Energy Efficiency Improvements in Electric Motors and Drives

Springer
Berlin
Heidelberg
New York
Barcelona
Hong Kong
London
Milan
Paris
Singapore
Tokyo

Paolo Bertoldi Aníbal T. de Almeida
Hugh Falkner (Editors)

Energy Efficiency Improvements in Electronic Motors and Drives

With 260 Figures
and 87 Tables

Springer

Dr. Paolo Bertoldi
European Commission
DG XVII
Rue de la Loi, 200
1049 Brussels
Belgium
E-mail: Paolo.BERTOLDI@cec.eu.int

Prof. Aníbal T. de Almeida
University of Coimbra
ISR - Dep. Eng. Electrotécnica
Pólo II
3030 Coimbra
Portugal
E-mail: adealmeida@isr.uc.pt

Dr. Hugh Falkner
ETSU
Harwell Didcot
Oxfordshire OX11 ORA
United Kingdom
E-mail: hugh.falkner@aeat.co.uk

ISBN-13: 978-3-540-67489-4 e-ISBN-13: 978-3-642-59785-5
DOI: 10.1007/978-3-642-59785-5

Die Deutsche Bibliothek - CIP-Einheitsaufnahme
Energy efficiency improvements in electronic motors and drives: with 87 tables / Paolo Bertoldi ... (ed.). - Berlin; Heidelberg; New York; Barcelona; Hong Kong; London; Milan; Paris; Singapore; Tokyo: Springer, 2000
ISBN-13: 978-3-540-67489-4

Cover-Design: Erich Kirchner, Heidelberg

SPIN 10724842 42/2202-5 4 3 2 1 0 - Printed on acid-free paper

Preface

The 1997 Kyoto Conference defined CO_2 emission targets for the developed regions of the world. The EU target of decreasing the emissions 8% below the 1990 level, by 2010, will require a very substantial effort covering basically all activities if such a target is to be reached. Energy-efficient motor systems can provide one of the most important opportunities to achieve electricity savings in a cost effective way, avoiding at the same time the emission of tens of millions of tons of carbon.

The reduction of energy consumption through improvements in energy efficiency is one of the major instruments for developed and developing countries to meet the Kyoto commitments. Energy efficiency is also a key element of the European Union (EU) energy policy, since it improves the efficiency of the economy, increases energy supply security, and decreases harmful emissions due to electricity generation.

Electric motor systems use over half of all electricity consumed in developed countries. Typically about 70% of the electricity which is used in the industrial sector and about 35% of the electricity used in the commercial sector in the EU is consumed by motor systems. In industry, a motor on average consumes an annual quantity of electricity which corresponds to approximately 5 times its purchase price, throughout its whole life of around 12 to 20 years .

Motor systems are by far the most important type of electric load. They are used in all sectors and in a wide range of applications, such as: fans, compressors, pumps, conveyors, mills, winders, elevators, transports, home appliances, and office equipment. As motors are the largest users of electrical energy, even small efficiency improvements will produce very large energy savings, across the EU.

Although individual components in electric motor drive systems perform relatively well, there are sources of losses throughout the power drive system. Motor electricity consumption is influenced by many factors such as motor efficiency, motor speed controls, power supply quality, harmonics, systems oversizing, distribution network, mechanical transmission system, maintenance practices, load management, and the efficiency of the end-use device (e.g. fan, pump, compressor etc.). In order to benefit from the available savings potential, in most cases it is necessary optimise the whole motor system.

Motor systems improvements produce substantial electricity savings contributing to improve plant efficiency, to reduce the use of fossil fuels, and to reduce greenhouse emissions. Additionally, the profitability and competitiveness of many economic activities can be significantly improved through the use of energy-efficient motor technologies.

The 2nd International Conference "Energy Efficiency in Motor Driven Systems" (EEMODS '99) held in London, 20-22 September, 1999, follows on from the very successful First International Conference *"Energy Efficiency Improvements in Electric Motors and Drives"*, Lisbon, 29-31 October 1996. There is evidence that the discussions and networking resulting from these motor conferences play an important step forward to better define research, development, standardisation, policies and programmes to promote energy-efficient motor systems around the world.

EEMODS '99 brought together around 200 prominent experts from all over the world, providing a variety of experiences and perspectives, to discuss the latest developments covering:

- New technology developments, covering both research results as well as innovative applications
- Policies and policy instruments, including labelling and minimum performance standards
- International programmes for market transformation and energy efficiency improvements
- Novel alternatives for financing and promoting energy-efficient motor systems projects

This book contains all the key presentations made in the plenary sessions , as well as the papers presented in the four workshops: *"Test Methods for Induction Motors"*, *"Examples of financing Energy Efficiency in Motor Systems"*, *"Minimum Standards and Motor Labelling"* and *"Pump Design and Applications"*. Summary papers of the posters on display during the conference are also included. It is hoped that the availability of this book will enable a large audience to benefit from the presentations made in EEMODS '99. Potential readers who may benefit from this book include researchers, engineers, policymakers, energy agencies, electric utilities, and all those who can influence the design, selection, application and operation of electric motor systems. The book is structured into the following chapters:

- Introduction
- Technology Developments
- Motor System Characterisation and Applications
- Pump Design and Applications
- Test Methods for Induction Motors
- Policies
- Minimum Standards and Motor Labelling
- Programmes
- Examples of Financing Energy Efficiency in Motor Systems

EEMODS '99 had as its main sponsors the SAVE II Programme (Specific Actions for Vigorous Energy Efficiency), which was adopted by the Council of the European Communities, and the UK Government's Energy Efficiency Best Practice programme.

We also wish to acknowledge the following organisations that helped by sponsoring the event: ABB AUTOMATION, ALSTOM, BROOK CROMPTON, CONTROL TECHNIQUES, EASTERN ENERGY, SIEMENS and the United States' Department of Energy. The support of all the sponsors and their generous funding was very important in ensuring the success of EEMODS '99, and is greatly appreciated.

The Editors

Aníbal T. De Almeida
Paolo Bertoldi
Hugh Falkner

S A V E

BEST PRACTICE PROGRAMME

ENERGY EFFICIENCY

Contents

Programmes

Examples of Financing Energy Efficiency in Motor Systems

Energy Efficiency in a Competitive Economy

Keynote Speech by Dr. Hans–Joachim Jähne

Thank you Dr. Bertoldi, Minister, Mr. Lambert, Mr. Hamilton.

Ladies and Gentlemen,

Today is a great honour for me on two counts; firstly to have been invited to attend this conference, and secondly to have the opportunity to share a few of my ideas with you. These are fundamental ideas on how a competitive economy can be linked to the current energy efficiency efforts being made within the framework of Agenda 21; in other words, energy efficiency in a competitive economy.

I am here mainly on behalf of the German Electrical and Electronic Manufacturers' Association. I also represent the European Manufacturers' Association CEMEP. I am a board member of the Drives Systems Sector, and last but not least I am President of a Siemens division.

My background lies mainly in semiconductors and electronics on a scientific basis as well as in the global business of drives.

I am very pleased to have this opportunity to speak to you. Despite many differences in detail, I can see a broad consensus between the European economy, the Commission and national governments in their attempts to make an active contribution towards reducing Carbon Dioxide emissions. As regards electricity, although there is no shortage of innovative ideas for its generation and transmission, perhaps our greatest combined potential would be to achieve the targets set in Kyoto, namely to use it far more effectively. It is our duty to deploy that potential as best we can.

On the consumer goods side there are already a number of supportive initiatives in place from a wide spectrum of active sources. On the capital goods side of industry, however, associated political measures are at a slightly earlier stage. This means that we must ascertain whether and where supportive measures are needed. It does not automatically indicate that less effort has been made on the capital goods side. On the contrary, the increasing use of automation technology, originally aimed at the qualitative aspects of end products, has in many cases become associated with improvements in the efficiency of industrial production.

Drive technology, for example, is having to cope with two different developments:

1. On one hand automation technology and drive technology are growing closer together

2. On the other hand there is a strong development towards decentralisation of drives.

The industry's answer to this is intelligent drive technology with totally integrated automation (TIA)

TIA provides a complete and uniform system module for manufacturing technology and process engineering.

The results of these activities are:

1. More intelligent features for field devices

2. Meeting real customer requirements - tailoring to their processes and expectations

And last but not least:

3. Process tailored functions and optimised control, which save energy throughout the whole system.

This can be seen clearly through the way in which energy consumption has become de-coupled from economic growth. In Germany for example, where I am obviously far more familiar with the situation, the proportion of industrial energy consumption has actually halved to around 25% since the 1950s, and the ratio of energy usage to production volume has fallen by more than 10% since 1990 in the West German electrical industry. I emphasise "West Germany" - our friends in the new eastern part have not been so successful.

These developments show that, even without state intervention, it is still possible to make considerable progress. However, although the trend is a positive one, it does not follow by any means that we should continue with "business as usual". The general trend does not hold true for all product areas, nor does it mean that we are already achieving the optimum in the successful product areas.

Let me prove this by using two examples:

Typical case 1:

The example of electric motors illustrates how initiatives by individual manufacturers have not necessarily led to success. Economy motors introduced during the 1970s by Siemens, along with other companies, were not accepted by the market. Now however, following a popular campaign initiated by the European Commission, an agreement between European manufacturers has been reached and the situation looks far more promising. This voluntary agreement, which will be presented to you during the course of this event, is an expression of the confidence of the European electric motor manufacturers and represents a breakthrough as a result of collective ideas.

Typical case 2:

The triumph of power electronics in the drives sector is impressive. Besides which, almost every time there is a further increase in market share it is accompanied by a greater saving potential in energy consumption, as I just mentioned. However, could the many existing applications have been opened up even more quickly? I will provide a few examples later.

These problems are being discussed very seriously by European industry, both with the European Commission and with national governments. The increasing number of cases of individual responsibility and voluntary agreements also shows how industry is ready to draw appropriate conclusions from the results of these discussions.

At the same time, the discussions have sharpened awareness of the factors needed to ensure that such discussions are objective and successful.

Allow me now to deal in a little more detail with some of the core points that might prove useful in explaining the Commission's position, and might be of additional importance.

1 Objective Measures Should Always Be Implemented at System Level

Just as a customer buying a refrigerator takes little interest in the energy-saving properties of the insulation or the motor, but only looks at the overall efficiency of the refrigerator as a self-contained system, so too are industrial customers less interested in machinery components than in the overall efficiency of a drive process. From this aspect, I would welcome a much greater concentration of this conference on the subject of drive systems than was the case in Lisbon. In comparison, agreements on components can only ever be of a supplementary nature.

2 Political Measures Must Be Aimed at the Global Economic Environment

European industry produces for and in a global economy. In this international environment, the national economies that have developed best are those in which the various economic units have been allowed a reasonable degree of freedom. Economies that have been dominated by government regulations, on the other hand, have fallen behind in international competition. Therefore, the great challenge is to achieve the just aims of protecting the environment and the world's climate by methods and measures, which have the least effect on degrees of freedom and do not devalue productive capital. Voluntary agreements in specific sectors appear to me to be the tool that promises the most success.

3 Energy Efficiency is an Important Product Criterion, but Not the Only One

There is no denying that the importance attached to the efficiency criterion can only be adjusted within certain limits. Competing criteria such as service life, customer support, price or even recycling cannot be manipulated at will. Those people who seek to apply the sustainable development principle also need to take due account of social and economic aspects.

4 Changes to the Existing Quality Demands of the Market for Political Reasons Must Take Due Account of the Time Factor

Customers have learned to attach certain priorities and values to product qualities. Some of these judgements differ very considerably between individual groups of customers and between individual firms. The shifting of these judgements in favour of individual factors leads to a positive or negative effect on the outlook of individual companies. This is only acceptable if sufficient time is allowed for these changes of outlook to take place. Imagine an Olympic decathlon competition where, in future, instead of the pole-vault results being rated at 10% the figure is increased to 15%, and the high-jump results are rated at 5% instead of 10%. Changing these ratings a few weeks before an Olympic Games would certainly lead to an uproar, even though they might be acceptable under other circumstances.

The more effectively that individual measures are structured and focused, the greater the influence on the competitive position of individual companies. Siemens can afford to see this development through calmer eyes than others — despite the fact that it might be a painful process — because, with the inherent reserves of a large organisation, it is able to recover from the political devaluation of some of its assets relatively well compared with small and medium-sized companies, who would find the situation much more difficult. Particularly for employment policy, the European Union will have to take into account the fact that changes made without allowing a reasonable transitional time will cause shifts in employment not only within Europe, but also on the international stage.

Despite all the restrictions that I have pointed out, there are still a number of sensible options for achieving further cuts while remaining true to the needs of competition. The German Electrical Manufacturers' Association is currently placing much more emphasis on providing better information for users. In a recent brochure, for example, the Association has drawn attention to several kinds of intelligent drives that are able to achieve lower energy consumption figures. This is intended to provide the basis for an objective longer-term dialogue between the Association and the various buyer groups.

We will be looking at this particular brochure in more detail later on during the conference.

Nevertheless, I would just like to mention two specific examples:

- Many coal-burning power stations still use coal mills that employ the bypass method. If this method were replaced by controllable clutches, it would be possible to achieve power savings of up to 29%. If frequency converters were also used, it would even be possible to reduce the power consumption by about 45%. This means that, every year, approximately 20 GWh could be saved in the boiler plant of a coal-burning power station by using modern, variable-speed electric drives.

- If, in the drive systems of the fans used in commercial and industrial buildings, the mechanical throttling dampers were replaced by frequency converters, the saving achieved with each drive would run into several megawatts per hour.

I know that these innovative ideas are increasingly making their mark in the market-place, but what sort of time-scale is involved, and is there any way in which it can be shortened?

As with electric motors, this is where I would like to see concerted discussions on a European level between the Commission and the sector committees responsible. I am certain that, especially at the equipment and system level, methods acceptable to market economies can be found to enable the energy-saving idea to be pursued more strongly and more effectively without necessarily involving individual obligation or official tagging.

I wish the conference every success in achieving its aims.

My message to the conference is:

Innovative energy saving systems will ensure our future survival on earth; there is no doubt about that. All other measures and political restrictions will only postpone our death!

Thank you.

Ethics & Answers
Engineering Efficiency for a Sustainable Future

Speech at EEMODS '99 2nd International Conference, London
by Richard J.M.Hamilton

President, Specialty Products, ITT Industries, Inc.

In ancient mythology, Icarus created wondrous wings of wax and feathers and, with this technology, was able to soar above the earth and fly to the heavens - only to soar too close to the sun and consequently plummet to his death. In the 20th Century, the technology of electricity has indeed provided the wings to enable modern man to soar. We must, however, beware that this exceptional marvel is not like the wings of wax for Icarus, with unforeseen consequences that lead to our destruction. Instead we recognize a need to do everything in our power to engineer a sustainable environment and secure the future for generations to come.

Today I would like to provide some perspectives from the USA, including a few facts on energy usage. Then I will share with you why my company, ITT Industries, sees the value of ethics in energy conservation, and what we are doing to provide leadership in developing answers to some of the issues. Finally I will discuss the concept of partnership, and the benefits that it can bring to your deliberations.

First, a few energy facts

The US Department of Energy estimates that consumption of electricity in North America has increased by about 23% in the last decade. Implementation of motor and system energy-efficiency practices would yield annual savings of from 75 to 122 billion kWh with a value of $3.6 to $5.8 billion[i] at current industrial energy prices.

Add to this picture the fact that fewer than one-third of American manufacturers are really reviewing energy usage.

In 1998 alone, US residential and commercial electricity use rose 3 percent. Utilities in the US increased the capacity of coal-fired plants with pollution equal to the exhaust of 1.5 million cars. Even with a trend away from coal fired power stations, carbon emissions from coal use over the next 10 years are projected to increase by 6% in the US, and almost 20% worldwide. The average coal plant called into service to meet these burgeoning needs loses almost two-thirds of its energy in waste heat and emissions[ii].

To address this situation, several initiatives have been driven by the 1992 US Energy Policy act, including the DOE Motor Challenge Program, and the National Energy Building Code. My company has chosen to play a leadership role in these programs, including the appointment of a key executive to coordinate our involvement.

Ethical behavior creates value

At ITT Industries we have found that ethical behavior towards our environmental responsibilities is not only personally satisfying, but ***It is good business to conserve energy and protect the environment.***

Our Environment Safety and Health Policy is an integral part of doing business, an integral part, not an afterthought. We work in accordance with ISO 14001 standards of environmental management and EMAS, the European Directive of "Eco-management and Audit Schemes." In fact we were the first to receive the certificate in Canada and among the first in the US.

We have found again and again, that making environmental planning as much a part of our company as cost analysis, saves us money both long and short term. There is no room at our company for "window dressing" environmental commitments. Our commitments must be real because they are the right things to do, and because they do make our company more profitable. For example, after one year of ISO 14001 certification, one of our Chicago units reduced hazardous waste by 92 percent, cut energy and disposal costs by $40,000, and cut water consumption by 1.4 million gallons.

Our company has made the commitment to implement best practices within our business units and to take a lead in encouraging others to follow.

- We educate our employees, suppliers and customers to meet the requirements of ISO 14001 and publish an environmental report each year on our progress.
- We actively invest to conserve natural resources and reduce waste.
- We implement measures and perform audits to reduce the company's environmental impact.
- We act in cooperation with authorities and industry associations to develop standards and programs to encourage enlightened behavior by others.
- Naturally, we also hope that our efforts will encourage others to buy our products.

Providing the answer

The second half of my talk's title promised some answers, so let's look at some specific applications where ITT Industries has worked to master energy efficiency. There are two ways to approach responsible energy efficiencies, both of which will be extensively covered during this conference.

- The first way is to examine actual product efficiency, in our case we look at the whole pumping unit, both the motor and the hydraulic end. We can make the motors do more, last longer, with less use and waste of energy. Our impellers are more efficient, and can maintain pressure and flow even under tough conditions, which will ultimately improve the total efficiency of the pump.
- The second way is to focus on system efficiency and the improved system design using a better match between component size and load, plus improving maintenance.

An interesting example of the first method is an elegant new impeller design for sewage applications that we have actually called the N-impeller, after the Greek letter η the international symbol for efficiency. This new pump design is engineered to be foul and clog resistant, thus decreasing start-stop cycles, reducing maintenance, while increasing life. From all of our test sites we have no reports of clogging and energy has been reduced by up to 50%. There are other applications for this type of pump. The paper industry, for example, can pump pulp mixtures of a much higher density than in the past. This new capacity cuts water needed for the process by a factor of two and a half and reduces the power requirements for pumping by about 60%.

There is no doubt, however, that it is easier to save energy and increase efficiencies by the second method - examining systems themselves. One analyst suggests that while there is a possible annual reduction of 24.6 billion-kilowatt hours from motor improvements, there is at least three times as much to be saved by systems' improvement[iii].

This is why we have taken a lead in a pump industry wide activity and together with the US Hydraulic Institute and Europump we are now promoting Life Cycle Costing - LCC. This concept focuses on all aspects of resource usage by a product in its lifetime. By far, the most dominant cost is energy. We believe that by studying the total cost of ownership, new and more efficient systems will emerge.

All over the world we see extensive and creative use of existing systems and technology. Quite recently we worked with the city of Kalmar in the South Eastern region of Sweden in upgrading its wastewater treatment plant to treat and process waste from food industries in the area, in addition to ordinary sewage, without building a new plant. 100 % of the heating and 50% of the electric power

needed for the plant is generated from the process in addition to gas for all gas driven vehicles and district heating for 1/6 of the population.
Farmers are given processed fertilizer in return for the manure they send to the plant. And pure water is returned to the sea.
This represents a move to a complete closed loop of production where everything is used and nothing is wasted.

A second example of the systems approach to savings comes from my own home area of Connecticut in the Eastern USA. In neighboring towns we found that pumping stations typically had pumps which cycled on and off as needed. This is expensive both in maintenance and energy. We proposed a new design with a small pump running constantly and a larger pump cycling only when required for peak loads. The results were truly amazing, 50% less maintenance and 21% to 43% less electricity. When you consider that U.S. sewage pumps use 3% of the nation's electricity, the impact becomes obvious. No glamorous new technology here, just good engineering and re-engineering.

This example is now a showcase project for the DOE Motor Challenge program.

For a further illustration of the systems approach, I would like to turn to the problems of the developing nations. Energy inefficiencies know no national boundaries. China, for example, is operating at up to 30% below western standards of energy efficiency and is suffering from smoke filled air across the country. These developing nations are increasingly aware of the risks of ignoring the environment. At the same time their capital is limited.

A possible answer here may well be one of scale. In the past, bigger was often considered better, resulting in large and expensive infrastructure projects. However, we have found in many instances, smaller is better. By building small water treatment plants, to serve 3000 - 5000 people, the job can be completed in a few months. The plants are made of concrete building blocks, prefabricated for easy combination and fast construction. The efficiencies of the water treatment and the ability to monitor and contain energy use are dramatic. Re-thinking scale can be an answer to new efficiencies in many contexts of energy and conservation. In a different application, the city of Zagreb found that installing individual variable speed pressure boosting systems in high rise apartments to be significantly more efficient than pressurizing the whole municipal supply system.

I cannot claim that these are the only answers, and many more must be found before our "wings of wax" soften further. We have found in our work with the DOE, Hydraulic Institute, Europump and others that partnerships are an effective way to go. It is our experience that partnerships between Industry, Government and Academia can:

- Create national recognition of the issues and initiate research programs of value to industry and the community.
- Increase overall awareness of the necessity to be more energy efficient.
- Help harmonize standards to create international competitiveness and speedier solutions.
- Capture opportunities to participate in the education of engineers and end users, thus creating goodwill for the institution, the company and the students.
- Build customer awareness of the benefits of LIFE CYCLE COST purchasing, which will lead to a realization that buying equipment based on the lowest initial cost can be a very expensive solution.

In short, in our role as leaders, we need to involve the emerging markets, make customers aware of the potentials of dramatic cost savings, educate manufacturers around the world of new efficiencies required to compete successfully, and support research and education. It is the ethical thing to do, and it is good business.

Remember that though we seem to have mastered the forces of nature, we must not abuse these gifts, nor take them for granted. We must recognize the consequences of squandering energy and ignoring the implications on the environment. We must acknowledge values, just as we must explore new technologies, re-examine our engineering of systems, and re-think issues of scale. Only then can we avoid the fate of Icarus who in his pride soared too close to the sun and was destroyed.

This is your challenge for this conference here in London. And as you meet here over the next three days, let me remind you of a view of London composed by William Wordsworth "Upon Westminster Bridge":

The city now doth, like a garment, wear
The beauty of the morning: silent, bare
Ships, towers, domes, theatres and temples lie
Open unto fields, and to the sky
All bright and glittering in the smokeless air"

Lets us hope that future generations can also enjoy "the smokeless air". Thank you for your efforts, and your kind attention

---

[i] Energy Savings Go Untapped, Quality, Wheaton, May 1999.
[ii] Joseph Romm, With Energy, We are simply too demanding, The Washington Post, 8/99
[iii] Richard L. Nailen, How the U.S Department of Energy Regards the electric Motor Market, Electrical Apparatus, July 1999.

Integration of Motors and Drives

Dr. D.C. Jackson

Brook Hansen, UK

Abstract. This paper examines the benefits of adopting a second-generation integrated motor and inverter. Removing the barriers to ensure that variable speed drives are more readily applied results in on-going cost savings to the user through energy savings plus process control benefits. In addition, the use of an integrated product instead of two separate components results in cost and time-savings to the installer. The simplification of integration, by transferring the guarantees of performance in efficiency, torque overload and stiffness, speed accuracy, noise and EMC compliance, allows optimisation by the design team to be realised by users and ease of application since the primary design team guarantees the product performance. The introduction of second generation compact product assists user conversion from present inefficient mechanical solutions. This technology is currently applicable in power ratings below 22kW, which includes the vast majority of practical applications.

1 Introduction

There are many industrial applications for fans, pumps and compressors that use mechanical means of flow regulation. Typically, for example, a pump may be driven at full speed by an induction motor, and the flow rate is varied by means of a valve. Whilst relatively straightforward this means of flow control is well known to be highly inefficient. There are a myriad of documented case studies that show remarkably short pay-back periods when such mechanical flow regulation is replaced by variable speed control of the driving motor. One of the traditional barriers to this technology has been the high cost of the speed controlling inverter. As inverters become more commoditised, the prices have fallen in real terms, and yet the adoption of variable speed has not been as rapid as one might expect. The reason for this is that there are other barriers that are perceived by the customer. In this paper I shall review these barriers, and explain how they are overcome by the use of integrated motor drives.

2 Barriers to Widespread Adoption of Variable Speed Motors

2.1 Barrier No 1: EMC Compliance

Due to the synthesised waveforms of the inverter EMC compliance comprises the control of the high frequency airborne radiated interference from the inverter to motor cable as well as the conducted noise of the supply current which should have a low harmonic current (Figure 1). To limit the radiated interference (RFI) the inverter to motor cable should be screened and earthed at both ends. The inverter to motor cable has capacitance that increases with length. So additional high frequency currents flow out of the inverter and these have the effect of increasing the inverter input current harmonic content. A range of input filters is therefore required to be available to suit likely inverter to motor cable lengths. This adds to the complexity of selection of inverter, motor and filters as well as compounding logistical and installation difficulties.

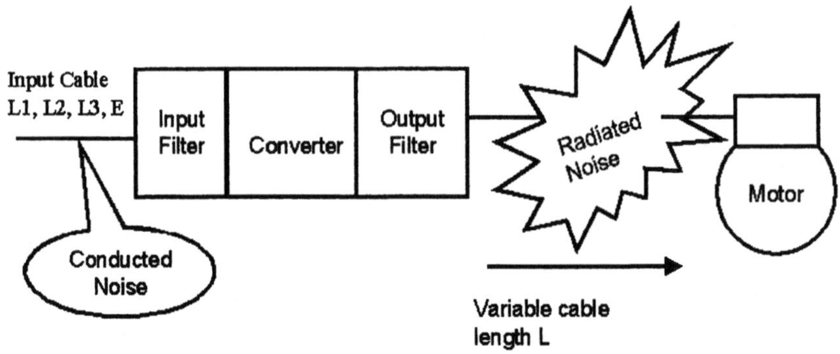

Figure 1: EMC Sources

2.2 Barrier No 2: Complexity of Installation

The installer requires to verify that the power supply matches the inverter capability and motor. Should the motor be connected in Star or Delta?. Are the links in the correct place to suit the desired phase voltage? Has the direction of rotation been established?

The control panel must provide a fused power supply that can be isolated but the application in which the drive is to be used decides the degree of local and remote control using start/stop switches, potentiometers and enclose when required, for example, PID controllers, PLC's or field bus cards. The installation complexity can vary significantly as well as the degree of protection of the panel, plus the panel cooling method.

2.3 Barrier No 3: Matching of Inverter and Motor

Customers are often concerned when matching separate motors and inverters. This is hardly surprising, as their expertise is in their own technology. Has the inverter been set up correctly to suit the motor? Will the inverter waveform make the motor hot? What quality of matching is needed? Set up simplification is a key opportunity to ease installation and remove a barrier for adoption.

Users concerns include

a) Obtaining the motor parameters: using default parameters or from the manufacturer, or from commissioning tests cold or hot.
b) Configuring the inverter to control the load to suit the application.
c) Optimising drive performance.
d) Ensuring that reliability results from avoiding mismatch of inverter to motor - insulation system or bearing type. This is also dependent on the length of the inverter to motor cable and controller type.

2.4 Barrier No 4: Flexibility

Users wishing to replace a fixed speed motor with a variable speed equivalent are often concerned that the replacement should fit in the same space as the predecessor and that normal motor features such as brakes, and forced ventilation cooling options are available.

2.5 Barrier No 5: Reliability

The initial perception of potential users of integrated motors is that several areas of reliability are suspect. Amongst the perceived apprehensions are

a) The electronics sits on top of a hot motor hence the electronics operates hotter than in a separate inverter.
b) The electronics must withstand application vibration.
c) The enclosure requirements IP55 could cause on product cooling condensation.

2.6 Barrier No 6: Initial Cost

Whilst the ability of variable speed drives to save running cost has been demonstrated in case study after case study, there remains the barrier of first cost, which acts to deter many potential users. Although inverter prices continue to fall, this barrier still remains.

3 Integrated Drive

In this section I will review how an integrated drive will overcome these barriers, but what is an integrated drive?

At first sight an integrated drive comprises a motor onto which is mounted a controller but that would only be a combined product mechanically joined with no added value intelligence built into the complete product. Integration is more than just the combination of suitable components. The integration involves the total product design comprising torque envelope, thermal appreciation, noise reduction, vibration acceptance, optimisation trade off's, life expectancy and many more aspects including the overall product appearance. In simple terms, the integrated product is a single complete product in itself. To a user, the integrated product has an inverter as an input interface that provides a shaft torque output where the speed is set by the control electronics and the physical location of the product is where a motor would be. This leads to the important differentiation of the integrated product being compatible with decentralised control where the power is supplied locally and the control can be remote aiding multiple communications to a diversity of integrated variable speed motors within a site. The communications may be RS485 or by a field-bus such as Profibus. An integrated motor can use alternative configurations of control from analogue to digital and computer driven.

3.1 Solution to Barrier No 1: EMC Compliance

An integrated motor has consistent short leads within the motor/inverter which are screened by the metal enclosure. The input and output filter of the integrated drive can therefore be designed to be correct avoiding EMC decisions during installation and providing straight forward compliance with the EMC directive. In effect this configuration guarantees the EMC compliance. The only decision is the choice of ordering Class A (industrial) or Class B (light industrial/domestic) EMC compliance. The integrated product has the benefit of reduced variety and logistical inventory of components giving error avoidance. Transferring the EMC compliance responsibility from the supply chain to the primary design team removes a barrier to variable speed motor adoption by simplifying in part the complexity of the installation and focussing responsibility.

3.2 Solution to Barrier No 2: Simplify Installation

Eliminating the motor cable removes all the concerns reference the star-delta links. The responsibility to combine the correct voltage inverter to motor is transferred to the original manufacturer avoiding supply chain errors. The local control panel need only provide a fused power supply that can be isolated. The simpler control panel does not need to consider the mounting and cooling of an internal inverter providing a substantial installation cost savings. The inclusion of a range of inputs - analogue, digital and serial communication or PID controller means these can be connected directly from the sources. To facilitate user control close to the integrated variable speed motor, keypad interfaces with differing degrees of intelligence are available.

A local operating panel can be readily connected that functions as a digital input. It provides basic functions comprising start/stop speed up/ slow down and jog with visual speed indication. It can be mounted onto the product or locally. This simplified unit precludes users from changing pre-set parameters and hence provides a degree of tamper proof requested by users.

Figure 2: Local Operating Panel

For more complex functionality, in addition to the features of the local operating panel a local control panel is generally available for most integral products. It has the facility of controlling operation, product set-up as well as indicating diagnostic advice. To meet the application enclosure requirements a plug kit is available for use to allow the handheld operation pad to be used elsewhere.

Figure 3: Local Control Panel

Setting up the inverter to operate with a motor as well as the application control configuration adds to the installation time.

These factors compared to separate motors and inverters lead to simpler installation requirements in space, time and cost for the integral product thereby removing a barrier for adoption. In reality for 80% of the applications there is not much additional connecting work compared to a motor alone.

3.3 Solution to Barrier No 3: Simplify Matching Inverter and Motor

3.3.1 Motor set up

The motor parameters are pre-programmed into the controller at manufacture, so set-up is greatly simplified. Furthermore additional matching intelligence is provided to ensure optimum performance at all speeds.

3.3.2 Application set up

As well as configuring the inverter to suit the motor, the individual application configuration requires to be established covering how to start, stop, set speeds, set direction of rotation, set maximum speeds, combine references etc. To simplify the application set-up many manufacturers pre-configure the default set up which caters for the majority of customers without change. To accommodate particular customisation technical handbooks are usually available describing the parameter choices available.

3.3.3 Performance

Since the integral product has been designed for optimum overall performance (motor and inverter designed together as a single product) the system performance is guaranteed. Separate figures for motor and controller need not be quoted as well as the need to state assumptions as to their performance.

A distinction between combined product (combined to fit together but not designed to work optimally together, first generation product) and integral product (designed together optimally, second generation product) should be made.
Combined products typically have a standard terminal box mating face onto which has been mounted an inverter. In this scenario performance, efficiency, noise and other parameters may not be optimal, although development costs may be lower and speed to market quicker. A fully integrated design may result in a superior product. At the time of writing this paper, users have an extensive choice of products with a wide spectrum of performance and features that require careful selection to meet initial expectations as well as future needs.

3.3.4 Insulation Life

A PWM waveform exhibits high frequency pulses that are applied to the inverter output to the motor. The pulse propagates along the inverter to motor cable. The pulses reflect at the motor end due to the impedance imbalance. On separate inverters and motors the reflected wave can combine with the next PWM pulse to produce a doubling effect, dependent on the length of the cable and PWM rate. The application of the high voltage stress combined with the dV/dt ages the motor insulation system. The motor manufacturer at a cost can address the concerns of insulation life due to high voltage and dV/dt as shown in Figure 4.

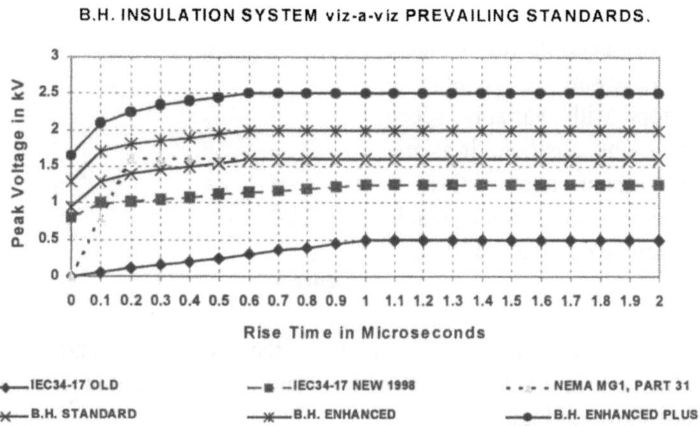

Figure 4: Insulation Capability Chart

An integrated product has short internal leads between the inverter and motor. The length of lead is so short that the propagation time and reflection is complete before the next PWM pulse so voltage doubling does not happen. The motor insulation system cannot therefore age as fast compared to separate components. This benefit enables integral product to accept a higher PWM frequency that, if selected, reduces noise at the expense of slightly higher losses.

In contrast, stand alone components, with random PWM or DTC can result in not just doubling but 3 - 4 times the voltage spikes which therefore can age the insulation system faster. In addition this waveform promotes larger zero sequence voltages which causes more high frequency current to earth which can reduce bearing life. So insulated bearings are essential with random PWM or DTC. This is not the case with normal PWM but provides increased bearing reliability.

Since insulation life doubles for every ten degrees Kelvin reduction in temperature (Arrhenius reaction rate equation.) the reliability of the motor to insulation degradation is better for higher efficiency cooler motors.

3.4 Solutions to Barrier No 4: Flexibility

The space taken up by a standard IEC Induction motor is shown opposite in Figure 5.

To obtain the benefits of variable speed it is necessary to be able to replace motors with variable speed types in the same position. It is not possible to mount and enclose a separate inverter in the space available to the enclosure degree required.

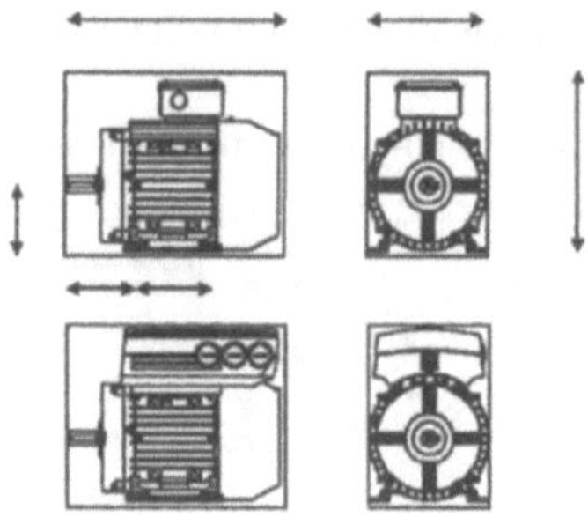

Figure 5

Ideally the variable speed motor must not exceed the envelope of standard motors in order to facilitate interchangeability of fixed and variable speed motors. Product examples are shown in Figures 6, 7 and 8.

Figure 6

Figure 7

Figure 8

The facility of a double shaft extension enables the choice of fitting gearboxes, brakes or forced ventilation units as standard motors.

Figure 9

Mechanical mounting adaptability is not impaired with integral products. (Figure 10). The flexibility of the multi-mount feature (Figure 11) enables compact installations to be satisfied.

Figure 10

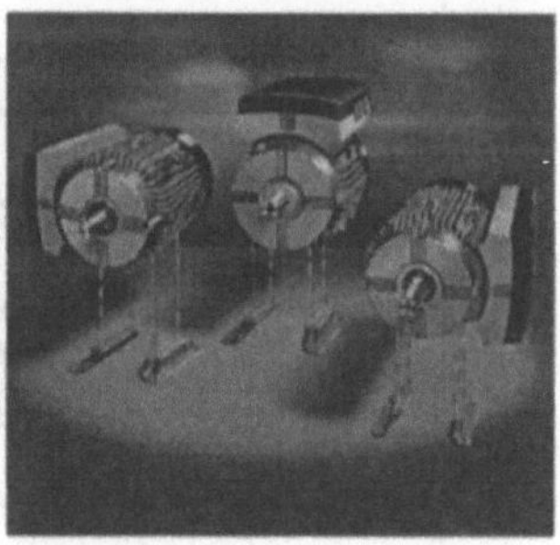

Figure 11

3.5 Solutions to Barrier No 5: Reliability

For a separate inverter the reliability is paramount at full power the maximum load condition. At this load maximum cooling is required and the normally hotter parts usually comprise the power components – input rectifier, chokes, dc link capacitors, and power switches. The life of the dc link capacitors is dependent on the ripple current that is controlled by the input chokes. Low cost, short life cooling fans in separate inverters have a life of 3-5 years unless operated under temperature control. At constant torque and reduced speed, even with the motor line current remaining constant, the input current reduces which results in less inverter losses.

Combining the motor and inverter has several effects. The inverter part usually has around 3% losses at full load and for motors in the range up to 7.5kW 10-20% losses. Since the motor has a larger thermal mass it dominates the thermal behaviour of the integrated product. The motor driven fan enhances the usual inverter cooling above half speed. However at below half speed the combined effect of the inverter sitting at the temperature rise of the motor plus ambient temperature along with the reduced inverter cooling causes the components within the inverter to be examined. To ensure reliable operation (of the normally hotter inverter components) they are placed in areas of maximum cooling and potted to reduce hot spot temperatures as well as provide resistance to vibration. These power components at rated load can in some integrated products be operating at lower temperatures than in a separate inverter enhancing the reliability at that load. This is necessary to ensure adequate reliability when the shaft driven fan reduces airflow cooling with speed. The continuous rating, for an ambient of 40C, is illustrated by typical curves IP1 for the lower output of a frame and IP2 for the higher output (Figure 12) where the current input and torque output is reduced. These curves unsurprisingly are not too distant from the class B rise of a separate motor since this is the temperature band limitation expected due to the thermal life/reliability constraints imposed by the inverter components.

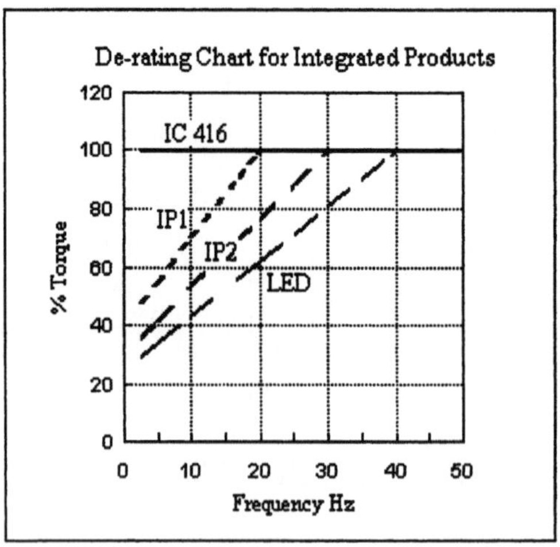

Figure 12

It is usually possible below half speed to maintain full load torque for well over an hour, in a 20C ambient, before the thermal overload condition results so low speed intermittent operation is permissible. Overloads up to the maximum limit of say 160% torque would reduce this time or duty cycle allowed. Most integral product controls monitor the product temperature and trip if their thermal limit is exceeded. To achieve continuous full load torque at low speed a long life separate cooling fan is used similar to separate units as indicated by IC416 in Figure 12. The thermal robustness of alternative manufacturer's products varies significantly. Those integral products with lower cooling capability or efficiency exhibit additional de-rating requirements as shown by curve LED in Figure 12. In addition the time to accept an overload or low speed load excursion is more limited. Protection of the inverter part to over temperature is therefore key to a reliable product.

The inverter part of an integrated product usually has many control similarities to a separate inverter in that it would instantaneously trip above say 160% current, and between 105% to 160% current to implement an I^2t counter to time out. In addition if over temperature within the inverter passes a warning threshold then a self protecting mechanism of reducing the PWM frequency can be initiated to reduce losses on some products. If the temperature is sustained beyond the warning level to the desired temperature limit for reliability then a thermal trip will result thereby protecting the product.

When separate motors and inverters are used the inverter is not often subjected to application vibration or environmental conditions, in contrast to an integrated product. To overcome these application vibrations integrated products have the larger components or circuit boards of the inverter part resiliently mounted or encapsulated. They are usually vibration tested over a range of frequencies and amplitude to ensure adequacy as well as shaken in 3 planes to 2.5G for 30 minutes.

Integrated products require to be mounted in any standard motor location. This means that they have to meet the enclosure requirements of IP 55 (Figure 13) or for more arduous applications IP66 indicating that the product can be washed down without detriment.

Figure 13

In addition precautions to avoid internal condensation, within the integrated product when the shaft is not rotating, can be made by using the motor windings as standby heaters thereby preventing condensation.

The performance window and life of integrated products can vary dependent on the product selected from first generation products to second generation, with life expectancies exceeding that described by bearings at 60,000 hours (Reference 1) to half of this (Reference 2) for first generation product. The number of operating hours for variable speed motors varies significantly (Reference 3) and it is difficult to establish a suitable typical figure. In Europe the products are self-certified to applicable standards whilst independent evaluation by Underwriters Laboratory (UL) can result in a UL listing which satisfies that the product is safe. At the time of writing I understand that the Brook Hansen Danfoss partnership product is the only European product to pass the UL tests.

3.6 Solution to Barrier No 6: Initial Cost

The energy savings offered by an integrated drive are identical to those obtained using a separate inverter plus motor. The purchase price of an integrated drive is typically 10% lower than that of separates since the components are designed to complement one another. Additional first cost savings are made because of the relative simplicity of installation - smaller control cabinet, simpler panel wiring and no expensive inverter to motor cable. The installation cost of a 7.5kW integral drive is typically a third of the installation cost for an equivalent drive composed of separate components. Overall, the first off cost is approximately 13% less for a variable speed pump. This lower first off cost is a positive step to enable variable speed drives to be adopted which will pull through energy saving benefits.

The steps to achieve the savings can be summarised: -

a) Select a higher efficiency integrated variable speed motor.

b) Size the application and review your needs with drive specialist's support or use software tools available.
c) Include additional sensors or controls to optimise the savings balanced with the investment.
d) On the application use an interface (key pad or RS 485 port with computer and dialogue software) to establish the actual power required, the kWh consumed then up grade the integrated motor to match the system.

4 Concluding Remarks

The European annual growth rate forecast by IMS in 1999 (Table 1) indicates that the benefits described earlier will aid sales penetration and increase adoption of variable speed motors.

Table 1:

Annual Growth Rate	1997	1998	1999	2000	2001	2002
	Base	77.6	119.2	78.6	62.0	40.0

To date the introduction of integrated products has been limited due to confidence in using them based on the perceived barriers. The use of integrated drives can reduce first off costs by 13%. Further education of the market is needed before the full potential of integrated products is realised. This is now being overcome by second generation products which match standard motor physical size and separate inverter attributes as well as having independent verification of their operation and safety through UL recognition. We believe that integrated drives will do much to promote the use of variable speed in pumps, fans and compressor type applications. As such they will certainly make a growing contribution to energy efficiency in Europe and the rest of the World.

5 Bibliographic References

[1] Jackson D. C. and Chang S.C.: Brook Hansen, UK.

[2] A second Generation Drive for Industrial Applications: PEVD 1998.

[3] Turner M. J. and Blake R. J.: Switched Reluctance Drives Ltd. Harrogate, UK. The development of a Multi-Purpose, Integrated, Industrial Switched Reluctance Drive. PCIM 1999.

[4] Almeida A.T. and Fonseca P. University of Coimbra, Portugal. Characterisation of the Electricity Use in European Union and the Savings Potential in 2010.

Design of Motors for Inverter Operation

Tapio Haring

ABB Motors OY, P.O.Box 633, FIN-65101 Vaasa, Finland

Abstract. This paper describes very practical principles of how an induction motor should be designed for converter application. The main focus targets the efficiency of the motor and drive. The results presented are based on actual test motors and FEM-calculation simulations. FEM-calculation together with a time-stepping function is a powerful tool for estimating magnetic flux densities, iron losses, current densities and corresponding losses in windings,in other words a tool for optimisation of the motor design. Time-stepping is rather time consuming because all the circuit equations must be solved for each time-step, but it provides a way to estimate the iron losses; hysteresis and eddy current losses as well as current distribution and current losses.
The calculation tool also provides the possibility to check if an existing motor is feasible for a converter drive. Alternatively if a motor is only to be supplied by a converter there are many more degrees of freedom in the electrical design and the motor may be optimised for that converter drive by incorporating rather simple design changes.

Additionally a design compromise, "a general purpose motor" useable for DOL and feasible for converter drive can be produced following the principles presented herewith. The converter types which are considered are indirect types and mainly voltage source converters since they are the most common on the market and are "general purpose converters" and providing a certain freedom to select the motor for the drive.

Current source converters require "matching" with the motor and therefore need a precise knowledge of the motor equivalent circuit, making the selection of the motor more complicated.

Keywords. Induction motor, converter drive, additional losses, efficiency improvements

1 Introduction

Induction motors are the most popular electric motor types in industry - real prime movers. They have gained this majority position by their excellent properties such

as good availability - thanks to standardisation - extreme simplicity in construction and very high reliability in use. They also have high efficiency and reasonable purchase price - important features for customers. Squirrel cage motors are also relatively easy to modify to satisfy special applications and environmental conditions.
As a matter of fact they have only two serious weaknesses: starting performance and the difficulty to control the speed smoothly over a wide range. Both of these problems can be solved by using frequency converters. Modern converters offer control properties which are superior to those of conventional d.c. drives. The speed control of cage induction motors has proved to be an excellent way to control different processes and to save electrical energy in various applications. There is a great variety of different drive applications where the reported annual energy savings have been from 20 % up to 70 %, even more in some regenerative drive cases - the realistic saving potential has been estimated to be 30 ... 35 % on average.
The mass-introduction of converter fed induction motors has, however, not taken place completely without problem. Additional motor losses due to harmonics in the motor voltage and current, generation of harmonic currents to the supplying network (EMI), bearing voltages and currents, additional acoustic and electromagnetic noise, torque ripple and increased voltage stresses on the insulation system of the motor are the risks and negative side of the otherwise excellent converter drive systems.
In this presentation the emphasis is laid on additional motor losses in converter drives since energy saving is the theme of this conference. A couple of University studies are presented and very practical examples are shown of how to tackle these issues in normal life.

2 Common Selection Practice

It is common practice that when a standard squirrel cage motor is selected for a converter drive the continuous output power is derated. With the lower output the temperature rise of the motor is maintained at the original level in spite of the fact that converter supply has caused additional losses in the motor. In other words the temperature rise and the maximun running temperature determine the limits of the continuous output capability (usually expressed in terms of torque as a function of the speed).

How much this reduction should be for different cases is really a difficult question. It depends on several factors in the total drive system, converter type and modulation, motor design, mechanical power transmission and control equipment. Each of these can occur in several variants, so a general answer covering all cases in not possible. According to converter manufacturer's own guidelines this derating varies from 0 % up to 20 % at nominal frequency.

Motor manufacturers have been forced to test their products with different converters /5/. In such studies we have found that the derating variation is very wide and in some cases there are chances for practically impossible combinations: the temperature limits being exceeded already at no load running!
The derating percentage, if accurate, gives an initial rough estimate of the additional losses in the motor although the temperature criteria is based on the stator winding temperature and not on the total losses of the motor /2/.

3 Distribution of Losses

The distribution of different loss components in a cage induction motor depends on the size, pole number and electrical design of the motor. Some very general rules may be given: in small motors - up to some kWs - and specially with high pole numbers, the current losses are the major part of all losses, iron, windage and friction losses playing a minor role. In bigger machines the distribution of losses is more even and also iron and windage losses are important.
In a converter drive friction and windage losses are easy to estimate at any speed if they are known at the base speed - on the other hand in high speed and very low speed applications a separate "constant speed" cooling system is commonly used. In order to hit the right target we have to know the distribution of different losses in the motor and additionally we have to know the main operating characteristics of the converter.

4 Methods of Analysing the Losses

Basically there are two different approaches to analysing the effects of the converter supply on motor losses; by making extensive and time consuming tests in a well equipped laboratory, or by using modern calculation and simulation tools to estimate the changes caused by different drive configurations. In the ideal situation, both methods are used: calculation simulations for rough selection of good candidates and then repeated with higher resolution for analysis of the short listed solutions and finally careful testing with real prototypes to verify the results and to improve the calculation accuracy.
In the analysis of a 37 kW cage induction motor /1/ the calculation of the motor is based on a time-stepping, finite element analysis of the electromagnetic field. The field is assumed to be two-dimensional and the three-dimensional end-region fields are modelled by constant end-winding impedances in the circuit equations of the windings. This (2D)approach does not cause any impairment of the calculation when the core length is relatively long. The iron losses are calculated from the results of the field analysis. The time-dependence of the field and the motion of the rotor are modelled by the Crank-Nicholson method. The non-

supplied by the converter is imposed on the formulation through the circuit equations of the winding, in other words the output voltage wave form and modulation of the converter must be known. This may be a problem with very modern, state of art, converters which do not have any fixed modulation since the switching of power components is carried out according to results of drive status calculations (e.g. Direct Torque Control, DTC).

5 Findings of the Studies

The simulation calculations and laboratory measurements show clearly that the greatest changes and increases occur in rotor current losses and stator iron losses (Figure 1 and 2) when motors are fed by PWM voltage source converters.

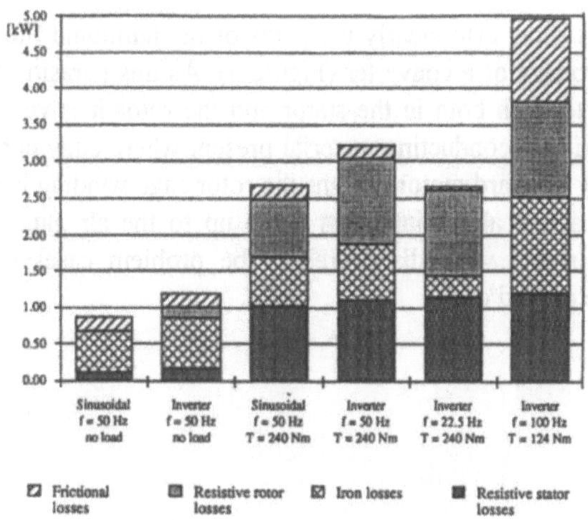

Figure 1: Magnitudes of various loss components of the unskewed 37kW motor at some typical operating points

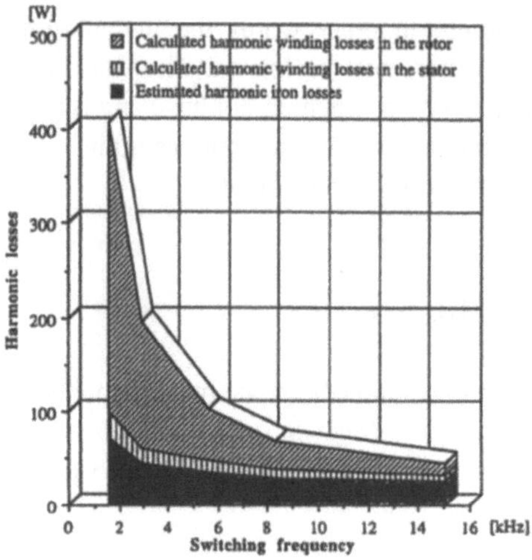

Figure 2: Subdivision of harmonic losses as functions of the switching frequency

FEM simulations show also clearly the paths of the harmonic flux caused by the non-sinusoidal voltage of a converter (Figure 3). As this parasitic flux is crossing over from teeth to teeth both in the stator and the rotor it is very important that there is no electrically conducting material present where eddy currents might be induced. In many standard motor designs the rotor cage winding is made from cast aluminium so that the aluminium bars come up to the air gap surface (fig. 4). These kinds of rotors were discovered to be problem cases even before the simulations were carried out !

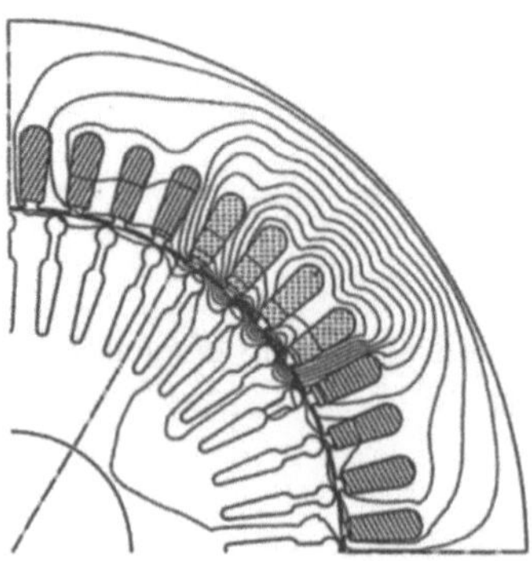

Figure 3: Typical distribution of the high-frequency flux caused by the inverter supply

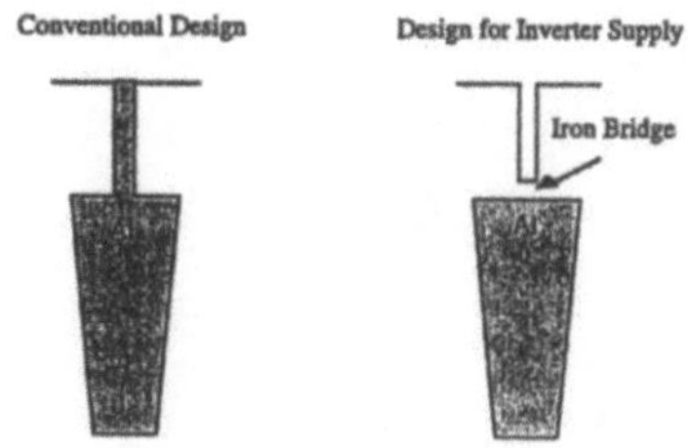

Figure 4: To the left of a typical slot of a standard motor, to the right a slot adapted to inverter-supply

The reason seemed to be high eddy currents and additional losses at the top of rotor bars. Since the rotor slot and bar design seemed to be one of the key factors in a good "converter proof" motor concept, several different rotor slot forms and bar shapes were studied (Figures 5,6 and 7).

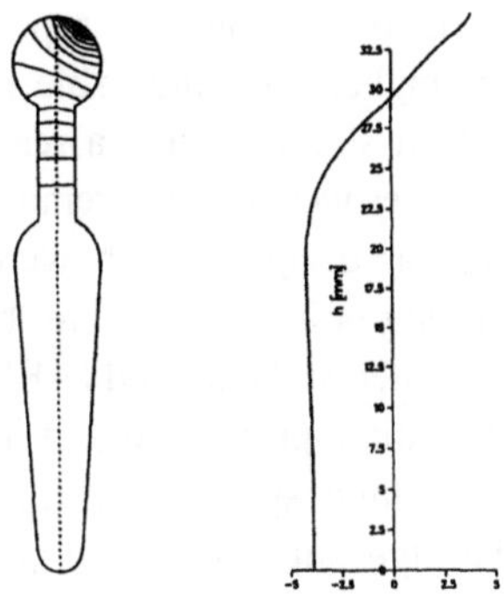

Figure 5: Typical current-density distribution on the cross-section of a rotor bar

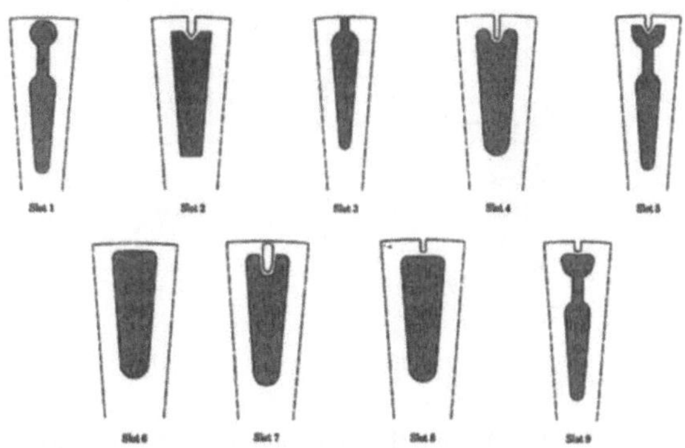

Figure 6: The cross-sectional geometries of the rotor slots

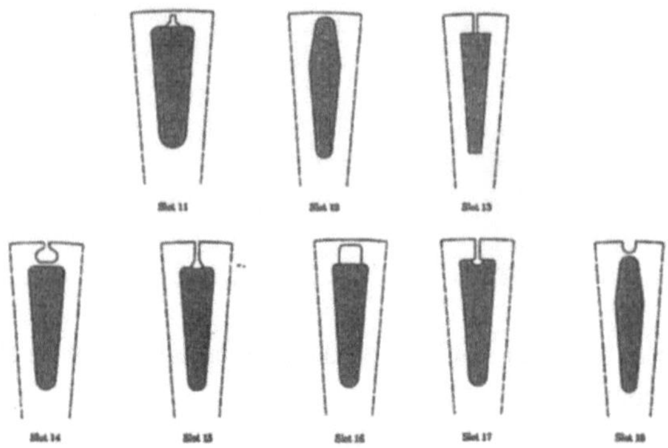

Figure 7: The cross-sectional geometries of the rotor slots

A very significant finding was made immediately: if the electrically conducting material (aluminium) was removed from the top areas of the bar where the harmonic flux densities are highest, a remarkable reduction of additional rotor current losses was achieved, in simple terms; a removal of material from those places where the additional losses were created removed the losses too!
Many of the rotor slot and bar designs which were studied are impractical or difficult from the manufacturing point of view but the so called U-shaped bridge slot (Figure 8) which was invented and patented (ABB) already in 1989 fulfills the usual requirements of the casting process and is used in some special motor designs such as traction and high speed applications. The fine tuning of the geometry of this U-shaped bridge slot is presented in the reference /2/.

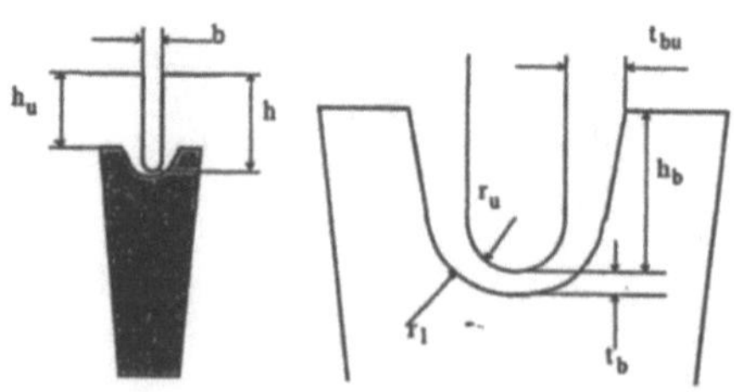

Figure 8: The U-shaped bridge slot and the main geometrical dimensions

Some other ways to improve the motor performance in a converter drive were also studied and the results are presented in fig. 9. Here the changes are compared with the original standard motor design as separate actions e.g. copper as cage winding material may reduce total losses some 8 ... 14 % and thinner lamination steel (from 0,65 ----> 0,5 mm) some 3 ... 12 %. If we make several of these changes in one motor e.g. better slot design (no 17) with copper bars and endrings, thinner lamination (0,5 mm), slightly increased air gap (from 0,8 to 0,9 mm), with chorded 5/6 winding and magnetic slot wedges the total losses can be reduced by up to 30 % at lower supply freguencies (up to 50 Hz) and by 40 % in the field weakening range (100 Hz). This is also clearly shown in the temperature rise measurements (Figure 10).

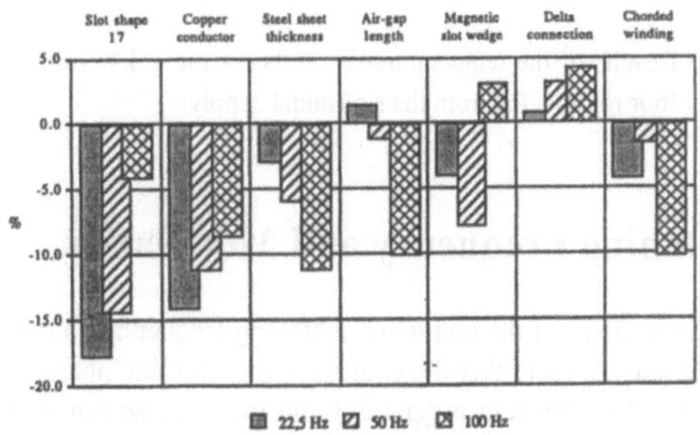

Figure 9: The changes in the total losses of the 37kW motor caused by variations in the construction

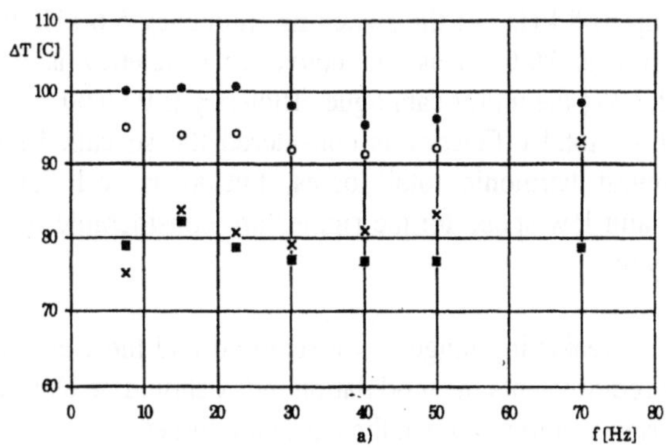

Figure 10a: Results of the temperature-rise tests for the 37kW motor equipped with the four rotors – supplied by the frequency converter

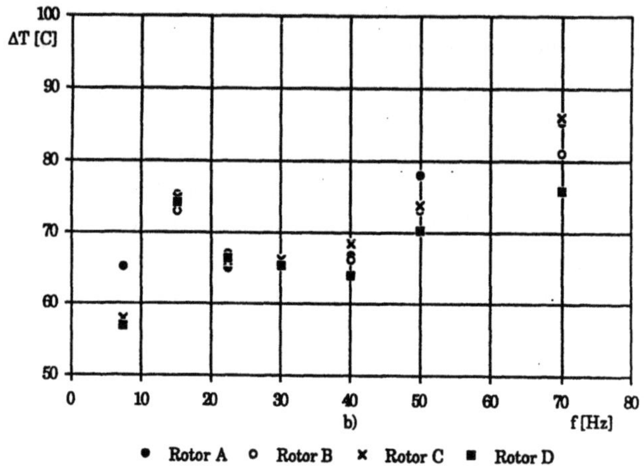

Figure 10b: Results of the temperature-rise tests for the 37kW motor equipped with the four rotors – fed from the sinusoidal supply

6 Switching Frequency and Modulation

In Figure 2 the effect of the converter switching frequency on harmonic losses in a test motor is shown. Note that the values are only for one phase, so the values have to be multiplied by three in order to obtain the 3-phase power of the test motor. The figure shows clearly that motor harmonic losses are reduced rapidly when the switching frequency is increased from e.g. 2 kHz to 6 kHz. On the other hand the converter losses will increase when the switching frequency is increased (Figure 11). The best total drive efficiency is achieved when the sum of losses reaches the minimum. In this case the optimum switching frequency range is rather large e.g. from 3 kHz up to 8 kHz. At the switching frequency 5.6 kHz the calculated total drive efficiency is 90,0 % as the converter efficiency is 97,6 % and motor efficiency 92,2 % (the motor catalogue efficiency is 92.9%)
If only the high total efficiency is considered the so called six-step-modulation gives the lowest harmonic total losses, but as is well known, the dynamic performance and low speed torque ripples are considerable disadvantages of this modulation type.

If a wide field weakening range is not required and there are no special dynamic performance requirements the modulation index should be close to 1 in any kind of PWM-converter, just to minimize the harmonic losses.

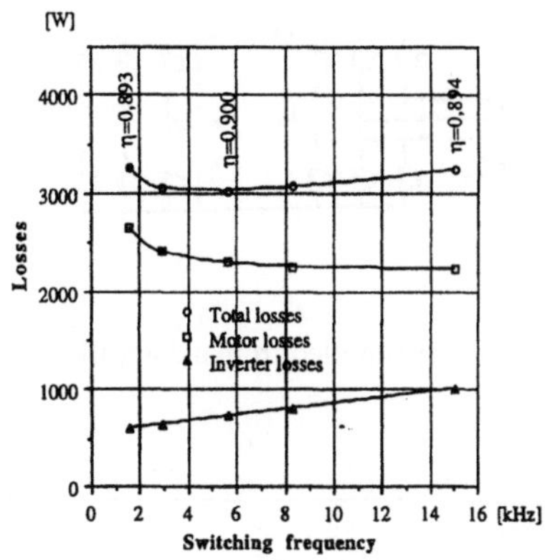

Figure 11: Total losses, motor losses and inverter losses versus switching frequency

7 Different Converter Types

All of the comments made above concern mainly voltage source frequency converters (VSFC) since they are the most common, "general purpose" converters - easy to apply because the only practical limitation for the motor is the output power. Current source frequency converters must be carefully matched with the motor (design) and therefore they are used only in special cases.
In Figure 12 two equivalent, complete drive systems are presented - one VSFC drive and the other CSFC drive /3/. Note that in both cases there are input filters to reduce the conducted interferences down to very low level (supply current total harmonic distortion THD < 1 %). In Figure 13 the total losses and efficiencies are presented as a function of output frequency. The studied case is a 2.2 kW constant torque drive and the Figure 13 shows clearly that VSFC has better efficiency specially at low frequencies. For high fundamental frequencies and at high switching frequencies the relationship is reversed.

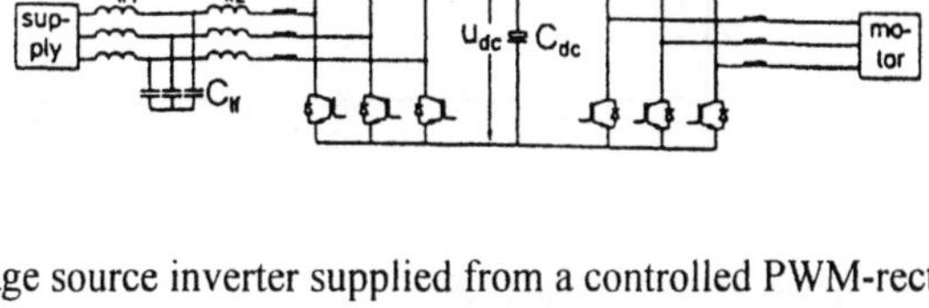

Figure 12a: A voltage source inverter supplied from a controlled PWM-rectifier

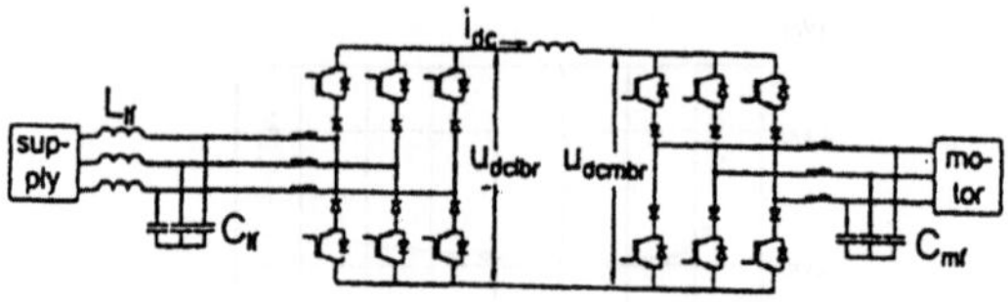

Figure 12b: A current source inverter supplied from a controlled PWM-rectifier

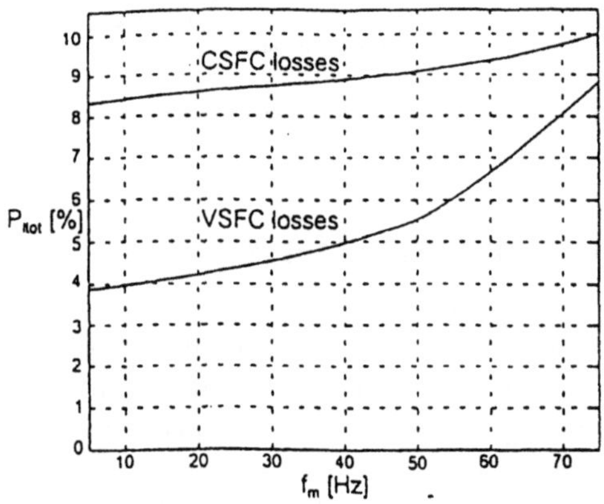

Figure 13a: Total losses $P_{\pi ot}$ of the VSFC and the CSFC

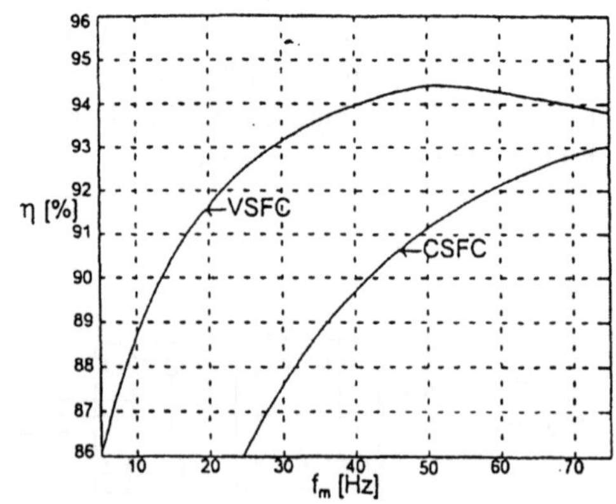

Figure 13b: Efficiency of the VSFC and CSFC

8 Filters

There are several reasons for using filters between the converter output and motor terminals; filters reduce the voltage transients and gradients (du/dt) at motor stator winding, they reduce the electromagnetic noise, they reduce the risk of bearing currents and as they reduce the harmonic currents in the motor they minimize the harmonic losses and torque ripple in the motor, and in general they have a positive impact on EMC.
Following the theme of the conference, a well designed filter eliminates more harmonic losses from the motor than it creates within itself ! - in other words the total efficiency of the drive is improved !

9 Conclusions

As generally known a converter voltage or current causes more losses in an induction motor than a sinusoidal voltage supply. These additional losses are usually compensated by derating the motor in the converter drive.
The energy savings gained with the speed control are in typical cases a decade greater than the loss increase in the motor caused by the converter voltage, resulting to pay-back-times typically from few months to a couple of years - and even greater environmental benefits; reduced pollution due to decreased energy production in electric power plants.
If a new installation is planned and the motors are to be converter driven, premium "converter proof" motors may be used. These are standard motor modifications -creating an improved design, tailor made for converter operation. A high efficiency motor is a high efficiency motor also in converter drive (the harmonic losses are relatively lower in high efficiency motors).
The world is bursting with induction motor driven applications where speed control improves the process and saves energy several tens of percent annually. The penalty which has to be paid for this great advantage is the converter price, installation cost and a small cost for decreased motor efficiency - peanuts compared with the total savings. The improvement of the system efficiency is multifold compared with the degradation of the motor efficiency. And with careful selection of the motor this drawback can be minimised.
In the future, variable speed drives may use motors designed for really high efficiency, motors with copper cage winding and synchronous type motors, such as reluctance motors and permanent magnet motors.

10 References

[1] Arkkio A."Analysis of a 37 kW cage induction motor". Report 30, Helsinki University of Technology, Laboratory of Electromechanics, Espoo Finland 1991

[2] Nee H-P."On rotor slot design and harmonic phenomena of inverter fed induction motors", Royal Institute of Technology, Department of Electric Power Engineering, Electrical Machines and Drives, Stockholm 1996

[3] Halkosaari T.,Tuusa H.,"The comparison of the efficiency in a current and voltage source inverter supplied from a controlled rectifier". Institute of Power Electronics Tampere University of Technology, Tampere Finland 1999

[4] IEC 60034-17, Technical Report "Guide for the application of cage induction motors when fed from converters"

[5] ABB test reports, "Motors fed with different converters". Internal reports 1972...1999

Die Casting Copper Motor Rotors: Mold Materials and Processing for Cost-Effective Manufacturing

Drs. Dale T. Peters[1], John G. Cowie[1] and Dr. Edwin F. Brush[2], Jr.

[1] Copper Development Association Inc
[2] Consultant to International Copper Association Ltd

Abstract. This project seeks to demonstrate mold materials for copper pressure die-casting that are cost-effective and practical for production use in die-casting copper motor rotors. The incorporation of die-cast copper for conductor bars and end rings of the induction motor in place of aluminum would result in attractive improvements in motor energy efficiency through reductions in motor losses ranging from 15% to 20%. Die-cast motor rotors are produced in aluminum today because rotor fabrication by pressure die-casting is an established practice. Lack of a durable and cost-effective mold material has been the technical barrier preventing manufacture of the die-cast copper rotor. This project tested H-13 steel die inserts that establish the baseline. Nickel-, tungsten-, and molybdenum-based high temperature alloys were extensively tested. Results indicate that substantially extended die life is possible using high temperature die materials, pre-heated and operated at elevated temperatures. Pre-heating and high operating temperatures were shown to be critical in extending the die life by decreasing the cyclic stresses associated with thermal expansion. Extended die life provides the opportunity for economically viable copper motor rotor die-casting.

1 Introduction

It is well known that incorporation of copper for the rotor conductor bars and end rings in the induction motor in place of aluminum would result in attractive improvements in motor energy efficiency due to copper's exceptional electrical conductivity.

Die-cast motor rotors are universally produced in aluminum today because fabrication by pressure die-casting is a well-established and economical method. Only small numbers of very large motors utilize copper in the rotors by mechanical fabrication. Such fabrication involves intensive hand labor and therefore is expensive. Die-casting, when it can be performed, is widely recognized as a low cost manufacturing process. For these reasons, die-casting has become the fabrication method of choice and aluminum the conductor of choice in almost all but the largest frame motors. Tool steel molds as used for the aluminum die-casting process have proved to be entirely inadequate when casting higher melting point metals including copper. Lack of a durable and cost-effective mold material has been the technical barrier preventing manufacture of the die-cast copper rotor.

An important study sponsored by the US Department of Energy found that motors above 1/6 Hp used about 60% of the electricity generated in the United States[1]. When extrapolated worldwide, the potential economic and environmental benefits of this project are substantial. Medium horsepower motors, 1-125 Hp (0.75 to approximately 100kW), use about 60% of the electricity supplied to all motors in the US. Because of the proliferation of electric motors in this horsepower range, the target of this project, the projected energy savings of the copper rotor motor is a significant national consideration. Efficiency increases (a function of motor size) from improved electrical conductivity are projected to result in total US energy savings in the year 2010 of 20.2 E+12 Btu/yr at only 10% market penetration and 143 E+12 Btu/yr at the expected market penetration of 50 to 70% (dependent on motor size). These numbers for the US are equivalent to the yearly output, respectively, of 0.5 to 3.5 600 MW generating plants operating at 75% of capacity.

This research project is funded jointly by the International Copper Association, Ltd. (ICA) through the US Copper Development Association Inc. (CDA), the U.S. Department of Energy (DoE) through their NICE[3] program, and the Air Conditioning and Refrigeration Institute (ARI). Additional sponsors include ThermoTrex, Formcast, and motor manufacturers. The objective of this research is to identify high temperature, thermal shock resistant materials, and then to design, fabricate and demonstrate molds and processing conditions designed to withstand the copper motor rotor die-casting environment for an economically acceptable life, i.e., thousands of casting cycles.

Several candidate die materials were identified; tungsten and molybdenum based refractory metals, beryllium-nickel, nickel-based superalloys, and one or more compositions in the tungsten-based composite family produced by a high-speed chemical vapor deposition (CVD) technique by the Trex Enterprises. This project in its *first* phase fabricated and tested a mold design (simulating the gate and volume of molten copper for a rotor mold) with inserts of the several mold materials for testing on a Buhler, 800-ton horizontal-shot-controlled pressure die-casting machine located at Formcast, Inc., in Denver, Colorado. In the *second* phase of this project, the most promising mold materials from Phase I are being fabricated into motor rotor molds and will be run for an extended number of shots at this same facility. For these runs, motor company partners will supply iron

---

[1] Classification and Evaluation of Electric Motors and Pumps, DOE/CS-0147, February, 1980

lamination stacks for appropriate motors designed to use copper rotor conductors. The motor partners will test the performance of the die-cast copper rotors.

2 Background

Motor manufacturers currently offer standard efficiency motor designs, and many offer one or more higher efficiency designs at price premiums. Operating energy efficiency improvements have been a strong emphasis since the late 1970's in the US, when many aluminum stator windings were replaced by copper. Newer designs have recently improved efficiencies further by increasing the amount of copper in windings, additional core and copper coil size, reduced windage losses, improved core steel, etc. However, the rotor remains die-cast aluminum because long-lived molds are not available.

Recent analysis by two U.S. motor manufacturers shows that the economics of motor operation and manufacture favor the use of copper in all classes of motors if the die life in the pressure die-casting process can be extended to the order of 20,000 shots.

Die-cast copper rotors can provide advantages to motor manufacture and/or performance in three ways:

- improvement in motor energy efficiency in operation
- reduction in overall manufacturing cost
- reduction in motor weight .

The motor manufacturer can accentuate one of the advantages at the expense of the other two. For example, in the case of a premium 10 Hp motor recently analyzed, the motor efficiency is 91.0%, as tested by IEEE Method 112-B. It should be noted that for a similar 10Hp (7.5kW) motor, the minimum efficiency levels, as tested by true input/output methods such as IEEE, or CAS, are 89.5% as required by the US Energy Policy Act (EPACT), and 91.7%, as currently recommended for minimum "premium" efficiency by the US Consortium for Energy Efficiency (CEE).

Three design scenarios using cast-copper-rotors (CCR) have been analyzed:
(1) seeking maximum efficiency improvement; (2) seeking maximum manufacturing cost reduction; and (3) seeking motor weight reduction.

Motors losses result from primary (stator winding) I^2R (usually 34% to 39%), secondary (rotor) I^2R (usually 16% to 29%), iron (core), friction and windage, and stray load[2] In addition to direct reduction in rotor loss with CCR's, designs achieve additional reductions from overall motor re-optimization of iron, strays, etc. CCR-based designs show overall loss reduction from 15% to 20%.

(1) If motor re-design efforts were devoted solely to improving efficiency, it is estimated that the new design with CCR could achieve 92.5% efficiency. This CCR example creates a "super" premium efficiency motor with an efficiency level (i.e., 92.5%) higher than currently available premium efficiency motors.

(2) If motor re-design efforts were devoted solely to reducing manufacturing costs for the current 91.0% efficient premium motor, it is estimated that the new design using CCR could be manufactured at a $36 reduction in overall manufacturing cost (15% of current $240 estimated manufacturing cost), maintaining exactly 91.0% efficiency.

(3) If motor re-design efforts were devoted solely to reducing motor weight, it is estimated that the new design could reduce weight by 5% to as much as 10%.

In addition, CCR could be utilized to reduce rotor I2R losses in an existing motor design, (replacing the existing die-cast aluminum rotor, not re-designing the motor to include more/better quality core, more stator windings, etc.—the existing methods to achieve higher efficiencies), in order to improve motor energy efficiency in operation.

CCR's can be used in specific motors to achieve a multiplicity of intermediate combinations of these design advantages. For example, where a smaller efficiency increase is required, the CCR could be used to achieve some reduction in manufacturing cost (stator winding, core, etc.) than would otherwise have been the case with traditional aluminum die-cast rotor technology.

The problem encountered in attempting to die-cast copper motor rotors is thermal shock and thermal fatigue of mold materials. Thermal cycling of the mold surface limits the mold life even in aluminum die-casting. However, cyclic thermal stresses are so severe in copper die casting that in at least one recent instance, a mold-gate-plate made of high strength steel (H-13, a die casting industry standard) being tested fractured in fewer than 10 casting shots. To be economically feasible, mold life must be measured in thousands of casting cycles.

---

[2] Steels for Laminations in Energy Efficient Motors, CMP/EPRI Report 9-11, June 91 Table 2-1

3 Innovation

A problem with common mold materials is that they lose strength at high temperature thus requiring low mean operating (and pre-shot surface) temperatures. A low initial temperature results in a large ΔT at the surface of the die, and thus the stress in the die, on each shot. The high melting temperature, high heat of fusion, substantial latent heat and high thermal conductivity of copper combine to maximize the thermal shock.

As a starting point, the solution to the thermal shock problem lies in the use of high temperature materials having thermal and thermo elastic properties conducive to minimizing thermally induced strain. Studies conducted by the International Copper Research Association (INCRA) in the 1970's confirm these expectations.

A major innovation in our program is in the process handling of these mold materials; we are die-casting with these high-temperature, high-performance materials when they are pre-heated (to elevated temperatures) to reduce this thermal differential and, thus, reduce the thermal shock to the molds within each cycle. Extended mold life appears feasible with several of our materials tested with this reduced temperature differential. In addition, the higher the pre-heat temperature (e.g. the lower the thermal differential) with several of our tested mold materials, the greater the reduction in heat-checking tendencies (or the longer in-service life before any heat-checking).

Tungsten and molybdenum were identified in the INCRA studies as good candidate materials for copper die-casting; however, they have not found extensive use in industry largely because of high base metal and fabrication costs. These refractory materials hold promise for long life (we included TZM, Anviloy, and Trex's tungsten Chemical Vapor Composite (CVC) materials in our testing). However, they formed oxides at our testing temperatures and may be still more expensive than necessary for all components of the mold system. Two promising candidates for mold material are the nickel-based super alloys (we tested 617, 625, 718, and 754) and the beryllium-nickel alloys. None of these materials has the low expansion of tungsten or molybdenum, but they do retain exceptional strength at high temperatures, and the Inconel alloys especially demonstrate good oxidation resistance at our operating temperatures.

Some of these high-temperature, high performance materials have been available for many years. Our efforts indicate that the large thermal differential when die-casting copper requires that the mold be pre-heated (reducing the thermal differential), in order to achieve a longer, economically viable mold life. In contrast, traditional die-casting uses molds to solidify, and reduce the molten metal temperature rapidly.

4 Results

An Inductotherm melting furnace was installed for *phase 1* at Formcast in Denver, Colorado, capable of continuously melting eight pounds of copper at 2-minute intervals for successive die shots for feeding to the 800-ton Buhler real-time, computer controlled, die-casting machine at this facility. A test mold was designed to simulate the action at one gate of a multi-gate mold. Die casting trials of candidate mold materials do not incorporate the iron lamination stack because of the high cost of lamination material involved in the thousands of shots of testing candidate mold materials.

Figure 1: Inductotherm melting furnace in operation

<u>H-13 Tool Steel</u>

Efforts were directed at evaluating a number of different high-temperature mold materials. Since the dies operated at very high temperatures, only materials designed for high-temperature service are under consideration for use in these molds (e.g., nickel-based super alloys and refractory metals such as tungsten and molybdenum). However, to establish a baseline for the test mold, the first copper die-casting trial was conducted using the H-13 steel die inserts (Figure 2). As expected, the dies generally degraded with increasing usage. Quite surprisingly, although some heat checking of the H-13 steel mold inserts and shot sleeve was evident after 20 shots, an extended run with copper against H-13 was accomplished. This success is attributed to the dry mold release system, the fast cycle time, which minimizes heat input to the mold, and through the use of advanced computer controlled casting equipment. Over 800 castings were produced using one set of dies, exceeding expectations by an order of magnitude. (Figure 3)

Figure 2: Copper die-casting

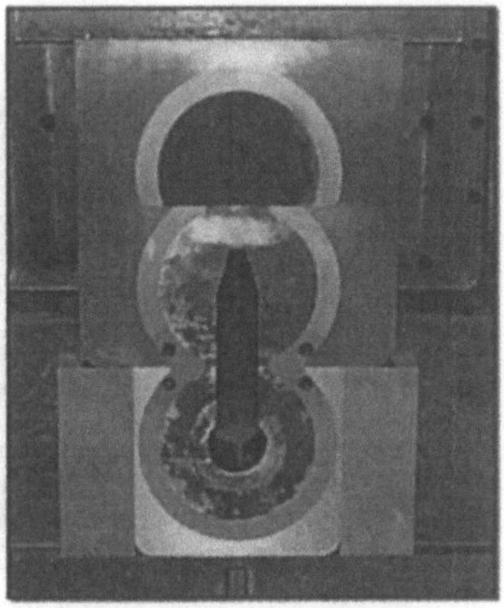

Figure 3: The H-133 steel test cavity tool set after the first several shots

Several copper castings were metallurgically, chemically and physically analyzed. The gate and runner macrostructures showed an outer columnar chill zone and a mixture of equiaxed and columnar grains in the bulk. The microstructures also showed the presence of an interdendritic phase most likely a eutectic copper-oxygen phase. Surface cracks and tears were found in the gate sections, in general the number and depth decreased with shot number. Internal defects resembling oxide films, macroscopic pores and slag type inclusions were also found and again decreased in size and frequency with shot number. A small amount of porosity was also present within the castings, but the overall microstructures are sound (Figure 4). The electrical conductivity measurements taken from the castings averaged 98% IACS and varied between 95% to 101% IACS. Samples were chemically analyzed for oxygen and iron contamination. The iron content varied from 10 ppm to 350 ppm and the oxygen levels from 0.06% to 0.15%.

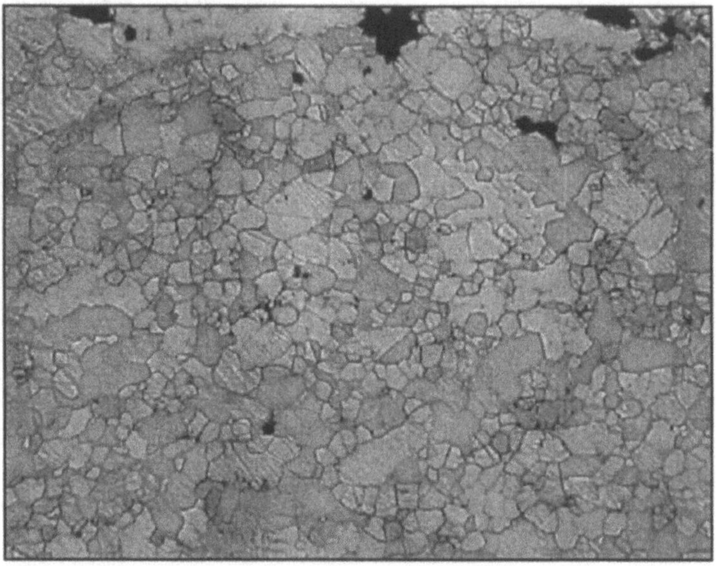

Figure 4: Photomicrograph of die-cast copper. Microporosity is a common defect associated with the die-casting process. 50X

Nickel-Based Superalloys

Die inserts were machined from Inconel alloys 617, 718 and 754. Over 250 die-castings were performed using these die sets, with the Inconel mold inserts pre-heated to over 300C. The Inconel 754 set began cracking very early (50 shots) into the run. This was somewhat surprising in that this particular alloy exhibits the highest strength at temperature of the three nickel-base alloys tested. But this alloy also has very low ductility at elevated temperature. The alloy with the lowest strength at temperature, Inconel alloy 718 began cracking after about 100 shots. The best performing alloy was the Inconel alloy 617, which exhibits the best combination of strength and ductility at elevated temperatures. Only minor craze cracking was evident on these die sets after 250 shots. These data provide an important clue toward solving this engineering problem, namely that a high fracture toughness at the service temperature may be important toward reducing the propensity to cracking and ultimately achieving extensive mold life.

Three copper castings were metallurgically, chemically and physically analyzed. Traces of iron, nickel and oxygen contamination were evident. Again, a small amount of microporosity was present within the castings, but the overall microstructures were sound. The electrical conductivity was better than that produced with the steel molds, nearly 100% IACS. To retain such high conductivity after melting in an open air environment and casting through a steel shot sleeve into nickel molds is very promising, as this will allow the copper scrap to be completely recyclable within the foundry.

A new set of Inconel 617 and 625 mold inserts was fabricated, and are presently being tested (targeting 650 C. preheat) for an extended run (over 1000 shots). The objective is to determine mold processing temperatures at which no mold degradation is evident after an extended run with these mold materials (the most promising of the Inconel mold materials tested to date).

CVD Tungsten

Tungsten inserts were fabricated by a high-speed chemical vapor deposition (CVD) technology developed by Trex Enterprises, San Diego, California. This technique offered the advantage of net-shape or near-net-shape fabrication on a graphite mandrel. An appropriate backing for the relatively thin tungsten or tungsten rhenium alloy shape remains to be developed. In the meantime the tungsten deposition was applied to a machined TZM molybdenum performs. Copper die-castings of the shape shown in Figure 2 were produced at Formcast. Tool life of the CVD tungsten molds has been disappointingly short at this point largely due to the columnar grain structure of the tungsten. Subsequent modifications to the mold set at Formcast to raise the tool temperature and efforts by Trex Enterprises to refine the grain structure are expected to considerably improve the life of tooling made by this technology.

TZM (Molybdenum Alloy) and Anviloy (Tungsten Alloy)

These alloys were considered in the early INCRA work on mold materials for pressure die-casting of copper, but their high ductile-brittle transition temperatures threaten survival of molds, especially in the first few shots. The solution is to pre-heat the molds by electrical resistance heaters.

TZM and Anviloy alloys were machined into die sets. They were heated to about 500 C before and during the die-casting trial. Over 500 shots were made with these die sets. There was some limited oxidation on both the TZM and Anviloy dies. However, no cracking due to heat checking was evident. The performance of these two die materials was unparalleled (Figure 5 - Left). Experience with these dies has shown that increasing the die operating temperature limits the thermal expansion and contraction, thereby decreasing the propensity to thermal fatigue cracking, more commonly known as "heat-checking." It is expected that further increasing the operating temperature of the dies will greatly extend their life by limiting the amount of cyclic strain associated with heating and cooling. It is predicted that operating the dies at elevated temperatures (just below the freezing point of copper) will limit the propensity for heat checking, thereby extending the life of dies to thousands of cycles.

Additional testing of these dies has recently been completed. Over 940 total die-castings have been produced. The TZM and Anviloy show minor cracking at the ejector pin locations, as shown in Figure 5 (Right); this cracking was due to inadequate allowance for thermal expansion in the steel ejector pins, and is able to be easily corrected. Otherwise, these molds showed long-life in die-casting service, without heat checking in the contours, gate areas, flat surfaces, etc., to a remarkable extent. With these results, further shots with these dies were not pursued, in order to install the new Inconel 617 and 625 insert set and begin that extended-life run, at higher pre-heat temperatures. Results to date show for these die materials that the higher pre-heat temperatures produce the substantially improved mold life, as anticipated.

Figure 5: Left: Moving half of die set containing the TZM and Anviloy materials. Right: Close-up of Anviloy die after 500 shots. No thermal fatigue cracking (heat checking) was evident.

Inconel Alloy Dies

The new Inconel insert set shows little degradation in the mold set with over 600 shots already run, using the higher pre-heat temperature (Figure 6). A run of 1000 shots is planned for the Inconel alloy die set. After 300 shots, there is no evidence of cracking. Oxidation in and around the dies appears self-limiting. It is becoming increasing clear that operating the dies at elevated temperature is absolutely essential toward improving the die life. The dies are being operated at elevated temperature (600 to 650 C) to reduce the thermal expansion and contraction associated with casting and subsequent cooling. Limiting the cyclic expansion and contraction is helping decrease the thermal fatigue.

Figure 6: Fixed die half with Inconel alloy die inserts after 300 shots. The dies have oxidized a little but have not cracked after 380 shots.

Rotor Die Set

A die set was purchased from Buhler in Switzerland. The die set is presently being machined to fit the Buhler die-casting machine at Formcast in Denver. (Figure 7) Drawings were made for machining die inserts for producing rotors.

Inductotherm constructed and delivered an induction-heating furnace capable of melting 20 pounds of copper in less than 8 minutes. This furnace upgrade is necessary to die-cast the copper rotors. (Figure 8)

Figure 7: One section of the die set purchased from Buhler for die casting copper rotors

Figure 8: Large Inductotherm melting furnace capable of rapidly melting over 20 pounds of copper.

5 Summary

Copper die cast motor rotors would result in attractive improvements in motor energy efficiency. Advances have been made toward the development of durable and cost effective mold materials, presently the last major hurdle preventing die-casting of the copper rotors. An extended run of copper against TZM and the tungsten alloy Anviloy was accomplished without major heat checking.

A major innovation in our program is its design to reduce the thermal differential through pre-heating to high temperature the mold material (raising the operating mold insert temperature), in order to reduce thermal shock when contacted by the copper. With this technique, several high temperature, high performance materials may have long-life in copper motor rotor die-casting.

A new, long run with Inconel 617 and 625 is underway,. and is anticipated to exceed 1000 shots and demonstrate no die degradation, using even higher die pre-heat temperatures (targeting 650 C.) to improve upon the earlier Inconel test results (conducted at die pre-heat over 300 C.). Increasing the operating temperature of the dies has shown to date that it extends their life by limiting the amount of cyclic strain associated with heating and cooling. The use of high pre-heat temperatures with these high temperature, high performance die materials significantly increased the mold life making possible the die-casting of copper and other high melting point materials. It is anticipated that a mold set incorporating a combination of nickel-base alloys and refractory alloys in the hottest portions will allow economical production of die-cast copper rotors.

Present plans entail the completion of the Inconel 617 and 625 long-run testing to demonstrate long-die-life-in-service, demonstrating the mold material and processing economic viability, and Phase II, namely die-casting several copper rotor designs for testing and analysis by the electric motor manufacturers. After which, the emphasis for this project will be to transfer this new die-casting technology to the manufacturers of electric motors.

Emphasis for this project will be to transfer this new die casting technology to the manufacturers of electric motors. Fabrication of prototype rotor die-casting for the largest motor manufacturers would expedite the transition to copper rotor motors. A meeting will initially be held with the electric motor manufacturers in order to announce the program. After which, the plan is to transfer technology to each electric motor manufacturer through demonstration of the die casting process.

Review of the System Compatibility and Ride-through Options for AC and DC Drives Including Multilevel Inverters

Annette von Jouanne[1] and B. Ben Banerjee[2]

[1] Power Electronics laboratory - Elect. & Compt. Engr. Dept. - Oregon State University - Corvallis, OR 97331-3211.
[2] Electric Power Research Institute - Power Electronics, Energy Delivery - 3412 Hillview Avenue - Palo Alto, CA 94304

Abstract. Adjustable Speed Drive (ASD) compatibility and ride-through issues have caused increased concerns due to the susceptibility of AC and DC drives to power disturbances, and the costly results of process disruptions. These losses can be avoided for critical production processes by using ASDs with ride-through capabilities. This paper assesses industrial ride-through requirements and application issues for AC and DC drives, including medium voltage (2300/4160V) multi-level inverter topologies. Ride-through alternatives are evaluated based on design, implementation and cost considerations in order to determine the most suitable solutions for various kVA ratings and time duration requirements.

1 Introduction

The application of Adjustable Speed Drives (ASDs) in commercial and industrial facilities is increasing due to improved efficiency, energy savings and process control. However, AC and DC drives are often susceptible to voltage disturbances such as sags, swells, transients (e.g. due to capacitor switching) and momentary interruptions (outages). According to survey reports, voltage sags of 10-30% below nominal for 3 to 30 cycle durations account for the majority of power system disturbances, and are the major cause of industry process disruptions [1].

Depending on the application, and the characteristics of the disturbance, the ASD controlled process may be momentarily interrupted or permanently tripped out. This can result in a significant loss in revenue and costly down time. For example in continuous process systems such as metal casters, paper machines, winders, extruders etc. any interruptions to that process can halt the entire manufacturing flow, with extremely costly implications. The cumulative cost estimates of power disturbances in the U.S. range from $20B - $100B/year, where industries have reported losses ranging from $10k to $1M per disrupting event [1].

These losses can be significantly reduced for *critical production processes* by using AC (Figure 1) and DC drives with ride-through capabilities. This paper will present ride-through alternatives and issues for AC drives, DC drives and then medium voltage (2300/4160V) multi-level inverter topologies.

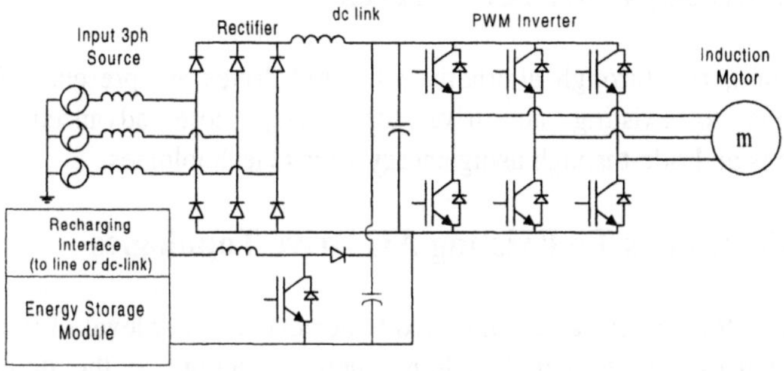

Figure 1: Use of Energy Storage as an Add-On Module for AC Drive Ride-Through.

1.1 The Industrial Power Environment

A piece of critical information necessary to determine the most cost-effective ride-through solution is the environment in which the critical ASD controlled process is expected to operate. Once this is defined, design objectives and mitigation devices can be specified. Recent advances in power-line monitoring technologies enable detailed analysis of the electrical environment. Thus, using statistics of average annual events, industry customers can predict downtime costs and make comparisons with the cost of additional ASD ride-through.

A number of power quality surveys aimed to define the electrical environment have been conducted in North America including: Sabin et al. (conducted by EPRI) [2], Dorr (conducted by the National Power Laboratory - NPL) [3], Hughes et al. (conducted by the Canadian Electrical Association - CEA) [4], Allen-Segal [5], Key [6], and Goldstein-Speranza [7], in addition to several European surveys [8]. From these power quality surveys and an ASD ride-through questionnaire conducted by the authors, it was determined that the most beneficial full power ride-through duration is 0.5 - 5 seconds, and should withstand a 50% sag. In addition, industrial ASDs are primarily process critical, requiring full speed and torque operation during a disturbance.

Another study conducted on the use of medium voltage (MV) ASDs reported that their application was also found to be associated with critical process areas of the plant, with system shut downs resulting in major process upsets and loss of production [9]. This study reported that manufacturers sold over 210 MV drives in 1995, over 400 in 1996 and over 500 in 1997. In addition, it was found that MV drives tend to experience ride-through problems during voltage sags below 85% of rated voltage for more than one cycle [9].

2 Ride-through for AC Drives

In this section, ride-through alternatives for AC drives are presented including modifications to existing ac drive topologies, more advanced hardware modifications and ride-through using energy storage technologies.

2.1 Modifications To Existing AC Drive Topologies

Existing drive topologies can be modified to achieve a higher level of immunity to line disturbances. These include adding more capacitors to the dc-bus, ride-through using load inertia, operating ASDs at reduced speed and/or load, and using lower voltage motors. An important distinction between each of the possible ride-through approaches is their ability to provide full power (full speed and torque) ride-through, which is required by many applications.

2.1.1 Additional Capacitors

By adding capacitors to the dc-bus, additional energy needed for full power ride-through during a voltage sag can be provided to the motor. A typical 460V, 60Hz, 10hp ac motor drive can be assumed to have a dc-link capacitance of C = 5000μF. The dc-link voltage V_{dc} (assuming continuous conduction) is given by:

$$V_{dc} = 1.35 * V_{LL} = 620V \text{ for } V_{LL} = 460V \qquad (1)$$

A typical ASD is set to trip if the dc-link voltage drops 0.9 times the nominal value, which is $V_{dc,trip} = 558V$. Also, the average dc-link current I_{dc} for a 10hp (P_o) load is:

$$I_{dc} = \frac{P_o * 746}{V_{dc}} = 12A \qquad (2)$$

Now, under short term power interruption, the filter capacitors must provide the power to the ASD motor. The ride-through duration t_r can be computed as follows:

$$t_r \approx \frac{C*(V_{dc} - V_{dc,trip})}{I_{dc}} = \frac{5000*10^{-6}(620-558)}{12} = 25.8ms \qquad (3)$$

A t_r of 25.8ms translates into 1.55 cycles at 60Hz frequency. Thus, 5000μF on the dc-link can only provide a full power ride-through of 10hp for 1.55 cycles. If the outage were to last for 0.5 seconds (i.e. t_r = 0.5, 30 cycles) the capacitance required to provide the ride-through can be calculated as:

$$C = \frac{I_{dc} * t_r}{V_{dc} - V_{dc,trip}} = \frac{12 * 0.5}{(620 - 558)} = 0.097F \qquad (4)$$

From Eqn. (4) it is clear that an additional 20 x 5000μF capacitors would have to be added to the dc-bus. Therefore, if the user were to use 2500μF caps rated at 400V connected in series, they would then need 160 capacitors, connected in 80 groups. Assuming the cost of each capacitor to be $40, the total cost amounts to $6,400 for a 10hp drive. The cost of enclosures, fuses, a precharge circuit and bus bars are additional.

Advantages:

- Simple and rugged approach, can provide limited ride-through for minor disturbances.

Disadvantages:

- Cost is relatively high.
- Large cabinet space, additional precharge circuits and safety considerations.

Approximate Cost - $600/kW, as demonstrated above.

2.1.2 Use of Load Inertia

The inverter control software can be modified such that when a power disturbance causes the dc-bus voltage to fall below a specified value (i.e. below 560V, on a 460V L-L system supporting a 620V dc-bus) the inverter will adjust to operate at a frequency slightly below the motor frequency, causing the motor to act like a generator [10]. The drive will absorb a small amount of energy from the rotating load to maintain the dc-bus at a specified level (i.e. 560V). Commercial drives are available on the market with this feature, and will maintain the specified dc-bus for 2 seconds during a dip that does not exceed 20%.

Advantages:

- No additional hardware is required, only a software modification in the inverter.
- Commercial drives are available on the market with this feature with 2 seconds of ride-through for sags to 80% nominal voltage.
- Since the drive and motor have been actively transferring energy during the power disturbance, no loss of phasing has occurred between the drive and the motor and the motor's magnetic field has not de-energized. Thus, there are no delays to start accelerating the motor as soon as the ac power line returns to normal, assuming the load can handle it [10].

Disadvantages:

- The motor speed is reduced and the torque is reversed which is acceptable for fan/pump type loads, but may not be acceptable for certain load types.
- The sustainable ride-through duration will be dependent on the load inertia.

Approximate Cost - Negligible, minor software modification [10].

2.1.3 Operate ASDs at Reduced Speed/Load

Since the dc-bus current varies with the frequency of the drive for variable torque loads, such as fans and pumps, a reduction in the motor speed will result in a reduction in the dc-bus current. Therefore, a fan and pump system running at 40Hz will draw less current than a system running at 60Hz and will therefore be able to operate for a longer period during a voltage sag situation [10]. The effect would be the same as increasing the capacitance, or energy stored, on the dc-bus. Similarly, if the motor load were reduced, the dc-bus current would be reduced and thus a longer ride-through duration could be achieved.

Advantages:

- No additional hardware is required.
- At 50% speed and load, would provide four (4) times the ride-through of a normal drive system.

Disadvantages:

- Application may not tolerate reduced speed/load operation.
- Only useful for variable torque (fans and pumps) loads.

Approximate Cost - None [10].

2.1.4 Use of Lower Voltage Motors

If a 230V ac motor were used with a 460V ac drive, the dc-bus voltage (nominally 620V) could drop to as low as 45% (to 280V) and still provide 230V ac to the motor. Note that the maximum V_{ac} output to the motor = V_{dc} x 0.82 [10]. Then, as the voltage drops, the inverter changes its duty cycle to maintain a constant 230V ac to the motor.

Advantages:

- No additional hardware is required.
- An increase of approximately 2.8 times the ride-through time of a normal drive system [10].

Disadvantages:

- The ASD rating is twice the hp rating of the 230V motor.
- A 230V motor with the same hp rating as a 460V motor will require twice the current at full load, and thus will have to be larger.
- The motor insulation must be capable of handling the higher voltages provided by a 460V ASD.

Approximate Cost - Cost of drive in $/kW for derating.

2.2 More Advanced Hardware Modifications

2.2.1 Boost Converter Ride-Through

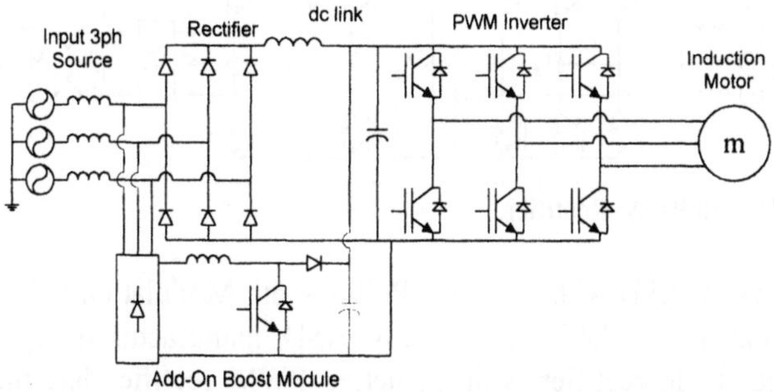

Figure 2: ASD with Boost Converter Ride-Through Device as an Add-On Module.

A boost converter can be used to maintain the dc-bus voltage during a voltage sag, and can either be integrated into new drives between the rectifier and the dc-link capacitors or retrofitted as an add-on module as in Figure. 2 [11]. The add-on module is used to retrofit existing drives with ride-through capabilities, or for multiple drives with a common dc bus, such as synthetic fiber drives. During a voltage sag, the boost converter will sense a drop in the dc-bus voltage and begin to regulate the dc-bus to the minimum voltage required by the inverter (i.e. 585V, which is user adjustable). In the case of a retrofit where a boost module is added to an existing ASD, proper coordination of fault protection logic is necessary.

Advantages:
- Can provide ride-through for sags up to 50%.
- The dc-bus voltage can be regulated as required by the inverter, and is user adjustable.

Disadvantages:
- Additional hardware required, which will have to be suitably rated due to the additional current drawn during a voltage sag.
- In the case of an outage, the boost converter will not be able to provide ride-through, and the drive will trip.

Approximate Cost - $100 - 200/kW [11].

2.2.2 Active Rectifier ASD Front End

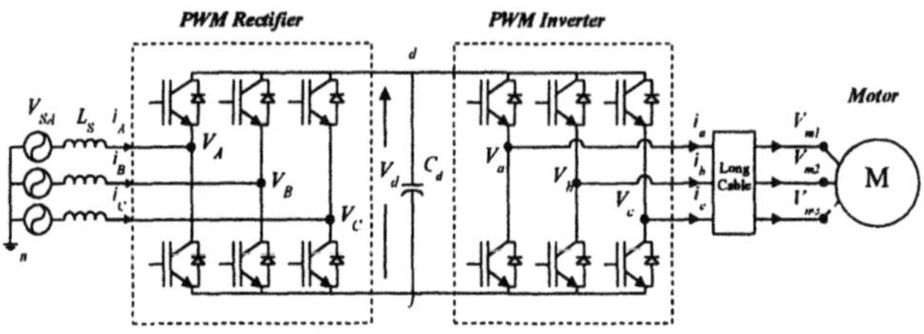

Figure 3: ASD with PWM Rectifier.

Figure 3 shows an ASD with an active Pulse-Width Modulation (PWM) rectifier. Such a system is available from many ASD manufacturers up to 500kW. Replacing the diode rectifier with an active PWM rectifier has the following advantages: regulated dc-bus which offers immunity to voltage sags and transients; low input current harmonics and compliance to IEEE 519 harmonic limits; and power flow in both directions enabling regenerative braking. The range of ride-through that this approach can provide is limited only by the current rating of the rectifier. By this method, ride-through for sags of up to 10% can be provided at full load. By derating the rectifier by a factor of 1.5, the ride-through capability can be extended to sags of up to 40% at full load [12].

Advantages:

- Clean input power in steady-state, unity power factor and compliance to IEEE 519 distortion limits.
- Active rectifier provides a regulated dc bus voltage, hence is self correcting under voltage sags. Suitable rectifier derating is necessary to provide full power ride-through under a sag.
- Power flow in both directions enables regenerative braking. This feature could add to improved efficiency is some applications.

Disadvantages:

- An ASD with an active PWM rectifier is nearly equivalent to two diode rectifier ASDs in cost.
- The ASD package is larger in size since in addition to the active rectifier hardware, three input filter inductors become necessary.
- Activerectifier operates the ASD with higher dc-link voltage; this results in higher differential mode dv/dt at the motor terminals. Also due to two PWM insulated gate bipolar transistor (IGBT) inverter stages the common mode dv/dt and EMI are higher.

Approximate Cost - Twice the cost of a regular ASD [13].

2.3 ASD Ride-Through Using Energy Storage Technologies

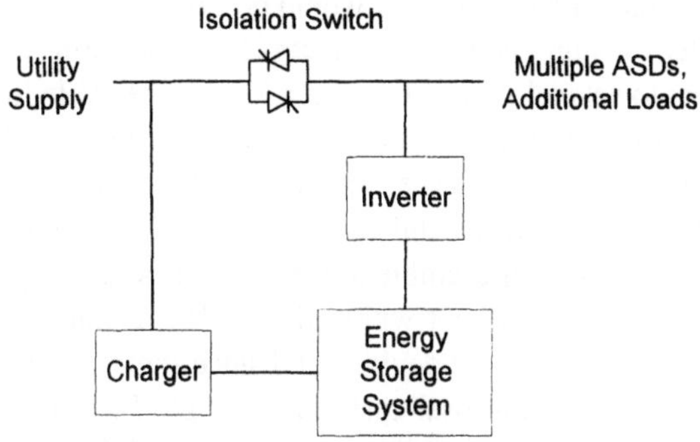

Figure 4: Energy Storage System as a Standby System for AC Line Power.

A variety of energy storage technologies are candidates for providing the needed full power ASD ride-through including: battery back up systems, super capacitors, motor-generator sets, flywheel energy storage systems, Superconducting Magnetic Energy Storage (SMES), and fuel cells. Table 1 gives a comparison of the energy storage characteristics [14-18]. Energy storage can be used as a standby system for the ac line power to supply a number of ASDs along with other loads as shown in Figure 4. A voltage sag mitigation approach termed as Dynamic Voltage Restoration (DVR) has been developed by Westinghouse as part of EPRI's custom power program for advanced distribution [19]. The DVR (Figure 4) is a solid state dc to ac switching converter that injects three single-phase ac output voltages in series with the distribution feeder and in synchronism with the voltages of the distribution system. The DVR is therefore capable of injecting missing voltage during a sag and can provide ride-through to sensitive loads and ASDs. A portable 2-MVA DVR with 1-MJ storage has been demonstrated by Westinghouse [20].

In addition, independent modules can be used to maintain the dc-bus of a single ASD (Figure 1), or for multiple drives connected to a common dc-bus. The charging devices in Figs. 1 and 4 are needed to power the energy storage system to normal levels during startup and after discharges. In Figure 1 a boost converter is used as an interface with the dc-bus for when a modest step-up is required. For ASD ride-through, power density, energy density, and efficiency characteristics are critical

2.3.1 Battery Back Up Systems

Battery back up systems operate similarly to adding capacitive energy storage, with the advantage that their energy per volume ratio is much higher than standard capacitors [14]. An ASD can be retrofitted with battery back-up as an add-on module on the dc-bus as shown in Figure 1. In order to maintain a dc-bus of

575V during an outage, 48 x 12V series connected batteries can be used. The required battery current will be dependent on the load. For example, for a 10hp motor, batteries capable of handling 15A would be sufficient.

Because batteries store energy in electrochemical form in "cells", their operation is subject to several limits. One such limit is the cycle life, which is the number of charge/discharge cycles possible for a given cell (see Table 1). Another limit is the depth-of-discharge, which is the fraction of stored energy that can be withdrawn. This will be dependent on the rate of discharge, and will affect the cycle life. In addition, the ambient temperature and the proper charging current must be monitored and kept within limits. The average footprint will be large (the floor area covered by the device), and since some depleted materials are considered hazardous waste, disposal costs can be high. Batteries are suitable for 5kW-10MW loads and can provide full power ride-through for up to 1hr.

Advantages:
- Can provide ride-through for deep sags and full outages.
- Batteries are easily obtained.
- Transfer time is almost instantaneous.

Disadvantages:
- Additional hardware and space required, though not as much as with standard capacitors.
- Relatively low cycle life.
- More maintenance required to ensure peak performance.
- The electrolyte is corrosive and may be hazardous to the application, and will need to be properly disposed of when depleted.

Approximate Cost - \$100 - \$200/kW [14].

2.3.2 Super Capacitors

Super capacitors (or ultracapacitors) offer substantial increases in energy density over conventional capacitors due to the choice and preparation of the electrode materials and increases in the effective capacitive plate surface area [14]. An ASD can be designed with integrated super capacitors, or an add-on module as shown in Figure 1.

A 100kW commercially available configuration [15] uses 8x 56V modules in series resulting in 448V, which is then boosted to the necessary voltage using a boost converter. Each module has a combined capacitance of 96.4F, and thus has an energy storage (E_{sm}) capability of:

$$E_{sm} = \frac{CV^2}{2} = \frac{96.4 * (56)^2}{2} = 151.2kJ \qquad (5)$$

The total energy storage (E_s) for a 100kW ride-through system is then:

$$E_s = 8 * 151.2kJ = 1.2\,MJ \qquad (6)$$

Eqn. (6) shows the total energy stored in the supercapacitors. Further, the stored energy is a direct function of the capacitor voltage. Therefore, a drop in 30% of the capacitor voltage from V to 0.7V amounts to the release of 50% of its stored energy. Using this as a design criteria, the ride-through duration (t_r) for a 100kW ASD load (under short-term power interruption) can be computed as follows:

$$00 * 10^3 * t_r = \frac{1}{2}\left[CV^2 - C*(0.7V)^2\right]*8 \qquad (7)$$

For C = 96.4 and V = 56V,

t_r = 6 sec (8)

Thus a maximum of 6 seconds ride-through can be obtained with this approach.

Advantages:
- Can provide ride-through for deep sags and full outages.
- Long cycle life and fast recharge rates.
- Easily monitored state of charge.
- Minimal maintenance needs.

Disadvantages:
- Additional hardware and space required, though not as much as with standard capacitors.
- Emerging technology.

Approximate Cost - $300-400/kW [14,15].

2.3.3 Motor-Generator Sets Using Written Pole Technology

Motor-generator (M-G) sets use their rotating mass to supply energy during a voltage sag or outage. Written pole technology is able to provide 15 seconds of ride-through for complete interruptions at full load [21]. This M-G set uses an electric-motor-driven synchronous generator that can output a constant 60Hz frequency, regardless of the speed of the machine, and thus could be used in the configuration shown in Figure 4. Typical units are available in 15kVA and 35kVA ratings. The generator can supply a constant output by continuously changing the polarity of the rotor's field poles. During a sag or outage, the generator rotor can continue to output 60Hz for 15 seconds under full load due to the mass and inertia of the rotor. For longer time periods, a diesel engine can be used to supply energy to the generator. A M-G set can also supply power directly to the dc-bus of the ASD through an ac motor-driven dc generator. In order to reduce the necessary size of the M-G set and the rotor mass, a flywheel can be used for the storage of kinetic energy as shown in Figure 5, which will be discussed in the next section [10].

Table I: Comparison of Energy Storage Characteristics

	Batteries	Super Capacitors	Flywheels	SMES	Fuel Cells
Efficiency	70-90%	90%	90%	95%	40-55%
Power Range (W)	5kW-10MW	5-100kW	1kW-10MW	300kW-1000MW	10kW-2MW
Energy Range (J)	0.1-600MJ	1kJ-10MJ	1-15MJ	1-500MJ	continuous
Cycle Life	2,000	100,000	10,000	10,000	continuous
Charge Time	hours	seconds	minutes	minutes-hours	continuous
Maturity	mature [14]	available [15]	available [16]	available [17]	available [18]
Capital Cost ($/kW)	100-200	500	300	700-1000	1500

Advantages:

- Can provide ride-through for deep sags and full outages.
- Very reliable, can provide 15 seconds of ride-through.

Disadvantages:

- Additional hardware and space required.
- Maintenance is required for the rotating components.

Approximate Cost - $200-300/kW [10,18].

2.3.4 Flywheels

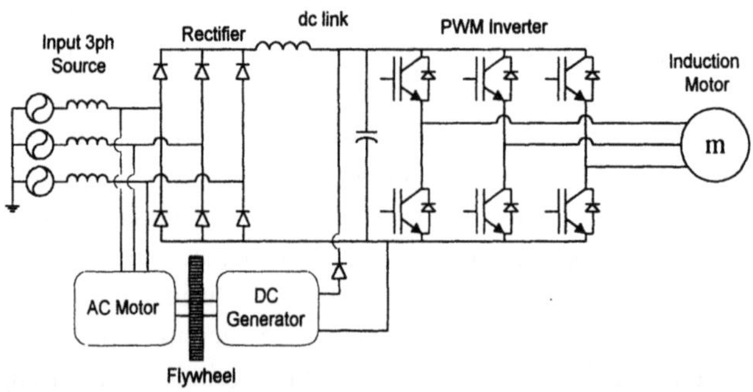

Figure 5: M-G Set with a Flywheel, Providing Power Directly to the ASD DC-Bus.

Flywheels, which store kinetic energy in a rotating mass, are also showing promise for ASD ride-through. A modification to the typical M-G set described in the previous section is shown in Figure 5, where an ac motor drives a dc generator, with a flywheel, providing power directly to the dc-bus of the ac drive [10]. Advancements in high strength composite materials have been important in the development of flywheels. In addition, high-temperature superconducting (HTS) bearings that are magnetically levitated to reduce bearing friction and drag losses, have recently become available [22]. Flywheels are suitable for 1kW - 10MW applications, and can provide full power ride-through for up to 1hr.

Advantages:
- Can provide ride-through for deep sags and full outages.
- Significantly reduces the size and necessary mass of M-G sets.

Disadvantages:
- Additional hardware and space required.
- Maintenance is required for the rotating components.

Approximate Cost - $200-300/kW [14].

2.3.5 Superconducting Magnetic Energy Storage (SMES)

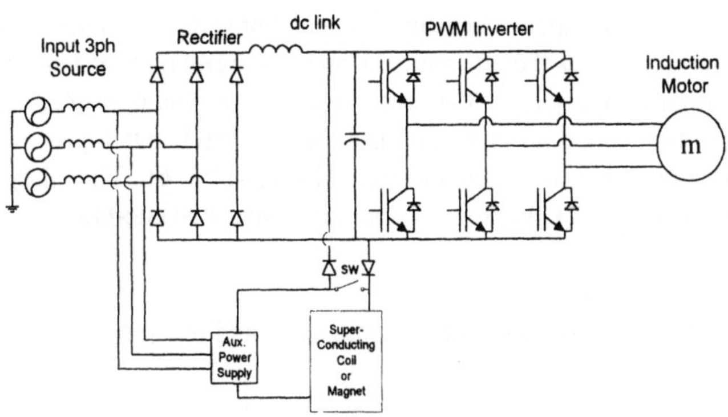

Figure 6: SMES Unit Providing Ride-Through Power Directly to the DC-bus of the ASD.

In a SMES system, a large amount of current is kept circulating in a superconducting coil or magnet, to be supplied to the system when needed. Since there are only negligible losses in the superconducting coil, the transfer of energy in and out of storage is highly efficient, and rapid. However, to remain superconducting, the coil must be cooled to cryogenic temperatures, which requires a fairly sophisticated refrigeration subsystem [14]. A SMES system can be interfaced with a power distribution system as shown in Figure 4, now taking into account that the SMES is a dc current source rather than a voltage source. A SMES unit can also be applied as shown in Figure 6, and directly connected to the dc-bus of an ASD or to a number of ASDs which share a common dc-bus. The auxiliary power supply shown in Figure 6 is needed to supply power to the part of the circuit that is outside the liquid helium chamber. When there is a sag or an outage, the switch in the outer circuit will open, forcing the circulating current into the filter capacitors of the drive to maintain the dc-bus [10]. Safety concerns have been addressed through the design of protection systems that dissipate the stored energy resistively, either internally or externally [14].

Advantages:
- Reliable, with little maintenance.
- SMES can handle rapid repeated discharging and charging without affecting its performance or life.

Disadvantages:
- Additional hardware and space required.
- Sophisticated cooling system required to maintain cryogenic temperatures and the associated power loss.
- High cost and safety concerns.

Approximate Cost - $600-800/kW [18].

2.3.6 Fuel Cells

Fuel cells provide continuous power through the consumption of hydrocarbon fuel; typically natural gas. Thus the operation is more similar to a battery in that electrochemical rather than electromechanical conversion occurs. A fuel cell could be interfaced with the ac line of a power distribution system as shown in Figure 4, similar to applying a M-G set. However, it would be more appropriate to operate the fuel cell at all times because the expense is much greater than a M-G set and it cannot be quick started from cold stand-by [14]. Thus they would not be as appropriate for back up power for an individual ASD dc-bus.

Advantages:
- Reliable and efficient, with little maintenance and few mechanically moving parts.

Disadvantages:
- Cannot respond rapidly to load changes.
- Additional hardware and high cost.

Approximate Cost - $1500/kW [18].

3 What Makes DC Drive Ride-through Different From AC Drive Ride-through

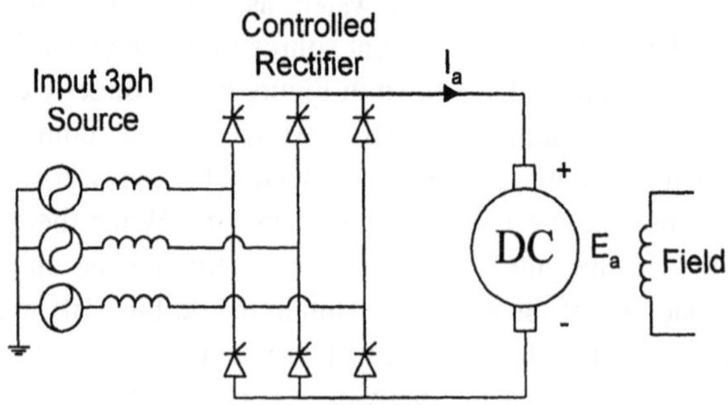

(a)

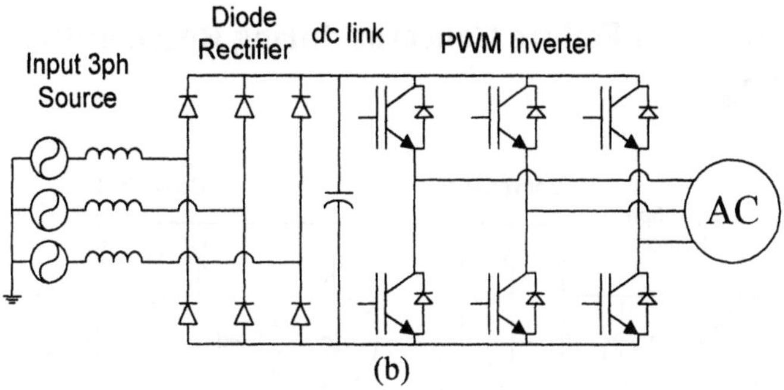

Figure 7: Typical a) DC Drive b) AC Drive.

The main differences between DC drive (Figure 7a) and AC drive (Figure 7b) ride-through stems from the front-end converter topology of the drive, the DC field winding source which also must be able to ride-through disturbances, and the issue of regenerative braking for speed reduction or reversal [23]. In some applications, such as in the steel and paper industries, regenerative braking and speed reversal are required frequently, which makes the process more susceptible to power system disturbances. Another major difference is that typical thyristor DC drives as shown in Figure 7a do not have energy storage (capacitors) on the DC link, which are seen in AC drives (Figure 7b).

Controlled rectifiers employing thyristors, i.e. silicon controlled rectifiers (SCRs), are generally used for speed control in industrial DC motor drive systems as shown in Figure 7a. The alternative form would be a diode rectifier (as with an AC drive front-end) followed by a DC-DC converter (chopper) as shown in Figure 8, which is widely used in lower power servo and high power traction applications. Since DC drives with chopper topologies have the same rectifier front end as AC drives, ride-through is also achieved in a similar way. Though for complete ride-through, the field excitation must also be robust enough to ride-through a power disturbance [23].

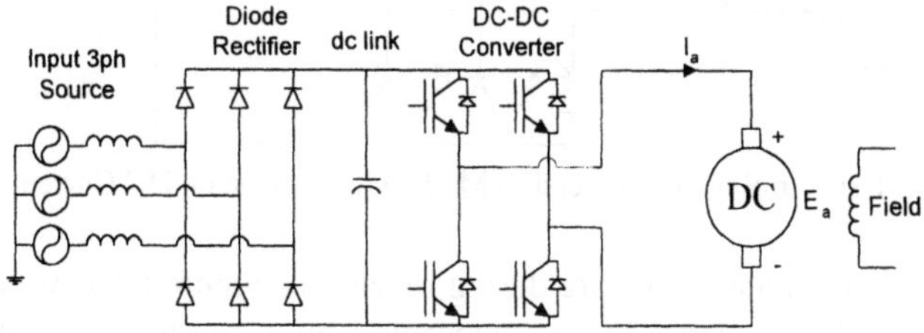

Figure 8: Alternative DC Motor/Drive Topology with a DC-DC Converter (Chopper).

3.1 Commutation Failure Concerns during Regenerative Braking

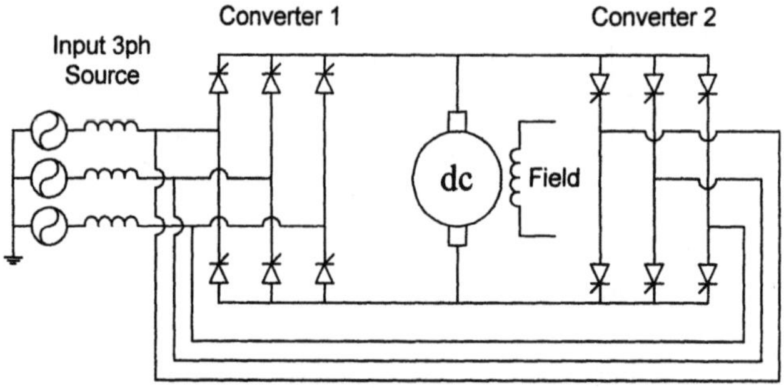

Figure 9: Regenerative DC Drive employing back-to-back thyristor converters enabling four quadrant operation.

For full control of the DC motor, two back-to-back connected thyristor converters can be used as shown in Figure 9, which allows four quadrant operation. During regenerative braking, the active converter in Figure 9 operates as an inverter and the DC motor acts as a generator, supplying power to the electrical system. If a power disturbance such as a voltage sag or an outage occurs during this inversion operation, there could be insufficient line voltage and commutation failure could occur resulting in blown fuses and interruption of the process [23].

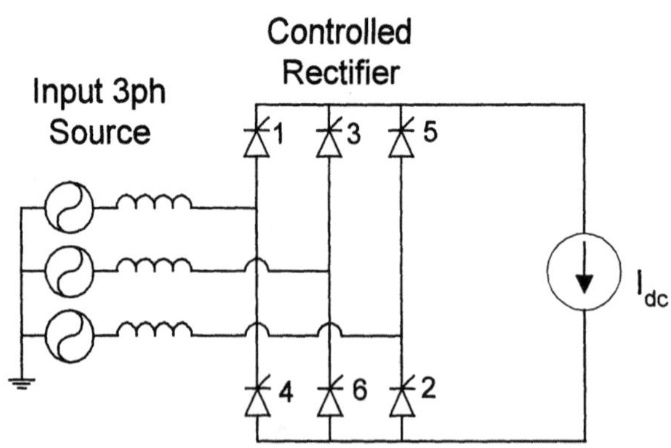

Figure 10: DC Drive in Regen or Inversion Mode with a Constant DC Current.

To understand commutation failure during regen mode more fully, we can assume the dc side of the converter to be represented by a constant current source as shown in Figure 10, which also shows the order of thyristor commutation. For a thyristor delay angle $90^0 \leq \alpha \leq 180^o$, the converter is in regen or inversion mode and the average value of the output voltage is negative. The extinction time interval (t_γ) during which the voltage across the thyristor is negative, must be greater than the thyristor turn-off time (t_q). (Note that thyristor turn on is controlled by the gate, and turn off begins to occur when the voltage across the thyristor goes negative) If $t_\gamma < t_q$, the thyristor will not be fully turned off when positive voltage is impressed across it causing premature conduction. Commutation failure, or the failure of current to commutate from one thyristor pair to another, will occur and cause a short circuit condition of the regenerative dc motor load, resulting in large destructive currents. Commutation failure can occur during regen mode for a power quality disturbance such as a voltage sag because the commutating voltage could be reduced to zero, or to a low value. Thus, DC drives are more susceptible to power disturbances during regenerative braking [23].

This condition can be avoided somewhat by limiting the range of delay angle α to be less than 165° or by modifying the regen-SCR converter by integrating a gate-turn-off thyristor (GTO) device in the converter dc side to assist commutation during regen operation.

Note that some industries employ DC drives with dynamic (resistive) braking. During dynamic braking, the armature is disconnected from the converter and a resistor is connected across the armature terminals. The DC motor acts like a generator, dissipating its kinetic energy in the braking resistor. Therefore, a power disturbance during dynamic braking is not a concern like it is with regenerative braking.

4 Ride-through for Thyristor Controlled Rectifier DC Drive Configurations

Since thyristor DC drives supply a variable DC voltage to the armature circuit for the control of torque and speed, complete ride-through during a power disturbance can entail either interfacing a controllable voltage source on the armature circuit such as through a buck-boost converter, or supporting the AC system on the line-side of the drive. Both of these approaches can take advantage of energy storage technologies. The use of lower voltage motors can also provide full power ride-through if suitable for the application. If full power (full speed and torque) ride-through is not required by the application, a flying restart algorithm can be used if the load condition allows.

Industrial surveys indicated that DC drives can appear to be less susceptible than AC drives to power disturbances [23]. This is due in part to the fact that when the DC motor requires less than rated voltage, a safety margin exists if a voltage sag does occur. Upon the occurrence of a voltage sag, the thyristor delay angle alpha decreases towards zero to maintain the appropriate armature voltage. This differs from AC drives in that the drive is commonly set to trip if the DC bus voltage falls below a set threshold value. However, if the thyristor firing angles go to zero, and the armature voltage is less than what the DC motor load requires, alternative ride-through techniques can be applied as will be discussed in the following sections.

4.1 Flying Restart Algorithm in DC Drive Control

Applications that do not require full power (full speed and torque) ride-through such as fan and pump type systems, can be modified with a flying restart algorithm. In this case, during a power disturbance the DC drive would take itself off-line and the motor will coast until it is determined that the power has been restored, at which time the DC drive would automatically restart. Note that this is not appropriate for some critical processes and thus some drives have a relay in the control circuit to prevent automatic restarting after a power disturbance.

Advantages:
- No additional hardware is required, only a software modification in the drive.
- This approach can be used while in either motoring or regen mode.

Disadvantages:
- The motor speed and torque will be reduced which may not be acceptable.

Cost - Negligible, minor software modification.

4.2 Use of Lower Voltage Motors

If a 230V DC motor were used with a 460V DC drive, the line voltage could drop as low as 50% (to 230V) and still provide full controllable voltage up to 230V DC to the motor. In this case, as the line voltage drops, the thyristor control angle α reduces towards zero to maintain the required voltage to the armature of the circuit of the DC drive. While this approach requires no additional hardware to provide full power ride-through, the disadvantages are that the DC drive is not fully utilized (derated by 50%), and that a 230V motor with the same hp rating as a 460V motor will require twice the current at full load, and thus will have to be larger.

Advantages:
- No additional hardware is required.
- This approach can be used while in either motoring or regen mode.

Disadvantages:
- The DC drive rating is twice the hp rating of the 230V DC motor.
- A 230V motor with the same hp rating as a 460V motor will require twice the current at full load, and thus will have to be larger.

Cost - Cost of drive in \$/kW.

4.3 Interfacing a Line Input Buck-Boost Converter on the Armature Circuit

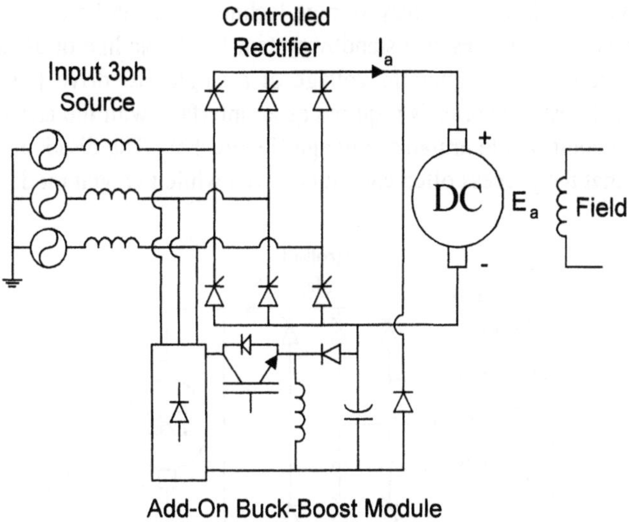

Figure 11: DC Drive with Buck-Boost Converter Ride-Through Device.

A buck-boost converter can be used to maintain the armature voltage during a voltage sag, and can either be retrofitted as an add-on module as shown in Figure 11 or integrated into new drive systems between the rectifier and the DC motor. During a voltage sag, the buck-boost converter will sense a drop in the armature voltage and begin to regulate the voltage of the armature circuit according to the demands of the load. The disadvantage of this approach is the additional hardware required, which will have to be derated due to the additional current drawn during a voltage sag. Note that a buck-boost converter can also be applied for DC drive ride-through using energy storage devices as the converter input/supply. This will be presented in the following section.

Advantages:
- Can provide ride-through for sags up to 50%.
- The armature voltage can be regulated as required by the load.

Disadvantages:
- Additional hardware required, which will have to be suitably rated due to the additional current drawn during a voltage sag.
- In the case of an outage, the buck-boost converter will not be able to provide ride-through, and the drive will trip.
- This approach will not be able to save the process if in regen mode.

Cost - $100 - 200/kW

4.4 DC Drive Ride-Through Using Energy Storage Technologies

For DC drive ride-through, energy storage technologies can be applied similarly as discussed with AC drives as a standby system for the ac line or as independent modules to maintain the armature voltage of a single DC drive [23]. For DC drives a buck-boost converter is required as an interface with the armature circuit to maintain a controllable armature voltage (Figure 12). The disadvantage of this approach is that the process often cannot be saved while in regen mode.

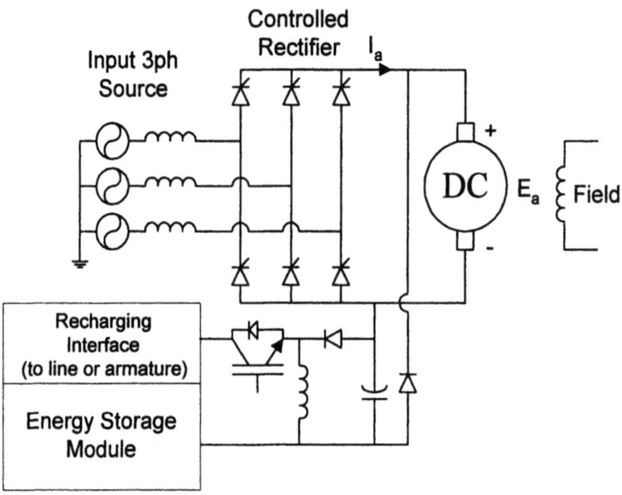

Figure 12: Energy Storage as an Add-On Module for a DC Drive Armature Circuit.

4.4.1 Ride-Through Considerations for the Field Winding of DC Motors

Industrial DC motors are typically separately excited and thus have a field winding that is supplied through either a diode rectifier (fixed field) or an SCR bridge, providing a variable field for speed control above base speed through field weakening. With a fixed voltage field winding supplied through a diode rectifier, energy storage techniques such as additional capacitors, battery back up or super capacitors can be used in a similar way as would be applied for ride-through in chopper topologies.

For higher performance DC motor drives with a wide speed range requirement (above and below base speed), controlled rectifiers are used to supply the field winding. For critical systems, to maintain control of the variable DC field voltage during a voltage sag, a buck-boost converter supplied by the line (as in Figure 11) can be used for sags no deeper than 50% nominal voltage. To ride through momentary interruptions (outages), additional energy storage in combination with the buck-boost converter can be used such as batteries or super capacitors.

5 Techniques for Medium Voltage Ride-through

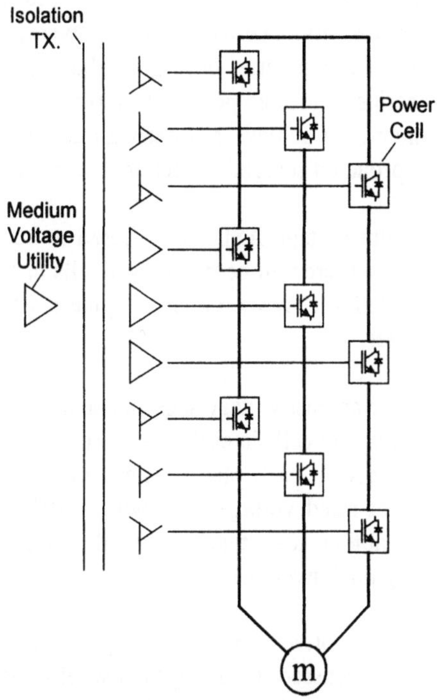

Figure 13: CI-ML-VSI for medium voltage ASDs [26,27].

A majority of ASDs for medium voltage (2300/4160V) drives address fan/pump type loads for energy savings and fall into two categories: current source inverters (CSIs) and voltage source inverters (VSIs). The CSI approach employs SCR/GTO devices and ride-through options can be implemented by employing flywheels and SMES systems as previously discussed [24].

The VSIs for medium voltage motor drives are based on multi-level inverter technology. These can be further classified into two types [24]: Cascaded inverter type multi-level voltage source inverters (CI-ML-VSI) as shown in Figs. 13-15 [25-27], and three-level neutral point clamped (NPC) inverters as shown in Figure 16 [28,29]. The CI-ML-VSI consists of several power-cells connected in series (Figure 13). Each power-cell is a single phase IGBT based VSI. A complicated input transformer becomes necessary to supply dc power to each power cell. The NPC inverter on the other hand is constructed with integrated gate commutated thyristor (IGCT) devices (Figure 16) [28,29].

5.1 Ride-Through for CI-ML-VSIs

In CI-ML-VSIs (Figure 13) often N+1 power-cells are used for redundancy and to enhance fault tolerant capability. The use of N+1 power-cells also enhances the ride-through capability for this inverter during a voltage sag disturbance. Due to the presence of many series connected power-cells, ride-through via load inertia is complicated to implement. To achieve ride-through during a power interruption (say 5 sec) the DVR approach discussed in section 5 seems to be appropriate.

CI-ML-VSI based medium voltage ASDs are offered at both 2300V and 4160V [26-27]. For a 4160V drive there will be five power cells in series, each capable of 480V pulse-width modulation (PWM) output per phase, with fifteen secondaries on the integral transformer.

Ride-through performance against voltage sags can be achieved with this topology by employing additional power cells in each phase. For example, in a 4160V drive if six power cells are added in each phase instead of five, then the system can operate the motor load at rated voltage even with a 20% loss of input voltage during a voltage sag. Another advantage of the additional module per phase is additional redundancy against inverter failures.

In this section, several alternative approaches to providing ride-through for CI-ML-VSI systems will be discussed including the use of boost converter modules, battery energy storage systems and super capacitors.

5.1.1 Boost Converter for CI-ML-VSI Ride-Through

A boost converter ride-through module can provide ride-through as shown in Figure 14, which is supplied by the ac line. This would be appropriate for a retrofit situation. An alternative implementation would be with the boost stage integrated into each rectifier stage. It is clear from Figure 14 that a boost type ride-through module becomes necessary for each cascaded VSI stage in a CI-ML-VSI. In a CI-ML-VSI based medium voltage ASD for 4160V applications it is typical to employ at least 5 cascaded VSI stages in each phase. Hence for ride-through of a 4160V CI-ML-VSI based ASD, a minimum of 15 boost modules become necessary.

Without any additional energy storage, as shown in Figure 14 where the ac line is the source, boost converters can provide ride-through for single-phase sags of up to 50% for up to 5 seconds. The voltages of the cascaded inverter stages can be regulated by the boost converters as required by the inverter and would be user adjustable. The additional boost converter hardware required would have to be suitably rated due to the additional current drawn during a voltage sag, and involves a cost of about $100-200/kW.

For ride-through during brief power interruptions (outages) appropriate energy storage devices such as super capacitors or batteries can be added to each boost module as shown in Figure 15. The applications requiring external energy storage modules as shown in Figure 15 would be critical continuous processes such as refinery extruders.

5.2 Ride-Through for NPC Inverters

In NPC type medium voltage inverters a single dc-link voltage powers the inverter as shown in Figure 16. Therefore all of the techniques discussed earlier to provide ride-through apply to this inverter as well. The best suited options would be load inertia ride-through, the boost converter approach and energy storage options such as flywheels and SMES.

6 Justifying ASD Ride-through Costs

In order to determine whether or not applying ASD ride-through technologies will be cost effective, interruption costs can be estimated. With recent advances in power-line monitoring technologies, detailed analysis of the electrical environment have been conducted and can be obtained by industrial customers. Thus, using statistics of average annual events, industry customers can predict downtime costs and make comparisons with the cost of additional ASD ride-through. Table 2 and Figure 17 summarize ASD ride-through characteristics and costs. Interruption costs can be estimated through the following general formula [30]:

interruption cost = value of lost production + outage-related costs - outage-related savings

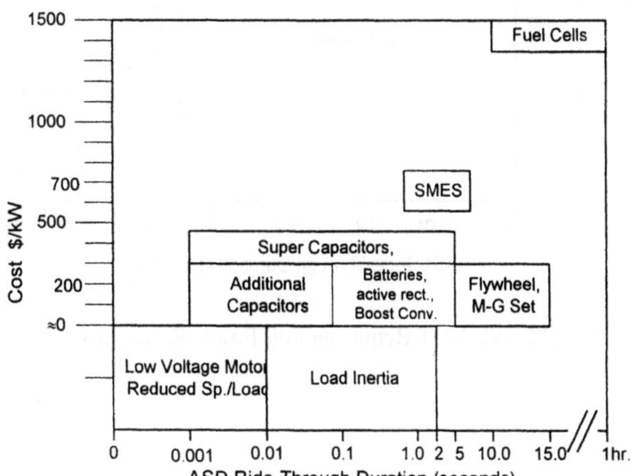

Figure 17: Summary of Ride-Through Time vs. Relative Cost.

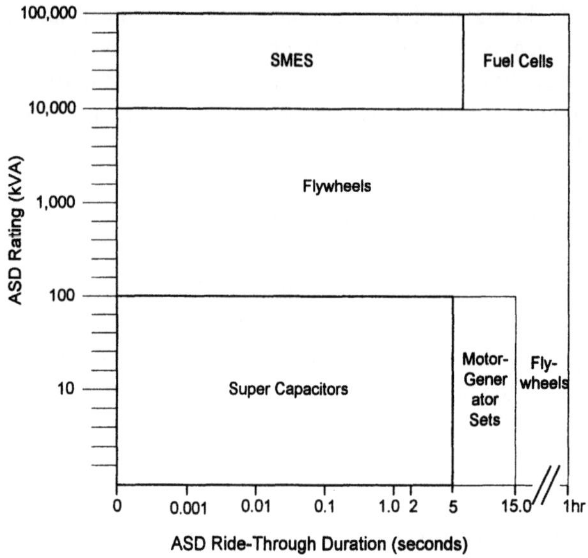

Figure 18: For Critical Loads Requiring Full Power Ride-Through.

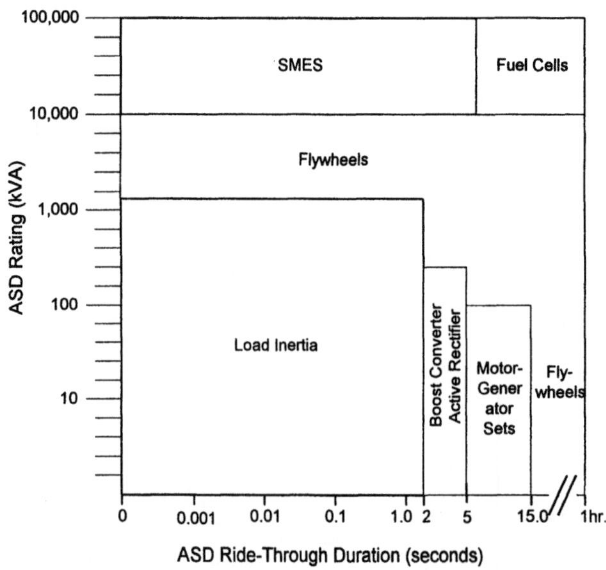

Figure 19: For Critical Loads NOT Requiring Full Power Ride-Through.

Table II: Summary of ASD Ride-Through Alternatives

ASD Ride-Through Alternatives	Cost $/kW	Ride-Through Duration Up To:	Power Range (W)
Additional Capacitors*	$600/kW	0.1 sec.	100kW
Load Inertia	≈0	2 sec.	1kW - 1MW
Reduced Speed/Load	≈0	0.01 sec.	5-10kW
Lower Voltage Motors*	≈0	0.01 sec.	5-10kW
Boost Converter**	$100-200/kW	5 sec.	5-200kW
Active Rectifier**	$100-200/kW	5 sec.	5-200kW
Battery Backup*	$100-200/kW	5 sec., 1hr.	5kW-1MW
Super Capacitors*	$300-400/kW	5 sec.	5kW-100kW
Motor-Generator Sets*	$200-300/kW	15 sec.	100kW
Flywheels*	$200-300/kW	15 sec., 1hr.	1kW-10MW
SMES*	$600-800/kW	10 sec.	300kW-1000MW
Fuel Cells*	$1500/kW	1 hr.	10kW-2MW

*provides full power ride-through
**can provide full power ride-through for single-phase sags < 50%

7 Conclusions and Recommendations for ASD Ride-through

The purpose of this paper is to address industrial ride-through requirements and application issues for AC and DC drives, including medium voltage (2300/4160V) multi-level inverter topologies. After analysis of the implementation and cost studies, the most cost effective and practical ASD ride-through solutions were matched with the most appropriate ASD ratings and ride-through durations and summarized for low voltage applications in Figs. 18 and 19. Figure 18 suggests the most appropriate ASD ride-through techniques for critical loads requiring *full power* (full speed and torque) ride-through. Therefore only the approaches indicated with an asterisk in Table 2 were possibilities. Figure 19 suggests the most appropriate ASD ride-through solutions for loads that do NOT require full power ride-through. Note that the boost converter and active rectifier techniques are included in Figure 19, since they can only provide full power ride-through for single-phase sags < 50% as indicated in Table 2.

Acknowledgment

Financial support from the Electric Power Research Institute (EPRI) for the funding of this research study is acknowledged.

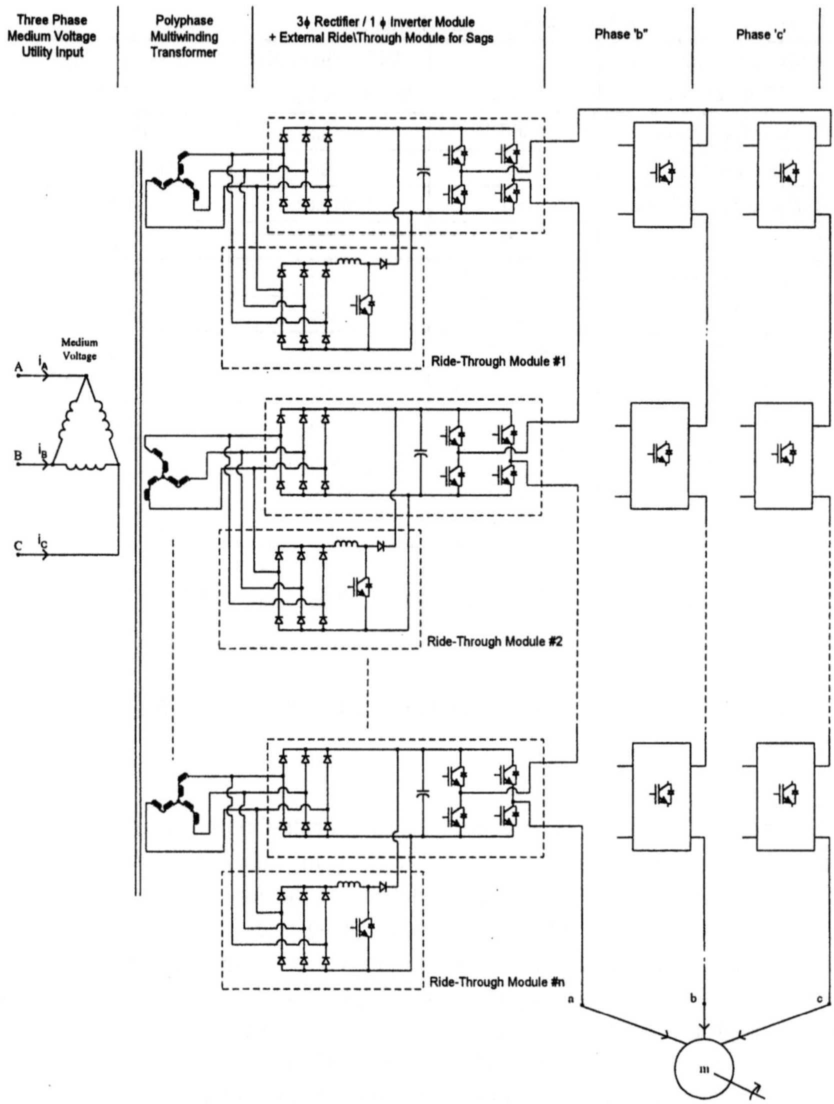

Figure 14: Application of Standard Boost Modules to Achieve Ride-through for a Cascaded Inverter type Multi-level Voltage Source Inverter (CI-ML-VSI).

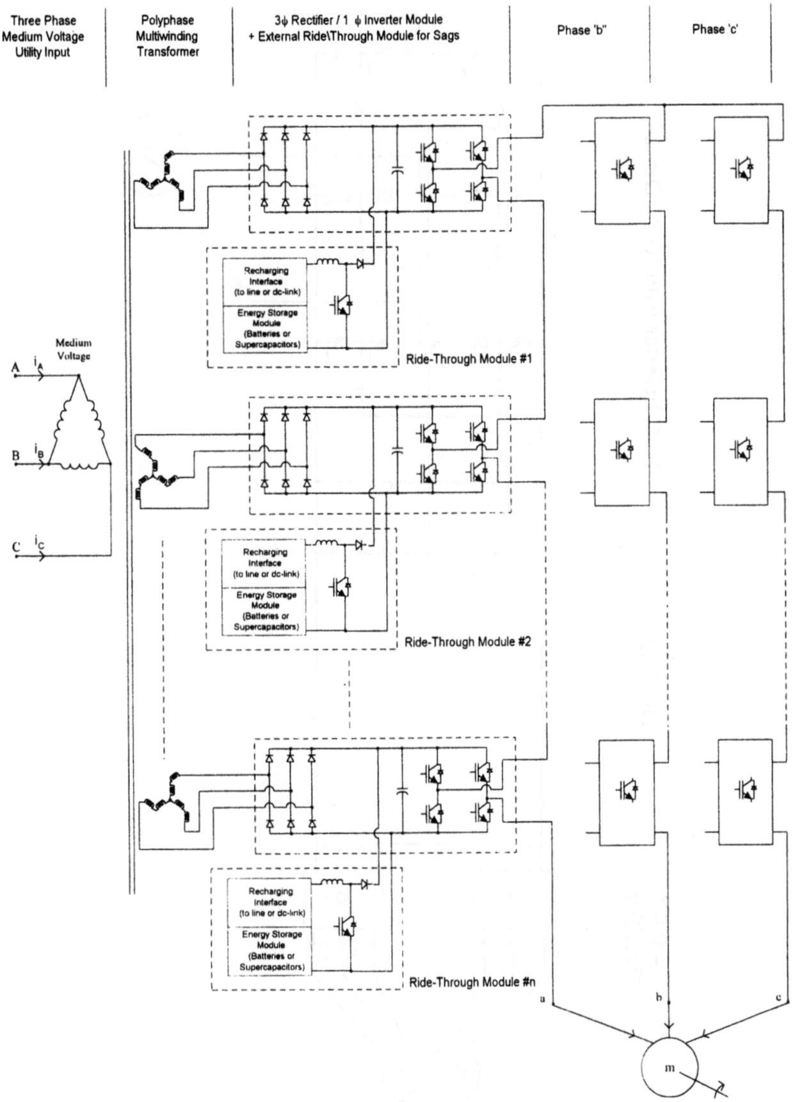

Figure 15: Application of Boost Converter Modules with Energy Storage (Batteries or Supercapacitors) to achieve Ride-through of a CI-ML-

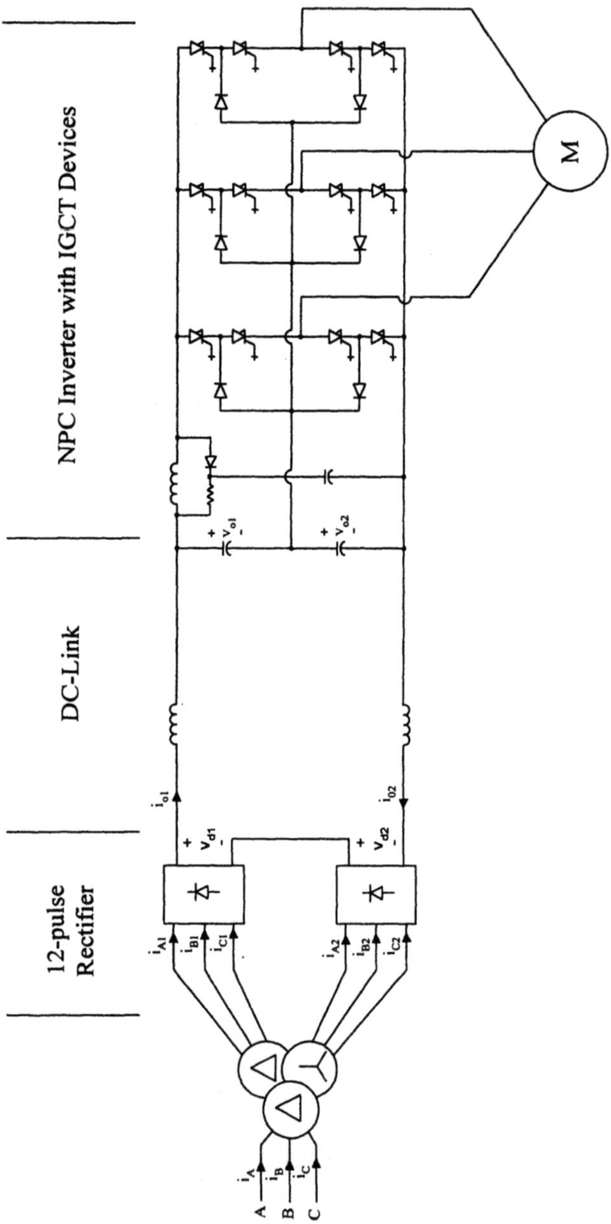

Figure 16: Three-level Neutral-Point Clamped (NPC) Inverter Topology for MV-ASD.

8 References

[1] A. von Jouanne, P. Enjeti, "ASD Ride-Through Technology Alternatives and Development", EPRI Final Report TR-109903, December 1997.

[2] D. D. Sabin, "An Assessment of Distribution System Power Quality", EPRI Final Report TR-106294-V2, May 1996.

[3] D. Dorr, "Point of utilization power quality study results," *IEEE Trans. Ind. Applicat.,* vol. 31, pp. 658-666, July/Aug. 1995.

[4] M. B. Hughes, J. S. Chan, "Canadian National power quality survey results," in *Proc. EPRI PQA'95*, New York, NY, May 9-11, 1995.

[5] G. W. Allen, D. Segal, "Monitoring of computer installations for power line disturbances," presented at the IEEE Winter Power Meeting, New York, NY, 1974, Paper WINPWR C74 1996.

[6] T. S. Key, "Diagnosing power quality-related computer problems," *IEEE Trans. Ind. Applicat.,* vol. IA-15, pp. 381-393, July/Aug. 1979.

[7] M. Goldstein, P. D. Speranza, "The quality of U.S. commercial ac power," in Proc. *IEEE INTELLEC*, 1982, pp. 28-33.

[8] D. S. Dorr, M. B. Hughes, T. M. Gruzs, R. E. Jurewicz, J. L. McClaine, "Interpreting Recent Power Quality Surveys to Define the Electrical Environment", IEEE Transactions on Industry Applications, Vol. 33, No. 6, Nov./Dec. 1997, pp. 1480-1487.

[9] R. A. Hanna, S. Prabhu,, "Medium-Voltage Adjustable-Speed Drives - Users' and Manufacturers' Experiences", *IEEE Trans. Ind. Appl.*, vol. 33, no. 6, Nov./Dec. 1997, pp. 1407-1415.

[10] F. Hoadley et al., "Motor Drive/Power Systems Interactions", IEEE Industry Applications Society Tutorial Course, 1997, pp. 3-1 - 3-15.

[11] K. Benson, J. R. Chapman, "Boost Converters Provide Power Dip Ride-Through for AC Drives", Power Quality Assurance Magazine, July/Aug., 1997, pp. 76-82.

[12] A. van Zyl, R. Spee, A. Faveluke, S. Bhowmik, "Voltage Sag Ride-Through for Adjustable Speed Drives with Active Rectifiers", IEEE Industry Applications Society Annual Meeting, Oct. 1997, pp. 486-492.

[13] www.baldor.com.

[14] S. M. Schoenung, J. M. Eyer, J. J. Iannucci, S. A. Horgan, "Energy Storage For A Competitive Power Market", Annu. Rev. Energy Environ. Report 21:347-70, 1997, pp. 347-370.

[15] Maxwell Technologies Systems Division DC Bus Ride Through System brochure, Oct. 1998, E-mail: pe@maxwell.com, Tel: 619-637-7410.

[16] Piller, Inc. information at: www.piller.com.

[17] American Superconductor information at: www.amsuper.com.

[18] S. M. Schoenung, "Power Conditioning Technology Assessment for the Power Quality Park Concept", EPRI Final Report WO4897-04, April, 1997.

[19] R. J. Nelson, N. H. Woodley et. al., "Requirements For Dynamic Voltage Restoration To Revive Distribution System Voltage Sags", Presented at the American Power Conference, April 1995.

[20] K. Berton, "Custom Power Equipment For Power Quality Enhancement", Paper Presented at the ONR PEBB Applications Workshop, Nov. 5-7, San Diego, CA, 1997.

[21] C. J. Melhor, A. Braz, P. Hofmann, R. J. Mauro, "An Evaluation of Energy Storage Techniques for Improving Ride-Through Capability for Sensitive Customers on Underground Networks", IEEE Transactions on Industry Applications, Vol. 33, No. 4, July/Aug. 1997, pp. 1083-1095.

[22] J. R. Hull, "Flywheels on a Roll", IEEE Spectrum, July 1997, pp. 20-25.

[23] A. von Jouanne, P. Enjeti, *DC Drive Ride-Through Technology Alternatives and Development,* EPRI Report TR-111759, December 1998.

[24] A. von Jouanne, P. Enjeti, *Medium Voltage (2300/4160V) Adjustable Speed Drive (ASD) Ride-Through Options and Implementation Methods,* EPRI Report TR-111760, November 1998.

[25] D.A. Paice, C.W. Charles, "High Voltage Modular Inverter and Control System Thereof", US Patent No: 4674024, June 16, 1987.

[26] www.robicon.com

[27] www.ticind.com

[28] GE Drive Systems at: www.ge.com/gemis/ds0.html

[29] ACS-1000 ABB medium voltage ASD at: www.abb.fi/vsd/mediumac.html

[30] M. Sullivan, T. Vardell, M. Johnson, "Power Interruption Costs to Industrial and Commercial Consumers of Electricity", IEEE Transactions on Industry Applications, Vol. 33, No. 6, Nov./Dec. 1997, pp. 1448-1458.

Introduction to the Permanent Magnet Motor Market

Toshihiro Sawa, Kaneyuki Hamada

Yaskawa Electric Corporation

Abstract. According to the Kyoto Summit on global warming (COP3) in December 1997, the green-house gas emission level has to be reduced to 92-94% of the 1990 green-house gas level by the year 2014-2018. This would require conserving energy. An efficient means of achieving this voluntary goal is by employing high-efficiency drives, since motors consume 70% of all electricity for industrial use in Japan.

As adjustable speed drives become popular, interior permanent magnet (IPM) motors, lately, have been recognized for high-efficiency performance. Due to the progress in permanent magnet technology combined with modern control methods, especially vector control with and without speed-sensors, the IPM motor is gaining in popularity. Compact size and high-efficiency performance is furthering the IPM motor as the preferred motor in many applications.

This paper describes the principle and operation of IPM motors and compares its performance with that of an induction motor. Important features and practical control methods for IPM motors are presented. Various application examples highlighting the advantages of employing an IPM motor system are discussed. The applications include, but are not limited to, machine tools, fans, pumps, elevators, cranes, etc.

1 Progress of Energy Saving in Japan

Though clean energy systems employing solar-energy and/or wind energy has already been developed in Japan to help the global environment, energy saving by high efficiency drive is still the most effective means to attain the voluntary goals set by the Kyoto proclamation mentioned above. It is believed that almost all of the manufacturing facilities have electric motors and that 70% of all electricity for industrial use are consumed by electric motors in Japan. The most popular electric motor used in the industry, unquestionably, is the induction motor. In its hundred year history, induction motor applications range from general-purpose line start fan and pump drives to high performance rolling mill drives. According to the Japan Electric Machine Association (JEMA) report of 1999, about 10 million induction motors are produced each year in Japan. Fan and pump drives are a large proportion of induction motor application in Japan.

After the oil crisis in the 1970's, many high power induction motors used in fan and pump applications that were in the 300-kW or higher power range were converted from direct across the line start to adjustable speed drives to achieve energy savings. Large power consumers that include paper mills and steel mills are very interested in adopting energy saving techniques and encourage use of high efficiency equipment. However, low capacity adjustable speed drives are still expensive and this has prevented the complete conversion from direct across line start to adjustable speed drives in these power ranges. Since the price of inverters is decreasing gradually, low capacity drives are increasingly being accepted in the industry. Since adjustable speed drives are gaining popularity, they are more than before, required to have high-efficiency characteristics. In this light, inverter driven permanent magnet (PM) motors are being increasingly considered for many applications.

The PM motor, including the surface permanent magnet motor and the interior permanent magnet (IPM) motor, utilize the permanent magnet flux and are known to be superior to the conventional induction motor in efficiency as shown in Table 1. However, PM motors use rare earth magnets and hence are more expensive to manufacture than the conventional induction motors. Therefore PM motors at present are mainly used as servos in machine tools, robots, and in the semiconductor industry.

Table 1: Comparison of Different Type s of AC Motors

	Induction Motor	SPM Motor	IPM Motor
Cross Section	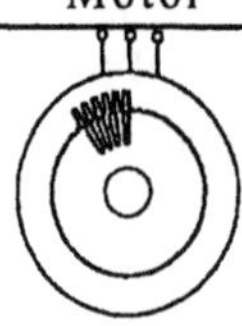		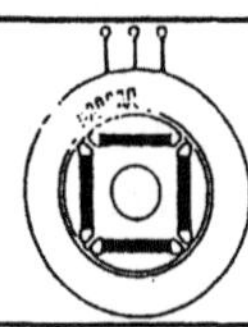
Cubic Volume	1.0	0.7	0.7
Efficiency	1.0	1.05	1.05
Power Factor	1.0	1.0	1.2
Constant Power Range	1.0	0.2	0.8
Max. Rotor Surface Speed	1.0	0.3	0.6

The permanent magnet characteristics govern the performance of PM motors. The introduction of sintered Neodymium-Iron-Boron (NdFeB) permanent magnet in 1983 provided a major boost to the use of permanent magnets in motors [1]. Rapid progress in controlling and applying these motors were made in the 1990s. Further, the above mentioned rare earth magnets were not cost prohibitive. By burying the magnets inside the rotor, the Interior Permanent Magnet (IPM) motor was developed. This reduced the rotor size and was instrumental in minimizing the overall size of permanent magnet motors. In addition to this, for attaining high performance from IPM motors, intelligent motor control technology such as flux observers and parameter tuning features made possible due to the progress in the area of microprocessors, were successfully implemented. These technological advances have made the IPM motor drives popular for industrial applications.

The industry recognized this fact and conferred the 1997 MITI award for the best energy saving motor drive equipment to Yaskawa Electric Corporation's (YEC) new IPM motor drive known as the "super energy saving drive VS-686PS5", which is shown in Figure 1. The award seems to have given impetus to improve the efficiency of general-purpose adjustable speed drives.

Figure1: IPM Motor Drive

However system manufacturers regard motors as tools to achieve high performance at low costs in their systems. Hence, high efficiency motor drives are regarded as an added cost to the system. From the system manufacturer's point of view, the motor drive manufacturer should make effort to include high efficiency motor as part of the drive. Hence, YEC has introduced many types and sizes of IPM motor drives for use in various industrial applications. Energy saving has already become a big selling point in home electronics equipment such as air conditioner and washing machine, and soon will become a big selling point in the industrial drive equipment market as well.

2 IPM Motor for Excellent Energy Saving

IPM motors have buried permanent magnets inside steel rotor cores as shown in the cross-sectional drawing in Table 1. According to Dr. Tom Jahns [2], this rotor structure is rather old and dates back to the 1950s when it was used embedded in squirrel caged induction motors for operation directly from utility-supplied AC power. These old structures employed Alnico magnets and were widely applied in the textile industry that required large numbers of machines operating at the same speed.

Figure 2 shows the principal of flux paths and stator currents in an IPM-motor. The torque developed by an IPM motor is made up of two distinct components. The first component is referred to as the electromagnetic torque and is produced rectangular stator current I_q along the q-axis.

The second component is referred to as the reluctance torque and is produced due to the difference in reluctance of the rotor structure as seen from the stator along the d-axis and q-axis. Orthogonal current components I_d and I_q produce orthogonal fluxes. The induced flux along the rotor d-axis has amplitude of $L_d I_d$ due to the d-axis stator inductance, L_d, and the induced flux along the rotor q-axis has amplitude of $L_q I_q$ due to the q-axis stator inductance, L_q. The interaction of the d-axis flux with stator q-axis current, I_q, and the q-axis flux with stator d-axis current, I_d, produce reluctance torque. The vector addition of the electromagnetic torque and the reluctance torque results in the total torque developed in an IPM motor.

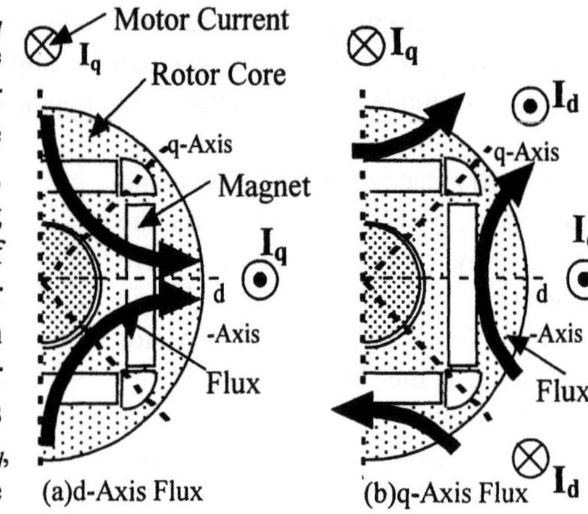

Figure 2: Principal of IPM Motor

To realize the optimum motor design, the latest CAE technology is useful. Figure 3 shows the simulation results of the cross-sectional flux distribution and the core stress in the rotor. These analyses using the simulation tool made possible the optimum design for the IPM motor shown in Figure 4. Figure 4 also compares the center height of an induction motor with that of an IPM motor of similar rating. As can be seen, IPM motors are 2 to 3 sizes smaller than conventional induction motors. This difference makes the IPM motor 20-55% smaller in cubic volume and 20-40% lighter in the motor mass. Hence, the IPM motor with its feature of small size and lightweight, is a good candidate to replace the induction motor resulting in saving energy and conserving natural resources.

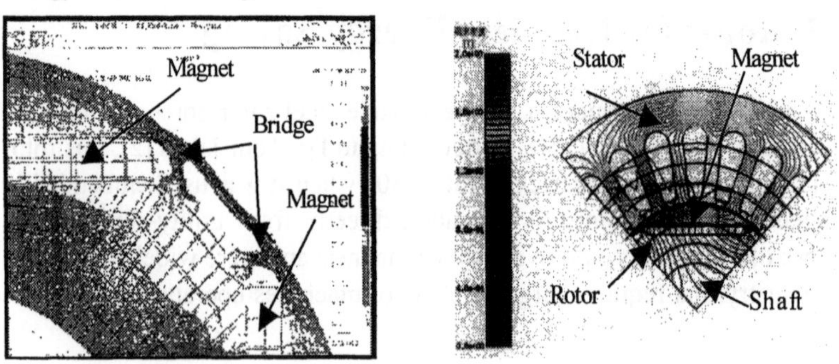

Figure 3: Simulation Results

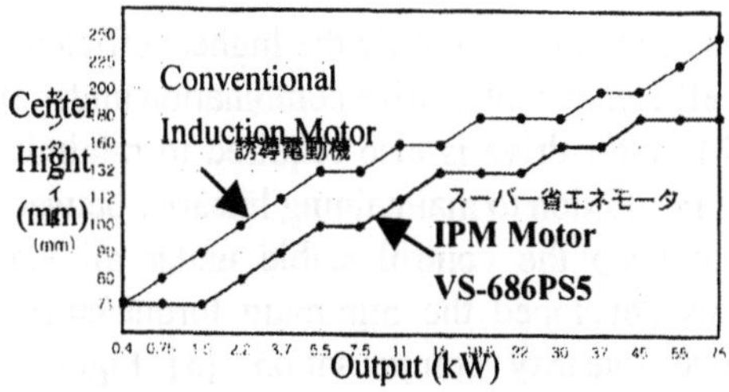

Figure 4: Comparison of the Motor Frame

From the point of view of drives, the IPM motor drive is more efficient than a general purpose Induction Motor drive of comparable size. The above fact along with smaller IPM motor size and weight results in total system cost reduction. Lately, IPM motor-drive combinations are being increasingly applied due to their energy saving and compact size features.

3 Controlling the IPM Motor

Though equivalent resistances, which generate eddy-current losses, exist in the rotor of an IPM motor, they are too small to have any damping effect. Therefore the IPM motor requires vector control. Basically the vector control algorithm of an IPM motor is similar to the vector control algorithm of that of an induction motor. Figure 5 shows the control block diagram of an IPM motor drive with an encoder. The difference in the control algorithm between induction motor and IPM motor is the phase angle detection. The former requires the estimation of slip frequency based on the torque producing current, I_q, and the flux producing current, I_d. The latter requires the information pertaining to the detected or estimated position of rotor flux. Hence it is easy to use general-purpose inverter for both induction motor and IPM motor with small software modification.

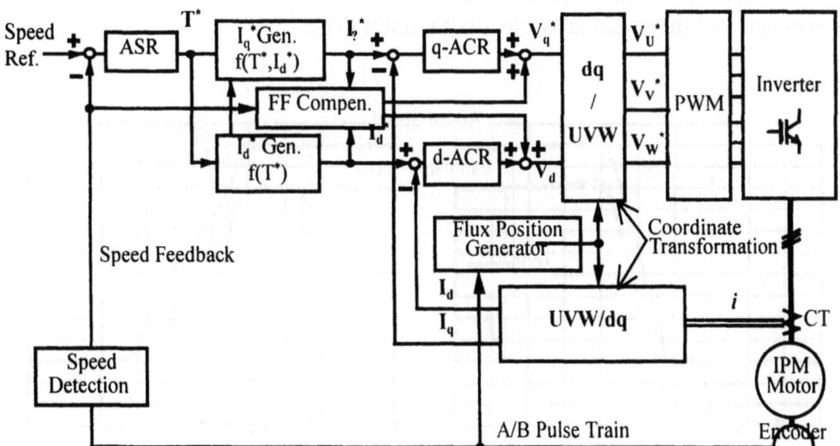

Figure 5: Block Diagram of IPM Motor Control

However in order to customize the IPM motor for each application several special control techniques have to be added to the IPM motor drive. In the case of energy saving using adjustable speed drives, users require not only the highest efficiency machine operation but also the highest efficiency motor drive combination under all load conditions. Accordingly the IPM motor drive is also required to minimize motor current for a desired torque level in addition to maintaining linearity between torque reference and actual torque to keep the control stable under all load conditions. For this reason, YEC has developed the minimum torque/current control algorithm with built-in torque linearity compensation. [3] Figure 6 illustrates the current vector locus for a given constant torque. The current vector for minimum motor current is 10% lower than the current vector for maximum motor efficiency. The choice of optimizing the current vector based on the new algorithm is superior to the choice of optimizing the current vector based on motor efficiency from a total system viewpoint since this minimizes inverter size and cost.

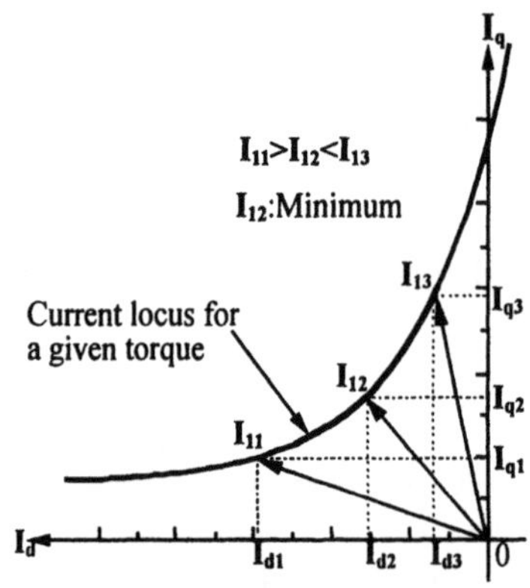

Figure 6: Minimum Torque/Current Control

In the application of fans, pumps and compressors line start induction motors with constant speed operation are used hitherto and do not need knowledge of rotor position and speed. Hence to employ IPM motor drive for these applications, position and speed sensorless control must be adopted. Instead of position and speed sensors, YEC has developed the flux position and rotor speed estimation methods using flux observes and rotor magnetic pole identification methods. For the magnetic pole identification method, YEC has introduced the detection method based on rotor saliency reversal in conjunction with the direction of permanent magnet flux. Figure 7 shows the speed estimation performance of the speed sensorless IPM motor drive.

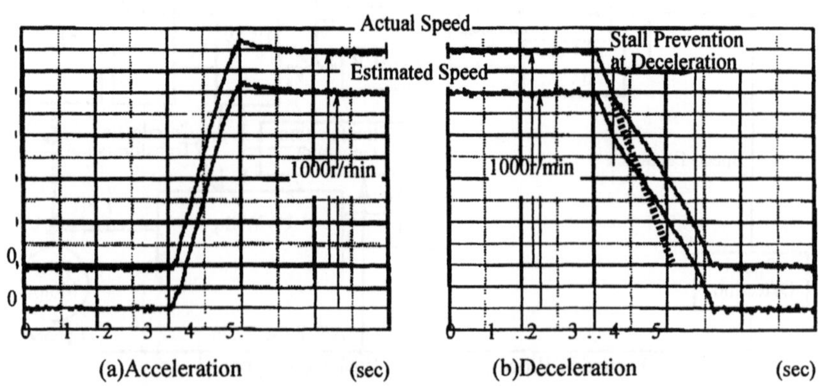

Figure 7: Comparison of Actual and Estimated Speed

The Figure shows the actual speed and the estimated speed under the acceleration and deceleration operation. There is scarcely any difference between the actual speed and the estimated speed. Even when stall prevention is engaged and the drive is decelerating, the estimated speed is very close to the actual speed.

4 Efficiency Characteristic of the IPM Motor Drive

A well-designed IPM motor is highly efficient. Figure 8 shows the efficiency characteristic of an IPM motor. It shows the excellent high efficiency characteristic of the IPM motor. At variable load levels ranging from 25% to 100%, the efficiency of the IPM motor is higher than 90%, while that of the conventional induction motor is generally lower than 85%. The power loss in an IPM motor is 65% of that in an induction motor of same size. The conspicuous difference is due to the fact that the rotor loss is extremely low in IPM motors. This makes it smaller in size, higher in efficiency and results in longer motor bearing life.

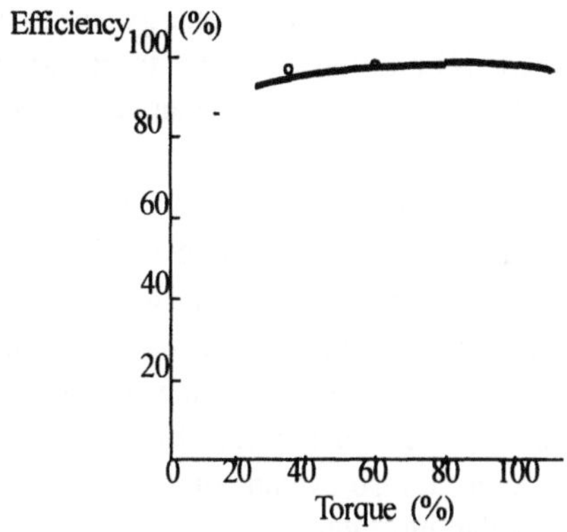

Figure 8: Efficiency of IPM Motor

Similar to the efficiency feature, the torque linearity feature with minimum torque/current control has to be verified. Figure 9 shows a plot of the actual torque versus the reference torque. The error is less than 2% of rated torque when the reference changes from zero to rated torque. This torque linearity characteristic is similar to that observed in DC motors with thyristor based dual converter scheme.

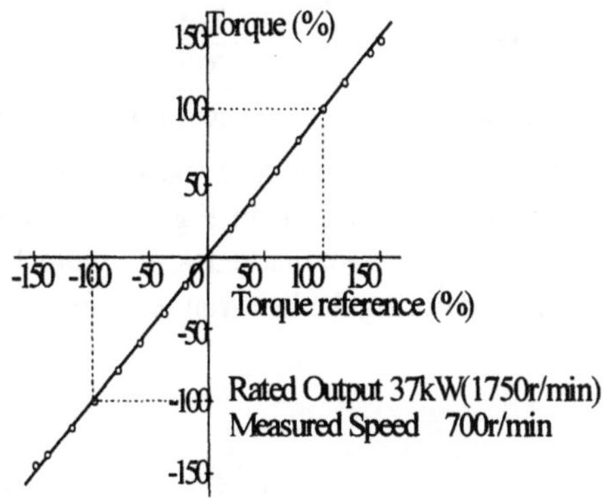

Figure 9: Torque Linearity Characteristics

5 Introduction to the IPM Motor Market

5.1 Machine Tool Application

YEC applied the IPM motor to the machine tool application at first. Machine tool application requires low loss, especially low rotor loss to maintain low thermal expansion of drive shaft for high precision performance. Low loss also saves energy. Since most machine tools are direct drive type and use motorized spindle for space saving and for avoiding mechanical vibration, the IPM motor fits well for this application.

The machine tool industry started to address energy saving issues in the early 1970s. In particular, the machine tool spindle drive was required to have regenerative capabilities to regenerate power back to the source during frequent decelerations. When DC motors were used, thyristor based dual converter schemes were employed [4]. Whenever energy was to be regenerated, the inverter thyristors were turned ON and the converter thyristors were turned OFF. More recently, spindle AC drives employ active front ends (PWM rectifiers) to regenerate power back to the source.

Generally speaking motors are capable of converting energy in both directions – electrical to mechanical and mechanical to electrical. In this process, there are losses and it is important to minimize these losses. As mentioned earlier, temperature variation of the rotor shaft can have profound effect in the machine tool industry where precision is key to operation. The IPM motor with its extremely low rotor loss is suitable for this application. It not only helps to maintain high precision; it also extends motor bearing life. Calculated losses for the IPM motor and induction motor, both designed for machine tool spindle drive is shown in Table 2 [5]. In the calculation, the eddy current loss in the rotor is neglected.

Table2 Comparison of Calculated Motor losses

Motor Speed	420r/min		1500r/min	
Motor Type	IM	IPMM	IM	IPMM
Stator Loss	2100W	2200W	1850W	1800W
Rotor Loss	1330W	0W	1050W	0W
Total Loss	3430W	2200W	2900W	1800W

* Rotor iron loss is not considered to loss calculation.

The simulated temperature distribution in an induction motor and in an IPM motor of same size is shown in Figure 10 while the heat run test results are shown in Figure 11. The spot marked A in Figure 10 is the position where the temperature is measured during the heat-run test. Figure 11 shows that the temperature rise of the IPM motor is less than one third that of the induction motor.

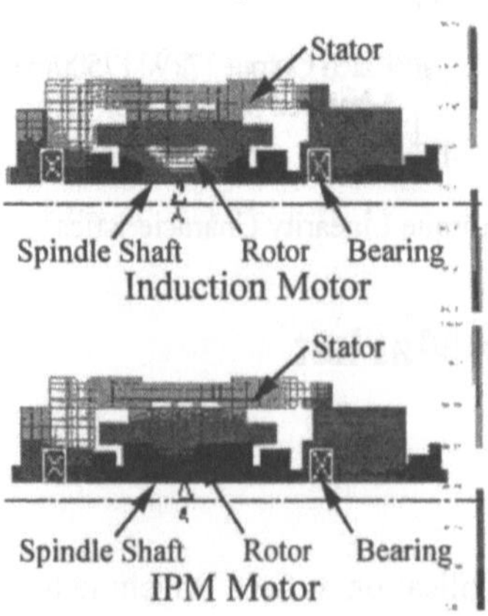

temperature Distribution

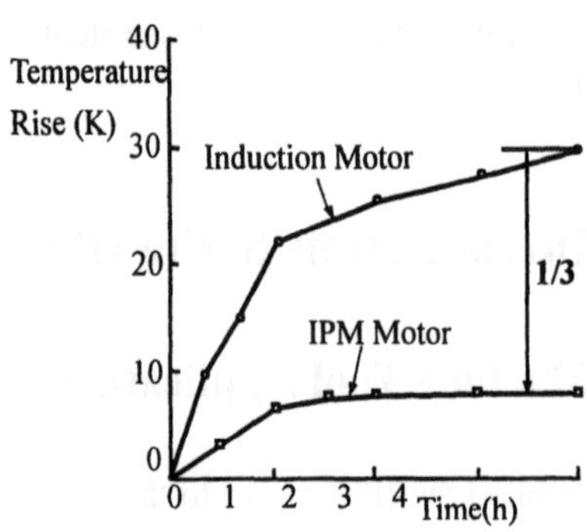

Figure11: Heat-run Test Results

Another requirement of the machine tool spindle drive is the wide constant power range and low torque ripple. Flux weakening control using the rectangular current control along with motor winding changeover technique has been shown to achieve constant power over a wide speed range of 1:12. The winding changeover technique shown in Figure 12 is to switch between low speed winding and high speed winding as the occasion demands and allows use of high efficiency spindle drive with wide constant power range for this application. Torque ripple is caused by cogging torque due to stator slots in the motor and is resolved by skewing the stator core. The speed fluctuation then becomes less than 0.5 r/min at 100r/min operation. This countermeasure along with direct spindle drive addresses the mechanical vibration problem.

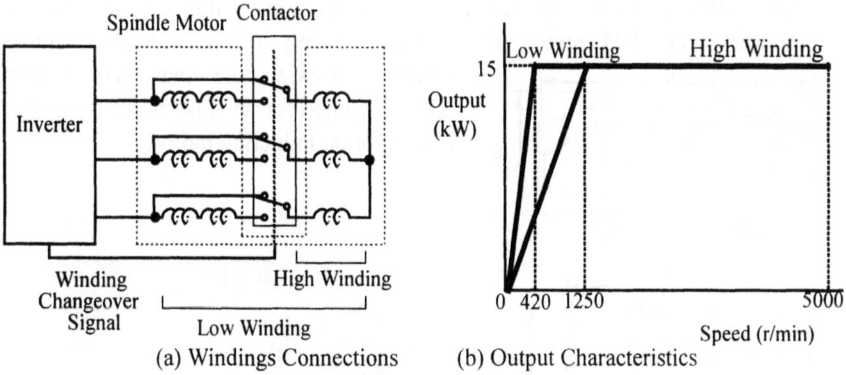

Figure 12: Winding Changeover

5.2 Fan, Pump and Compressor Applications

Though the total amount of fans, pumps and compressors produced in Japan reaches several million per year, the amount of them with adjustable speed drives are not so many. In fans, pumps and compressors speed control is more efficient than valve or damper control. In the case of fan the required energy is in proportion to the positive cube of air volume when speed control is used. However, the required energy is in proportion to the negative cube of air volume when damper control is used. These results indicate the necessity of adjustable speed drives. Accordingly, YEC proposes the IPM motor drive with special features for centrifugal load users, fan, and pump manufacturers.

Figure 13 shows the overall compressor system efficiency. An IPM motor drive has 4.1% higher efficiency than an induction motor drive of the same size at rated compressor output. In the case of submersible pump application (Figure 14), the IPM motor not

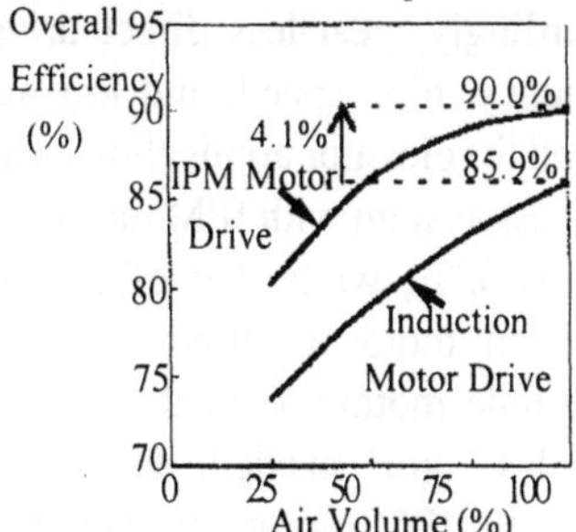

Figure 13: Efficiency Comparison of Compressor

only has higher efficiency compared to an induction motor drive, it is 50% lighter and 30% smaller than an induction motor of the same rating.
Figure 15 shows an example of a drainage pump, which falls under submersible pump application category. The Ministry of Construction in Japan has decided to apply the IPM motor drive to drainage pumps as a countermeasure for floods because of their higher efficiency and smaller size compared to other motor drives. Since, sensorless (open-loop) inverters control induction motors for centrifugal loads, the IPM motor drives have to use position and speed sensorless vector control to be comparable. Furthermore, the minimum torque/current control method discussed earlier has also to be employed for energy saving. By implementing the modifications discussed, the IPM motor drive becomes ideal to replace the induction motor drive in the submersible pump application. It is easy to commercialize the IPM motor drive using existing induction motor control inverters because the sensorless vector control needed for an IPM motor is similar to that for an induction motor.

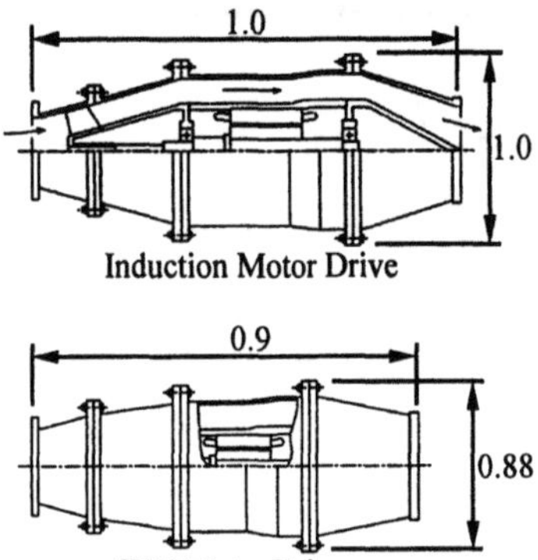

Figure 14: Comparison of Pump Size

Figure 15: Drainage Pump

5.3 Elevator Application

The elevator application needs high-speed operation. Further, the motor drive combination should occupy the least possible space. An application example is shown in Figure 16. Accordingly, gear-less direct drive having high efficiency, high speed, and low audible noise is needed for elevator application. The efficiency of gear-less system with IPM motor drive reaches as high as 82%, while that of geared system with induction motor is only between 60-68%. As multi-pole motors, typically having greater than 16-poles, are needed to achieve gear-less operation, IPM motors are superior to induction motors, in this application as well.

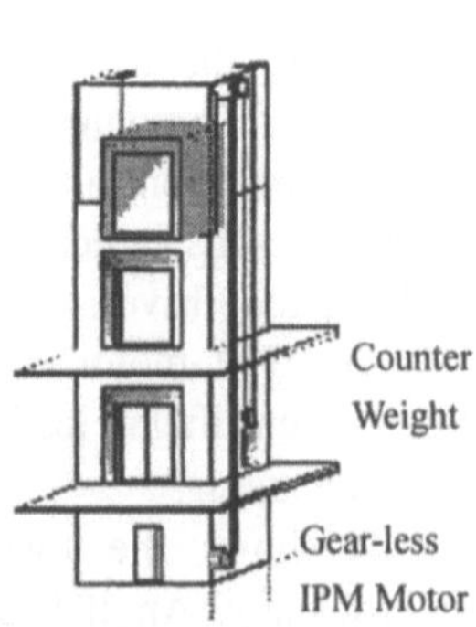

Figure 16: Machine Room-less Elevator

In addition to this, motor room-less elevators require compact motor for space saving and construction cost reduction. The feature of extremely low rotor loss suits well for this requirement of low motor axial depth. Figure 17 compares the dimensions of induction motor and IPM motor of similar ratings for gear-less drives. As evident, the motor has to be compact to be set up in the elevator pit. The depth of an IPM motor is two-thirds of that of an induction motor. Table 3 indicates the characteristics of both motors and shows that induction motors are not suited for direct drive elevator application.

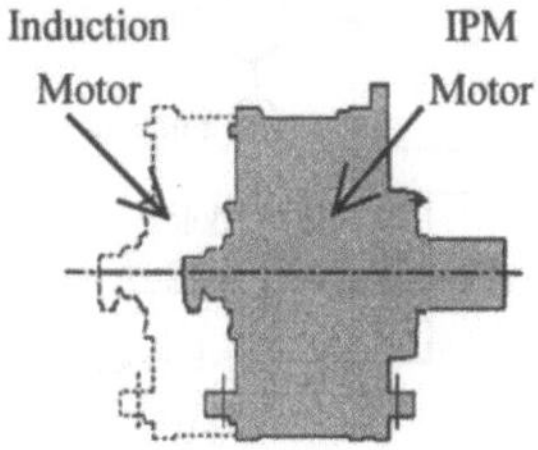

Figure 17: Comparison of Elevator Motor Dimensions

Another requirement for elevator application is the robustness of the control algorithm. First, true line regeneration using PWM rectifier has to be used. Secondly, the "S" shaped acceleration and deceleration control with auto floor position control for smooth and comfortable operation has to be employed. In conclusion, suffice it to say that well-designed IPM motors for direct drive elevator application is increasing year by year.

Table 3 Characteristics of Elevator Motors

Motor Type	IPM Motor	Induction Motor
Output (kW)	9	9
Speed (r/min)	150	150
Max. Output (%)	200	200
Power Factor (%)	**90**	70
Efficiency (%)	**90**	88
Volume (%)	**80**	100

5.4 Crane Application

New application of IPM motors in the crane industry is emerging. These cranes are called "container carriers" and "rubber tire gantry (RTG) cranes". Motor drives in cranes are needed for hoists and gantries. Main requirements in the crane industry are high efficiency, small size, and low maintenance (high reliability). Figure 18 shows the single line drawing of RTG crane using IPM motors. Since the hoists and gantry are actuated by electricity from an on-board engine generator, the need for high efficiency motor drives like IPM motor drives is truly felt to minimize the capacity of the generators. The IPM motor reduces the volume by 23% and weight by 14% compared to an induction motor of comparable size for crane application. Hence, it contributes towards space saving as well. Moreover, the IPM motor's low temperature rise extends the bearing grease life.

Figure 19 shows measured data for hoist operation. By using vector control and constant power control, smooth operation is achieved. These kinds of cranes, in the near future, will become motor driven instead of engine driven for high efficiency and to reduce noxious exhaust gas.

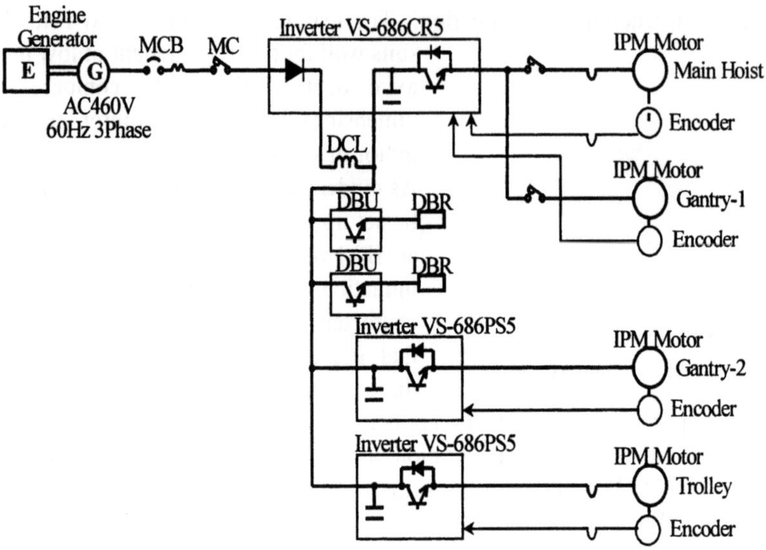

Figure 18: RTG Crane System

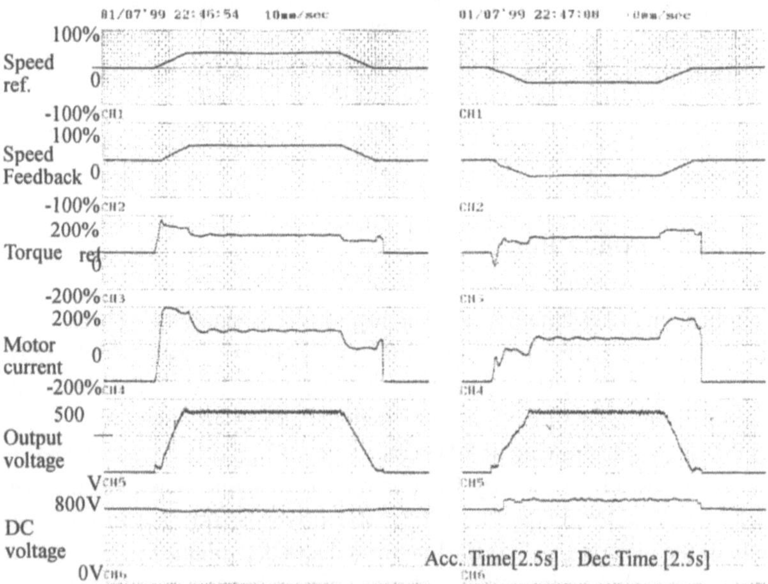

Figure 19: Hoist Operation Data

6 Future of the IPM Motor Drive

In this paper, the principal and operation of IPM motor drives has been introduced. Control and application issues of IPM motor drives have been compared with that of conventional induction motor drives. Various application examples where IPM motor drives have made significant contribution have been highlighted. YEC, the leading manufacturer of industrial IPM motor drives in Japan, has commercialized many kinds of IPM motor drives suited for various applications. IPM motor drives with position and speed sensors ranging from 0.4kW to 300kW for cranes, elevators, etc. are currently in use in the field. The sensorless drives ranging from 0.4kW to 75kW for fans, pumps and other centrifugal loads are also in use in the industry. High-speed IPM motor drives operating at speeds higher than 10000r/min have also been put to practical use in machine tools, such as lathes and machining centers. Owing to the tremendous demand from various system manufacturers for the past several years, the IPM motor drive has established its own application niche in the industry. The high efficiency performance of these motor drives is highly respected in the field.

There are applications where the performance of IPM motor drives can further be improved. More work needs to be done in the area of low speed sensorless operation. Further, new application areas should be investigated to expand its use. This will contribute to saving more energy and natural resources. The authors believe that, in the not so distant future, IPM motor drives may take the place of induction motor drives.

Acknowledgement

The authors would like to thank the support of Dr. Tsuneo Joe Kume and Dr. Mahesh Swamy, Yaskawa Electric America Inc., in helping prepare this document.

7 Reference

[1] G. R. Slemon (1997): Chapter 2 – Electrical Machines for Drives, *Power Electronics and Variable Frequency Drives – Technology and Applications,* edited by Bimal K. Bose. *IEEE Press*, ISBN 0-7803-1061-6.

[2] T. M. Jahns, G. B. Kliman, and T. W. Newmann (1986): Interior Permanent Magnet synchronous Motors for Adjustable-speed. *IEEE Trans. Ind. Appl.*, vol. IA-22, 738-747

[3] N. Shibata, T. Idemitsu, and T. Kamei (1998): New Adjustable Speed Control Method for Interior Permanent-Magnet Synchronous Motors Capable of Operating at High Motor Efficiency. Record of Metal Industrial Society, IEEJ, vol. MID-98-12, 7-12, (in Japanese).

[4] S. B. Dewan and A. Straughen (1975): Chapter 5 – Controlled Rectifiers, *Power Semiconductor Circuits*, John Wiley and Sons,; ISBN 0-471-21180-X.

[5] M.Sawamura, E. Yamamoto, Y. Miyamoto, and T. Sawa (1997): Application of Interior Permanent Magnet Motors for Machine Tool Spindle Drive System. Record of Semiconductor Power Conversion Society, IEEJ, vol. SPC-97-122, 37-42, (in Japanese).

High Efficiency Motors – Performance, Economy & Reliability, by Optimisation

P. Caselotti[1], A. Conchetto[1], P. J. Tavner[2]

[1] MarelliMotori SpA
[2] FKI Engineering

1 Introduction

The TEFC machine is designed to standard dimensions and outputs with a tough enclosure and blow-over cooling to give a rugged, interchangeable motor that can be used in a wide variety of applications.
Historically manufacturers have perfected their products to achieve requirements as economically as possible. This has resulted in machines with improving power-weight ratios, as shown in Figure 1, taken from Glew, Ref1, so electric motors have become progressively cheaper.
However, the figure shows that the power-weight of TEFC motors has not increased at the same rate as larger machines, in recent years there has been a tendency for the power-weight to deteriorate. This is due to the poor heat transfer in TEFC machine, compared to modern large machines, which have air-blown internal active parts. It is also due to the limitations of standards that confine outputs to particular frame sizes. Therefore it has been harder for manufacturers to make TEFC motors more competitive, by increasing the power-weight further, without deteriorating the noise and efficiency of the machine.
The challenge of the efficiency initiatives in North America and the European Union is to improve performance without reducing power-weight by adding materials and increasing cost.
This has been the subject of other papers, Haataja et al Ref2, where the key area has been identified as reducing loss.
This paper suggests that addressing the fundamental problem of heat transfer in the TEFC machine, in addition to reducing loss will make it possible to improve performance and still provide competitive products.

2 Heat Transfer & Temperature Rise in TEFC Machines

The TEFC motor has evolved from earlier geometries, which compromise the designer's need for cooling air in the active parts with the user's need to protect the motor from its environment. The present geometry has been adopted in the majority of motors in the range 0.75-75 kW. The problem with the geometry is that all heat from the active parts, the core, rotor and stator windings, must be extracted from the interior of the machine, passed to the ribbed barrel and removed by convection by the blow-over air. There are four weaknesses in this arrangement:

- The internal air does not remove much heat from the active parts.
- Heat from the rotor to the ribbed barrel crosses 3 interfaces, the air gap, the stator insulation and the core/barrel interface.
- Production factors affect these interfaces and increase their thermal resistance.
- The convective heat transfer is inefficient.

In consequence the bulk heat transfer coefficient of the TEFC motor, relating total losses to stator winding temperature rise and motor surface area, is typically $15 W/m^2K$, much lower than for larger electrical machines at $300\ W/m^2K$. There are some references that address one or all of the weaknesses in TEFC machines, for example Pickering et al Ref3.

3 Interaction between Losses, Efficiency & Power-Weight

The losses in a motor can be classified as follows:

i. Load Dependent Losses including:
   - Joule losses in the stator winding;
   - Joule losses in the rotor winding;
   - Stray losses.

ii. Constant Losses of load, including:
   - Iron loss;
   - Mechanical losses including fan, bearing and shaft seal.

In small and medium sized motors at normal load the Joule losses represent the significant proportion of total losses, typically around 50%. Therefore their reduction must be one of the principal objectives in raising motor efficiencies.
Haataja in Ref3 implied the reduction of losses is the only way to develop high efficiency motors and this must increase the volume of active parts. He quotes Applebaum et al, Ref4, who show increasing efficiency for increased volume of active materials. But this is only so if currents, fluxes and materials remain constant. No self-respecting manufacturer improves his machines by keeping everything the same.
This paper is arguing that one can increase power-weight by raising the bulk heat transfer coefficient in the machine and increase efficiency. This could increase loss density, reducing efficiency, but only if the materials and design are not changed to challenge that increase.
To investigate this point one can compare motors of standard and high efficiency designs, selected from the catalogues of European manufacturers. We consider

identical constructions, using aluminium frames, to eliminate variability due to structural materials, exposing the variation due only to the active parts.
Results of a comparison between standard designs and an early high efficiency 2-pole design are shown in Figure 2. This demonstrates, for this range, higher efficiencies by reducing power-weight by at least 20%. Figure 3 compares the same design with another, optimised, high efficiency design. This shows optimised results achieved without the same reduction in power-weight.

4 Production Factors & Optimisation

To reach higher performance competitively one must optimise:

i. Performance parameters such as:
   - Losses & efficiency.
   - Noise.
   - Starting current & torque.
ii. Production factors
   - Improved processes to reduce loss.
   - Choice of core frame diameters.
   - Cost of production.

The cost consists of three components:

i. Material.
ii. Direct labour.
iii. Overheads.

The first two are influenced by the process technologies and the third is a function of the organisation of the Company. To reduce those first two, design and process choices must be made that minimise cost but still focus on performance. It is also necessary to improve the variability of processes and materials. The following areas for optimisation have been identified:

i. Design calculations;
   - Electromagnetic;
   - Thermal;
   - Ventilation.
ii. Materials;
   - Use improved steels;
   - Low loss bearing seals;
   - Reduce material variance.
iii. Processes;
   - Better impregnation to improve heat transfer;
   - Limit tool wear to reduce loss & wear variance;
   - Improved stator winding to reduce loss;
   - Improved rotor cage construction;
   - Improved core/barrel fit.

5 Temperature Rise, Insulation & Reliability

An important performance area, linked to loss and efficiency, is temperature rise in the insulation system. Modern insulation materials are capable of withstanding higher temperatures. Most Class F insulation systems are now capable of withstanding Class H. If a Class F series of motors were operated to Class H an immediate improvement in power-weight could be achieved but at the expense of higher Joule and iron losses and a potential decrease in efficiency.
On the other hand a high efficiency motor has lower losses than a standard motor and has less thermal problems. Reducing temperature rise enables ventilation losses to be reduced, which in the case of 2 pole motors are a significant proportion of the total, enabling further increases in efficiency and a significant reduction of noise. A reduced temperature rise also confers increased reliability, as predicted by the Arrhenius Law. This suggests that high efficiency machines with modern higher temperature insulation systems will have an extended life.
On the other hand, for a motor supplied from an inverter, one must consider the dielectric withstand of the insulation on the life of the motor. It is noted that PWM inverters with IGBT technology have increased the speed of commutation. This imposes voltage spikes on the winding, whose severity are related to the commutation rise time and the length of cable feeding the machine, see IEE Ref5. Therefore the insulation withstand needs to be improved for inverter supply. When taken with the lower temperatures obtainable in motors of high efficiency this could convert into increased reliability but motors on inverter supply experience additional heating that may efface that advantage.

6 Industrial Experience of Reliability

The reliability of a motor is defined by the Mean Time Between Failures (MTBF) as:

MTBF (Motor) hrs =1/(Sum of Failure Rates for each Subsystem of the Motor)

The IEEE conducts surveys on industrial electrical plant reliability and has established a standard, Ref6. This describes how to carry out a reliability survey and how equipment can be designed to achieve high reliability. Ref6 also contains reliability surveys for electrical equipment including motors, published in the IEEE Transactions, for example Refs7-8. Their data gives us the opportunity to consider the reliability of TEFC Induction Motors. We have reviewed those surveys and identified relevant results in Table I showing MTBFs of over 150000hrs. The Arrhenius Law gives a stator winding insulation life at Class F temperatures of 100000hrs. An analysis of failure modes in service, Table II, taken from three sources, shows that bearings rather than windings are the principal failure mode. The higher MTBF figures found in the surveys therefore show that real winding life must be much longer than 100000hrs. So, although higher temperatures would reduce insulation life, the failure modes show that this will have little effect in reducing the life of a motor, compared to the bearings. Similarly a reduction in reliability due to commutation pulses is only going to be significant if it substantially reduces winding life. In other words raising motor temperatures is both feasible from a winding point of view and is unlikely to significantly reduce reliability from current typical figures of 150000hrs.

7 Conclusions

Improving the efficiency of TEFC motors does not necessarily mean increasing the volume of material and therefore cost, but does necessitate optimising design and production factors. An important byproduct of such work could be reduced winding temperatures and a potential increase in the reliability of the high efficiency motor. In fact operating such motors at higher temperatures is unlikely to reduce motor reliability below 150000hrs. However, reliability may be compromised when the motor requires an inverter supply.

Acknowledgements

This paper is published with the permission of the Directors of FKI Engineering and of MarelliMotori SpA.

8 References

[1] Glew N, Design and manufacture of energy efficient and environmentally friendly large machines, IEE Colloquium 1999, pp1/1-1/5.

[2] Haataja J, Pyrhonen J, Improving three phase induction motor efficiency in Europe, the challenge for manufacturers, IEE EMD Conf, Cambridge, 1997, Publn 444, pp190-194.

[3] Pickering S, Lampard D, Hay N, Roylance T F, Heat transfer from the stator end windings of a low voltage concentric wound induction motor, IEE EMD Conf, Durham, 1995, Publn 412, pp 477-481.

[4] Appelbaum J, Fuchs E F, White J C, Optimisation of three-phase induction motor design, IEEE Trans, Vol EC-2, 1987, No 3, pp 407-422.

[5] Various papers, Effects of high speed switching on motors and drives, Birmingham, June 1999, IEE Colloquium 99/144.

[6] IEEE Std 493-1997, Recommended Practice for Design of Reliable Industrial and Commercial Power Systems.

[7] O'Donnell P, Report of large motor survey of industrial and commercial installations, Part I, IEEE Trans IA, Jul/Aug 1985, pp 853-872.
[8] Thorsen OV, A survey on induction motors in offshore oil industry petrochemical industry gas terminals and oil refineries, IEEE Trans IA, Sept/Oct 1995, pp 1186-1196.

Table I: Industrial Surveys

Machine	Type	Source of Information	Unit-Years Surveyed	No. of Failures	MTBF, hr, Surveyed, Industrial
AC Motors	Induction 0-1000V	IEEE 1985, Ref7	1080	89	106311
AC Motors	LV 11-50 kW	IEEE 1995, Ref8	15658	832	164972
AC Motors	LV	Sum of above	16738	921	159021

Table II: Comparing Proportions of Failures

Subsystem	Proportion of Failures, FKI Predicted	Proportion Of Failures, Customer Measured	Proportion Of Failures IEEE, Ref8 Measured
Bearings	75%	95%	51%
Stator Winding	9%	2%	16%
Rotor Winding	6%	1%	5%

Fig 1 Weight per unit output as a function of year, Ref 1

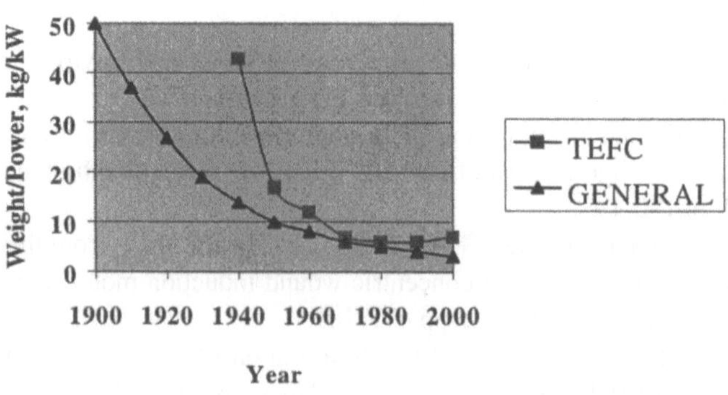

Fig 2 Improved vs Standard Motors

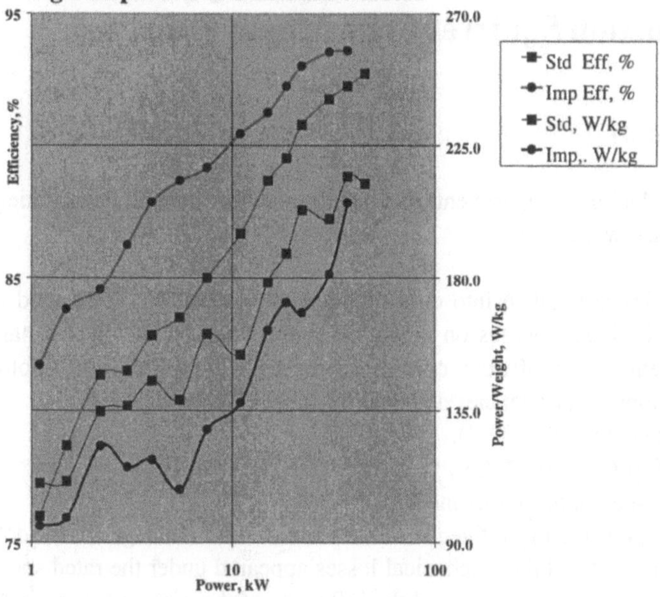

Fig 3 New Design vs Standard Motors

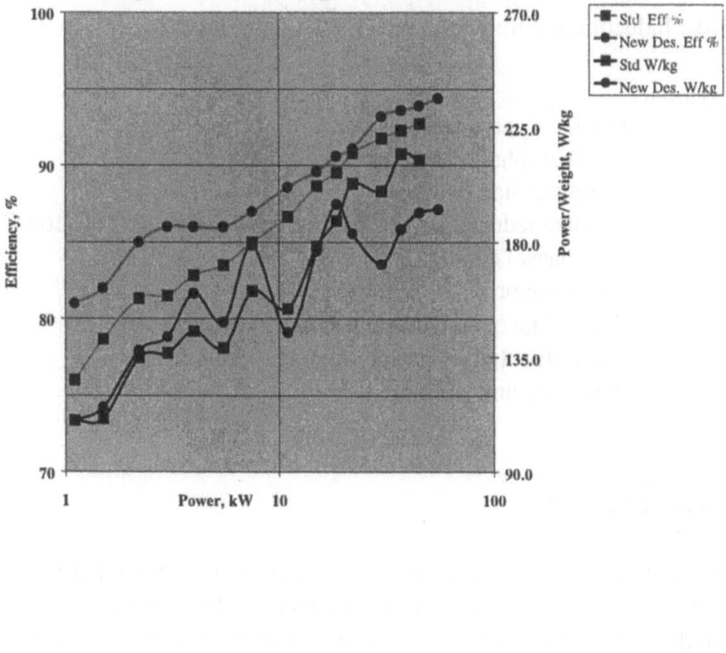

The Influence of Select Losses Components on Induction Squirrel-Cage Motor Efficiency

Konrad Dabala

Electrical Machines Department, Electrotechnical Institute, ul. Pozaryskiego 28, 04-703 Warsaw, Poland

Abstract. There is taken into consideration the influence of measured core and friction and windage losses on induction squirrel-cage motor in this paper. This paper presents a way of exact core losses determination in full-load motor. There are also compared efficiencies obtained for three core losses values:
- from no-load test (IEC 34-2);
- from draft IEC 2G/102/CDV;
- from presented in this paper method.

This paper presents the influence of the friction and windage losses determined from no-load test and the mechanical losses appeared under the rated speed on the efficiency, too. There is compared the influence of the core and mechanical losses sum on the efficiency dependently on the way of this losses components determination.

Keywords. Induction squirrel-cage motors, efficiency, losses

List of symbols

U (U_{ph})	terminal (phase) voltage
U_i (U_{iph})	terminal (phase) electromotive force
U_{aph} U_{bph}	phase auxiliary voltages
U_r	terminal reduced voltage according to project IEC 2G/102/CDV
I (I_{ph})	line (phase) current
$\cos\varphi$	power factor
R (R_{ph})	line-to-line (phase) stator winding resistance
X (X_{ph})	line-to-line (phase) stator winding reactance
j	imaginary unit

1 Introduction

In many standards [1, 2, 3] the core losses determined for the rated voltage and mechanical losses from no-load test are assumed to determine efficiency of the motor. In draft [4] is proposal to compensate for the influence of the voltage drop in the resistance of stator winding and to determine core losses under load on the basis of reduced voltage. This paper presents experimental estimate of different losses determination and efficiency methods.

2 Methods of Core Losses Determination

2.1 The Method According to IEC 34-2

Core losses are determined on the basis of no-load measurements for rated voltage.

2.2 The Method According to Draft IEC 2G/102/CDV

Equation (1) is accorded to draft IEC 2G/102/CDV and presents reduced voltage U_r on which are determined core losses (based on the no-load measurements). In this method is compensated for the influence of the voltage drop in the resistance of stator winding only.

$$U_r = \sqrt{\left(U - \frac{\sqrt{3}}{2} IR\cos\varphi\right)^2 + \left(\frac{\sqrt{3}}{2} IR\sin\varphi\right)^2} \qquad (1)$$

2.3 A Method Presented in this Paper

On a basis of Figure 1 we obtain equations (2) - (4) and final equation (5) which determined terminal electromotive force U_i. For this value of U_i we can determine core losses on the basis of no-load measurements.

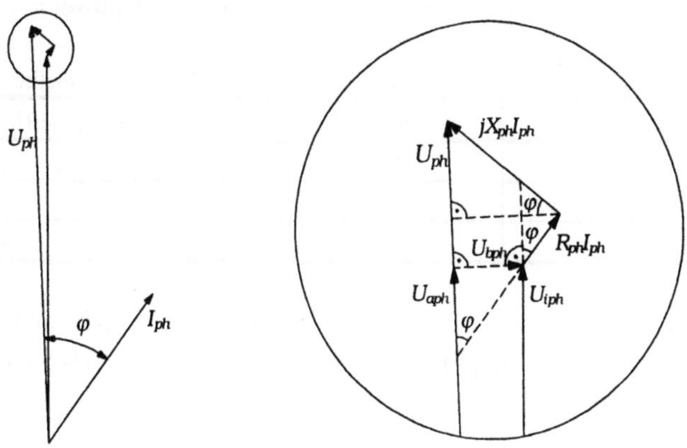

Figure 1: Per-phase sinor diagram for full-load squirrel-cage induction motor.

$$U_{aph} = U_{ph} - I_{ph} R_{ph} \cos\varphi - I_{ph} X_{ph} \sin\varphi \tag{2}$$

$$U_{bph} = I_{ph} X_{ph} \cos\varphi - I_{ph} R_{ph} \sin\varphi \tag{3}$$

$$U_{iph} = \sqrt{U_{aph}^2 + U_{bph}^2} \tag{4}$$

$$U_i = \sqrt{\left(U - \frac{\sqrt{3}}{2} IR\cos\varphi - \frac{\sqrt{3}}{2} IX\sin\varphi\right)^2 + \left(\frac{\sqrt{3}}{2} IX\cos\varphi - \frac{\sqrt{3}}{2} IR\,\text{si}\right.} \tag{5}$$

In this method is compensated for the influence of the voltage drop in the resistance and reactance of stator winding. The value of reactance X we can determined from load test measured values or from designed calculations of the motor.

3 The Influence of Different Methods on the Losses Values and Efficiency

Investigated motor had following rated values: power 12,45 kW, voltage 380 V, frequency 50 Hz, speed 1460 r/min.

3.1 Core Losses

The results of calculation of core losses and efficiencies by different methods are presented in table 1. The calculations were made for following values: R = 0,563 Ω X = 1,067 Ω I = 26,2 A cosφ = 0,811 rated input power = 13975,7 W.

Table 1: Core losses and efficiencies determined according to different methods

Methods	voltage [V]	core losses [W]	efficiency [%]
IEC 34-2	380,0	387,3	89,08
draft IEC 2G/102/CDV	369,7	366,6	89,22
presented method	355,7	339,4	89,42

The difference between efficiency by method IEC 34-2 and draft IEC 2G/102/CDV is equal 0,14 point of % but between IEC 34-2 and presented method even 0,34 point of %.

3.2 Mechanical Losses

According to IEC 34-2 mechanical losses from no-load test are taken to determine efficiency of the motor but in full-load condition rated speed is different than in no-load test. So there were made measurements of mechanical losses versus speed under sinusoidal supply. The points from measurements were approximated by 3-rd degree polynomial and the results of calculation for two speeds are presented in table 3. The difference between efficiency by method IEC 34-2 and presented method is equal only 0,04 point of %.

Table 2: Mechanical losses and efficiencies according to different methods

Methods	speed [r/min]	mechanical losses [W]	efficiency [%]
IEC 34-2	1499,0	69,4	89,08
presented method	1459,4	64,5	89,12

3.3 Sum of Core Losses and Mechanical Losses

The influence of sum of core losses and mechanical losses on efficiency is presented in table 3.

Table 3: Sum of core losses and mechanical losses and efficiencies determined according to different methods

Methods	sum of core and mechanical losses [W]	efficiency [%]
IEC 34-2	456,7	89,08
draft IEC 2G/102/CDV	436,0	89,22
presented method	403,9	89,46

The difference between efficiency by method IEC 34-2 and presented method is equal 0,38 point of %.

4 Conclusion

The results of losses and efficiency determination by presented in this paper method have proved that the 'real' efficiency of investigated motor is bigger than determinated by draft IEC 2G/102/CDV about 0,24 point of % and by IEC 34-2 even about 0,38 point of %. I'll suppose that the differences of efficiencies in smaller motors will be more bigger.

5 References

[1] IEEE Standard Test Procedure for Polyphase Induction Motors and Generators.IEEE Std 112-1991

[2] Energy Efficiency Test Methods for Three-Phase Induction Motors. CSA Standard C390-M 1985

[3] Rotating electrical machines Part 2: Methods for determining losses and efficiency of rotating electrical machinery from tests (excluding machines for traction vehicles). IEC 34-2 (1972)

[4] Method for determining losses and efficiency of three-phase cage induction motors. IEC 61972, 2G/102/CDV (1988)

Inverter Fed High-Speed Solid-Rotor Induction Motors for Industrial Applications

Huppunen, Jussi[1], Pyrhönen, Juha[1] and Alamäki, Jarmo[2]

[1] Lappeenranta University of Technology (LUT), Finland
[2] Rotatek Finland Oy, Lappeenranta, Finland

Abstract. An inverter fed 250 kW, 9000 min^{-1} solid-rotor induction motor drive for an industrial compressor application is introduced. New designing methods for the electric motor have made it possible to create a high efficiency high-speed solid-rotor induction motor. The results of the research work are new motor structures that are also easy and economical to manufacture. This technology is very reliable and economical for compressor and pump applications in power range from 100 kW to 1000 kW.

Keywords. Solid-rotor, high speed machines, induction machines, sine filter.

1 Introduction

The main advantages of using the motor in the high-speed range are the decreasing motor size and the absence of a mechanical gearbox and a coupler. When using appropriate materials the volume per power ratio and the weight per power ratio are nearly inversely proportional to the rotating speed in the medium speed range. Thus, when the motor speed is near 10000 min^{-1} the motor size and weight will decrease to about 1/3-part of the size of a conventional network frequency motor for 3000 min^{-1}.

When the load is directly attached to the rotor shaft and elevated speed is used, the solid-rotor construction is still capable of achieving sufficient mechanical strength and avoiding balance fluctuations and vibrations that might damage the bearing system.

It is usually believed that a solid-rotor induction motor has poor electrical properties. According to our experience, however, very good properties can be reached when a motor and a drive are properly designed. In addition, high-speed motors always need a frequency converter as a supply. This gives also all the advantages of the speed control for the drive. Because of its good control properties the system provides us with high energy efficiency. Converters, however, cause extra heating problems even in normal speed machines and thus a careful design to combine an inverter with a solid-rotor machine is needed.

2 Applications

Typical applications of elevated speed motors are process industry fans, compressors and pumps on 100 ... 1000 kW power range. The theory and usage of high-speed machines has shown that the optimal speed range for this power range applications varies between 6000 and 30000 min^{-1}.

This system construction is extremely suitable for the replacing of low pressure liquid ring pumps which are constant speed driven and controlled with a throttle. The nominal efficiency for a liquid ring pump without a throttle valve is about 60%, but under throttle control the efficiency is between 20 and 50 %. The efficiency of the high-speed compressor is around 85 %, and while this new system is speed controlled the efficiency stays high in a wide speed range. The new technology saves huge amounts of energy and its pay-back time is typically less than two years.

3 The System Construction

The system consists of a three-phase solid-rotor induction motor, a commercial PWM-modulated frequency converter with a special sine wave filter and one or two compressors which are designed for the speed of 9000 min^{-1}. The compressors are directly attached on both ends of the motor shaft, without any gear or coupler. Therefore, the bearings have uniform load without any axial loading.

3.1 Solid-Rotor Induction Motor

The stator of this solid-rotor induction motor has a two layer 5/6 short pitch three phase winding. That winding type minimises the winding harmonics since it gives the minimum value for the leakage factor [1]. Stator slot openings are specifically designed to avoid permeance harmonics [1]. Although a copper induction cage in the solid-rotor is very efficient and gives particularly good properties for the solid-rotor motor, it is not essential in high power applications unless the smallest possible motor size is required [2]. The same power can be attained without a cage winding when the rotor size is slightly increased. In many applications the copper cage winding may also be too expensive to manufacture. Thus an axially slitted solid-rotor with copper end-rings is used.

The motor has oil lubricated ball bearings. That bearing system works hydrodynamically at high speeds. Therefore, the bearing system has very low friction losses and a long maintenance cycle.

The motor has an open structure and an air-cooling system. There is an air channel in the middle of the stator core. This channel leads the cooling air into the air-gap.

On both sides of the shaft there are fans which suck the cooling air through the stator core and the air-gap.

3.2 Filter Structure

A passive filter is employed between the motor and the inverter that smoothens the voltage and current waveforms very close to a sinusoidal form. Therefore the motor mmf will smoothen and harmonic losses will be reduced and it can be attained a motor efficiency level that is similar to that reached by a normal motor in direct network supply. Because of almost sinusoidal waveforms, the filter permits the use of long cable between the inverter and the motor. The filter also compensates the magnetising current of the motor and thus reduces the effects of the slightly poor power factor of the solid-rotor induction motor. The inverter output current is also reduced remarkably. Thus the inverter does not need over-sizing despite of the poor power factor of the motor. The filter structure is a passive LC-circuit. The drive configuration is shown in Figure 1 [4].

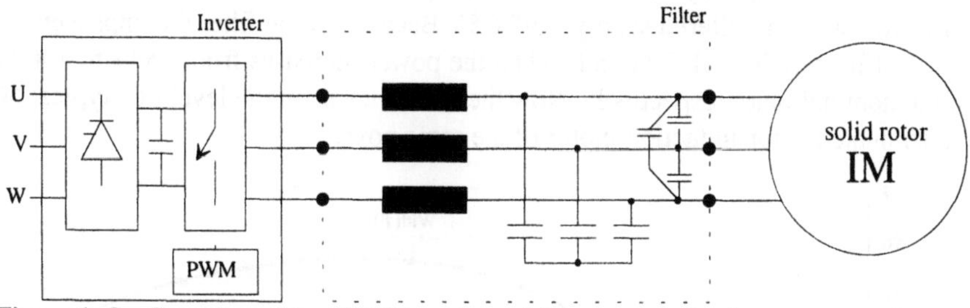

Figure 1. Configuration of the PWM-inverter and filter on an induction motor supply.

4 The Drive Performance

4.1 The Effects of the Filter on the Inverter Output Waveforms

Figure 2 shows the frequency converter output voltage waveforms without filter and with filter at nominal load using 6 kHz switching frequency, respectively. As it can be seen, the frequency converter generates a large amount of harmonics near the switching frequency. The filter smoothens the waveforms extremely effectively [3].

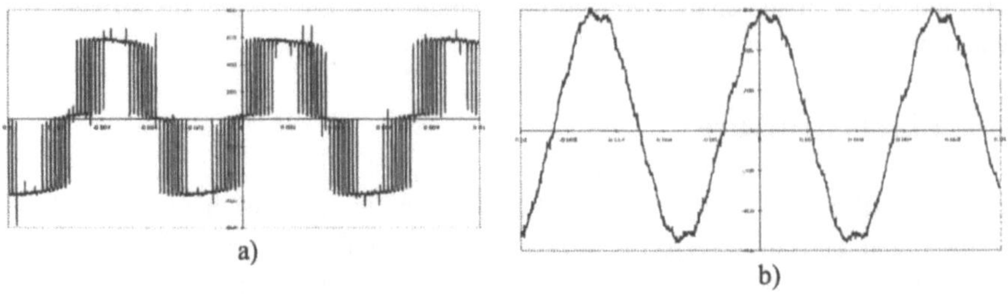

Figure 2. Inverter output a) voltage and b) current waveforms at 140 Hz supply frequency and 6 kHz switching frequency.

4.2 Loaded Running Motor Tests

Measurements were done in a laboratory test bench and also extensive field tests were carried out. In Figure 4 the motor efficiency is drawn at nominal supply frequency. The filter improves the motor efficiency remarkably despite of the reduced power capacity caused by the voltage drop in the filter. The maximum efficiency with the filtered supply is 95.6 %. Because of the filter the improvement in the efficiency is 5...1.5 %-units when the power increases from 25 % to 100 % of the nominal load, respectively. Now the efficiency is on the level of a typical 50 Hz laminated-rotor-induction-motor of the same power.

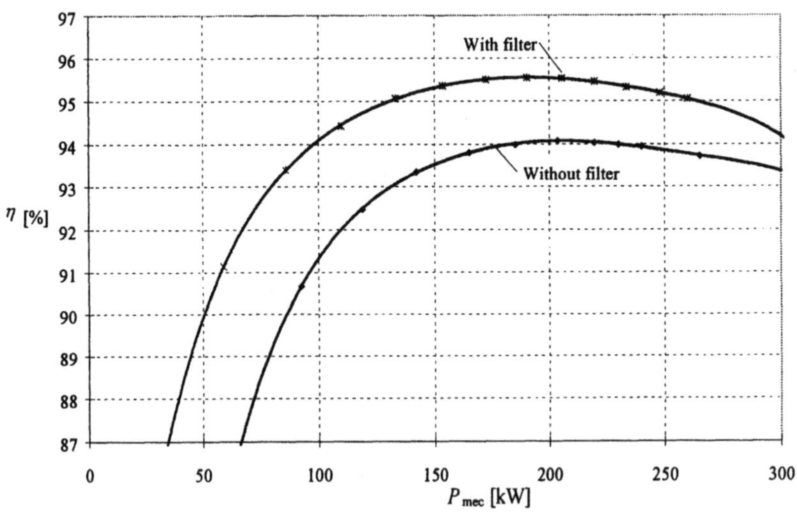

Figure 3. The motor efficiency versus motor output power at nominal supply frequency.

5 Applications' Economy

The efficiency of the high-speed compressor is 85 %. The inverter gives speed control for the drive, thus the total efficiency stands high in a wide speed range, unlike in throttle or shutter valve controlled constant speed systems where the efficiency drops deeply while they are working at partial load, Figure 4. In the following case the efficiency of the new technology is compared to a liquid ring pump systems in a vacuum creating system for which this new technique is very suitable. In this case, for 250 kW vacuum creating system, the new high speed technology gives 220 kW power saving and 24 m^3/h water saving. For example, in 6000 h full operating time this leads to 1320 MWh/a energy saving and 144000m^3/a water saving, Figure 5.

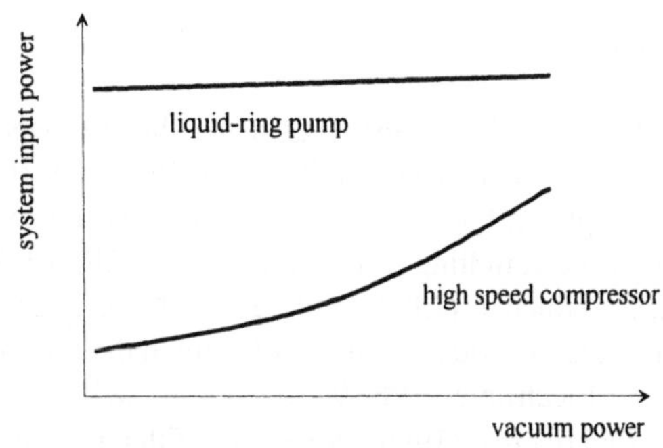

Figure 4. Power demands of two vacuum creating systems.

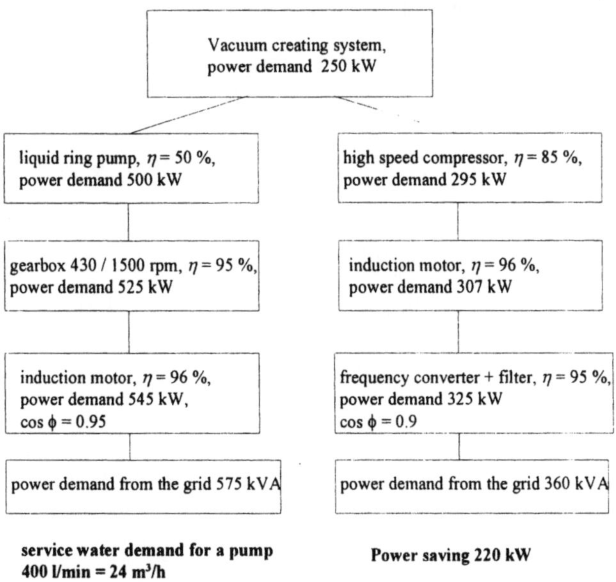

Figure 5. A comparison between a high-speed compressor and a liquid ring pump as a vacuum creating system.

6 Conclusions

It is well known that suitable speed for industrial compressors in the power range of 100 ... 1000 kW varies between 6000 and 30000 min^{-1} depending on the pressure needed. In the case of a compressor it is very profitable to integrate the motor and the compressor and to use a standard frequency converter to make a system economical in space and weight. Furthermore, the system gives also very promising results for energy economy. Because in the high speed range the manufacturing of the system does not require extremely sophisticated technique and the bearing system is well-known engineering, there are no technical obstructions for the systems to penetrate into the market.

7 References

[1] Pyrhönen J., Kurronen P. (1994): Increasing the Efficiency of High Speed Solid-Rotor Induction machines, ETEP vol. 4, No. 2, pp. 101-107.

[2] Pyrhönen J., Huppunen J. (1996): A new medium speed solid rotor induction motor for a high speed milling machine, Proc. SPEEDAM -96, pp. B5.1-8

[3] Huppunen J., Pyrhönen J. (1997): Filtered PWM-inverter output waveforms in a medium speed (< 30000 rpm) solid-rotor induction motor drive. Proc. European Power Electronics EPE -97. pp. 3.813-3.818.

[4] Huppunen J., Pyrhönen J. (1999): Sine wave filter for PWM-inverter driven medium-speed (< 30000 rpm) solid-rotor induction motors. Proc. European Power electronics EPE -99.

Switched Reluctance Drive for Mass Application

Dr.M.G.Bitchkov, Prof.N.F.Ilinski, Dr.V.A.Sementchuk

Moscow Power Engineering Institute, Russia

Abstract. Variable Speed Drives (VSDs) are the most effective means for energy saving in many practical applications particularly in pumps, fans, compressors, conveyors, etc. Up to now the only wide used VSD is a frequency controlled electric drive with AC induction motor. A Switched Reluctance Drive (SRD) having noticeable advantages such as simple construction, low cost, high reliability, high efficiency, etc. still does not compete with traditional AC drives mainly because of the following unsolved problems: complexity of proper control, necessity of a rotor position sensor, poor acoustic characteristics. Results of R&D project realised in Moscow Power Engineering Institute in co-operation with Jaroslavl Electric Machine-Building Plant allowed to overcome the above mentioned problems and to obtain reliable low cost sensorless SRD with high efficiency and acceptable acoustic characteristics. The original sensorless SRD version is based on current measuring and exploiting a specially organized observer that provides a stable drive operation within broad speed-torque ranges. Acoustic SRD characteristics were improved by means of proper current pulses shape control. Small increasing the time of current pulse trailing edge allowed to decrease the noise level to 70 dB(A) and brought the SRD to a vary smooth operation even at a low speed.

1 Introduction

Switched Reluctance Drives still did not find wide application in technologies where their advantages (low cost, high reliability, high efficiency etc.) seem to be the most noticeable- pumps, fans, compressors, etc. These spheres are occupied by traditional frequency controlled AC induction drives in spite of at least twice more expensive AC motors and more complicated and less reliable frequency converters with approximately the same price. This can be explained not only by final users conservatism but mainly by still existing SRD negative features - a rotor position sensor that complicates the motor construction and brings some practical inconveniences, and poor SRD acoustic characteristics due to discrete character of the drive behaviour.

The R&D project undertaken at Moscow Power Engineering Institute, Electric Drive Department, in co-operation with Jaroslavl Electric Machine-Building Plant has been aimed to overcome above mentioned SRD disadvantages. Previous to the project the general SRD theory has been carried out that allows to design optimal SRD configurations and parameters for any given requirements. This optimal design procedure is presented in the form of a dialogue computer program. It has been used for working out some SRD prototypes: 8/6 poles, 380V, 9 kW, 3000 rpm (general-purpose version), 12/16 poles, 36 V, 160 W, 200 rpm (electric bicycle), 6/4 poles, 220 V, 1.2 kW, 4000 rpm (domestic equipment). SRD control principles have been established and corresponding hardware and software have been worked out.

All theoretical results have been proved by laboratory experiments carrying out on specially designed and produced computerized testing equipment allowing to make on-line investigation of any processes in SRD.

The main results of the project concerning sensorless SRD version and acoustic characteristics improvement are given below.

2 Sensorless SRD Version

A rotor position sensor normally used in SRD is rather undesirable element of the drive for mass application as far as it noticeably affects the simplicity of the drive construction, so the elimination of a rotor position sensor is the necessary condition for wide SRD application.

This problem has been solved by implementation into the drive a specially designed original observer that controls the phase currents and uses this on-line information for nominating the angles of switching the phases on and off in any operation modes. This technical solution is based on the same principles that have been presented at [1,2] and differs in some important details.

The effectiveness of this technical solution is illustrated in Figure 2.1 where one can see practically identical phase currents while SRD is starting with (Figure 2.1a) and without (Figure 2.1b) rotor position sensor. This identity exists for all modes of drive operation.

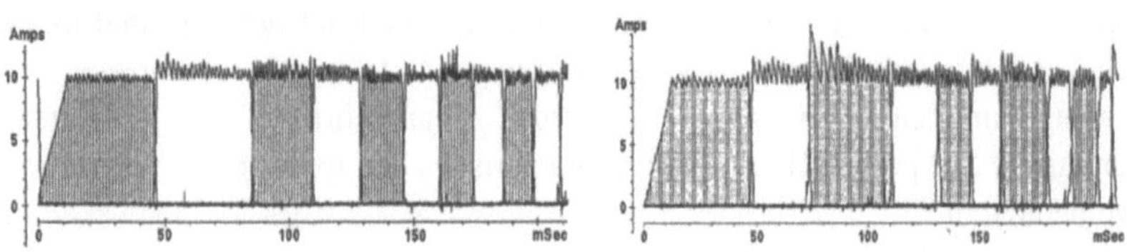

a) b)

Figure 2.1: SRD starting currents with (a) and without (b) rotor position sensor

The described sensorless SRD control method can be realized on a very simple controller with 8-bit microprocessor like MOTOROLA MC68HC908MR24. The simplest sensorless SRD is completely workable and has per-unit speed – torque and efficiency – power characteristics shown on Figure 2.2 for the 9 kW, 3000 rpm sample. The speed-torque characteristics, if required, can be transformed by any traditional method into the desired form.

The efficiency-power characteristic has practically horizontal part for power 0.6 – 1.2 of nominal value. It is to be noted that the efficiency of 90-92 per cent has been obtained in a nonoptimized drive and can be increased in an optimized version.

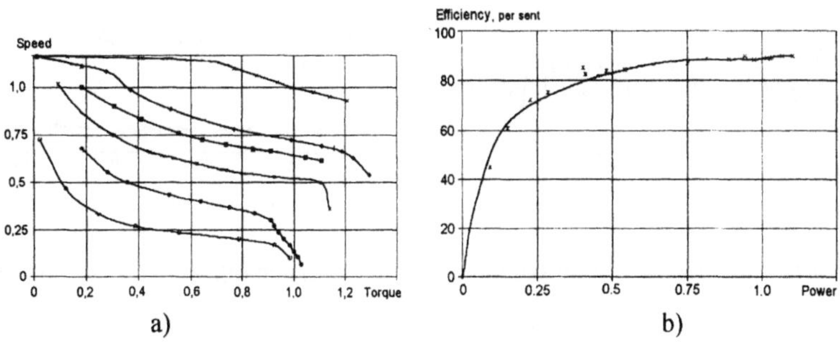

Figure 2.2: SRD characteristics

The sensorless SRD control can be effectively realized for the drives of kilowatts and more where the per-unit winding resistance is relatively small. For the drives of less capacity (hundred Watts) the resistance changing with the temperature rise is to be taken into account to organize the proper control algorithm. In other way the drive happens to be out of control.

The proper software for eliminating this disadvantage has been carried out for low voltage SRD (36 V, 160 W, 200 rpm) for electrical bicycles. This drive worked satisfactory even at very low speed (up to 50 rpm). The switching curve transforming by means of the properly organised control algorithm allows to obtain stable low speed characteristics on wide torque range.

3 Acoustic Characteristics Improvement

The main source of motor excessive noise and vibration has been defined when only one phase operates (Figure 3.1a): the front edge of the current pulse does not create the vibration while current decreasing is followed by strong current oscillation and noise. So the main noise and vibration source is the rare edge of current pulses.

Finding out the source of the noise allowed to eliminate it by means of the proper current pulse shape control. Corresponding delay of the current pulse rare edge allowed to decrease the noise and vibration up to the acceptable level as it is shown in Figure 3.1b (four phases operation). In practice this problem has been solved on the base of power controller with Intel 8xC196MH processor.

Another noise and vibration source can be the improperly selected PWM frequency that also has to be taken into consideration in SRD design.

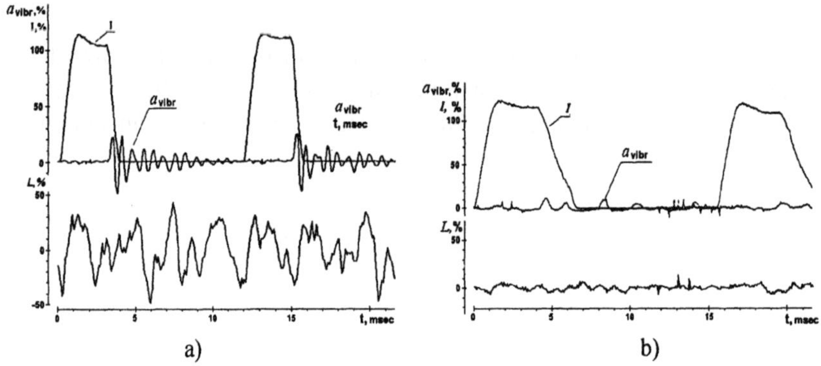

Figure 3.1: Current pulses, noise and vibration: initial version (a), corrected version (b)

4 Conclusions

Well-known SRD advantages: low cost, high efficiency, high reliability make this type of variable speed electric drive the most suitable for general purpose mass application.

Eliminating by the proper control the main SRD negative features – rotor position sensor and excessive noise and vibration leads to a much wider SRD application in different technologies.

5 Bibliography

[1] Arefeen M.S., DiRenzo M., Beierke S. (Texas Instruments) (1998): DSP for Switched Reluctance Motor Drives. PCIM Europe 5, 276-280.

[2] US Patent N 5043643 (Int. Cl.[5] HO2P 6/02).

The Written-Pole™ Motor: High Efficiency - Low Start Current

Brian Beck[1], Dale Friesen[2]

[1] C.Eng. Precise Power Corporation, Bradenton, Florida
[2] P.E. Manitoba Hydro, Winnipeg, Canada

Abstract. Written-Pole™ technology is a patented machine technology, which changes the magnetic polarity of the rotor field in a rotating machine, while the machine is operating. The number of poles is thereby changed, resulting in a constant frequency - variable speed machine. When operating as a motor, a Written-Pole machine has inherently low starting current and high operating efficiency.

1 Introduction

Precise Power Corporation of Bradenton, Florida developed a three-phase Written-Pole motor, originally as part of a motor-generator design. The Electric Power Research Institute (EPRI) of Palo Alto, California, the National Rural Electric Cooperative Association (NRECA) of Arlington, Virginia, and their member companies recognized the easy start capability of the Written-Pole motor as a potential significant benefit. In particular, there is a need for large horsepower motors to be used on single-phase distribution lines. A single-phase development was commissioned and is currently part of an on-going EPRI project, funded through the collaborative efforts of EPRI, NRECA and participating electric utilities. The project's objective is to commercialize a range of single-phase motors up to 75 hp. The first series of designs was for motors with an external rotor. Written-Pole motors exhibit excellent starting characteristics. They can be used to accelerate high inertia loads and also have start currents less than two-times full load current. Under steady-state conditions they operate as synchronous machines at unity power factor and a high operating efficiency, comparable to three-phase machines of the same rating. Subsequently, a range of internal rotor single-phase machines has been introduced and designs are in progress for three-phase motors of ratings up to 1000 hp.

2 Written-Pole Technology

The patented Written-Pole technology consists of a rotor design that has a layer of ferrite magnetic material that can be changed magnetically to any desired pole pattern while the machine is operating. An electromagnetic field from a stationary exciter coil is used to change the magnetic polarity of the revolving layer of ferrite, effectively inducing the poles as the machine rotates. Hence the name, "Written-Pole". In conventional machines the rotor poles are mechanically located and the magnetic pattern cannot be altered after the machine is assembled. The ability to electrically place the magnetic poles in any position allows a synchronous motor to be built that has a very low per-unit direct line starting current, high operating efficiency, and very high inertial load starting capacity. Also it allows for generators where the output frequency is independent of rotational speed.

3 Construction of the Written-Pole Machine

Conventional synchronous machines have a fixed, even number of magnetic poles, which causes a fixed relationship between shaft speed and frequency.

The Written-Pole machine has a continuous layer of "permanent" magnetic material on the surface of the rotor. (Figure 1) Typically this material is a highly anisotropic S_r O·n Fe_2 O_3 composition ferrite ranging in thickness in the oriented direction from 15 to 75mm depending on machine size. The main windings of the stator are similar to those of conventional machines, with an additional concentrated winding around an exciter pole. The magnetic layer can be magnetized into any desired pattern by the exciter pole while the machine is still operating.

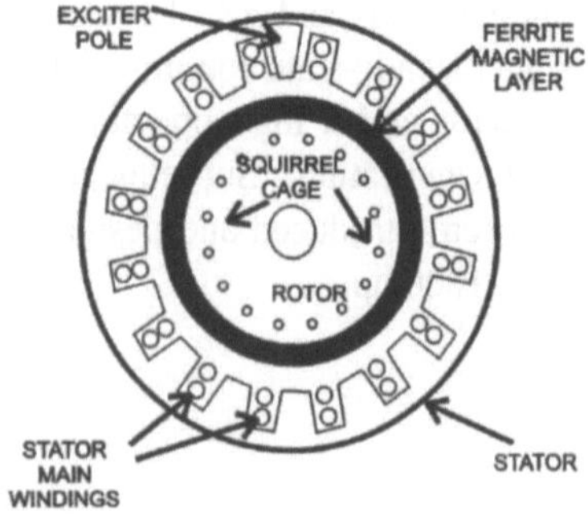

Figure 1: Internal Rotor Construction

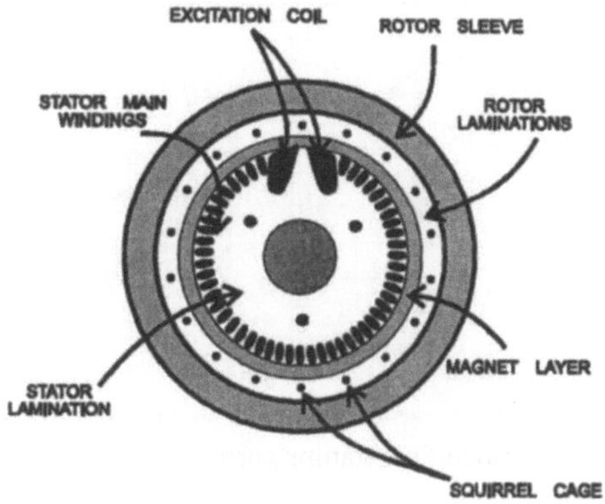

Figure 2: External Rotor Construction

4 Starting Characteristics

The Written-Pole motor is a synchronous motor that inherently develops starting torque. The rotor magnetics are designed to allow the stator winding starting current to influence the magnet layer and provide hysteresis torque. A squirrel cage is added under the magnet layer in the rotor for additional starting torque. (See Figures. 1 and 2.)

The high resistance rotor cage is designed for 20% slip or more and with the inherent hysteresis torque brings the rotor to partial speed before the motor exciter coil winding is energized. The exciter coil is energized with alternating current at line frequency inducing a pole pattern in the rotor magnetic surface that is magnetically synchronous with the stator rotating field.

The phase relationship of the induced poles and the rotating magnetic field of the stator causes the rotor to accelerate with maximum torque to synchronous speed. At this point, the exciter can be turned off and the motor will continue to operate as a permanent magnet synchronous motor. (Figures 3 and 4).

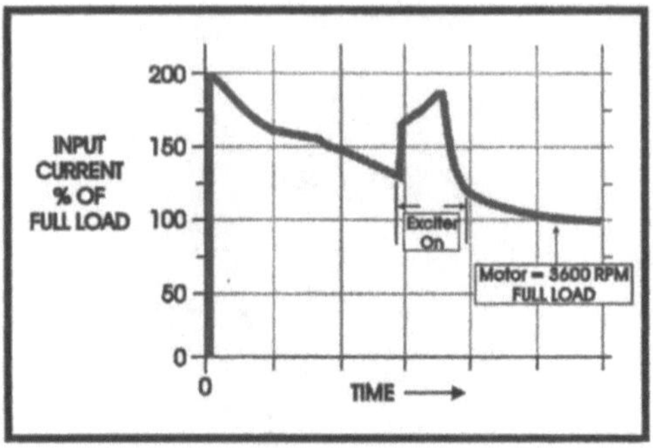

Figure 3: Typical Plot of Written-Pole starting current

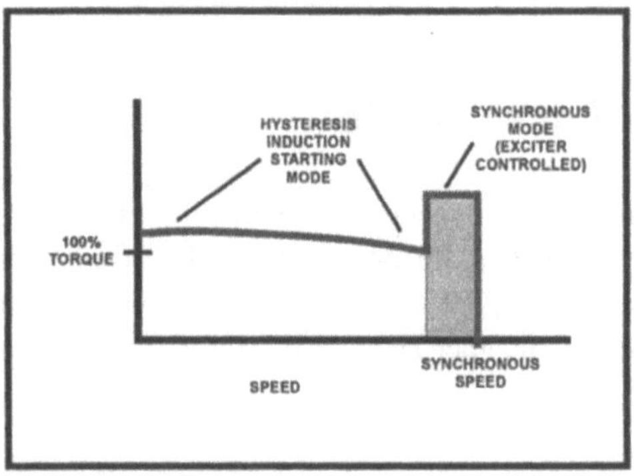

Figure 4: Typical Plot of Torque/Speed.

This starting method results in a starting current of 1/3 or less of that required by an equivalent induction motor. This low starting current makes the Written-Pole motor attractive for single-phase applications. Voltage levels on long distribution feeders are often negatively impacted by the starting current requirements of large single-phase motors. Many utilities impose starting restrictions on large single-phase motors, typically when the associated voltage reduction is greater than 3%. Tests performed by Manitoba Hydro on a 20 hp single-phase Written-Pole motor indicated that the distribution voltage reduction on starting was 1.4% (the calculated value was 1.6%).[1]

---

[1] Independent Laboratory and Field Tests of a Written-Pole™ Motor. D. Friesen, P.E. and D. St.George, P.E. Manitoba Hydro

5 Efficiency

Recent tests at an independent test facility in Canada illustrated efficiency in excess of 94% for a single-phase 30-hp prototype internal rotor machine.[2]

Laboratory personnel conducted a wide array of tests to verify the prototype's operating efficiency, starting performance, and locked rotor characteristics. Since conventional three-phase test standards are not applicable to this single-phase design, a modified input/output test was used to evaluate the prototype's operating efficiency over a range of loads varying from 25 to 135% of rated load. The prototype Written-Pole motor exhibited exceptional operating efficiency with a measured efficiency of 94.2% at rated load and 94.5% at 115% of rated load. The input power factor at rated load was 0.997 (leading) and 0.990 (lagging) at 125% of rated load. The high operating efficiency and near unity power factor contributed to the motor's low operating temperature with a measured temperature rise of 45°C at rated load.

Locked rotor tests and starting tests were used to verify the prototype's starting performance. The locked rotor current was measured at 144 amps when rated voltage was applied to the motor terminal with the shaft locked. Locked rotor torque was measured at approximately 85 N-m or 72% of rated torque. While the locked rotor torque is significantly lower than comparable three-phase designs, the dramatic reduction in locked rotor current when compared to conventional three-phase designs yields significant reductions in temperature rise during starting cycles. This characteristic was particularly evident after a prolonged locked rotor test in which the rotor was locked for a period of three minutes with rated voltage applied to the terminals. An initial temperature of 55°C was recorded at the start of the test (after performing 12 successive start tests over a period of 30 minutes). Upon completion of the test, the measured winding temperature was 107°C, yielding a temperature rise of 52°C

The prototype's peak starting current was recorded when the exciter was energized at approximately 1575 rpm The peak current recorded during this period was approximately 168 amps with rated voltage applied to the motor terminals. This value represents a peak starting current of 1.7 per unit or 1.7 times the rated current of the motor.

---

[2] Continuing Improvements in Written-Pole™ Technology – the internal rotor motor. J. Roesel and B. Beck, Precise Power Corporation, D. Friesen, Manitoba Hydro

Table 1

30 hp, 4-pole, 240 Volt single-phase internal rotor

Load %	pf	Efficiency %
125.37	0.979	94.2
115.06	0.984	94.5
99.75	0.994	94.2
75.42	0.996	92.7
50.01	0.998	88.8

20 hp, 4-pole, 240 volt, single-phase external rotor

Load %	pf	Efficiency %
124.20	1.000	91.8
115.10	0.996	92.1
99.10	0.983	92.0
74.7	0.947	90.7
50.40	0.855	87.2

The external rotor design with its increase in friction and windage suffers from degradation in efficiency of approximately 2 percentage points. Table 1 shows test results for the 30 hp internal rotor motor and earlier results from tests on a 20 hp external rotor motor.

Calculated efficiencies for a three-phase machine, internal rotor design, indicate for the comparable rating there is a percentage point improvement in efficiency over the single-phase machine value.

6 Applications

The combination of low starting current and synchronous operation allows loads with very large inertia to be easily accelerated to synchronous operating speed. Typically Written-Pole motors can safely accelerate inertial loads of at least 10-20 times that practical with a conventional high efficiency squirrel-cage induction motor. Conventional motors driving such high inertia loads require special electrical or electronic starting systems to avoid high temperature winding or rotor damage during starting or re-starting. The Written-Pole motor can drive such loads without electronics in the power path, which achieves better reliability, efficiency and ruggedness.

Written-Pole motors constructed with an external rotor are able to support rated load for periods up to 15 seconds, even in the case of a full voltage interruption. Written-Pole technology utilizes the inertia, the flywheel effect, of the external rotor, continuously re-magnetizing the ferrite layer as rotor speed decreases. On resumption of full voltage, the motor draws no more than the initial start current to re-accelerate the rotor.

The first commercial use of three-phase Written-Pole motors was to meet the requirements for a specific high inertia motor-generator (M-G) set application. The three-phase, 42 HP motors for these M-G sets drive a rotor assembly with very high inertia in order to obtain "ride-through" time (flywheel effect) after an input power interruption. The 28 kW, 35 kVA, three-phase Written-Pole generator in these M-G sets maintains a constant frequency, phase-locked output during the speed change.

7 Economics

Table 2 illustrates costs of (A) a 40 hp Written-Pole single-phase motor project; (B) a three-phase 40 hp motor with phase converter, and (C) a three-phase 40 hp motor with a new length of 3-phase line over a distance of 3 miles.

Item	A	B	C
Initial cost	$9,500	$6,800	$80,000
Efficiency	92%	79%	90%
Power factor	1.0	.85	.85
Inrush	2.0 p.u.	7.0 p.u.	7.0 p.u.
Energy Cost	$4,261	$4,963	$4,357
Motor life	15	15	10
Equivalent annual cost	$4,894	$5,416	$8,440

Assumptions: Motor duty cycle = 30%; Electricity cost = $0.05 per kWh; Three-phase line cost spread over 20 years.

8 Conclusions

The Written-Pole motor exhibits low inrush starting current without the need for electronic input control or reduced voltage starters. It exhibits very stable synchronous operation with high efficiency and power factor. Torque is controlled so that the motor is not easily pulled out of synchronism. In the case of a heavy shock overload the Written-Pole motor increases the pull-in torque and re-synchronizes easily.

The Written-Pole motor has an inherent ability to start and synchronize loads with very large inertia. It is also capable of multiple starts in a short time without overheating. Automatic controls in the Written-Pole motor instantly write new poles that are in phase with the line current.

Independent testing of the Written-Pole motors has demonstrated performance characteristics to be within reasonable tolerances of calculated values. Efficiency values at full load for a 20 hp and 30 hp single-phase motor have been shown to be 92% and 94% for the external rotor and internal rotor designs respectively. Associated starting currents were less than 2.0 p.u.

High efficiency and low starting current are the main attributes of the Written-Pole motor, differentiating it from other new motor designs.

Motor and Drive Integration with Passive Components Realised Using the Stator Core

S.D. Garvey

M. D. of Insight_M Ltd., Aston Science Park,
and Reader in Dynamics at Aston University, UK

Abstract. There is an inevitable trade-off between overall energy efficiency of any motor-drive pair and net capital cost. If the marginal cost of certain necessary components can be reduced, greater expenditure is possible in other areas with the potential of increased efficiency for a given cost. This paper investigates just such a proposition.

Keywords. Integrated Passive Components, Integrated Motor-Drives.

1 Introduction

Useful quantities of both inductance and capacitance can be realised in the stator cores of many classes of machines at minimal additional cost through judicious use of the stator core as one or more independent toroidal inductor cores and/or as one or more independent multi-plate capacitors. Any given section of stator core can simultaneously serve to provide both capacitance and inductance.

The "goodness" of the machine alone (taken in isolation from the drive) may or may not be slightly diminished as a result of this increased utilisation of the stator core. Some additional volume would result from the introduction of additional dielectric between laminations and from the necessity to insulate the core electrically from the frame. Some slot-space is also lost to accommodate windings which drive circumferential (toroidal) flux. Against this, the main windings of the machine may be much better protected from the ravages of rapid voltage transients appearing at the machine terminals with the result that insulation thickness might be reduced with the twofold benefit of improved thermal conduction and higher slot fill factors.

Moreover, the very real possibility exists that the need for any major capacitors and inductors in the drive is completely obviated.

The concepts of realising capacitance and inductance from the stator core are applicable to induction machines, switched-reluctance machines, permanent magnet and wound-rotor synchronous machines and others.

2 Realising Discrete Inductors within Stator Core

The entire back of stator core can be drawn into use as the magnetic circuit for one or more discrete inductors by incorporating toroidal windings which act to drive alternating magnetic flux in complete loops around the back of stator core encircling the machine axis and generally parallel to the airgap.

Figure 1a shows a machine stator carrying both main machine windings (1) and toroidal windings (2). Figure 1b shows that the stator core can be divided axially into portions to form different inductors.

Figures 2a and 2b show flux plots for a fictitious permanent-magnet machine with and without the toroidal field. The penalty paid on main machine field due to the presence of small levels of toroidal flux is second-order.

Energy densities in the order of **2 - 5 kJ/m^3** are achievable in the back of core inductors with the right conditions. When transformer function or operation of a choke on a DC line are of interest, the value of the integrated inductors can be far greater than the above Figures suggest.

If the magnetic characteristics of the iron in the back of core were linear, the toroidal flux field and the main machine flux field would be completely decoupled. Of course, saturation occurs and this causes a coupling between the two fields. Figure 3a shows a set of magnetisation characteristics for the main machine flux of a small switched-reluctance machine. Each curve on this figure corresponds to a different value of net toroidal MMF. Figure 3b shows the corresponding magnetisation characteristics for the toroidal field at various levels of net main-machine MMF.

Figure 4 shows an image of a small induction machine stator carrying the toroidal windings. In this case, the toroidal windings provided are several times more substantial than they would normally be in application. Spacer pieces to hold the toroidal windings away from the ends of core are evident (brown).

3 Realising Discrete Capacitors within Stator Core

Virtually all stator cores for rotating AC machines are currently laminated (Figure 5). Individual laminations are electrically conducting and they are insulated from their neighbours by thin layers of insulation. It is obvious that capacitance can be achieved by connecting the plates electrically.

Figure 6 shows that three-phase "star" or "delta" capacitor configurations can be achieved by appropriately-ordered connections to the plates.

Energy densities in the order of **200 kJ/m^3** of dielectric are achievable if suitable care is taken to ensure that the lamination surfaces are smooth. With, say, 1% of stator core volume being devoted to interlaminar insulation (dielectric), the average electrostatic stored energy density over the stator core may be in the order of **2 kJ/m^3**.

For low-voltage machines, a single layer of interlaminar insulation can be sufficient to withstand line-to-line voltage if it has been chosen and treated carefully. For higher voltage machines, it becomes sensible to share the voltage drop between several different interlaminar layers. Although this reduces the actual capacitance value, the net potential for storing electrostatic energy is unchanged.

4 Applications

The main applications for the integrated components include :

- Incorporation of DC link components into motors
- Passive filters for protecting main machine windings
- EMC filters at the front ends of integrated motor-drive pairs
- Passive components for Resonant Link Inverter Designs
- Passive components for Static VAR compensation
- Mutual-inductors acting to suppress undesirable spatial components of current in multi-phase (> 3-phase) machines.

Application notes are available (from 1/10/99) on three of the above at :- http://www.insight-m.demon.co.uk.

5 References

[1] S.D. Garvey. *New Design for Motors and Generators Enhances Machine Utility*. PCT Patent. WO 98 / 40957.

[2] S.D. Garvey. *Electrical Machine and Drive System Incorporates Passive Components in the Machine*. UK Patent. GB 9911635.5

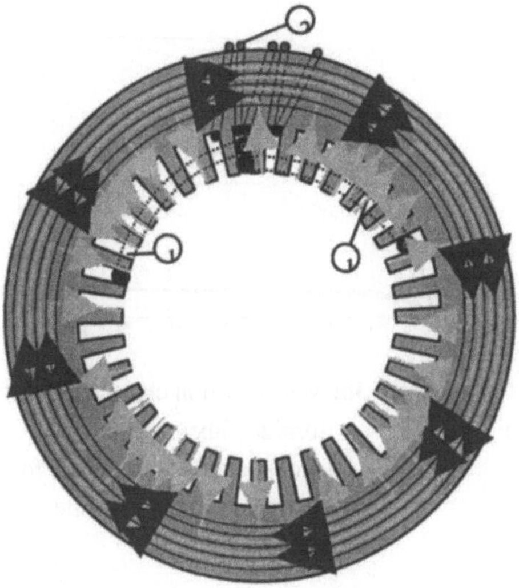

Figure 1a: Stator core carrying main-machine~ and toroidal windings

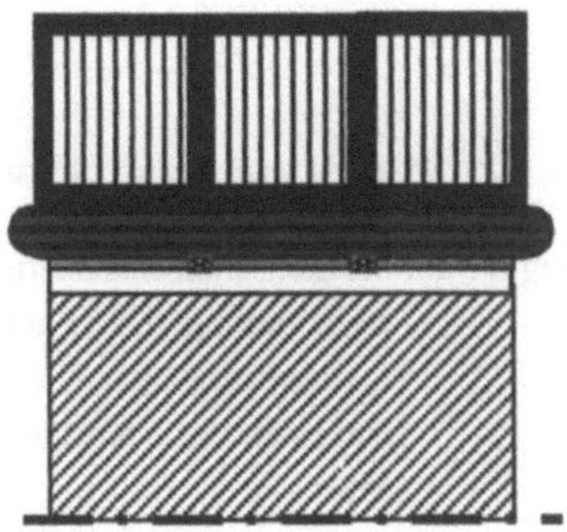

Figure 1b: Stator core divided axially into portions

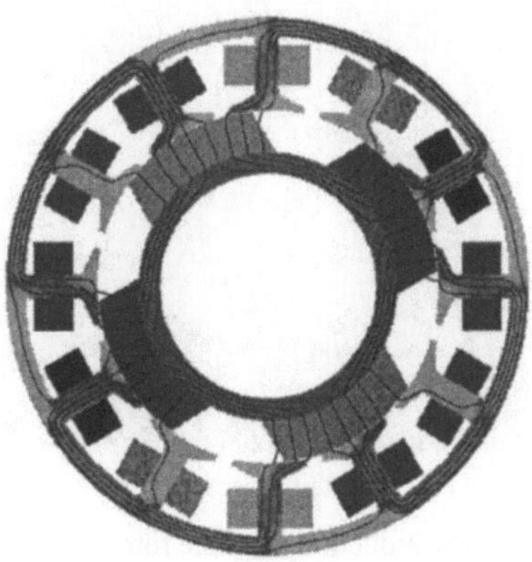

Figure 2a: Permanent magnet machine with no toroidal MMF (Fictious Machine Design)

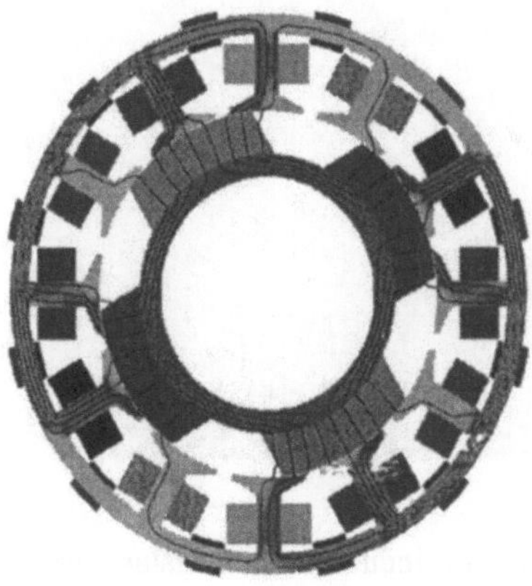

Figure 2b: Permanent magnet machine with some toroidal MMF (Fictious Machine Design)

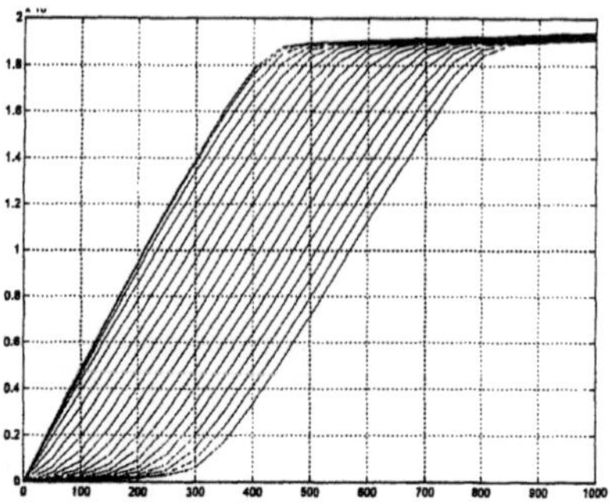

Figure 3a: Magnetisation characteristics for main machine flux vs net main machine excitation. Distinct curves for distinct net toroidal MMFs

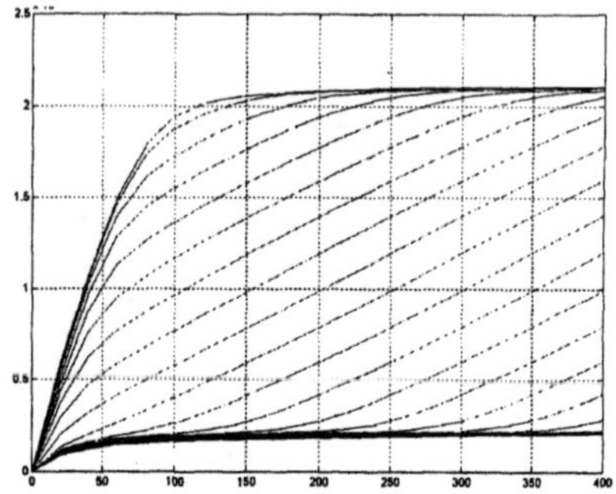

Figure 3b: Magnetisation characteristics for toroidal flux vs net toroidal excitation. Distinct curves for distinct net main machine MMFs

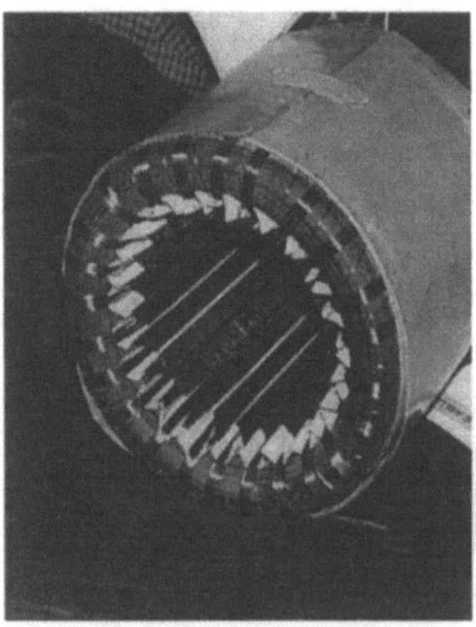

Figure 4a: Induction motor stator carrying toroidalwindings

Figure 4b: Induction machine rotor

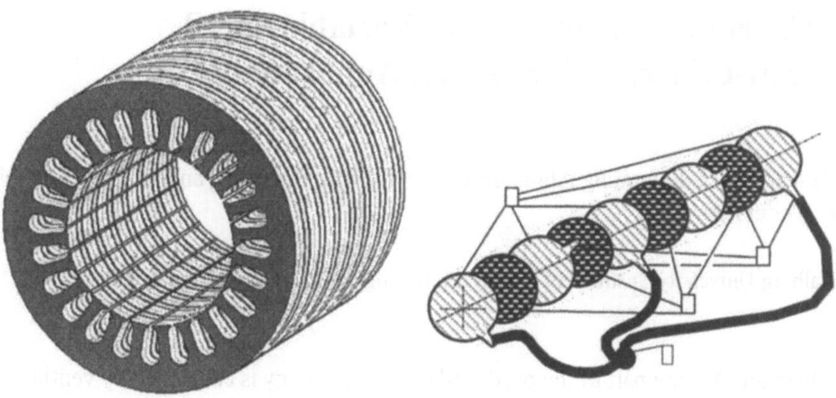

Figure 5: Every laminated stator is potentially a capacitor

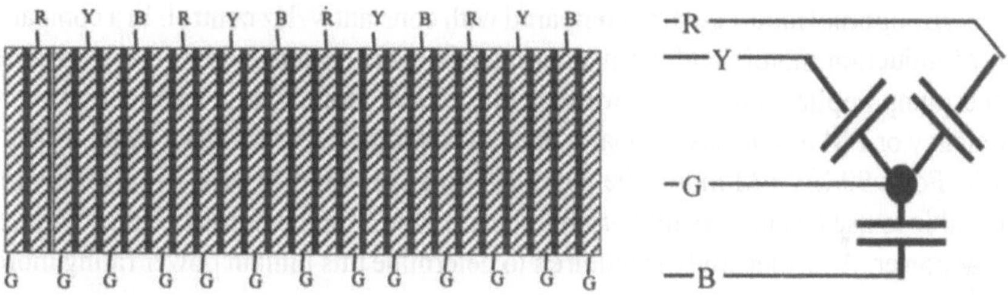

Figure 6a: Achieving 3 capacitors in "star"

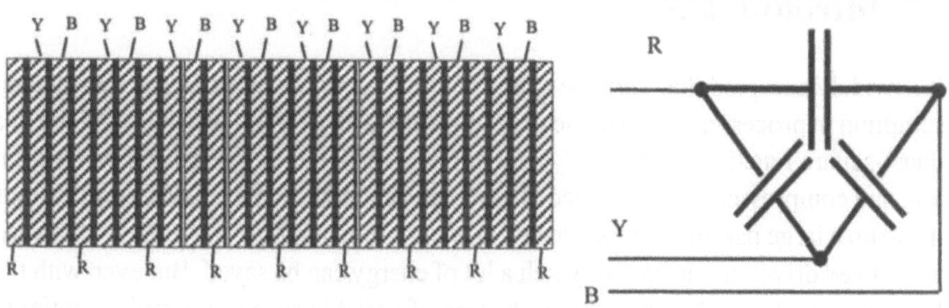

Figure 6b: Achieving 3 capacitors in "delta"

Efficiency Improvement of Variable Speed Electrical Drives for HVAC Applications

Research Ass. Flemming Abrahamsen, Prof. Frede Blaabjerg and Assoc. Prof. John K. Pedersen

Aalborg University, Institute of Energy Technology, 9220 Aalborg East, Denmark

Abstract. A large part of the produced electrical energy is consumed by ventilators, pumps and compressors, the so-called HVAC applications. A lot of this energy can be saved by speed control, but even with the large saving obtained alone by introduction of variable speed, it is still essential to optimise the control of the variable speed drive and to optimise the electrical machine with respect to efficiency. Experiments are made with energy optimal induction motor control on a 2.2 kW variable speed pump system. It is demonstrated that 10 % of the consumed energy can typically be saved by energy optimal motor control compared with constant V/Hz control. In a comparison of induction motors and permanent magnet synchronous motors for a variable speed pump application it is shown that for 2.2 kW motors an investment in high-efficiency or PM motors are typically paid back within 2.5 years and 7 years respectively. For a 90 kW PM motor the pay-back time would be 24 years. It is today not profitable to use PM motors for variable speed HVAC applications above 2 kW rated motor power. A further study is required to determine this limit in power rating more precisely.

1 Introduction

It is widely accepted that variable speed is a key factor for reduction of energy consumption in processes that load the drive machine with a torque which is a square or a quasi-square function of speed. Typical applications of this type are pumps, ventilators and compressors, the so-called HVAC applications. And as HVAC applications make up a large part of the total electrical energy consumption, it is evident that variable speed drives are means by which a lot of energy can be saved. But even with the large saving obtained alone by introduction of variable speed, it is still essential to optimise the control of the variable speed drive and to optimise the electrical machine with respect to efficiency.

By investigation of statistical material on energy consumed by electrical motors it is found that most energy is used by small machines (approximately 1 - 50 kW). In [1] and [2] it was estimated that electrical motors with ratings below 52 kW consume 73% of the total energy used by electrical motors, and dissipate 89 % of the total electrical motor losses. So with regard to saving energy by proper control and good motor construction, the main effort should be concentrated on reducing the losses in drives of that size.

The contribution of this paper is to analyse the aspects of energy efficiency of variable speed control in HVAC applications: 1. control of the motor drive, 2. motor type. In both cases the analysis is done with basis in a traditional voltage source frequency converter driving a three phase induction motor. The first analysis is based on experiments where energy savings in a variable speed drive obtained by energy optimal control are compared with constant V/Hz control in a realistic 2.2 kW pump system. The second analysis is a comparison between using induction motors and permanent magnet synchronous motors, and the purpose is to quantify the profitability of using a high-efficiency induction motor or a permanent magnet synchronous motor instead of a standard induction motor.

2 Energy Optimal Control of Variable Speed Pump Motor Drive

When an induction motor is fed by a frequency converter it is possible to reduce the flux in the machine at low load. The reason to do that is to reduce the core losses and the stator current, so the overall losses in the drive are reduced. A side-effect is that the acoustic noise is reduced as well. This control principle is especially applicable in HVAC applications where the load is reduced with speed, and where the controlled process (pump, ventilator, ...) operates with reduced output most of the time.

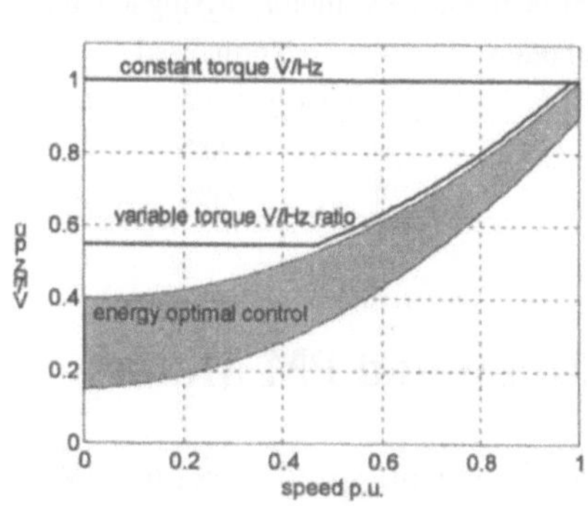

Figure 1: Different strategies of V/Hz-ratio control for induction motors in HVAC.

The traditional way to realise flux reduction in HVAC-drives is to use a predefined variable torque V/Hz-ratio curve which is speed dependent, see Figure 1. Because the same V/Hz is used at a given speed, no matter how the load changes, the flux reduction must be made quite conservative to avoid instability, and it does not provide full minimisation of drive losses. A more sophisticated solution is energy optimal control, where the V/Hz-ratio depends on both load and speed, see Figure 1. In that way the losses can come closer the loss minimum. Control strategies such as power factor control, model-based control and search control have been discussed in many papers, including [1].

Energy optimal control strategies that ensure maximal efficiency in any load case have been developed and tested experimentally. The load of the 2.2 kW motor was a realistic water-pump-system, see Figure 2a. The motor drive energy consumption was measured during a six minute test cycle with typical variations of water pressure and water flow, see Figure 2b. The measurements were done with constant V/Hz and with energy optimal control. The main results are shown in Table 1. In this case, where the average motor loading is 20 %, the consumed energy was reduced with 10 % by energy optimal control. If the average loading is higher than 20 % the percentage of saved energy will diminish.

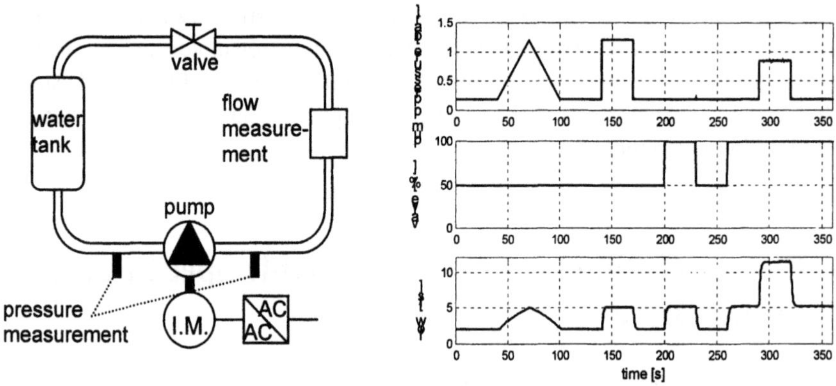

a) b)

Figure 2: a) Scheme of experimental 2.2 kW pump system. b) Pressure difference over the pump, valve opening and liquid flow during the six min. test cycle.

Table 1: Energy consumption for a six min. test-cycle of a 2.2 kW motor, driving a pump.

	Energy consumption [kWs]	Normalised energy consumption
Constant V/Hz	208	100%
Energy optimal control	186	90%

3 Comparison of Induction Motor and PM Motor

Standard induction motors are most often used in variable speed drives for HVAC applications. A way to improve the drive efficiency is to use a high-efficiency induction motor or a permanent magnet synchronous motor instead. A permanent magnet motor has a higher nominal efficiency than its corresponding induction motor, but on the other hand the efficiency of the induction motor can be increased at low load by flux reduction. In HVAC applications the motor will often operate at low load, and therefore it is not trivial to analyse further which motor is best suited for HVAC drive applications.

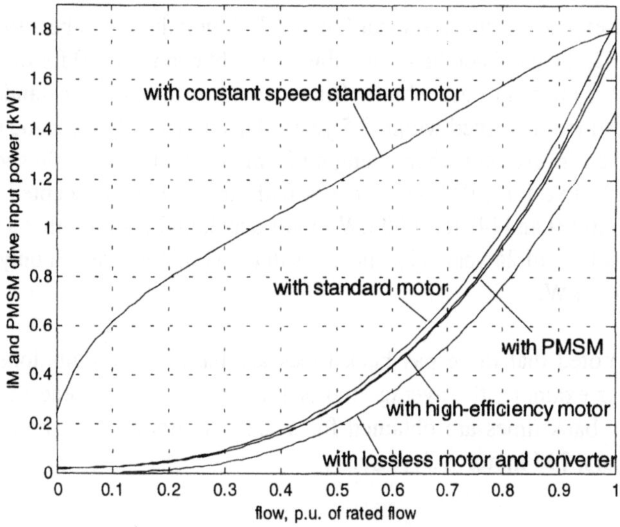

Figure 3: Induction motor and PM motor drive input power as function of pump flow.

The starting point is that the drive is equipped with a 2.2 kW standard induction motor driving a pump without head. The objective is to estimate how much better the drive will be if a high-efficiency motor or a PM motor is used instead, and how long the pay-back times are for the investment in more expensive motors. The high-efficiency motor used here is a standard motor which is improved with a longer stack, better iron and more copper. The PM motor is made from a standard motor stator frame and constructed with surface mounted neodynium-iron-bor magnets on the rotor surface. The induction motor losses are calculated by use of experimentally verified motor models, and the PM motor losses are calculated by use the SPEED motor designer software. In all cases the motors are controlled with optimal drive efficiency.

A calculation of how much power the drives consume as function of pump flow (proportional to speed) is shown on Figure 3. The high-efficiency and PM motor drive performances are almost equal, and better than the standard motor drive. As references are also shown the curves for a motor connected directly to the grid and for a fictitious motor drive without any losses.

The curves on Figure 3 are used to calculate the annual energy consumption when the pump operates with a typical share between high and low load, the average motor load is 30 %. The calculated energy consumption is listed in Table 2, converted into Euro and assuming an energy price of 0.0705 Euro/kWh. The numbers demonstrate the same as Figure 3: the PM motor drive is slightly better than the high-efficiency motor drive, and both are much better than the standard motor drive.
The energy expenses, however, have to be related to the extra prices of the PM motor and the high-efficiency motor. From the prices of magnets it is estimated that a PM motor is 250 Euro more expensive than a standard motor. A high-efficiency motor is

70 Euro more expensive than a standard motor. Relating these numbers to the annual costs in Table 2, the pay-back times in Table 3 are obtained, see [3] for details. These show that if it is chosen to use a high-efficiency motor instead of a standard motor, the extra investment is earned within 2.5 years. This is typically the same as the pay-back time for an investment when changing from constant to variable speed. The 7 years pay-back time for a PM motor is normally more than is accepted. A similar study is made in [3] and [4] for a 90 kW motor drive, and it shows that a PM motor has a pay-back time of 24 year. This indicates that PM motors are still only profitable in sizes above 2 kW.

It should be noted that these pay-back times are based on assumptions of motor loading. If, for example, the average motor loading is 25 % instead of 30 %, the following pay-back times are obtained for a 2.2 kW motor drive: high-efficiency motor: 3.5 years, PM motor: 9.5 years.

Table 2: Typical annual energy cost for a 2.2 kW pump drive, and annual cost related to the standard motor drive.

Loss less drive	PM motor drive	High-efficiency motor drive	Standard motor drive	Constant speed drive
410 Euro	496 Euro	505 Euro	532 Euro	862 Euro
23 %	6.7 %	5.2 %	0 %	-62 %

Table 3: Pay-back time for choosing another motor than a standard induction motor.

	2.2 kW motor	90 kW motor
High-efficiency motor	2.5 years	-
Permanent magnet motor	7 years	24 years

4 Conclusion

It is proved that energy optimal control is definitely a means for optimising the efficiency of a variable speed drive for HVAC. For a 2.2 kW motor driving a pump 10 % of the consumed energy can typically be saved compared with constant V/Hz control. For the same drive system the extra cost of a high-efficiency motor instead of a standard motor is typically paid back within 2.5 years. For a PM motor the pay-back time is 7 years. It is today not profitable to use PM motors for variable speed HVAC applications above 2 kW rated motor power. A further study is required to determine this limit in power rating more precisely. As the costs of permament magnets still continues to decrease the limit is also changing over time.

5 References

[1] Abrahamsen, F., Blaabjerg, F., Pedersen, J. K., Grabowski, P., Thøgersen, P. (1998): On the Energy Optimized Control of Standard and High-Efficiency Induction Motor in CT and HVAC Applications. IEEE Trans. on Ind. Appl., Vol. 34, No. 4, Jul/Aug 1998, pp. 822-831.

[2] Abrahamsen, F. (1998): Energy Optimal Control Strategies for Electro Motors, technical report (268 pages), Aalborg University, Institute of Energy Technology, Pontoppidanstraede 101, DK-9220 Aalborg East, ISBN 87-89179-23-4.

[3] Abrahamsen, F., Birk, J., Blaabjerg, F., Pedersen, J. K. (1999): Comparison of Induction and PM Motor Efficiency in HVAC Applications. Proceed. of EPE'99, Lausanne, Sep. 7-9.

[4] Abrahamsen, F. (1999): Energy Optimal Control of Induction Motors, Ph.D. dissertation, Aalborg University, Institute of Energy Technology, Pontoppidanstraede 101, DK-9220 Aalborg East, ISBN 87-89179-26-9.

Higher Efficiency Drive for Fluid Treatments

Aldo Boglietti, Mario Lazzari and Michele Pastorelli

Politecnico di Torino - Dipartimento di Ingegneria Elettrica Industriale
C.so Duca degli Abruzzi, 24 - 10124 Torino, ITALY

Abstract. In the paper is proposed a converter topology to be employed when the dynamic performances, required by the mechanical load, are not so pressing, but, on the contrary, good energetic motor performances are desired. Typical applications are the for all the drives devoted to fluid treatments (pumps, fans, etc.), where the motor is supplied by a VSI (Voltage Source Inverter), without current feedback. The proposed converter is based on a variable output voltage ac/dc converter followed by a unity modulation index PWM (Pulse Width Modulated) inverter. Experimental measurements of the efficiency, with motor supplied by the proposed converter or by a standard converter, are presented and compared. In addition, the possibility to reduce the supply voltage, at speed lower than the rated, is analysed. The experimental results confirms the better motor efficiency that can be reached with regulated dc bus voltage, unity modulation index, PWM-VSI inverters.

1 Introduction

The use of PWM inverters can be today considered the standard solution in electrical drives when speed and/or torque control of the mechanical load is required. The market has imposed the choice of the PWM inverter because of the its simplicity and flexibility. Unfortunately, from the motor losses point of view, the PWM inverter cannot be considered the best solution, since modulated voltage supply produces heavy increase of the motor iron losses, compared to the sinusoidal supply. This problem has already been investigated by the authors in [1, 2]. The iron losses increase heavily depends on the modulation index, the relationship between iron losses, at constant fundamental voltage and switching frequency, but with different modulation indexes and dc bus voltage is shown in [2]. If the voltage vs. frequency law is followed by contemporary modifying the frequency and the amplitude of the modulation waveforms, with constant dc bus voltage, the motor iron loss increase (with reference to sinusoidal supply) can reach 80% at 25-35 Hz around [2]. On the contrary, if the voltage vs. frequency law is followed using a constant modulation index (close to the unity) and a variable dc bus voltage the iron losses increase can be reduced under the 40 % [2]. Unfortunately the second solution requires a more complex and expensive converter with variable voltage ac/dc converter, realised with a thyristors rectifier bridge or a diode bridge rectifier followed by a dc/dc converter.

In this paper, the solution with full wave thyristors rectifier, half control is analysed from the motor energetic point of view, taking into account typical loads where the dynamic performances are not so pressing, such as electrical drive for fluid treatments i.e. fan and pump applications.

2 Mechanical Characteristics the Fans and Pumps

If the motor load is a fan or a centrifugal pump for fluid treatments, the output mechanical characteristics requires zero torque at zero speed, and then the torque increase with the speed following the law $T = k^*w^n$ with n equal to 1 with centrifugal pumps and n equal to 2 for fans.

3 Choice Methods of the Motor Rated Power for Industrial Applications

In order to understand the advantages of the proposed drive, it is interesting to analyse the effective use and the choice criterion adopted for defining the rated power of electrical motors employed in industrial applications. An inquire that took place in some mechanical firms, carried out by the authors [3], has shown the following results: a) the electric power required by the motors is a fraction of the total installed power (about one third); b) the electrical motors are over-sized with detriment of the motor efficiency and power factor. To confirm this trend, Figure 1 reports the electric power, required in a division of a mechanical firm, measured at ten casual instants, during the working time.

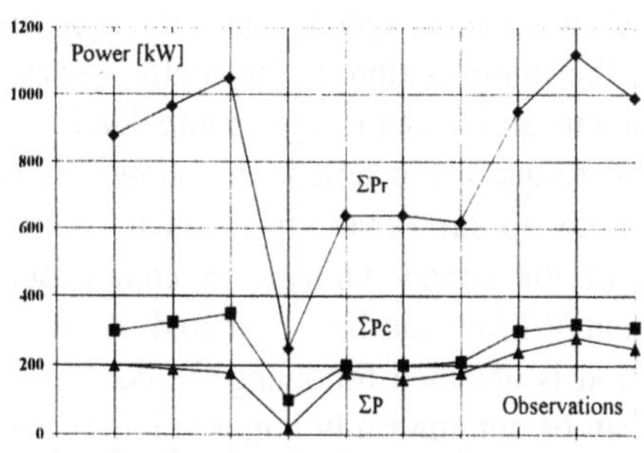

Figure 1: Total installed power, mean power required in a working cycle and absorbed electrical power in a division of a mechanical firm vs. casual observations instants.

With reference to Figure 1, it follows:

- Σ Pr is the sum of the rated powers of the supplied machines at the observation time, without taking into account their load condition;
- Σ Pc is the sum of the adsorbed average powers during a work-cycle of the machines specified at the previous point;
- Σ P is the effective electric input power measured at the casual observation instants.

The experimental data confirm that the required electric power is only a small fraction of the installed power and, even if those results have been obtained on motors employed on machine tools, the former considerations are worth for other kind of industries. In particular, the power measurements have been confirmed that during the system design process, every designer takes a safety coefficient on the required powers, that can lead to a final choice of the motor rated power 30%-50% greater than of the really requested one. This way of choice of the motor rated power does not allow the motor best employ from the energetic point of view, since the efficiency is optimised around the rated power. The role of the no-load losses, that are moreover incremented in case of PWM inverter supply, is important for the energy saving consideration during the whole process.

4 Proposed Higher Efficiency Drive for Fan and Pump Applications

On the basis of the previous discussion and the results reported in [1] and [2], an higher motor efficiency can be reached if a supply converter, based on a variable voltage ac/dc converter, followed by a PWM – VSI inverter, with nearly unity constant modulation index, is employed. This technique allows to produce the requested fundamental voltage by regulation of the dc bus voltage while the output frequency is regulated modifying only the modulation function frequency. In addition, if a $T = k*w^n$ load mechanical characteristic is considered, since the motor supply allows, below the rated speed, constant torque operation, the supply voltage regulation, for the minimisation of the motor losses, can be taken into account. In Figure 2 a fan and a centrifugal pump load torque characteristics, together with a constant torque characteristic are shown. It is clearly evidenced that at low speed the motor torque capability is much higher than the requested one, thus a reduction of the supply voltage, in comparison with the voltage requested by the constant V/f law can be considered. As well known, the motor constant torque regulation is obtained following the constant motor voltage over frequency law, which can be automatically imposed by the inverter supply. Since a reduction of the speed corresponds to a reduction of the torque, a constant torque regulation is not strictly necessary. As a consequence, at speed lower than the rated a reduction of the motor voltage (motor flux weakening) can be applied until a motor current increase is verified. The minimum supply current point allows the minimisation of both the copper losses and the iron losses and this kind of regulation can be easily automatically performed by the control circuitry of the drive control.

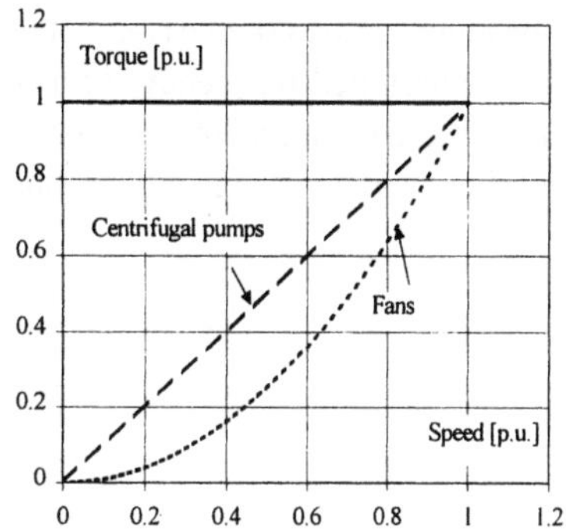

Figure 2: Constant motor torque characteristic and mechanical characteristics for fans and centrifugal pumps vs. speed (p.u. quantities)

5 Experimental Set-Up

In order to experimental verify the performance of the proposed drive a full wave half control converter, followed by an IGBT inverter was employed. Both ac/dc converter and the inverter are common industrial devices. The laboratory modifications are related to an interface board, that process the protection signals from the two converters, and to the modulation function generation board. The main ac/ac converter features of this converter are summarised in the following:

- Rated power: 60 kVA;
- Input voltage: 380 V, 50 Hz;
- Switching frequency regulation between 1 kHz up to 20 kHz;
- Full independent regulation of the modulation index, dc bus voltage, output fundamental frequency.

6 Energetic Comparison Between Standard and the Proposed Drive

In order to experimentally evaluate the energetic performance of the proposed drive a common industrial induction motor rated 15 kW, 50 Hz, 4 poles was tested on a bench provided with torque and speed transducer and dc generator. The load mechanical characteristic was supposed proportional to the square of the speed as in the case of a fan. The motor was supplied by the same ac/ac converter, but controlled as a common PWM inverter or as a variable dc voltage, unity modulation index, PWM inverter. All the efficiency measurements were carried out when the motor has reached the steady state temperature. Figure 3 shows the comparison of the motor efficiency with the two supply conditions, it is evident the efficiency increase in the case of a constant modulation index and variable dc bus voltage, in particular an efficiency greater than six percentage points around 30 Hz has been reached. In Figure 4 the comparison between the motor efficiency, using the rated voltage, at the assigned frequency, or a reduced voltage supply is shown. Also in this case the efficiency increase is well evident. In particular at low frequency (f < 30 Hz) the efficiency increase is higher than 200%.

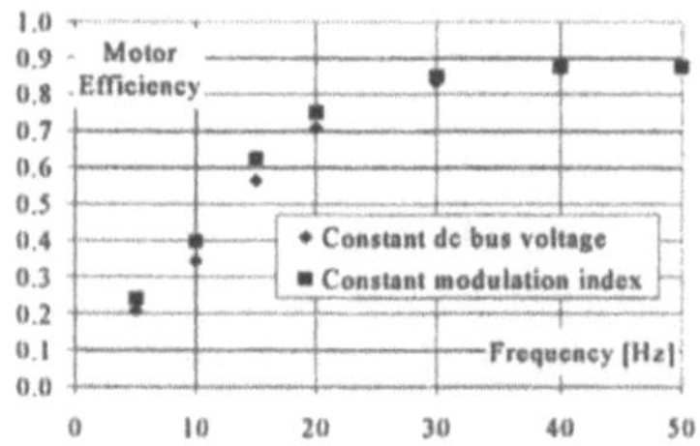

Figure 3: Efficiency comparison between standard and the proposed higher efficiency PWM inverter supply on a 15 kW, 50 Hz, 4 poles induction motor.

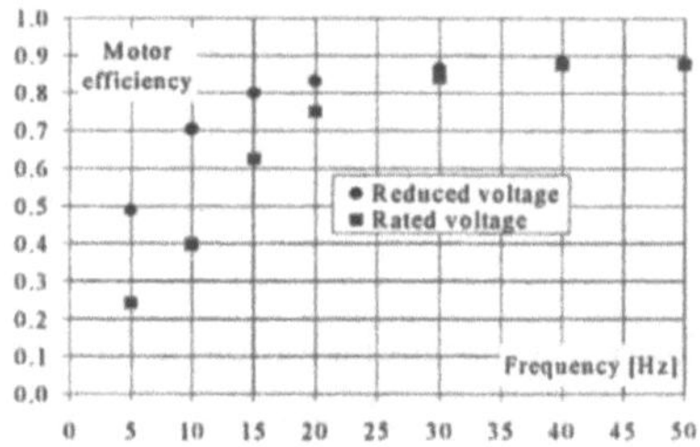

Figure 4: Efficiency comparison using rated or reduced voltage supply on a 15 kW, 50 Hz, 4 poles induction motor.

7 Conclusion

The adoption of a dc variable voltage, unity modulation index, PWM – VSI inverter is experimentally analysed from the point of view of the motor efficiency for general purpose drives devoted to fluid treatments. The additional energy savings looks to be interesting especially if reduction of the motor rated supply voltage is adopted, at speed less than the rated. Due to the modulated voltage waveform obtained the method shows interesting advantages also from the point of view of the stresses in the winding isolation, electromagnetic conducted interference and acoustic noise.

8 References

[1] A. Boglietti, P. Ferraris, M. Lazzari, "Power derating for inverter fed induction motors", IEEE-IAS 1994, Oct. 2-6 1994, Denver, USA, pp. 55-61.

[2] A. Boglietti, P. Ferraris, M.Lazzari, M. Pastorelli, "Influence of the inverter characteristics on the iron losses in PWM inverter fed induction motor, IEEE Trans. on Industry Applications, Vol.32, No. 5, Sept./Oct. 96, pp. 1190-1194.

[3] A. Boglietti, P. Ferraris, M. Lazzari, F. Profumo, "Analysis of the magnetic material quality influence on the induction motors energetic performance", ICEM' 86, 8-10 Sept. 1986, Munchen, Germany, pp. 219-222.

Characterisation of EU Motor Use

Aníbal T. De Almeida and Paula Fonseca

Department of Electrical Engineering, University of Coimbra, 3030 Coimbra Portugal

Abstract. This paper presents the main results of the field characterisation of the electricity used by motors in the European Union, in the scope of the project *"Improving the Penetration of Energy-Efficient Motors and Drives"*, recently carried out. This project was supported by the European Commission – DG XVII, under the SAVE II Programme.

The key objectives of this project were:

- To characterise the motor electricity end-use in the main industrial sub-sectors and in the tertiary sector, in order to deal with the lack of end-use data in the area of electric motors.
- To identify and estimate the potential energy savings through the use of efficient motor technologies, namely Energy-Efficient Motors and Variable Speed Drives.
- To evaluate the cost-effectiveness of the identified savings potential opportunities.
- To develop a database of end-user applications of electric motors in the industrial and tertiary sectors.
- To design DSM programmes targeted at the exploitation of cost-effective energy savings opportunities.

The most important industrial sectors in the EU have been characterised, namely non-metallic mineral, paper pulp and print, food and tobacco, chemical, iron and steel, and machinery and metal. These industrial sectors are responsible for 72% of the total industrial electricity consumption in EU.

1 Introduction

The study *"Actions to Promote Energy Efficient Electric Motors"* supported by the SAVE Programme of the European Union and carried out by the Motors Study Group in 1996, showed that electric motors consume a significant amount of electricity in the industrial and in the tertiary sectors of the European Union, reaching 69% and 36%, respectively, of the total electricity consumption.

The above mentioned study, shown that there was a substantial lack of accurate information about motor electricity consumption, installed motor capacity, number of operating hours, load diagrams, oversizing, variable speed requirements, distribution by horsepower range, etc, in industry and in the tertiary sector, in the European Union. Additionally the Motors Study Group could identify attractive opportunities that significantly improve the efficiency of motor systems, leading to large cost-effective energy savings.

In this context, under the scope of the SAVE II Programme, the project *"Improving the Penetration of Energy-Efficient Motors and Drives"* was carried out to improve the characterisation of the motor electricity consumption, which is a very important step in order to tap the huge savings potential and to design policies and DSM actions which can significantly improve the penetration of energy-efficient motor technologies.

The main objectives of the project were the field characterisation of motor electricity use in the most significant sectors in industry and in the tertiary sector, the identification of the potential savings with the application of Energy-Efficient Motors and Variable Speed Drives in 2015, and evaluation of their cost-effectiveness, as well as the identification, in the field, of the main barriers for the introduction of efficient motor technologies. After collection of data on motor use, a database for the European Union was prepared. The design of DSM guides addressing the most relevant energy-saving options was also carried out, including electronic variable speed drives, oversizing, labelling, procurement, education and training, rewinding and energy-efficient motors. Moreover, dissemination brochures, describing the key aspects of efficient motors in use in the characterised sectors, were prepared for the energy managers and technical staff of the industries which were characterised.

This paper focuses on two of the most relevant tasks of this project: the field characterisation and potential energy savings opportunities. The field characterisation was the main challenge of the project because it was dependent upon the collaboration of third party institutions and because it was a very complex and lengthy process. Another constraint was the lack of statistical data, and in same cases there were discrepancies among the scarce available statistics on the electricity consumption for the different countries.

2 Characterisation Approach

In order to obtain as much information as possible, there was a distribution of the most important industrial sectors in the EU among the countries of the study. This distribution was based on the importance of each industry in their country. Based on the sectorial characterisation in each country and on the total electricity consumption by sector in the EU countries, the information collected in the field characterisation was then extrapolated for the E.U.

Some of most significant sectors (in industry and in the services sector) have been characterised with field studies. It has been considered a division of industries according to the international NACE classification in the European Union.

Table 1 shows the sectors characterised in each country of the study and respective industrial Code:

Table 1: Sectors that have been characterised by country and respective international NACE code

Country	Industry Sector	Code
Denmark	Machinery and Metal Food and Tobacco	28 15
France	Paper, Pulp and Print Chemical	21 24
Germany	Tertiary sector (air conditioning, ventilation, lifts, circulating pumps)	
Italy	Chemical Food and Tobacco	24 15
Portugal	Paper, Pulp and Print Non-Metallic Mineral Sector	21 26
U.K.	Chemicals Iron and Steel	24 27

It is important to note that these surveyed industrial sectors represent 72% of the total industrial electricity consumption in EU, as it can be seen in Figure 1. Other sectors, which were not characterised, include non-ferrous metals, transport equipment, mining and quarrying, wood and wood products, construction, textiles and leather.

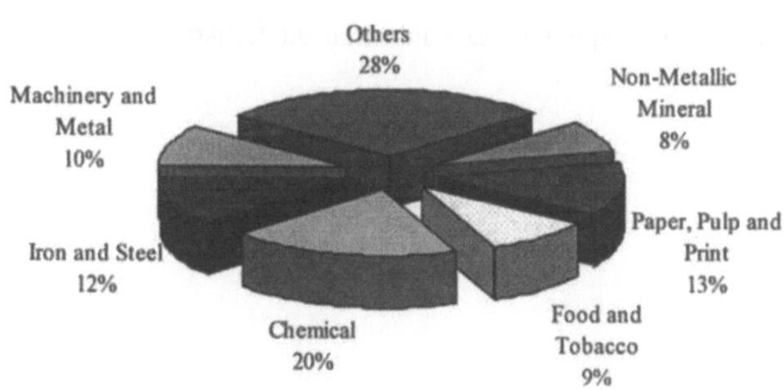

Figure 1: Disaggregation of the industrial electricity consumption by industrial sector

The tertiary sector was also addressed in this project. The electricity used by electric motors in the tertiary sector in EU countries is larger than several of the largest industrial sectors combined together. Modern tertiary sector buildings use a large amount of drive-power, namely for ventilation, air conditioning, circulating pumps and lifts. Additionally, all trends point in the direction of a significant increase in the air conditioning and ventilation loads in the tertiary sector. Taking into consideration that lighting and office loads are the main responsible for other loads in the tertiary sector, whose efficiency is continuously increasing, it can be concluded that motor loads are one of the factors responsible for the average growth of the load, making this part of the project even more pertinent.

Figure 2 shows the importance of the motor electricity consumption in the surveyed industrial sectors, and Figure 3 shows the share of motor electricity consumption in the tertiary sector in the EU.

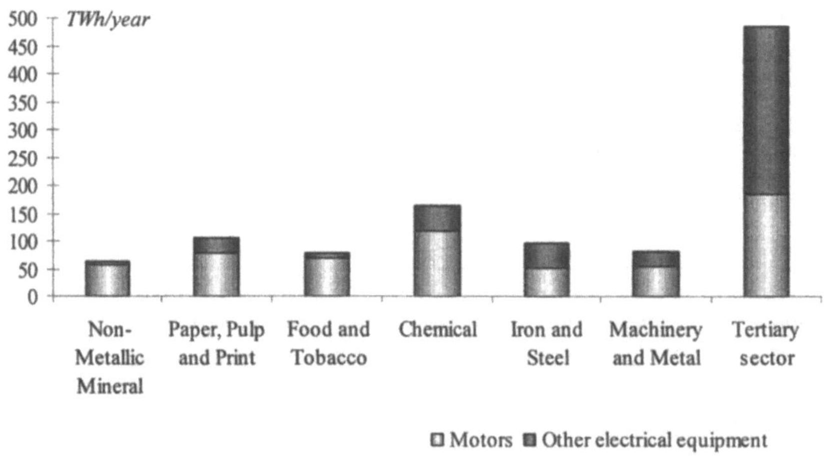

Figure 2: Electricity motor consumption in each industrial and tertiary sectors

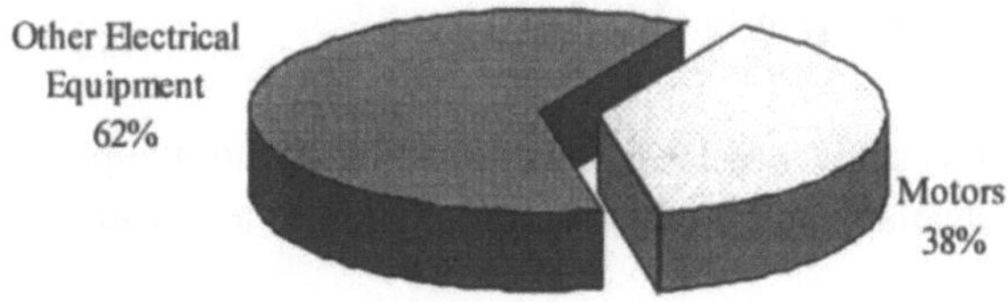

Figure 3: Motor consumption in the tertiary sector

To achieve the objectives of this project, a methodology was developed to collect the data from the field, through both audits and site-visits or questionnaires:

Audits:

- Each country selected a number of companies to be audited. The sample of companies should be representative of the sector in each country.
- The sample of motors to be subject of measures, in each audit, should enable to estimate detailed information on the motor system electricity use within the company.

Questionnaires:

- In the tertiary sector, a questionnaire has been filled through personal interviews, in different institutions. A questionnaire has also been prepared for the industrial sector, in order to complete the field characterisation.

To generalise the results for other countries than the one surveyed and characterise the EU situation, it is assumed that the difference in motor application structure between countries is small. Each partner surveyed the most important sectors in his country. The distribution of sectoral surveys over different countries therefore yields considerable savings and synergies for all countries. This approach is most realistic in the process industry because the used production equipment is produced by a small number of manufacturers.

3 Data Collection Methodology

The base year for this study was 1996, because at the start of the project it was the most reliable data available to characterise motor end-use applications.

As it was already mentioned, each partner of the study characterised the most relevant industry sector in his country. The selection of enterprises to be audited or visited was responsibility of each partner.

There was not a single process to collect the motor end-use data. The field characterisation was based on audits, on site checks, interviews and on already existing data from field surveys. In order to standardise the information and to better integrate all the data from each country, some guidelines have been agreed among the partners. The following guidelines helped to define with accuracy what type of information should be collected in each industry and tertiary sectors.

- *Distribution of the electricity consumption by types of load in the industrial and tertiary sectors*

The loads to be considered are: motor load, lighting and other loads

- *Distribution of the motor electricity consumption by end-use in the industrial sectors*

The considered motor end-use applications are pumps, fans, air compressors, cooling compressors, conveyors and other motors (typically: mills, mixers, stirrers, etc)

- *Distribution of the motor electricity consumption by end-use in the tertiary sector*

The considered motor end-use applications are pumps, fans, refrigeration, air conditioning, conveyors and other motors (typically:

- *Kind of motors considered*

AC induction motors, because they represent more than 90% of the motor electricity consumption.

- *Power ranges considered*

The power range should be 0,75-7,5 kW; 7,5-37 kW; 37-75 kW; 75-375 kW; 375-750 kW.

For each power range the following information was sought:

- *Lifetime*
- *Number of motors in operation by power range*

- *Number of operating hours by power range*
- *Average power by power range*
- *Load diagrams*
- *Oversizing*
- *Variable speed requirements*
- *Installed capacity by power ranges*
- *Load factor*
- *Efficiency by power range (η)*

The average EURODEEM (European Database of Energy-Efficient Motors) efficiencies were confirmed through the values obtained in field efficiency checking of motor data.

The above collected data enabled each partner to determine the motor electricity consumption in each industrial sector, by load type and by power range. The following information was prepared:

- *Estimates of the electricity savings potential* with:

Through the application of Energy Efficient Motors, for the same power ranges, in the industrial and tertiary sectors (percentage value and GWh value), for the time horizon 2015.

Through the application of Electronic Variable Speed Drives, for the same power ranges, in the industrial and tertiary sectors (percentage value and GWh value), for the time horizon 2015.

Through the combined application of Energy Efficient Motors and Electronic Variable Speed Drives, for the same power ranges, in the industrial and tertiary sectors (percentage value and GWh value), for the time horizon 2015.

Other sources of information

Field characterisation was by far the most important source of information. However, some recent statistics have been the basis for the extrapolations for the EU, namely the IEA (International Energy Agency) statistics for the year 1996. Another important sources of information were the previous SAVE I study

"Actions to Promote Energy-Efficient Electric Motors", the EURODEEM motor database and the IKARUS database.

4 Overall Assessment for the European Union

The countries of the study represent 72% of the total consumption in industry in EU. Based on the individual country assessments on motor electricity use in each country and on electricity consumption statistics in the base year 1996 we were able to extrapolate the data and characterise the EU situation in what concerns the motor electricity end-use in the industrial and tertiary sectors.

For the data extrapolations, two different situations are considered:

- *When an industry sector was characterised by only one country*

In this case the data is extrapolated from one country to EU:

Number of Motors (N):

$$N_{EU}=K_{sector}*N_{sample}$$

Installed Capacity (Cap):

$$Cap_{EU}=K_{sector}*Cap_{sample}$$

Motor Consumption (C):

$$C_{EU}=K_{sector}*C_{sample}$$

Where:

$$K_{sector} = \frac{Consumption_{EU}}{Consumption_{sample}}$$

Operating hours $_{EU}$ = Operating hours $_{sample}$

Load factor $_{EU}$ = Load factor $_{sample}$

Efficiency $_{EU}$ = Efficiency $_{sample}$

- *When there was more than one country to characterise the same industry sector*

In this case, the number of motors, the motor electricity consumption and the installed capacity were primarily aggregated by the sum of the values for each individual country and then extrapolated for EU:

$$X_s = \sum X_i$$

Where X_s represents the values in the sample and X_i represents the values in each country.

The duty (load factor times operating hours) and motor efficiency were calculated by weighting the average values of each country by the number of motors in use:

$$X_s = \frac{\sum N_i X_i}{\sum N_i}$$

The method used to extrapolate the data for EU is similar in both situations. It has been used a factor to extrapolate the data in each sector, which represents the relation between the EU consumption and the country consumptions.

4.1 Collected Data for each Surveyed Sector

Table 2 shows the total electricity consumption for the surveyed industrial sectors and the share of electricity used by electric motors in each sector. The total electricity consumption in the EU is based on IEA statistics and the electricity consumed by motors in each sector is based on field surveys.

Table 2: Electricity consumption by sector

Sectors	Total Electricity Consumption (TWh)	Share of Motor Electricity Consumption
Non-Metallic Mineral	63	90.9
Paper, Pulp and Print	107	75.1
Food and Tobacco	78	89.8
Chemical	165	71.9
Machinery and Metal	98	51.7
Iron and Steel	82	66.3
Other Industry	242	59.9
Total Industry	834	69.0
Tertiary	519.7	38.3

Considering the main end-use applications of electric motors in the industrial and in the tertiary sectors, Table 3 shows the distribution of motor electricity consumption by application in the surveyed industrial sectors. Pumps fans and compressors account for a very large percentage (62%) of the motor electricity consumption in the surveyed sectors.

Table 3: Share of motor electricity consumption in the six surveyed industrial sectors

	Pumps	Fans	AirComp	CoolComp	Conveyors	Other
Non-Metallic Mineral	6.8	21.4	16.8	0.0	6.2	48.8
Paper, Pulp and Print	56.9	21.7	13.2	0.4	0.9	6.9
Food and Tobacco	9.8	11.5	8.7	30.3	*	39.7
Chemical	26.4	10.6	28.1	5.7	2.6	26.6
Iron and Steel	8.5	15.0	14.1	0.0	5.3	57.1
Machinery and Metal	1.4	18.3	17.2	0.0	0.0	63.1

* Conveyors are included in the other motors.

Table 4 shows the distribution of motor electricity consumption by end-use application in the tertiary sector. Pumps, fans and compressors also account for the largest share of motor electricity consumption. Because the structure of the tertiary sector is different from the industrial sector in what concerns motor applications, a different type of loads has been considered for the assessment of motor end use consumption in this sector. The motor loads that have been considered in the tertiary sector were pumps, fans, refrigeration, air conditioning, conveyors (including lifts and escalators) and other motors.

Table 4: Share of motor electricity consumption in the tertiary sector

	Pumps	Fans	Refrigeration	Air Cond.	Conveyors	Other
Tertiary	15.8	24.0	26.1	16.6	10.9	6.6

All the data has been disaggregated by motor power range. The distribution of motors by motor size is important because the efficiency difference between standard and EEMs is larger in the smaller motor sizes than in the large sizes. This difference in motor efficiency is most relevant in the power ranges]0-0.75[and [0.75-4[. This efficiency gap is very important when calculating the electricity savings potential using Energy Efficient Motors.

Figure 4 shows the motor electricity consumption, the installed capacity and the number of operating hours by power range, in the Industrial sector in EU.

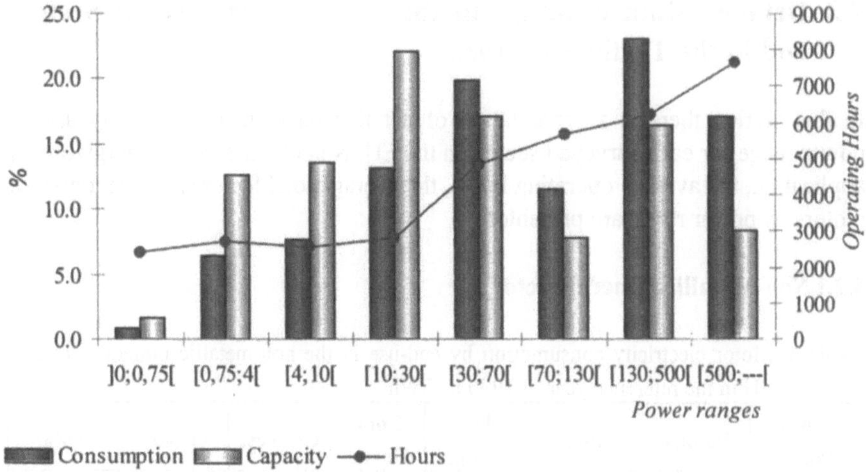

Figure 4: Motor electricity consumption by power range in the industrial sector

Figure 5 shows the motor electricity consumption, the installed capacity and the number of operating hours by power range, in the Tertiary sector in EU.

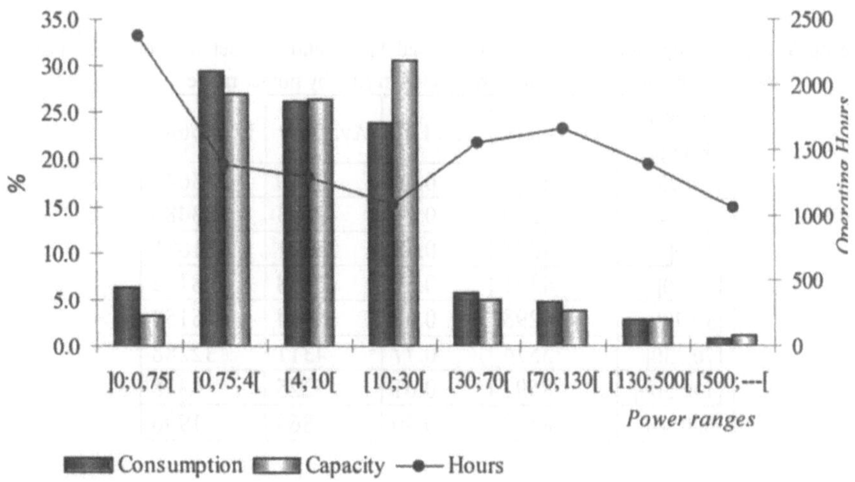

Figure 5: Motor electricity consumption by power range in the tertiary sector

4.2 Detailed Data Findings in the Surveyed Industrial Sectors and in the Tertiary Sector

In this section there is a presentation of detailed data on motor energy use by power range for each surveyed sector, in the EU. Namely, the motor energy use by application, the average operating hours, the average load factor and the number of motors by power range are presented.

4.2.1 Non-Metallic Mineral Sector

Table 5: Motor electricity consumption by end-use in the non-metallic mineral sector in EU in the reference year of 1996 in GWh:

Power Ranges	Pumps	Fans	Air Comp.	Cool. Comp.	Convey.	Others	Total
]0;0.75[	20	188	0	0	289	106	603
[0.75;4[	429	840	12	0	1513	474	3268
[4;10[	255	1135	88	0	373	1499	3351
[10;30[	984	1661	264	0	554	1727	5191
[30;70[	1310	2691	1026	0	298	6053	11377
[70;130[	664	385	5669	0	336	5608	12662
[130;500[	234	2210	2515	0	162	1089	6210
[500;---[	0	3083	0	0	0	11205	14287
Total	3895	12192	9575	0	3525	27761	56949

Table 6: Average operating hours, average load factor and number of motors by power range in the non-metallic mineral sector in EU, by power range

Power Ranges	Av. Hours	Av. LF	Av. Duty	Nº Motors
]0;0.75[	4999	0.60	3010	313647
[0.75;4[	4279	0.61	2593	583480
[4;10[	4075	0.58	2371	181636
[10;30[	4381	0.59	2568	108122
[30;70[	4298	0.68	2934	85153
[70;130[	5574	0.77	4311	32288
[130;500[	8592	0.61	5225	4418
[500;---[	8592	0.80	6869	1996

"Other motors" in the non-metallic mineral sector are mainly mills, extruders, rollers and large production machinery. Those motors are responsible for about half of the total motor consumption in this sector, especially in the higher power range. This consumption is mainly due to mills. Since most of the factories operate continuously, the average number of operating hours is high.

4.2.2 Paper, Pulp and Print

Table 7: Motor electricity consumption by end-use in the paper, pulp and print sector in EU in the reference year of 1996 in GWh

Power Ranges	Pumps	Fans	Air Comp.	Cool. Comp.	Convey.	Others	Total
]0;0.75[	6	5	0	0	19	26	55
[0.75;4[	827	1246	317	2	96	116	2604
[4;10[	1422	773	1366	5	97	171	3835
[10;30[	5456	2964	863	2	117	282	9684
[30;70[	9503	4009	1346	63	172	896	15990
[70;130[	9108	4102	907	259	65	540	14982
[130;500[	18085	3959	3152	0	0	1356	26553
[500;---[	1247	374	2654	0	131	2073	6478
Total	45655	17432	10605	330	697	5461	80181

Table 8: Average operating hours, average load factor and number of motors by power range in the paper, pulp and print sector in EU, by power range

Power Ranges	Av. Hours	Av. LF	Av. Duty	N° Motors
]0;0.75[	4496	0.53	2378	49196
[0.75;4[	4424	0.55	2420	317047
[4;10[	6007	0.49	2951	131038
[10;30[	6192	0.50	3120	175739
[30;70[	6241	0.56	3484	111472
[70;130[	5272	0.63	3335	44547
[130;500[	5623	0.60	3389	34598
[500;---[	6868	0.51	3504	2020

Pumps are by far the most important type of motor load in the paper industry, accounting for 57% of the total motor consumption. They are mainly used to pump pulp, clean water and wastewater. The second largest share of electricity consumption is for fans, which are mainly used to air blowing. In the pulp production sector, fans are used to extract waste gases, which result from the chemical processes involved in the pulp production. Because of these chemical gases, the environmental conditions in Paper plants are very arduous. Most of the cooling compressors found in this sector are used to filter the air and control the indoor temperature of the switch board rooms and control rooms, in order to avoid oxidation in the contactors.

The average number of operating hours is high, especially in the higher power range, because the paper industry operates continuously.

4.2.3 Food Beverage and Tobacco

Table 9: Motor electricity consumption by end-use in the food and tobacco sector in EU in the reference year of 1996 in GWh

Power Ranges	Pumps	Fans	Air Comp.	Cool. Comp.	Others	Total
]0;0.75[	214	252	192	662	867	2186
[0.75;4[	1127	1320	1004	3485	4556	11491
[4;10[	841	986	747	2596	3395	8566
[10;30[	606	709	542	1870	2451	6179
[30;70[	2370	2776	2110	7315	9570	24140
[70;130[	346	406	307	1063	1392	3514
[130;500[	1384	1618	1230	4266	5581	14079
Total	6888	8067	6132	21257	27812	70156

Table 10: Average operating hours, average load factor and number of motors by power range in the food and tobacco sector in EU, by power range

Power Ranges	Av. Hours	Av. LF	Av. Duty	N° Motors
]0;0.75[	4320	0.39	1704	2851101
[0.75;4[	4644	0.46	2136	2815377
[4;10[	3287	0.52	1714	815913
[10;30[	5404	0.58	3149	129292
[30;70[	7742	0.65	4994	95865
[70;130[	5137	0.70	3584	12028
[130;500[	4640	0.87	4024	19603

Motor electricity consumption in the Food and Tobacco sector presents a slightly different disaggregation by end-use application. "Other motors" include: mixers, conveyor belts, packaging machines and other motors that do not belong to any of the considered end-uses, typically processing machines.

Cooling compressors and other motors are responsible for about 30% and 40% of the total consumption, respectively. Motors in the power range]0, 4kW[are significantly over sizing, being good candidates to be replaced with smaller Energy-Efficient motors, when they fail. These motors represent a particularly attractive opportunity for electricity savings.

Gross oversizing leads to significant drop in efficiency and gives rise to a low power factor with all the associated problems.

4.2.4 Chemical

Table 11: Motor electricity consumption by end-use in the chemical sector in EU in the reference year of 1996

Power Ranges	Pumps	Fans	Air Comp.	Cool. Comp.	Convey.	Others	Total
[0.75;4[	821	226	25	0	769	130	1971
[4;10[	3600	630	106	0	908	908	6154
[10;30[	3707	270	338	0	1386	1386	7086
[30;70[	5483	2300	1271	0	0	5767	14822
[70;130[	4488	3832	1823	0	0	3541	13684
[130;500[	8207	4671	5293	0	0	16816	34987
[500;---[	4915	648	24351	6750	0	2890	39553
Total	31221	12578	33207	6750	3063	31438	118257

Table 12: Average operating hours, average load factor and number of motors by power range in the chemical sector in EU, by power range

Power Ranges	Av. Hours	Av. LF	Av. Duty	N° Motors
[0.75;4[	4089	0.66	2699	429763
[4;10[	4893	0.67	3278	313115
[10;30[	4711	0.75	3534	125293
[30;70[	5553	0.58	3221	107056
[70;130[	5870	0.69	4050	37635
[130;500[	6680	0.53	3540	39554
[500;---[	7681	0.73	5604	8925

Pumps, fans and compressors represent about 71% of the total motor electricity consumption. This high share of controllable loads with the application of VSDs is responsible for the large savings potential, which has been identified in this sector. Many of the motors used in the Chemical plants are very large and, if operated within their design limits, have very long lifetimes. Large motors, in the range [130, ---[, are responsible for 63% of the total motor consumption.

The "Other motors" end-uses include mixers, stirrers, agitators and dispersers. Using the latest design of paddles, large energy savings can be made in those motors.

Oversized motors were not seen as a particular problem on chemicals, with the exception of stirrers and agitators.

4.2.5 Machinery and Metal

Table 13: Motor electricity consumption by end-use in the machinery and metal sector in EU in the reference year of 1996 in GWh

Power Ranges	Pumps	Fans	Air Comp.	Cool. Comp.	Others	Total
]0;0.75[	14	156	149	0	545	864
[0.75;4[	92	1240	1162	0	4271	6764
[4;10[	120	1601	1502	0	5518	8741
[10;30[	340	4583	4299	0	15795	25018
[30;70[	177	2330	2182	0	8025	12714
Total	744	9909	9293	0	34155	54101

Table 14: Average operating hours, average load factor and number of motors by power range in the machinery and metal sector in EU, by power range

Power Ranges	Av. Hours	Av. LF	Av. Duty	Nº Motors
]0;0.75[	1231	0.26	319	5146563
[0.75;4[	1157	0.37	427	5282090
[4;10[	1527	0.42	647	2197129
[10;30[	1783	0.52	931	1416746
[30;70[	2452	0.62	1520	220383

"Other motors" in the Machinery and Metal sector are responsible for 63% of the total motor consumption. These motors are machine tools, robotic systems, cranes, and process machinery.

All motors especially motors bellow 0.75kW are characterised by low load factors, resulting in correspondingly lower efficiencies.

It should be noted that field characterisation has only targeted light metal industry. This is the main reason for the absence of motors above 70 kW. In larger production plants larger motors are used, particularly in air compressors.

4.2.6 Iron and Steel

Table 15: Motor electricity consumption by end-use in the iron and steel sector in EU in the reference year of 1996

Power Ranges	Pumps	Fans	Air Comp.	Cool. Comp.	Convey.	Others	Total
[0.75;4[	214	264	22	9	1088	316	1912
[4;10[	526	412	96	0	720	918	2671
[10;30[	850	588	118	0	869	1112	3536
[30;70[	1483	524	150	0	0	4701	6858
[70;130[	626	939	448	0	0	3053	5067
[130;500[	511	1181	2693	0	0	13006	17391
[500; ---[	125	3702	3620	0	0	5907	13354
Total	4334	7609	7147	9	2677	29014	50788

Table 16: Average operating hours, average load factor and number of motors by power range in the iron and steel sector in EU, by power range

Power Ranges	Av. Hours	Av. LF	Av. Duty	Nº Motors
[0.75;4[	4377	0.66	2889	388665
[4;10[	4164	0.67	2789	160058
[10;30[	4745	0.75	3559	62017
[30;70[	6097	0.58	3536	45079
[70;130[	7393	0.69	5103	11047
[130;500[	7435	0.53	3941	17715
[500;---[	7558	0.73	5516	3020

There are two types of operation within the Steel sector: steel making and forming.

"Other motors" in the Iron and Steel sector account for the largest share of motor electricity consumption (57%) and are mainly used for specialist activities related to steel forming in rolling mills, sanding, casting and finishing.

Modern AC drives can not only improve performance but also save electricity in comparison with old DC drives used in rolling mills. In steel making pumping for cooling and dust extraction are relevant motor loads.

There is almost lack of refrigeration equipment in the Iron and Steel sector. Fans, which are good candidates for VSDs, are the second largest energy users. They are used for dust and waste gas extraction and for air blowing. Motors above 70kW are responsible for about 70% of the total motor consumption, and have a large number of operating hours.

4.2.7 Tertiary Sector

Concerning motor end-use application, the tertiary sector is different from the industrial sector. Therefore, a different type of motor end-use application has been considered in the field characterisation.

Table 17: Motor electricity consumption by end-use in the tertiary sector in EU in the reference year of 1996 in GWh

Power Ranges	Pumps	Fans	Refrige-ration	Air Cond.	Convey.	Others	Total
]0;0.75[	9583	1105	0	350	0	705	11743
[0.75;4[	12410	19850	10107	7824	56	4527	54774
[4;10[	5576	10141	17382	5383	5845	4323	48650
[10;30[	1821	6619	12025	10008	12876	1056	44405
[30;70[	0	2496	3932	2418	1409	251	10506
[70;130[	0	2756	3109	2351	31	579	8826
[130;500[	0	1279	1638	1919	0	357	5194
[500;---[	0	317	298	555	0	351	1521
Total	29389	44564	48491	30807	20216	12151	185619

Table 18: Average operating hours, average load factor and number of motors by power range in the tertiary sector in EU, by power range

Power Ranges	Av. Hours	Av. LF	Av. Duty	N° Motors
]0;0.75[	2381	0.47	1122	15791566
[0.75;4[	1389	0.54	746	31687249
[4;10[	1292	0.56	729	10217309
[10;30[	1095	0.55	597	3636098
[30;70[	1561	0.57	890	303419
[70;130[	1664	0.61	1011	95423
[130;500[	1397	0.59	826	23886
[500;---[	1068	0.52	550	4423

Conveyors in the tertiary sector include lifts and escalators. Most motors, especially small motors, are over sized.

Pumps, fans and compressors are by far the most important loads in the tertiary sector, accounting for about 83% of the total motor electricity consumption. Motors up to 30kW are responsible for 86% of the total motor consumption. This share shows the importance of small motors in the tertiary sector when evaluating the energy savings with Energy-Efficient Motors.

5 Potential Savings

This section presents the identified potential electricity savings in the industrial in the tertiary sectors, which can be achieved by the application of energy efficient motors and electronic variable speed drives. Studies show that the electricity savings potential would be much larger if other motor system improvements were considered, namely, good design of the motor system, high efficiency end-use devices (pumps, fans, etc.), efficient transmission, good quality repair, improved maintenance, proper electric supply, correcting motor oversizing, etc. However, it is difficult to quantify these potential savings with accuracy, and the estimates of electricity savings potential that are drawn for the industrial and for the tertiary sectors are therefore very conservative. For example, in what concerns motor repair, a recent study estimated that the potential for energy savings in the EU is around 8 TWh, if a good quality repair is carried out.

5.1 Methodology

In order to estimate the electricity savings potential, it was assumed a time horizon of 2015. According to *"European Energy to 2020"*, Special issue of Energy in Europe, Spring 1996, in the "conventional wisdom" scenario, the annual average growth rates of the electricity consumption up to 2015, in the industrial and tertiary sectors are estimated to be 1.2% and 1%, respectively. Table 19 shows the estimated motor electricity consumption in the industrial and tertiary sector, by power range in 2015, assuming frozen efficiencies for induction motors.

Table 19: Estimation of motor electricity consumption by power range, by 2015 based on the average growth rates, assuming 1.2% and 1% growth rates in the industrial and in the tertiary sectors, respectively

Motor electricity consumption forecast by 2015 [TWh]		
Power Ranges	**Industrial sector**	**Tertiary sector**
]0;0,75[	6.2	14.2
[0,75;4[	46.9	66.2
[4;10[	55.8	58.8
[10;30[	95.0	53.6
[30;70[	143.9	12.7
[70;130[	83.6	10.7
[130;500[	166.3	6.3
[500;---[	123.5	1.8
Total [TWh]	721.3	224.2

For the assessment of the electricity savings potential with the application of Energy-Efficient Motors and Electronic Variable Speed Drives, two different conditions have been considered: the economic savings potential and technical savings potential.

Technical Potential
The technical potential represents the energy savings that can be achieved applying the Energy Efficient Motors and Variable Speed Drives to all the available opportunities, irrespective of the cost-effectiveness of the measure.

Economic Potential
The economic savings potential represents the energy savings that can be achieved when the efficient technologies are only applied to cost effective applications.

One of most sensitive factors for the cost effectiveness of EEMs and VSDs is the number of operating hours. Cost-effectiveness analysis in this study has been based on the cost of saved energy. The input data for the cost benefit analysis was the field survey.

Cost effectiveness analysis show that it is cost effective to introduce EEMs in all power ranges. In what concerns VSDs, there are some situations in which the introduction of variable speed control is not attractive, especially in the lower power ranges.

5.1.1 Potential Savings with the Application of EEMs

In the previous SAVE I Study "Actions to Promote Energy Efficient Electric Motors", it was found out that motors in the range 0–7.5kW last on average 12 years, motors in the range 11–75kW last on average 15 years and motors above 75kW are estimated to last 20 years. If mandatory minimum efficiency standards were applied, in 20 years time almost all the motors would have been replaced by more energy efficient models. The evolution of motor stock has been evaluated in another SAVE II Project, *"Study on technical/economic and cost/benefit analyses of energy efficiency improvements in industrial three-phase induction motors"*, recently carried out, for four different scenarios. Three levels of efficiency for motors were considered: Class_III, being average typical efficiencies (according to the EURODEEM database), Class_II and Class_I, are the established minimum efficiency levels for motors in European Union CEMEP (Motors Manufacturers Association) agreement, as seen in Table 20.

Table 20: Average Class_III and Class_I efficiencies and minimum efficiency for Class_II for 4 pole motors

Power Ranges	Class_III	Class_II	Class_I
]0;0.75[	0.610	0.670	0.790
[0.75;4[	0.785	0.805	0.840
[4;10[	0.846	0.859	0.894
[10;30[	0.886	0.898	0.920
[30;70[	0.910	0.924	0.938
[70;130[	0.930	0.940	0.950
[130;500[	0.940	0.950	0.960
[500;---[	0.950	0.960	0.970

In the most ambitious scenario in the above-mentioned study mandatory minimum efficiency standards are progressively applied. After January 2003 no Class III motors can be sold, and after January 2006 only Class I motors will be sold. Based on the average number of operating hours the use of Class_I motors leads to paybacks of under 3 years for most applications both in the industrial and in the tertiary sectors. According to this scenario, the motors distribution by efficiency level in 2015 will be 5%, 3% and 92% in Class_III, Class_ II and in Class_I, respectively.

5.1.2 Potential Savings with the Application of VSDs

The suitability of Variable Speed Drives for motor loads is generally high, but some applications, which could benefit from the speed control, may already have installed a VSD. Some loads are not suitable to apply speed control. To determine the potential electricity savings from VSDs an applicability factor of 80% was considered, based on the survey. The existing loads which already have electronic variable speed control, by motor end use application have also been considered, in the industrial and in the tertiary sectors respectively.

For the calculation of the potential electricity savings with the application of VSDs, it has been assumed 25% average savings for pumps, fans and compressors (air compressors, air conditioning and refrigeration), and 15% average savings for conveyors and other motors. These values are average savings based on a variety of case studies.

5.2 Potential Savings in Industry

Table 21 shows the estimated economic and technical potential energy savings by power range in industry in the EU, by 2015, considering the application of EEMs and VSDs. The joint application of EEMs and VSDs has also been estimated and is shown in the table.

Table 21: Estimated economic and technical electricity savings for the industrial sector

Total Industry	Technical Potential TWh			Economical Potential TWh		
	EEMs	VSDs	EEMs+VSDs	EEMs	VSDs	EEMs+VSDs
]0;0.75[	0.9	0.8	1.6	0.9	0.4	1.3
[0.75;4[	4.1	6.3	9.8	4.1	3.2	6.9
[4;10[	2.4	7.6	9.7	2.4	4.1	6.4
[10;30[	2.3	12.8	14.8	2.3	6.4	8.5
[30;70[	3.1	19.9	22.5	3.1	12.0	14.8
[70;130[	2.3	12.2	14.1	2.3	8.7	10.7
[130;500[	5.4	23.2	27.8	5.4	16.3	21.1
[500;---[	3.6	18.2	21.2	3.6	13.3	16.5
TWh	***24.1***	***101.0***	***121.5***	***24.1***	***64.4***	***86.1***

In Figures 6 and 7 the technical and economic potential electricity savings by power range for the application of EEMs and VSDs, in the industrial sector in EU can be seen.

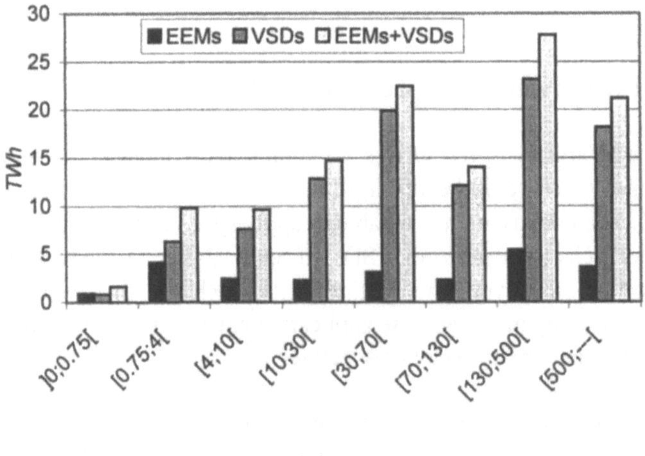

Figure 6: Technical Potential for the Industrial sector

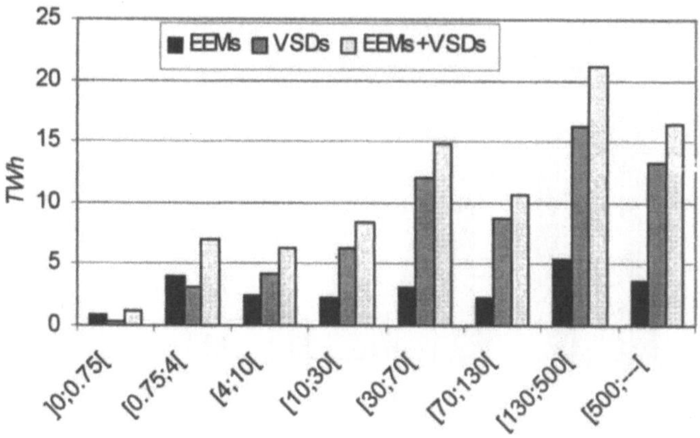

Figure 7: Economical Potential for the Industrial sector

5.3 Potential Savings in the Tertiary Sector

Using the same methodology described for the industrial sector, the potential savings for the tertiary sector have been estimated. Next Table shows the technical and economical electricity savings potential in the tertiary sector by power range in EU, with the application of EEMs and VSDs, by 2015.

Table 22: Estimated economic and technical electricity savings for the tertiary sector

Tertiary Sector	Technical Potential TWh			Economical Potential TWh		
	EEMs	**VSDs**	**EEMs+VSDs**	**EEMs**	**VSDs**	**EEMs+VSDs**
]0;0.75[	2.1	2.5	4.3	2.1	0.9	3.6
[0.75;4[	3.7	12.0	15.0	3.7	4.6	9.5
[4;10[	2.7	10.2	12.4	2.7	4.3	7.7
[10;30[	2.2	8.9	10.7	2.2	3.7	5.8
[30;70[	0.5	2.3	2.7	0.5	1.1	1.5
[70;130[	0.2	2.0	2.2	0.2	1.0	1.1
[130;500[	0.1	1.2	1.2	0.1	0.5	0.6
[500;---[	0.0	0.3	0.3	0.0	0.1	0.1
TWh	***11.5***	***39.4***	***48.8***	***11.5***	***16.2***	***29.9***

In Figures 8 and 9 it can be seen the technical and economic potential electricity savings by power range for the application of EEMs and VSDs, in the tertiary sector in EU.

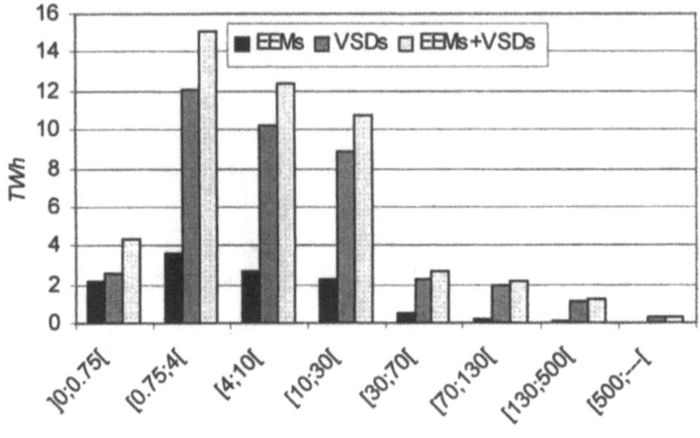

Figure 8: Technical Potential for the Tertiary sector

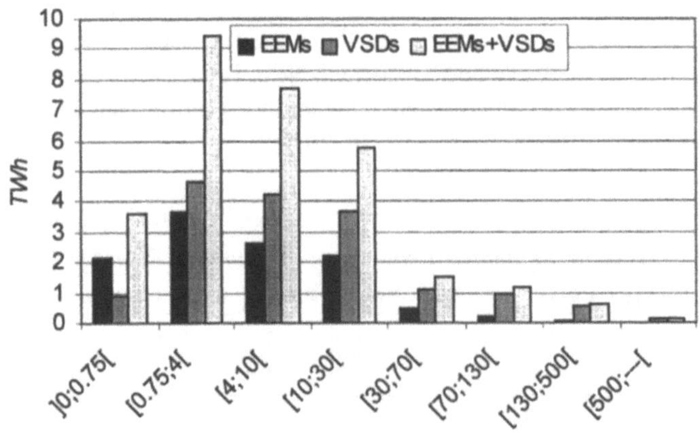

Figure 9: Economical Potential for the Tertiary sector

Motors up to 4kW in the tertiary sector are responsible for about 50% of the total energy savings with the application of EEMs. The main reason for this high potential for energy savings is the efficiency difference between standard and energy efficient motors in the ranges]0, 0.75[and [0.75, 4[. This shows the importance of small motors in the tertiary sector, which need be taken into consideration with more attention in the future.

6 Conclusions

The field characterisation showed that there is a considerable potential for energy savings in the industrial and tertiary sectors in EU.

Table 23 summarises the estimated technical and economic savings potential in the industrial and in the tertiary sector with the application of EEMs and VSDs.

The application of VSDs can lead to the achievement of a huge electricity savings (140TWh). The wide application of EEMs, although its impact is not so impressive as VSDs, still leads to substantial electricity savings (35TWh).

Table 23: Potential Electricity savings in the year 2015

	Technical Potential [TWh]			Economical Potential [TWh]		
	EEMs	**VSDs**	**EEMs+VSDs**	**EEMs**	**VSDs**	**EEMs+VSDs**
Industry	24.1	101.0	121.5	24.1	64.4	86.1
Tertiary	11.5	39.4	48.9	11.5	16.2	29.9
Total [TWh]	35.6	140.4	170.4	35.6	80.6	116.0

Cost-effectiveness analysis, based on the Cost of Saved Energy, showed that the application of Energy Efficient Motors in Industry and in the Tertiary sector is cost-effective in the vast majority of cases. Therefore, the economical potential is equal to its technical potential, which amounts to 24 TWh in industry and to 11.5 TWh in the tertiary sector. In what concerns Variable Speed Drives, there are situations, especially in the lower power ranges, in which the application of speed control is not cost-effective. In the industrial sector with a large number of operating hours, the fraction of application in which VSDs are cost-effective is larger than in the tertiary sector, where the number of operating hours is significantly less. The higher cost of electricity in the tertiary sector partially compensates the reduction in operating hours.

While the savings potential available in industry are concentrated in large motors, the savings potential available in the tertiary sector are concentrated in small motors (below 30kW), due to their relative importance in the use of electricity.

Most motors, especially small motors, both in industry and in the tertiary sector are oversized. The replacement of those motors when they fail with smaller motors would bring significant electricity savings. Those savings would be much higher if the motors were replaced with Energy-Efficient models.

Minimising Efficiency Loss Caused by Motor Rewinds

Professor D.G. Walters OBE, C.Eng

Abstract. It is estimated that rewind efficiency reductions for induction motors up to 250 kW cause annual energy losses in Europe of between 1500 and 2000 GWh/annum. This represents an increased energy cost to European Industry of up to 135M Euro/annum.

Work done in the USA and later in the UK shows that much of the technology already exists to reduce rewind losses almost to zero. This work has been reviewed and extended in a recent EU 'SAVE' Project (yet to be published) entitled "Barriers Against Energy Efficient Motor Repair".

The fragmented nature of the European Repair Industry (most repairers are SME's, some very small) coupled with a lack of awareness of any problem by motor users are considered the main barriers to improvement. It is recommended that simple qualification schemes for repairers be introduced which would include customer awareness, training and third party accreditation elements; also that some funding be allocated from future carbon tax revenues to set up such schemes.

Because motor repair is an essentially local industry the exact requirements in different countries may differ. However, it is recommended that a scheme being introduced by the UK Repair Trade Association (the AEMT) with help from the UK Government be monitored carefully and if successful be used as a model for other schemes elsewhere in the future.

1 Introduction

Some of the processes involved in motor repair, mainly stripping out the old winding and putting in a new one can result in a reduction of motor efficiency. Typically figures of about 1% efficiency reduction/rewind (and occasionally as high as 4%) have been quoted in previous papers on this subject [Refs 1, 2, 3] although a few have indicated that much lower figures are possible.[Refs 4, 5] In many cases the whole or the majority of the efficiency reduction has been put down to iron loss increase. Almost all the early work published originated in the USA and Canada but in the early 1990s the Association of Electrical & Mechanical Trades (AEMT), the Trade Association representing the majority of motor repairers in the UK, did further work which demonstrated that given certain conditions rewind efficiency loss could be reduced almost to zero in small/medium size motors. These results were published in a Good Practice Guide in 1996 [Ref 6]. More recently an EU 'SAVE' project entitled "Barriers Against Energy Efficient Motor Repair" which is not yet published, has yielded further valuable information and for the first time has attempted to quantify the scale of the problem in Europe in terms of energy loss and cost to industry.

The contents of this paper are derived mainly from the AEMT research work and the results of the EU 'SAVE' Project of which the author was a participant.

2 The Scale of the Problem

It is estimated that rewind efficiency reductions for induction motors up to 250kW cause annual energy losses of between 1500 and 2000 GWh/annum in the EU which represents a cost to EU industry of between 100M EURO (£67M, $US 93M) and 135M EURO (£90M, $US 125M) per year for wasted energy. The motor lifetime energy losses and costs are considerably higher - approximately 30TWh over 15 years costing 1800M EURO (£1232M, or $1654M) at current prices and exchange rates.

It is important to understand that rewind efficiency reduction is not a phenomenon applying only to Energy Efficient Motors (EEMs). It can apply to **all** motors whether they be of high or low efficiency. The only factors which have a significant effect on it are some of the original materials used by the motor manufacturers and the repair processes used by the rewinder.

Further information contained in the SAVE Report referred to above relates to motors in the output ranges ≥1.0 - 7.5 kW, > 7.5 - 75 kW, >75 - 250kW and indicates:

a) The average life of motors in each range is

≥ 1.0 - 7.5 kW	12 years
> 7.5 - 75 kW	15 years
> 75 - 250 kW	20 years

b) Statistically motors are rewound between 2.0 and 2.5 times during their lifetime.

c) The number of motors rewound/year in EU is approximately

≥ 1.0 - 7.5 kW	308,000
> 7.5 - 75 kW	1,667,000
> 75 kW	208,000

The relatively low number of motors rewound in the ≥ 1.0 - 7.5 kW category is accounted for by the fact that it is cheaper to replace rather than repair motors of this size in many parts of Europe.

d) Data from various sources (manufacturers, repairers etc) indicates that the "Weighted average" kW rewound in each category is approximately:

Category	"Weighted Average" kW rewound
1.0 - 7.5 kW	5.5 kW
> 7.5 -75 kW	20.8 kW
> 75 - 250 kW	124 kW

Again the "Weighted average" kW rewound for motors in the ≥ 1.0 - 7.5 kW category is near the top end of the range because of the large number of smaller motors which are replaced rather than rewound.

e) Based on this information the total rewind losses/year in MW are calculated as shown in Table 1 below:

Table 1

Power Range	**Ave Power Output (kW)**	**Ave Effy ***	**Ave Power Input (kW)**	**No.of motors Rewound per year**	**Rewind losses for 1% effy reduction (MW)**
≥ 1.0 -7.5 kW	5.5	83.9	6.55	308,000	20.2
>7.5 - 75 kW	20.8	91.25	22.79	1,667,000	380.0
> 75 - 250 kW	124.0	94.9	130.66	208,000	270.0
				Total	**670.2**

*Source - EuroDEEM database - 4 pole motors
22 kW data for 11-75 kW range.
Average of 110 kW and 132 kW data for 75-250 kW range

f) The energy loss depends upon the number of hours for which the 'average' motor is run each year. An earlier EU study [Ref 7] indicated the average annual running hours for motors in each group to be approximately:

≥ 1.0 - 7.5 kW	1213 hours
11 - 75 kW	1886 hours
> 75 - 250 kW	2053 hours

Since that study both the author and some other workers have indicated that because of increasing utilisation of industrial plant they now believe those hours to be too low, perhaps by as much as 50% or more. Thus for the purposes of this paper the average running hours are taken as:

≥ 1.0 - 7.5 kW	1819
11 - 75 kW	2829
>75 - 250 kW	3080

Assuming that each rewind **does** cause a 1% efficiency reduction the annual rewind energy loss for the motors in each group obtained by multiplying power loss by running hours is shown in Table 2.

Table 2

Power Range KW	Annual Rewind Energy Loss GWh
≥ 1.0 - 7.5 kW	36.8
11 - 75 kW	1075.0
> 75 - 250 kW	831.0
Total	**1942.8**

These calculations inevitably contain a number of assumptions and estimates; however it seems reasonably safe to assume that the energy lost by European industry as a result of the "rewind efficiency reduction" of induction motors up to 250 kW is between 1500 - 2000 GWh per annum. At an average electricity cost of 0.06 Euro per kWh the cost to industry in the EU is thus between 90M Euro (£60M, $82.5M) and 120M Euro (80M, $110M) per year.

Based on a weighted average motor size of 28 kW and a motor life of 15 years including two rewinds*, one after 5 and the other after 10 years the motor lifetime rewind energy losses and costs **for a 1% efficiency reduction/rewind** could be:

Table 3

Period	**Rewind Energy loss Twh**	**Cost**		
		Euro (M)	**£M**	**$USM**
Years 1-5	0	0	0	0
Years 6 - 10	10.03	600	400	550
Years 11 - 15	20.05	1200	800	1100
Total life	**30.08**	**1800**	**1200**	**1650**

Both the annual and "motor lifetime" figures are large sums of money which justify action to reduce "rewind efficiency losses" on both economic and environmental grounds.

Furthermore an efficiency loss of 1%/rewind will approximately halve the lifetime energy savings which might have been expected from the newer EEMs. This too is a serious matter and adds impetus to the need for action.

However, it should be noted that the bulk of the rewind energy losses occur in motors between 11 - 250 kW and it is on this range that action needs to be focussed.

3 Rewind or Replace?

The majority of motors sent to repairers can be repaired satisfactorily; however there will be some which have suffered irreversible damage and which even if rewound will not run for very long before further failure occurs; these motors will certainly run at reduced efficiency. When such a motor is received for repair, a replacement should be supplied if possible.

* Derived from data collected from UK Repairers in 1994 and 1998 by the AEMT.

Irreparable damage can include:

- Major damage to either stator or rotor core - pieces missing or fused together or widespread damage to the airgap surfaces of the stator or rotor such as that caused by a severe core rub.

- Uniform blue or brown discoloration of the stator and/or rotor core which indicates a prolonged period of overheating such as that caused by stalling. This may have damaged the interlaminar insulation on motors other than those having L3 type inorganic/organic insulation. If damage is suspected the cores can be stripped and the laminations re-insulated but this is generally not economic.

- All over discoloration of the stator winding - not in itself irreparable but may indicate motor operating on continuous overload - supply new larger motor

If a replacement motor is supplied consideration should always be given to supplying an energy efficient motor (EEM). Whether or not this is economic will depend upon:

- The price differential (if any) between the EEM and a standard efficiency motor
- The hours run by the motor each year
- The cost of electricity

There are now some EEMs on sale which have either zero or very low price premiums compared with standard efficiency motors. It is obviously helpful if the Repairer is also a stockist of these motors (as a number are) because then the Repairer will be able to help his customer make the right choice in an unbiased way.

4 The Causes of Rewind Energy Losses and Ways of Reducing them

Many rewind shops are small and do not appear to the outsider as very sophisticated workshops. Nevertheless the repair process they undertake is complex although fortunately only a few of its elements can have any significant effect on motor efficiency. Those that do are shown in Figure 1 and each is reviewed in the following paragraphs.

a) Cutting off one endwinding

In some repair shops this is done by hand using a hammer and chisel which renders the stator core liable to damage if the chisel slips. Such mechanical damage may result in iron loss increase.

By far the best method of cutting off one endwinding is to use a jig located circular saw, an example of which is shown in Figure 2. This method also has the advantage of being much quicker as well as safer.

b) Releasing the varnish from the old winding

In some small motors it is possible to remove the winding without pre-heating but in most cases some heat is applied to soften the varnish which is used as part of the winding insulation system. Other methods of varnish release (eg chemical) have become less common in recent years because of environmental problems.

Heating the windings inevitably results in some heat being transmitted to the stator core. The safe temperature to which the stator core may be heated depends upon a number of factors including:

- The type of interlaminar insulation used

- The size of motor
- Whether or not flaws in the interlaminar insulation allow hotspots to develop

As a general rule, if stator cores are heated to more than 300°C (360°C for smaller motors up to about 30 kW) some damage may occur to the interlaminar insulation.

Laminations insulated with inorganic/organic varnishes of the 'L3' type will withstand much higher temperatures without damage to the interlaminar insulation but higher temperatures are not recommended because of the possibility of distorting other mechanical components (eg stator frame spigots).

There are three methods generally used for heating:

- Hand held torch (normally propane or natural gas but may be acetylene)
- Heating table fixture
- Controlled temperature burnout oven

The hand held torch with the flame directed on to the remaining endwinding is both the cheapest and on small motors the quickest method of heating the insulation prior to removing the windings. It is almost universally used on small motors (up to 7.5 kW) and on many medium sized motors also but its use on large motors (say above 37 kW) is less common because the heat output of the torch is often insufficient to get all the winding warm in a reasonable time.

Used carefully there is no evidence that the torch method does any damage to the stator core; however the process is completely under the control of the operator and if the torch flame is accidentally allowed to play directly on to the core it can quickly damage the interlaminar insulation. From a motor efficiency viewpoint this method, although in widespread use, must be regarded as unsatisfactory except perhaps for motors in the ≥ 1.0 - 7.5 kW range and then only when done by an experienced operator who has received appropriate training.

Heating tables are fixtures in which the stator with one endwinding cut off is held vertically with the remaining endwinding pointing either upwards or downwards. Gas jets are directed on to this endwinding and heat is transmitted to the slots via the conductors. A heat insulating shield prevents the gas flames impinging directly on the stator core.

Whilst this equipment is more expensive than the hand held torch it is **much** less expensive than a controlled temperature burnout oven and it does concentrate the heat where it is needed. Tests conducted by the author have shown that it is quite possible to get over 200°C temperature differential between the wire in the slot and the adjacent core. The remains of the old slot insulation does much to prevent heat transfer from the copper winding to the iron core. Figure 3 shows an

11kW stator prepared for burnout (note the white coloured thermal shielding round the end of the core) and Figure 4 shows the actual temperatures recorded during burnout using a thermal imaging camera. It should be noted that in this installation the endwinding being heated is at the top of the core. This allows heat to be **conducted** down into the slots whilst surplus heat is **convected** away from the core to an exhaust extractor. Also with this arrangement the heating table can be combined with a coil puller which saves handling time. Similar equipment is obtainable commercially both in the UK and Germany (although some models have the endwinding to be heated at the bottom). This should become the burnout method of choice for most motors.

Controlled temperature burnout ovens are the most expensive option. Even with these, hotspot damage in the core can occur but the main problem is that their cost (about 15000 Euros in the UK for a $1m^3$ oven complete with fume extraction equipment) puts them outside the price range of many smaller repairers. Because they heat the whole mass of material in them equally (including stator frames etc) they are also the slowest of the three methods and can take several hours. Because of this there is sometimes a temptation to raise the thermostat setting to 400°C or more which reduces the cycle time but also causes damage to the interlaminar insulation which will reduce efficiency. Unless ovens are full, which is often not the case in a repair shop, they are not themselves very energy efficient. They are, however, widely used in the repair industry and may be the best method for the larger motors.

c) Cleaning the Core

When the old winding has been pulled out of the core a considerable amount of debris – mainly old insulation – is left, particularly in the stator slots, and this has to be removed before re-insulating and rewinding the stator. A variety of methods are used.

One of the most common (and potentially most damaging) is the use of a rat-tailed file to scrape out the insulation. Unfortunately this will almost certainly burr together the edges of adjacent laminations in the slots and increase both the mains frequency iron loss and the harmonic iron losses (often regarded as part of stray loss). A wire brush does less damage but is often less effective in removing insulation debris.

Another commonly used method is the so-called "oxygen lance" which projects a narrow oxy-acetylene flame at about 1500°C up each slot in turn, reducing the old insulation to ash which can then be brushed out. The operation is very quick and it is possible that the temperature of the stator teeth does not rise excessively.

A more recent Italian development [Ref 8] uses an oxy hydrogen flame at 3500°C. Recent tests on 15 and 30 kW motors fitted with core thermocouples indicated that the core temperature using the oxy-hydrogen flame could be over 100°C lower than when using more conventional and cooler oxy-acetylene (See Tables 4 & 5)

Table 4: 15 kW Motor - Slot Insulation Burnout

	Oxy-acetylene flame 1500°C	**Oxy-hydrogen flame 3500°C**
Teeth temperature on airgap surface	160-190°C	100-180°C
Core Temperature	120-130°C	100°C

Table 5: 30 kW Motor - Slot Insulation Burnout

	Oxy-acetylene flame 1500°C	**Oxy-hydrogen flame 3500°C**
Teeth temperature on airgap surface	240-280°C	100-150°C
Core temperature	220-310°C	130-160°C

The difference in temperature between the 15 kW and 30 kW motors may be significant and further tests on larger machines need to be done before the full benefits of the oxy-hydrogen flame are known. It should be noted that in the trials (and usually in practice) both the oxy-acetylene and oxy-hydrogen lances were hand held and are therefore capable of misuse (ie heating one slot for longer than is necessary). Careful training procedures need to be established for operators using this type of equipment.

The oxy-hydrogen equipment (shown in Figure 5) currently has a much higher first cost than the oxy-acetylene equipment although running costs are claimed to be up to 85% lower. Whilst oxy-hydrogen equipment can be used to advantage in other repair processes first cost will probably have to be addressed before its use becomes widespread in the repair industry.

d) Rewinding

Most repairers do a "copy" rewind – ie the winding they put in is a copy of the one they take out. The problem is that some "copies" are better than others. Time is usually of the essence in motor repair and it is quicker to put in a winding which is an easy, rather than tight fit in the slots or which has longer rather than shorter overhangs.

Thus, there is a temptation for some repairers to:

i) Reduce the conductor cross section slightly
and/or
ii) Increase the endwinding length slightly
and/or
iii) Reduce the number of turns

Actions (i) and (ii) increase the stator copper losses; (iii) can reduce the stator copper loss but will increase the flux and the iron losses unless the winding pitch factor is altered to compensate. Usually this is not done because increasing the flux will also increase the torque; if the repairer thinks the motor may have failed because of insufficient torque he may feel that a little more torque will solve the problem.

There can also be problems when a repairer attempts to replace a concentric winding (favoured by most manufacturers for small/medium sized motors because automatic winding machinery is usually designed for it) with a hand wound lap winding. It is perfectly possible to replace most concentric windings with a short pitched lap winding without increasing copper loss (the exception is the three tier concentric winding); however, not all repairers know how to calculate the equivalent lap winding correctly.

Not everything the repairer does when rewinding need make the motor efficiency worse. It is generally possible to put in a winding by hand which has a shorter endwinding than most machine wound concentric windings – this will reduce copper loss. Best of all though is to replace the two layer concentric winding (used by most manufacturers for 4 pole motors) with a three layer concentric winding – this significantly reduces the copper used by reducing the average coil span and thus reduces the copper losses. In theory this can yield an efficiency gain of more than 0.5% but practical considerations of endwinding dimensions mean that the gain achieved is more usually about 0.3% (and some of this may be offset by efficiency losses arising from other parts of the repair process). A further advantage of this winding configuration for repair shops where motors are usually hand wound is that all coils in the same phase can be in the same layer of the winding. Thus, the whole phase can be wound without interconnections which saves time overall and reduces the risk of errors.

Coil group configurations for 4 pole 36 slot 2 layer and 3 layer windings are shown in Figures 6a and 6b.

As far as winding is concerned the main barrier to good practice is not cost but time and the lack of understanding of simple winding theory by many repairers.

e) Insulation and Impregnation

(i) Insulation

There is a commonly held but not necessarily correct belief that if a motor runs hotter following a rewind the life of the insulation will be reduced. Insulation life will only be reduced significantly if the motor is run at an operating temperature above the temperature class of the insulation used. Most motors sold in Europe today have Class B temperature rises with Class F insulation; this coupled with the fact that the average load factor is something between .66 and .75 means that most motors run much cooler than the limiting temperature of their insulation class and a small increase in operating temperature will not significantly affect the life of the insulating materials.

That said, it is always possible that hot spots will form in the motor core and the likelihood of these is probably increased by a repair. A repairer following best practice will normally use the highest quality insulation materials available eg Nomex/polyester film/Nomex composite sheet materials in preference to polyester paper/polyester film/polyester paper composites even though both these materials have Class F temperature ratings.

(ii) Impregnation

Impregnating the winding with varnish, which is subsequently baked hard at temperatures up to 150°C, serves two purposes. It bonds the individual conductors into a rigid cohesive mass less likely to be affected by vibration and damp; the varnish also replaces the air in the spaces between conductors in the slots and improves the heat transfer from the conductors to the motor core. This is important in keeping winding temperatures (and hence joule losses) down.

Some repairers merely pour varnish over the endwindings – this is an unacceptable practice since varnish applied in this way usually fails to serve either of its main purposes properly. Other methods include:

- Dip, drain and bake
- Trickle impregnation
- Vacuum impregnation

Done properly, any of these methods is satisfactory although trickle or vacuum impregnation methods which use solventless or water based varnishes are generally regarded as the best for obtaining maximum varnish fill in the slots and thus best heat transfer from conductors to core.

Care needs to be taken if trickle impregnation is used on a three layer winding - it is difficult to impregnate the middle layer of the winding properly and more work needs to be done to develop best practice in this respect.

f) Re-barring and brazing rotors

Nowadays most induction motor rotors have diecast cages which are rarely damaged and even if they are, are replaced with a spare from the motor manufacturer; alternatively a new motor is supplied.

A minority of new motors and many older ones still in service have cages and endrings fabricated from copper, brass or phosphor bronze bars and occasionally these suffer fatigue failure due to vibration.

Most of the bar sections used by most manufacturers are available (at a price) from specialist suppliers but delivery is not always immediate. There may occasionally be the temptation to replace an unavailable copper conductor with, say, an available phosphor bronze one having the same cross section – this will increase the rotor resistance and slip losses and will thus reduce efficiency.

If the correct material for bar and endring sections is not available best practice is to replace the motor.

g) Skimming the rotor surface

Many types of motor failure, particularly bearing collapses, result in some damage to the surface of the rotor core. If the laminations in the damaged area are burred together this can cause a local hotspot because of eddy currents induced at slotting frequencies (typically 23rd harmonic) in the damaged surface of the rotor.

To try and overcome this problem and also to make the rotor surface look better, some repairers put the rotor in a lathe and take a skim cut across the whole surface. This can have several effects:

- The airgap will be increased – so also will the torque and the magnetising current – the latter reducing efficiency.
- If not done carefully <u>all</u> the laminations may be burred together and stray loss will be increased.
- Paradoxically, if the laminations are not burred together, increasing the airgap up to a point might actually reduce stray loss on some motors.

Much more research is required in this area – research which might be stimulated in Europe if current IEC proposals for direct measurement methods on motor test are adopted. However, for the moment the best advice to repairers is not to skim rotor surfaces but to ensure that laminations in damaged areas are not burred together. If skimming is considered essential two light cuts in opposite directions should be used in preference to a single cut in one direction only.

h) Bearing replacement

Most repairers replace the motor bearings whenever they do a rewind. Bearing friction losses are very small and provided that the replacement bearing is of the correct grade (ie size **and** clearance) there is no reason why any bearing from a reputable manufacturer should increase losses. Care should be taken not to overgrease bearings - remember the old adage "Half full is too full".

i) Seal replacement

Seal friction can be a problem, particularly on smaller motors. Most small motor seals are either of the face contact or spring loaded garter types. Face seals have generally lower friction losses than garter seals although the latter are generally regarded as giving better protection against the ingress of moisture and dirt. However, the garter seals must be properly fitted and must be adequately lubricated. A dry running seal can reduce the efficiency of a small motor (4 kW say) by 2 or 3% and apart from this will cause local heating which may damage the bearing which the seal is supposed to protect.

j) Fan replacement

Sometimes cooling fans and covers get broken and need to be replaced. There is no standardised dimension on induction motors for the shaft diameter at the fan seat and this has led to the growth of a small industry supplying “universal” fans with hub inserts to fit different shaft sizes. Often these have larger blades than the original fan fitted by the motor manufacturers and thus increase windage loss.

On small/medium sized motors this is not a problem because fan losses are fairly low; however, on larger motors (say > 50 kW) particularly 2 poles, fitting an oversized fan can reduce efficiency by up to 0.5%.

5 Encouraging the Take-up of Best Practice by the Motor Repair Industry

Knowing how to repair motors without loss of efficiency is one thing; disseminating this knowledge to the repair industry and persuading them to use it is quite different. In the EU much of the motor repair industry is made up of small companies often with less than 30 employees and each serving a local area. Their strength lies in their flexibility, which enables them to respond quickly to a wide variety of customer demands. However, unless these companies are members of a strong trade association which is prepared to undertake the work of disseminating knowledge of best practice on behalf of its members it is very difficult to carry out this task effectively.

Repair trade associations do exist in France,[Ref 9] Germany [Ref10] and the UK. The UK trade association is called the Association of Electrical and Mechanical Trades (AEMT) and has already begun work in this field. The most important tasks are seen as:

- The development of standards for best practice
- The development of simple qualification schemes for repairers which will be recognised as valuable by motor users
- The provision of training facilities and materials
- The provision of third party assessment for repairers
- The provision of awareness campaigns for motor users

(a) The development of standards for best practice

The AEMT in the UK has already produced a best practice guide entitled "The Repair of Induction Motors - best practice to maintain energy efficiency". This presently covers induction motors up to 22 kW but it is hoped soon to carry out further research which will increase the power level up to 250 kW or higher.

In addition to this there is an IEC Working Group (IEC.TC2/WG25) which is considering a standard on "rewinding and refurbishing of electrical machines". The scope of this standard is wider than that of the AEMT Good Practice Guide, however it is recommended that the standard should make reference to the Guide in a section dealing with minimising the potential loss of efficiency during a repair.

(b) The development of simple qualification schemes for repairers

The repairing of motors with minimum or no loss of efficiency is of no value to the repairer unless the benefits are recognised by his customer, ie the motor user. To help publicise the benefits of good repairs it is recommended that an energy efficient brand label for motor repair be introduced. This would lead to easy recognition of good repairers by customers and would complement the efficiency labelling scheme about to be introduced by manufacturers of new motors.

In the UK a pilot scheme for repairers using the Energy Saving Trust's logo is being considered and is likely to be introduced in the year 2000. Because of the essentially localised character of the repair industry other branding schemes may be more appropriate in other EU member states; however it is clear that such initiatives are only realistic if national trade associations are prepared to back them - as is the case in the UK.

(c) The provision of training facilities

In most of the EU there is no recognised qualification for young people entering the repair trade and the geographic diversity of the workforce, with relatively small numbers in any one location, would make it financially impractical for local technical colleges to provide courses even if a qualification existed.

In the UK most apprentices in the repair industry study part time for electrical or mechanical technician's qualifications - the drop out rate from such courses is fairly high because they are seen by many trainees as largely irrelevant to motor repair. In an attempt to overcome the problem the AEMT, with help from the UK Energy Efficiency Best Practice Programme, is constructing a specialised workshop for motor repair training and plans in due course to run residential short courses for apprentices. More immediately this training facility will be used by the AEMT to provide short course training in energy efficiency rewind practices for experienced employees of UK repair companies.

It is recommended that this initiative be closely monitored and if successful be replicated elsewhere in the EU.

(d) The provision of third party assessment of repairers wishing to be members of any qualification scheme

For any qualification scheme to be seen to have value by motor users it almost certainly needs to have some form of third party assessment. The main concern is that to be practical, third party assessments should be kept to the absolute minimum commensurate with achieving their objective of ensuring that repair processes which can seriously and adversely affect motor efficiency are carried out properly.

Some larger repairers have ISO 9000 certification and this can almost be certainly used where it exists with suitable additional pages added to the certification documents. However, for the majority of repair shops without ISO 9000 certification a much simpler and more focussed scheme needs to be developed. The AEMT are presently discussing such a scheme with two UK companies active in the quality assessment field.

(e) The provision of awareness campaigns for motor users

This is probably the most difficult task of all and will require co-ordinated efforts by Governments (through Energy Agencies), by trade associations and by individual companies.

It will inevitably be a fairly long term programme and the best method of approach will vary from country to country. A campaign such as this will also cost money and it is recommended that the EU consider using some carbon tax revenues for this purpose from the year 2001 onwards.

Any campaign should be linked to an accredited qualification scheme which includes employee training and must direct customers towards those repairers who are members of that scheme.

6 Conclusion

Much work has been done by some motor manufacturers in Europe to develop lower cost energy efficient motors and the European Commission has recently announced a voluntary agreement with CEMEP (the European Motor Manufacturers Trade Association) for efficiency labelling of new motors. In the USA and Canada the position is more advanced with minimum efficiency legislation already in place.

Because motors are (statistically at least) repaired more than twice during their lifetimes a significant part of the lifetime energy saving, which might have been expected from an EEM, will be lost if efficiency is reduced by 1.0% each time it is rewound.

Work done over recent years has shown that rewind efficiency losses can be reduced almost to zero (and in some cases efficiency might even be slightly improved) if the best practice outlined in this paper were to be widely adopted. Using best practice repair techniques does not mean the repair need take longer (time is often of the essence in the repair industry) but it may require some repairers to upgrade some of their equipment and all repairers will need some training. The geographic diversity of the Repair Industry and the comparatively small size of most individual repair shops make this difficult; in any case repairers are unlikely to be willing to invest in any improvements unless their value is recognised by the motor users who form the repairer's customer base.

To make this happen some form of simple qualification scheme for repairers which would include training, third party accreditation and customer awareness elements is recommended. Initially some pump priming revenues would be needed for any such scheme and it is suggested that some revenues from future carbon taxes (which are supposed to be recycled into energy efficiency improvement measures) be used for this purpose.

In the UK, the AEMT is well advanced with plans to introduce such a qualification scheme which will probably be linked to the UK's Energy Saving Trust. It is recommended that the development of this scheme be monitored carefully and if successful that it be used either in whole or in part as a role model for future schemes elsewhere.

Acknowledgements

The author would like to thank the following for their contributions to this paper and apologise to any he may have inadvertently omitted to mention.

The Association of Electrical & Mechanical Trades, Nottingham, UK.
The Dept. of Environment, Transport & the Regions (DETR) UK.
Professor AT de Almeida - University of Coimbra, Portugal.
Professor F Parasiliti - University of L'Aquila, Italy.
Oxyweld snc, Maggiore, Italy.
Rotary Electrical Ltd, Sheffield, UK.
Whitelegg Machines Ltd, Dorking, UK.
Wire Electric Supplies Ltd, Wolverhampton, UK.

7 References

[1] Montgomery, D (1989) Testing Rewinds to avoid motor efficiency degradation, Energy Engineering Vol. 86, No. 3,USA
[2] Ontario Hydro (1991) Rewound Motor Efficiency, TP-91-125, Ontario
[3] BC Hydro (1993): Rewound High Efficiency Motor Performance, M101. British Columbia
[4] Electrical Apparatus Service Association (1984): Core Iron Study, USA
[5] Hydro Quebec (1998): "Evaluation of Electric Motor Procedures Guidebook", (CEA 9205 U 984)
[6] AEMT (1997): The Repair of Induction Motors - Best Practice to maintain energy efficiency
[7] De Almeida AT & Fonseca P (1997): Energy Efficiency Improvements in Electric Motors & Drives, Springer, pp 19-36
[8] Oxyweld snc, Maggiore Italy
[9] Syndicat des Industries de la reparation et de la maintenance electrotechnique et electronique, Paris
[10] Zentralverband der Deutschen Elektrohandwerke, Frankfurt

- Cutting off the old endwinding
- Heating the old winding to release the varnish
- Cleaning out old slot insulation
- Rewinding
- Impregnation
- Rotor repair
- Seal or fan replacement

Figure 1: Repair Processes which can Affect Efficiency

Figure 2: Stator Cut Off Machine
(Whitelegg Machines Ltd UK)

Figure 3: 11 kW Stator Awaiting Burnout
(Rotary Electrical Ltd, Sheffield, UK)

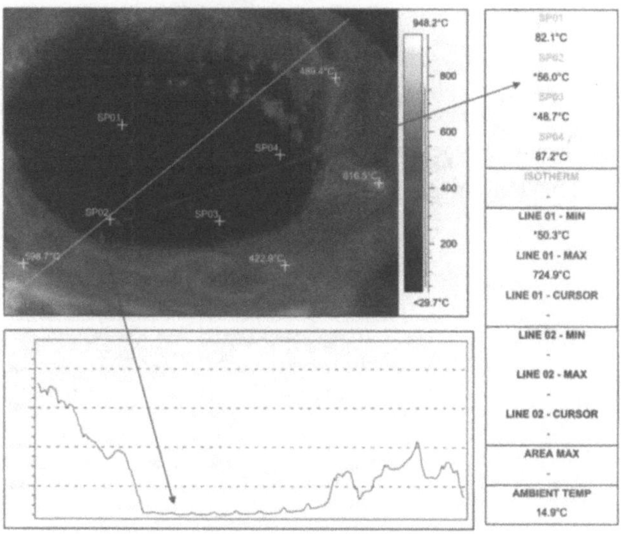

Figure 4: Thermal Image of 11 kW Stator being Burned Out
(Rotary Electrical Ltd., Sheffield, UK)

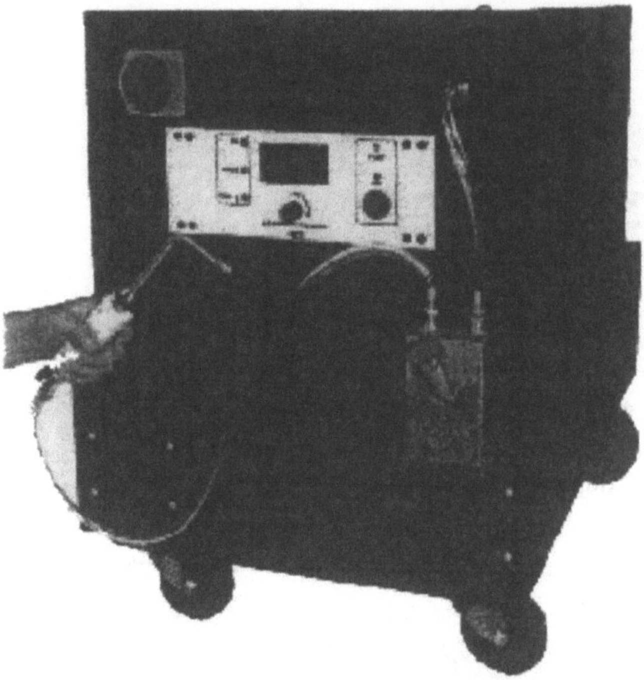

Figure 5: Oxy hydrogen generator
(Oxyweld snc, Maggiore, Italy)

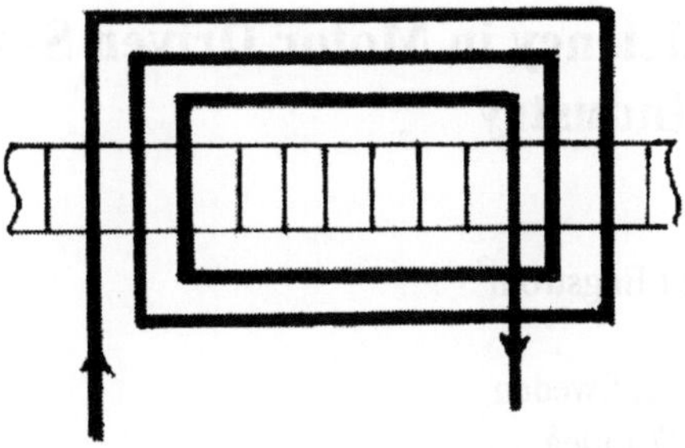

Figure 6a: Coil groups for 2 layer concentric winding
Coil spans 11, 9, 7 slots
Average Span = 9

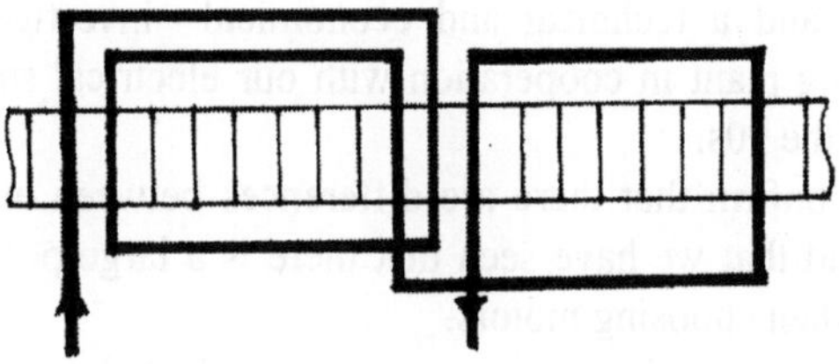

Figure 6b: Coil group for 3 layer concentric winding
Coil spans 9, 7, 7 slots
Average Span = 7.66

Improving Efficiency in Motor Driven Systems – The Iron Ore Industry

Lennart Mukka[1], Bernt Engström[2]

[1] LKAB, 981 86 Kiruna, Sweden
[2] Vattenfall AB, 971 77 Luleå

Abstract. This paper presents test results and examples of improvements in electrical energy savings through increasing the efficiency of motor driven systems at LKAB, Kiruna.
The first section presents results from a Motor Project consisting of laboratory tests on motors and a technical and economical investigation carried out at LKABs pelletizing plant in cooperation with our electrical supplier Vattenfall, in the beginning of the 90s.
Briefly we can confirm that there are differences between motors from different manufacturers and that we have seen that there is a large potential for improving energy savings when choosing motors.
The second section describes improvements in selected motor drive systems for example conveyor drives and belt driven systems.

1 Introduction

1.1 LKAB

LKAB is one of the eight largest iron-ore producers in the world, with two underground mines in production at Kiruna and Malmberget. The former is the largest in the world and has three associated concentrators and pelletizing plants. The ability to use the energy efficiently in our operations is a major factor in maintaining our competitiveness and the means our of survival.
LKAB's yearly consumption of electrical energy is 1.6 TWh which is about 1% of the total electrical energy consumption in Sweden, and of that about 90 % is used to supply electrical motors.

Electrical energy forms about 11% of the total product costs.

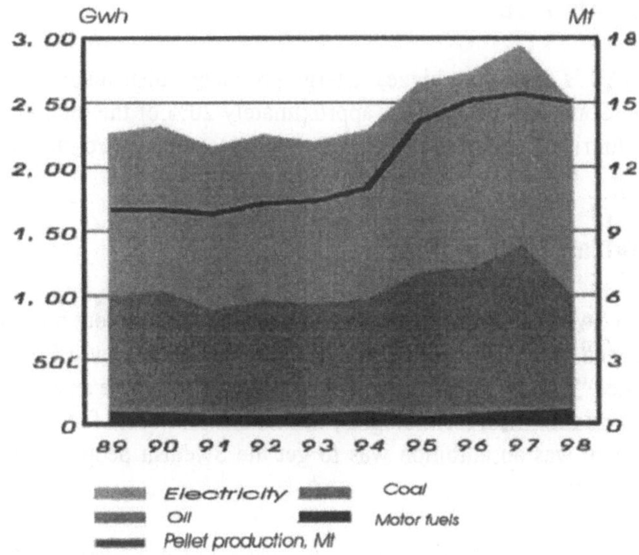

Figure 1.1: Energy consumption

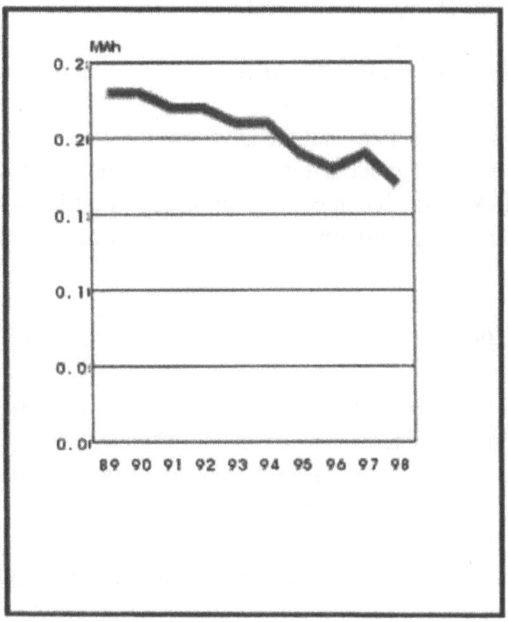

Figure 1.2: Energy consumption per tonne pellets

1.2 Vattenfall AB

Vattenfall AB is one of the biggest energy producers, distributors and suppliers in the Nordic Countries, generating approximately 20% of the total electricity used in these countries. Most of the electricity is produced by hydro- and nuclear power plants.

1.3 Uppdrag 2000 in Brief

In 1984 the Swedish Government made a decision to phase-out nuclear power. As a consequence of this, Vattenfall, at that time, the "State Power Board" created a project called "Uppdrag 2000", the aim of which was to give answers to questions concerning possibilities, technologies, limitations and costs for the situation . In addition, there was an ambition was to get the Swedish people to use electricity more efficiently.

Arising from this, and in support of the theme of this conference, the present paper describes some examples from electrical energy improvements carried out in the electrical motor driven systems during the period 1980-1998, and others in progress.

2 The Electric Motor Project

2.1.1 Vattenfall AB the Energy Company

Vattenfall AB was before 1992 and the advent of the "free electricity market", the "State Power Board" and as such , totallyresponsible for energy supply in Sweden. Today Vattenfall AB is active in several European Countries, where we lectricity, heat and other qualified energy services are offered to customers.

2.1.2 Uppdrag 2000

The guidelines were drawn up in 1984 and the project was finished in 1991. The scope of the work was intended to cover the whole of society. The experiences gained later became a qualified base for evaluating added values in the electricity industry and were extended to other energy services.

The following statements give an idea of the important parts of the assumptions for the project.

"Starting at 10 years from now, Sweden will commence start a nuclear power phase-out."

"If we do not build any more productions plants, which technologies can be used, how much can Sweden's need for electricity, be reduced, and to what costs, by smarter and more efficient usage?"

"Encourage the efficient use of electrical energy, throughout the country."

Areas for the project.
Single family houses
Multi family houses
Small and medium sized industries
Heavy and energy intensive industries.
Price and marketing considerations.
Method:
Specific investigations (approx 4000 cases)
Measurements, evaluation, analyses
Test installations
Interviews

Resources: Approx. 80 own personnel + consultants
Costs: ***435 MSEK (40 MGBP)***
Technologies:
Heating
Heat Storage
Heat Pumps
Electric Motor Drives.
Refrigeration, - food shops, ice rinks,
Compressed air.
Lighting, (types, control systems, etc.)
Air- and water transportation.

2.2 Motor Project in a Heavy Industry

2.2.1 The Electric Motor Study within Uppdrag 2000

For the electric motor-system project within heavy industry, LKAB was the co-operation partner. At that time, LKAB was already well aware of many electric motor issues. They had more than 4 000 motors in operation, in all categories, and, very important, the had very supportive company managers , the perfect partnership !

To ensure realistic and true motor and system tests and evaluation of measurements, good facilities were needed, and for this the resources of the Royal Institute of Technologies in Stockholm was utilized.

2.2.2 Managing and Sponsoring

The Project "Study of electric motors in a heavy industry plant" was managed by three sponsors, LKAB, STEM, Swedish National Energy Administration (former Swedish National Board for Industrial and Technical Development NUTEK) and Vattenfall AB.
A reference group of engineers from three other heavy industries and the Royal Institute of Technology, also took an active part in the project.

The project was divided in two parts,

Part 1. Laboratory Motor Tests at KTH (the Royal Institute of Technology)

Tests on a test bench for confirmation (?) of rated technical data
Tests for comparing motors from different manufacturers.
Tests for comparing Standard and High Performance motors of today.
Form an opinion of which improvements in efficiency, that can be made using High Performance motors.

Part 2. Measurements and Evaluation in the LKAB Pelletizing Plant.
The choice of motors in a real (10 years old) plant.
The energy and economy potential if the motor choice is optimal.
The potential for reduction of reactive power
The potential for reduction in investment costs.

2.3 Results

2.3.1 Results from Part 1, "Laboratory Motor Tests"

The tests were performed on a test bench, using motors with rated powers of 4, 11 and 37 kW, from manufacturers with good reputation, both standard and HP-motors.

The differences in efficiency between data sheets and measured data were found to be a maximum of 4.3, 2.7 and 1.1 percent respectively for standard motors, and a difference of 3.0, 4.0 and 2.0 percent respectively between standard and HP-motors from the same manufacturer. Differences in cos φ were also noted.

One possible reason for the differences between rated performance data and the test values can be due to the accepted IEC presentation form of data. Performance data may be based on efficiency data, calculated from losses. (However comparison showed that all manufacturers presented the data in the same way, so the differences between them should be correct.)

2.3.2 Content in Part 2, Study in the Pelletizing Plant

The motor-study in the 10 year old pelletizing plant started in 1988 by carrying out an inventory. The production in the pelletizing plant was 3 600 million tons in 1990, , or approximately 80% of planned production capacity.
The plant has 638 motors in total, but only 368 of economical significance. (>1 kW and >7000/kWh/a.). From that group a "controlled" random selection of 80 motors was made.

Measurement were made at the power supply points, during a seven day period. The active and reactive power was registered and in parallel, the associated production, inorder to be able to exclude non-relevant data at a later stage. The measurement equipment was good with an error of less than 2%.

The evaluation of collected data included an assessment of "traditional" technical data, the degree of load during the period, giving average and maximum values for every motor. Beside that, an economical evaluation was made to determine , for the given circumstances "Which motor is the best choice, technically and economically, taking into account the current technical and economical parameters?"

Technical assessments such as load and operation time were considered to be basic data , but not starting torque (see below), ambient temperature and degree of surface dirt collection was not possible to register for every motor. Economical facts were, for example, energy price, investment costs and internal economical requirements.

2.3.3 Results from Part 2, Measurement and Evaluation of Motor Installations in an Industrial Plant

The assembly of collected data, calculations and evaluation shows that, foı actual production situation in the pelletizing plant, only a few motors are optimal in all aspects! (20 years ago High Performance Motors did not exist and, at that time the company plans were to expand the production).

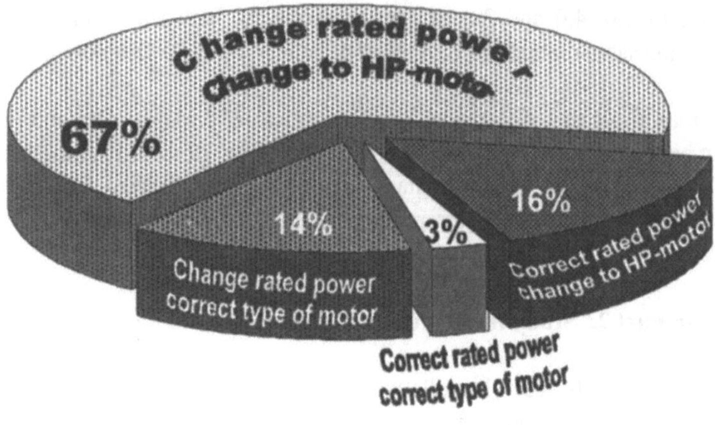

Figure 2: The motor choice; Rated Power – Type of Motor.

2.3.4 Other Observations, - there Are a Lot of Differences within the Framework of the IEC Standard

For many years a standard has existed for Electric Asynchronous Motors, IEC 34. Most motors from manufacturers all over the world (apart from North America), comply with this standard.

Very simply, we can say that the standard recommends mechanical sizes and power outputs, which technical performance shall be tested and how, etc. But this standard has no quality limits for efficiency and is therefore no guarantee,for example, for high efficiency. Therefore, we have differences! In other quality aspects, ex electrical safety, all motors are good motors.
Motors can have large performance differences, despite the fact that they comply with the same standard! Important factors can differ a lot, for example:

- Efficiency	**- Starting torque**
- Cos φ	**- Temperature classes**
- Noise	**- Service**
- Stock levels	**- Winding protection**
- Price	

The issue of motor efficiencies has now been adressed by the classification of motor efficiencies introduced within the European Community !

2.3.5 A Practical Example, Concerning the Starting Torque for a Belt Conveyor (from LKAB)

Demands: Starting torque; at least 3200 Nm (break-away torque)
Normal operation torque, 950 N

Table 1

	Manufacturer 1		Manufacturer 2		Manufacturer 3	
Rated Power KW	Rated Torque Nm	Start Torque Nm	Rated Torque Nm	Start Torque Nm	Rated Torque Nm	Start Torque Nm
160	1036	1761	1028	2364	1030	2678
200	1294	1811	1286	2829	1290	3225
250	1618	2265	1604	2566	1610	3381
315	2039	2854	2024	3440	2020	4242

From the table we can see that a 200 or 250 kW-motor can be a good choice, if we choose manufacturer 3, but that the other motors are not suitable for the application

2.3.6 Is the LKAB-Study Representative?

Are there more investigations carried out in other industries? The answer is Yes ! For example the result from a large Swedish paper mill, MoDo Husum, shows the following result from a study on 78 motors.
(There was no economical evaluation)
We consider that the LKAB-study is representative.

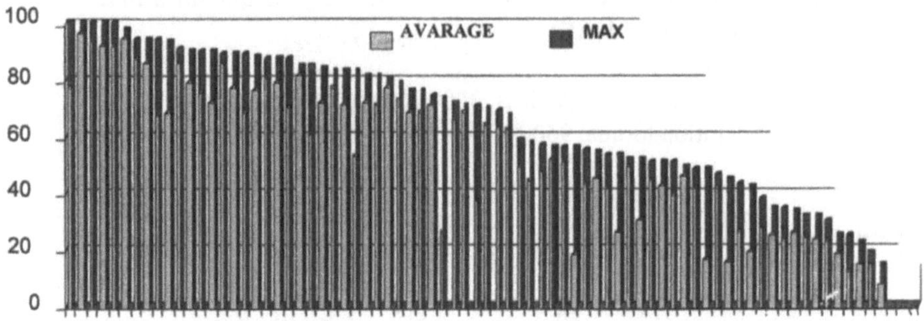

Figure 3: Individual motors, degree of load %

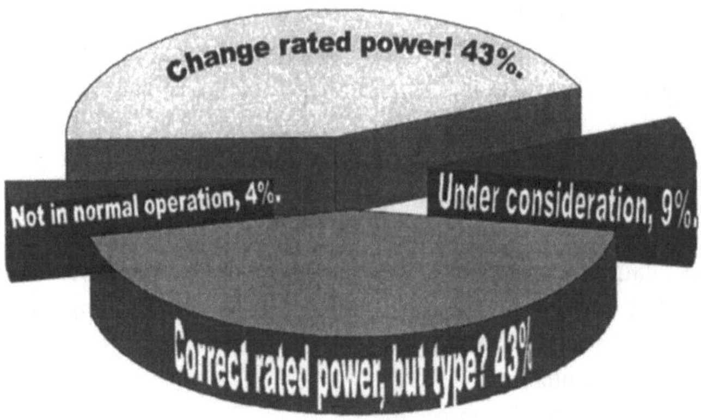

Figure 4: The motor choice, only related to rated power

2.4 Economy

2.4.1 Are there Any Significant Monetary Savings to be Made? And if so, How Much?

The answer is Yes! From the LKAB-study we know that a better motor choice can;

Reduce	Energy	4-8%
	Reactive power	up to 20%
	Investment costs	up to 20%

And if we, in addition, have variable speed drives ? It will be more…!

2.5 Conclusion

Traditionally the dimensioning and selection of electric motors for the industrial installation under study seems not to have been so good, for many reasons. By using better specifications of the real needs, moderate safety margins and better motor selection, costs can be reduced and reliability increased.

But a motor is only one component in a “Drive System”! There is no benefit in making things too complicated for nothing. But it is sometimes advantageous to know more of (all?) parameters involved.

For optimal efficiency a lot of components must co-operate. The advantages of an investment in a HP-motor can be destroyed by in-accurate installation of belts and wheels, for example. (But there are solutions even for that!) Try to make a complete overview when installing a new drive system! System solution, motor-selection, installation, start–up controls, operation, maintenance and repair. It is also a question of reliability!

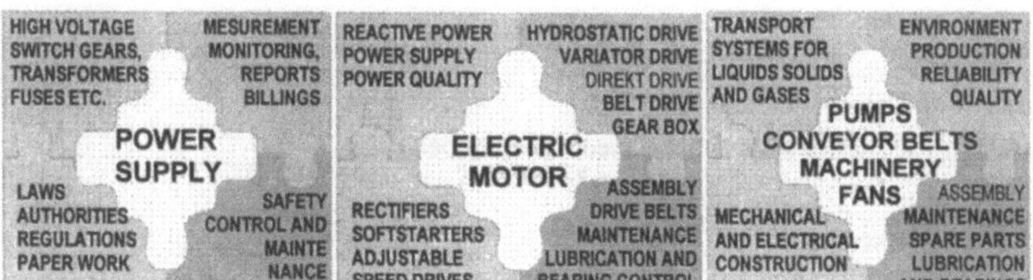

3 Belt Conveyors

Following a request from the maintenance personnel at our sorting plant we measured a conveyor drive powered by two parallel mounted 6kV motors with an installed power of 2x450kW.
The reason for the measurement was that the electrical foreman was interested to compare the load of the motors with the signal measuring the weight of the iron ore.
Measurements of the input power to the conveyor was carried out during normal load for several days.
The measurement showed that the power need for the drive was only 370kW.

3.1 Results

The result of this was that we were able to dismount one of the motors, which resulted in decreased energy and reactive power consumption and of course also lower maintenance.There were eight similar drive systems at this actual plant which meant a lot of energy costs to be saved.

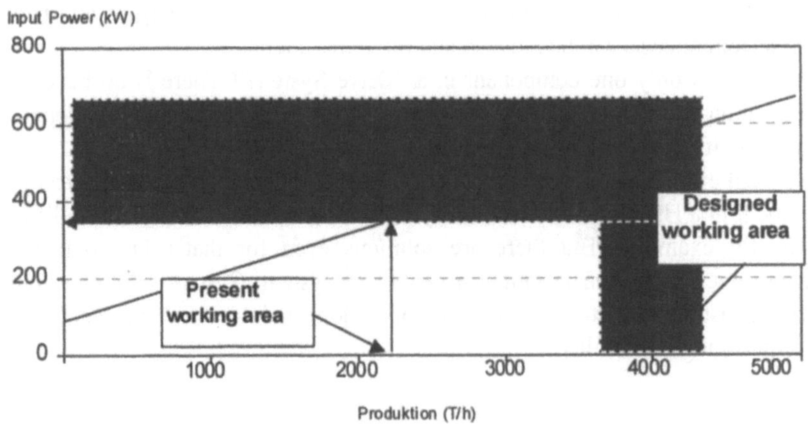

Figure 3.1: Reason for over dimensioned conveyor belt drive

3.2 Conclusions

The explanation for the situation was that the sorting plant and its belt drives had been designed and constructed for a higher production level than that existed at the time of the investigation, and that which we have today. But no-one had reflected about the over dimensioned drives.

4 Belt Drive Systems

Almost by an accident, when we were examining a belt driven pump drive to see whether it was possible to monitor the condition of the pump by measuring the input power, the flow and the density of the liquid, we obtained interesting results which were new for us.

4.1 Introduction

The pump drive 315kW, 1500rpm was powered through an inverter which allowed us to vary the speed and flow.
The normal production is 600 tons per hour.
The belt tension was varied higher and lower than that which was recommended by the belt manufacturer.
The angle between the driving and the driven pulley was also varied.
The belt tension in the pump drive we examined was easy to change (and if necessary, relace) since, the motor was mounted on a GEMEX shelf, a system which LKAB has been using for several years..

Figure 4.1: Motor mounted on a GEMEX shelf

4.2 Measurements

The diagrams shows input power in kW on the y-axes and mass flow in tons per hour on
x-axes.

The following diagrams show the results of the measurements when changing belt tension and angular error between the pulleys.

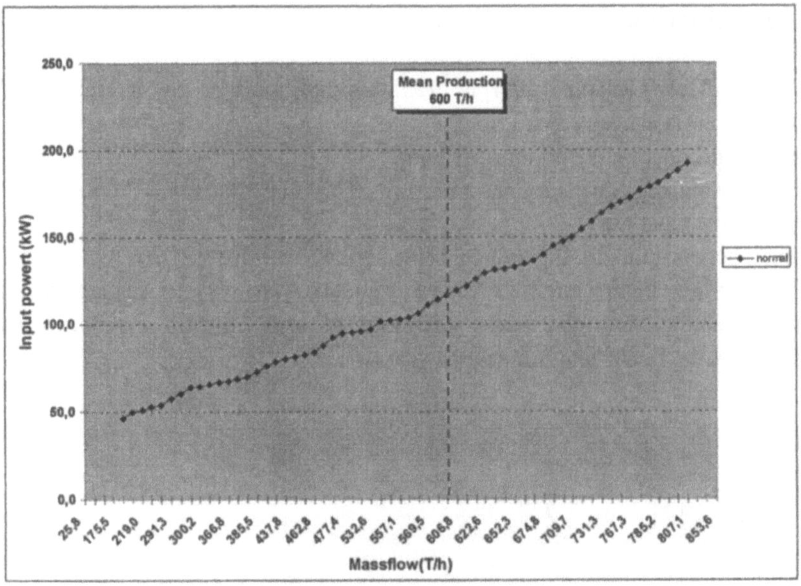

Figure 4.2: Normal belt tension (ie recommended by the belt manufacturer) and the input power is 120 kW.

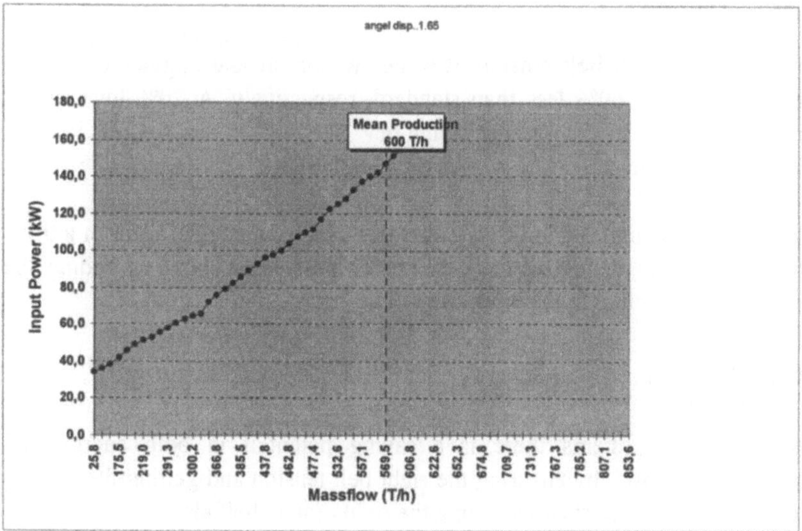

Figure 4.3: An angular error of 1.65 degrees between the pulleys and the input power increases to 160 kW

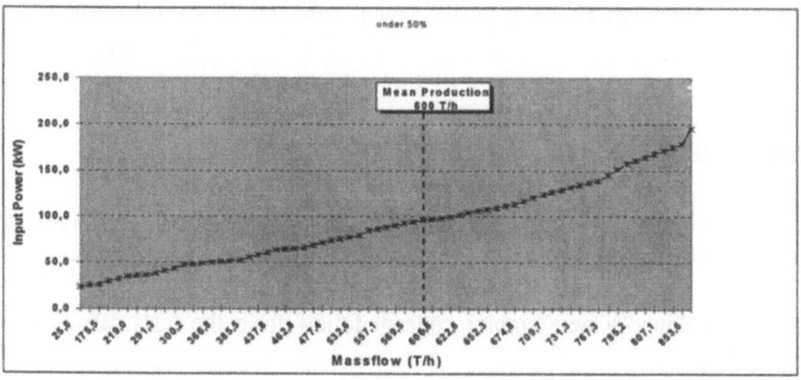

Figure 4.4: The belt tension is 50 % below recommended value and the input power is 100kW

4.3 Results

The input powers with belt tension 50% below normal and angular error 1.65 degrees are 20% and 60% less than standard, respectively. A 20% lower input power means for this drive an yearly saving of
20(kW) x 8000(h/year) x 0.25(SEK/kWh) = 40.000 SEK.

What the measurements also indicate is that if we have an angular error in a drive like this, an increase in production will not be possible without upgrading the whole pump drive including the power supply.

4.4 Conclusions

- In belt drive systems there is a saving potential in both energy and maintenance costs by ensuring the right belt tension and getting the pulleys in parallel and in most cases having the right motor shelf also.

- We are going to make an inventory of all belt drives in the whole LKAB combine.
  A training course will be arranged for all personnel who work with belt drives.
  Then the belt drives will be optimized. An inventory has already started and a rough estimation shows a yearly saving potential of several millions of SEK.

5 Results

I do not believe that Industry in Sweden is, in these respects, different to anyother countries within the EU.

LKAB has developed a new motor concept follow from the tests carried out during the 90s.

In the years following that investigation we have made it possible for most of the employees to obtain an education in awareness of energy consumption, and we have also improved the measurement technology to make it possible to detect if motor driven systems can be improved.

Optimising Fan Systems for Energy Efficiency

W. T. W. Cory

DEng, MSc, CEng, FIMechE, MCIBSE, MIAgrE, FRSH
Technical Director, Woods Air Movement Limited

1 Introduction

For Global warming and many other reasons, the spotlight of efficiency has been directed towards the reduction in energy consumption of all types of machinery, but none more so than in fluid or turbo-machinery such as fans, pumps and compressors. In addition, savings in running costs can usually justify a small increase in first cost, even for the humble fan.

Forthcoming changes to the Building Regulations, will also encourage the installer to design Air Conditioning or Mechanical Ventilation systems to meet defined energy targets. We even have to contend with a new found enthusiasm for 'natural' ventilation. Points will in any case be awarded for plants which incorporate efficient means of flowrate control such that supply and demand can be more closely matched at all times.

This paper attempts to review a number of the factors affecting the efficient utilisation of energy in fans, their systems, their prime movers and their controls. All or some of the following strategies should be considered:

i) Ensure that the plant is only in use when required.
ii) Use some form of capacity control to match the flow to requirements.
iii) Prime movers to be of high efficiency and matched to demand.
iv) Keep plant and motors in largest possible units, consistent with i) and ii) above.
v) Reduce system resistances to a minimum.
vi) Make no unnecessary energy conversions.

2 System Theory

Before attempting to improve the fan, we must firstly consider the system to which it is attached, for on this will be determined the system resistance, and the corresponding pressure to be generated by the fan.

A fan system will consist of straight ducting together with a number of fittings such as bends, change shape pieces, diffusers, branches, inlet or outlet grilles, cowls and guards. These are known as fixed elements. To pass a given air quantity or any other gas through such a system will require a certain pressure equal to the 'system resistance'. The more air which is passed the more pressure will be required. It is observed that for fully turbulent flow this pressure will vary as the air velocity squared. Thus for each element the loss in total pressure p_f equals:

$$p_{t1} - p_{t2} = k\,0.5\rho v^2\ \mathrm{Pa}$$

v(m/s) is the average velocity in the section indicated. k includes allowances for the influence of the component on duct losses downstream and for the duct length occupied by the component. The system loss is the sum of all the $k\,0.5\,\rho v^2$ items. Losses for typical duct fittings are shown in Figures 1 to 6 and emphasise the need for 'easy' bends, gradual tapers and minimum obstructions.

A curve of system resistance can now be drawn where $p \propto v^2$ which for a fixed system where:

$$v = \frac{\text{flowrate } Q}{\text{cross sectional area } A} \quad : \quad \text{then } p \propto Q^2$$

The point of intersection of this system curve with the fan characteristic will determine what flowrate will actually result (Figure 7).

It should be remembered that certain types of system such as fuel beds in boilers or grain in silos do not follow a $p \propto Q^2$ but that $p \propto Q^n$ where n lies between 1 and 2. The accurate analysis of such systems requires reference to fan and control performance curves. Over a limited range, however, they may be compared with fixed element systems.

To repeat, the resistance of a ductwork system is a function of air velocity squared. Obviously velocity should, therefore, be kept to a minimum consistent with the space available. There may be some plants, such as pneumatic conveying where a minimum velocity is essential, to maintain the solids in suspension. These, however, are in the minority. Reducing duct velocity need not necessarily increase the ducting cost unduly. To take the example of straight ducting:

$$p_L = \frac{\rho f L v^2}{2D_h} \text{ where}$$

ρ = air density kg/m^3
f = friction factor
v = air velocity m/s
L = length of duct m

D_h = hydraulic diameter m

$$= \frac{4 \times (\text{cross sectional area of duct})}{\text{periphery of cross section}}$$

Q = flowrate m^3/s

A = cross-sectional area m^2

For a given flowrate this becomes: $$p_L = \frac{\rho f L Q^2}{2 D_h A^2}$$

In circular ducting this can be simplified to:

$$p_L = \frac{\rho f L Q^2}{2 \times 4 \times \frac{\pi d^2}{4} \times \frac{1}{\pi d} \times \left(\frac{\pi d^2}{4}\right)^2} \quad : \text{ where d = duct dia m}$$

$$= 8 \frac{\rho f L Q^2}{\pi^2 d^5}$$

Thus for an increase in duct diameter of 10% then

$$p_L = \frac{8 \rho f L Q^2}{\pi^2 \times 1.1^5 \times d^5}$$

and the resistance, and therefore the power absorbed at constant efficiency, has been reduced to 62% of the previous value for this item.

$$\left[\frac{1}{1.1^5}\right]$$

Considering now a rectangular duct, whose sides have a 2:1 ratio, handling the same flow and having the same metal content/price/periphery then

$$p_L = \frac{\rho f L Q^2}{2 D_h A^2} \qquad = \frac{\rho f L Q^2}{2 \times 4 \times \frac{2a^2}{6a} \times 4a^4}$$

where a = dimension of short side of duct m

$$= \frac{3\rho f L Q^2}{32\ a^5}$$

$$\text{but} \quad \pi d = 6a \quad \text{or } a = \frac{\pi d}{6}$$

$$\text{and} \quad p_L = \frac{3\ \rho f L Q^2 \times 6^5}{32 \times \pi^5 d^5}$$

$$= 2.3822\ \frac{\rho f L Q^2}{d^5}$$

This should be compared with the result for circular ducting where:

$$p_L = 8\frac{\rho f L Q^2}{\pi^2 d^5} \qquad = 0.8106\frac{\rho f L Q^2}{d^5}$$

i.e. the pressure loss and power absorbed in 2:1 rectangular ducting of the same metal content and with the same flowrate will be $\frac{2.3822}{0.8106}$ or 2.94 times as high.

This emphasises the need for ensuring the maximum diameter of ducting consistent with space and money available, and the desirability of using circular cross-sections at all times. Referring also to the section on ducting fitting losses, bends should be of large radii, inlets to systems should be provided with bellmouths and exits should be fitted with diffusers.

3 Derivation of Fan Efficiency

From the preceding sections, it may be deduced that the fan work equals the system resistance. For customer satisfaction this equality should be at the design flowrate.

It can be shown that the fan pressure has the dimensions of work per unit volume. Thus the system resistance may also be regarded as the work required per unit volume of gas. This makes it easy to compute the powers required to maintain the flow of Q m^3/s through a system where resistance is p_s kPa.

$$\text{Power kW} = Q\ (m^3/s) \times p_s(\text{kPa})$$

As an alternative we may say:

work = force x distance

We define power as the rate of work development per unit time and hence:

power	=	force x velocity
	=	pressure x area x velocity
	=	pressure x volume per unit time

The ratio of this 'air' power to the power required to drive the fan is the fan efficiency.

4 The Importance of Aerodynamic Efficiency

Efficiency is a diagnostic parameter in fan design and a high value indicates a satisfactory design for a given duty. Apart from displaying a satisfactory maximum, it is equally important that the fan efficiency curve is maximised over as wide a range of flow as possible, such that fans can be selected to operate within the zone of acceptable efficiency.

The larger the fan the more crucial is the requirement for efficient energy conversion. There are three primary reasons for optimising aerodynamic design:

a) To enable the smallest possible fan to be used for a given duty. For a particular fan type, the manufacturing cost increases at least with the diameter squared.

b) To enable the minimum motor size to be selected.

c) To ensure the most economic running costs.

a) and b) are directly related to the first cost of a fan unit, the importance of c) may be illustrated by two examples:

i) There is an estimated 2.5 million kW of fan power installed in Britain; if the average unit is assumed to operate 8 hours/day and 300 days/year and the average generating cost is £0.04/kWh - a unilateral increase in efficiency of 3% would save:

$2.5 \times 10^6 \times 2400\text{hrs} \times £0.04 \times 0.03 = £7.2\text{m p.a.}$

ii) 100kW of fan power running continuously would consume 0.876×10^6 kWh p.a. At a commercial tariff of £0.05/kWh a 3% improvement in efficiency, with a corresponding reduction in motor power would save:

$0.876 \times 10^6 \times £0.05 \times 0.03 = £1{,}314$ p.a.

These two examples are purely illustrative. The choice of fan type will ultimately be determined by the capital cost, projected operating costs and maintenance costs. Regretfully in the past, the first of these has received undue prominence.

5 Size and Type Limitations to Good Efficiency

High operational efficiencies are only achievable with certain types and sizes of fan. Indeed, the definition of 'good' or 'high' efficiency depends on the class and quality of the fan being considered. Clearly, a 'good' public utility fan will have a peak static efficiency unachievable by a basic HVAC fan. Conversely, what is considered 'good' for a HVAC fan may be moderate or poor for an industrial or large utility fan.

For a particular type of fan the best efficiencies will be achieved by higher specific speed fans of backward curved, backward inclined or aerofoil bladed design, with the fans being medium or large diameter and operating at Reynolds numbers in excess of 20×10^5, where:

$$\text{Fan Reynolds number} = \frac{DU\rho}{\mu}$$

where:

D	-	Fan diameter
U	-	Tip velocity
ρ	-	Fluid density
μ	-	Dynamic viscosity

Low specific speed fans operating at low Fan Reynolds numbers and of paddle, radial or forward-curve bladed design will always exhibit reduced peak efficiencies. These constraints on efficiency are illustrated in Figures 8, 9 and 10. The use of fans of only moderate efficiency must now be questioned, and the trend to backward bladed designs encouraged for all clean air centrifugal applications. Even in dust collecting plant, a change from open paddle designs to shrouded radial fans, could give efficiency improvements of 10 percentage points.

Specific speed N_s for a given fan at a given rating is the speed at which a geometrically similar fan would need to run to produce a fan flowrate of unity and a fan (total) pressure of unity at unit gas density and the same point of rating (often the same efficiency).

Specific (impeller) diameter D_s is the size of the geometrically similar fan referred to above. The figures illustrate the relationships between specific speed, specific diameter and efficiency for various types of fan. It will be noted that if one variable is fixed, then for optimum efficiency, the other variable is determined. It perhaps emphasises that where direct drive is preferred (see later) then the manufacturer must have an arsenal of ranges available, if he is to put forward a unit with the best possible efficiency (Figure 11).

6 Fan Design Parameters

Continuous development by industry has led to a progressive increase in the maximum achievable total efficiencies, from about 75% in 1950 to about 89% currently. It is considered by many that 91% total efficiency is the limit: friction losses, inertial losses and mixing losses can be reduced but not eliminated. The aerodynamic performance of centrifugal fans is dependent on a number of design parameters. Torque is transmitted to the fluid by the impeller blades and it is the blade design features which are most significant. The energy transfer characteristics of an impeller are governed by blade shape rather than blade number, although every impeller will have an optimum blade number for minimum energy loss. In practice, there are so many secondary interactions between the various aspects of design that any impeller needs to be proven by an aerodynamic test.

7 Impeller Loss Mechanisms

Improvements in fan design are monitored by increased efficiency, through a reduction in impeller losses. The causes of energy loss are distinct and various, and it is the reduction of these which forms the basis of fan technology.

The principal impeller loss mechanisms for centrifugal fans are listed below, noting that individual mechanisms may affect pressure development, inlet volume flow or fan absorbed power:

i) Entry losses incurred in changing the predominant airflow direction from axial to radial. Sensitive to inlet design and shroud lip curvature.

ii) Impeller inlet losses caused by a mis-match between blade angle and relative flow angle. Sensitive to blade inlet angle α_1. Can also lead to a substantial reduction in throughflow if a stall zone blocks the inlet and diverts the substantially radial flow velocities, see Figure 12.

iii) Clearance losses incurred by the leakage of air between the inlet and the impeller. Power losses are due to the re-entry of discharged fluid back through this running clearance; pressure and volume flow losses may be caused by an interaction between the recirculating flow and the mainstream inlet flow.

iv) Diffusion losses as the kinetic energy of the discharged air must be efficiently converted to static energy within the volute. Pressure losses are caused by flow separation within the casing. A volute shape is only optimum at one flowrate and for fully turbulent flow conditions the diffusion loss increases with the square of a representative flow velocity.

v) Skin friction losses related directly to the formation and growth of boundary layers on the impeller surfaces.

vi) Friction losses within the impeller arising from the passage of the fluid. These losses are due principally to flow separation from the negative pressure blade faces, he formation of eddies within the fan-entrained fluid volume and turbulent mixing within the throughflow.

With axial flow fans, experience has shown that there is an optimum hub/tip ratio for each value of specific speed. High pressure fans require high ratios, whilst low pressures require only low ratios. The optimum number of blades may also be a function of specific speed. Blade profile should preferably be as thin and as smooth as possible indicating the superiority of die castings. The axial fan casing should be cylindrical over the blades with clearance between the blade tips and the casing skin the minimum practical, allowing for the requirements of any growth due to operation at temperatures above ambient. A spun casing and computer generated arms will enable these clearances to be closely controlled.

Fan selection has been governed for too long by considerations of price and size. Selection of higher efficiency types with non-overloading power characteristics will repay the difference in first cost many times over during their working lives. They are essential where the system resistance cannot be accurately assessed and where a close match between fan and motor is desired. The pressure/flowrate curve of such fans is relatively steep so that the effect of errors or variations in pressure demand has a small effect on flow.

8 Selection of the Prime Mover

Having calculated the power absorbed by the fan, we are then in a position to select a suitable driving motor. With small units this is usually straightforward as there will be no difficulties in starting. In the analysis which follows, however, we will consider all the various factors which can affect the selection.

a) Power absorbed by the fan

This will be obtained from the duty requirements of air/gas volume flow, pressure to be developed, and known air/gas conditions at fan inlet. It is also necessary to consider how all these factors may vary during fan operation. For example, it is usually difficult to assess accurately the fan pressure. The system designer often, therefore, adds a 'safety margin' to his calculated pressure to ensure that he achieves the design flow. If he can subsequently add in additional resistance by orifice plates or similar to bring the flow back to specification, then we believe there will be no problem. Alternatively he may be able to partially close a damper in the system to dissipate the unwanted pressure. It should, however, be realised that this so called balancing is wasteful in power and must be avoided wherever possible. If resistance cannot be added and the speed cannot be changed, then the fan will handle more air and this may affect the power consumption. With 'non-overloading' fans such as the backward inclined, backward curved or aerofoil the volume flow against power curve is relatively flat over the working range, i.e. an increase in capacity with reduced pressure has only a small effect, if any, on the power absorbed. With impellers having blades radial at the outlet, i.e. shrouded radial, open paddle, backplate paddle and radial tipped, the power increases uniformly with capacity. The forward curved impeller has a flow versus power curve which increases ever more rapidly towards the 'free air' or zero pressure condition. For this reason larger margins have to be added to the fan absorbed power simply to cater for the inevitable errors in system resistance calculations.

It is also important to know if the power absorbed can vary with time. In a ventilating system with a fan handling 'outside' air, the only variation will be that due to a variation of air density with changing barometric pressure or ambient temperature. Calculations of both fan duty and system resistance are normally made under 'standard' conditions, i.e. with air having a density of 1.2kg/m^3. Typically this would correspond to air at a temperature of 16°C 100kPa barometric pressure and 62% relative humidity. Between summer and winter there will be variations in both temperature and barometric pressure and these will affect the air density. The air temperature at the fan inlet could fall to -3°C (270°K) and barometric pressure could rise to 105kPa. The effect on air density would then be:

$$1.2 \times \frac{105}{100} \times \frac{273+16}{273-3} = 1.35\text{kg}/\text{m}^3$$

i.e. an increase of 12%. If such variations in conditions do occur, then the necessary margin has to be allowed. This could result in the motor efficiency under normal load being reduced. A possible alternative is for the motor to be 'overloaded' for short periods of time. This is not necessarily a danger where the motor is in the airstream as the performance (usually determined by winding temperature) can improve at low temperatures.

b) Electric motor design

Whilst a very small number of fans may be driven by prime movers such as steam turbines or petrol engines, the vast majority - in excess of 98% - are driven by electric motors. With axial flow fans, it is common for the fan impeller to be mounted directly on the motor shaft extension. Centrifugal fans, may, of course, be vee belt drive, directly driven or through a flexible coupling with or without an intermediate gearbox (common in the UK on large mine ventilation fans).

Again, with the majority of fans, electric motors are of the totally enclosed squirrel cage induction form suitable for a three phase supply. Single phase motors are usually limited to fractional horsepower outputs.

The AC induction motor can be easily adapted to variable speed output by inverter control and this is perhaps the most popular form of flowrate variation. According to the type of control and its effect on the electrical wave form, this may be distorted sufficiently from the ideal sinusoidal shape that motor noise may increase and efficiency reduce as the speed is decreased. Selection of the relationship between voltage and frequency is of especial importance particularly where the fan is of lightweight design, or where it is built onto a flimsy structure. It is usual to select the so-called fan torque variant where $v \propto f^2$. If the transmission efficiency also varies with speed (as in a vee belt driven fan) this may not, however, be ideal.

c) Selection of correct motor speed and type

During the early post war years, when energy was cheap, it became accepted for motors to be designed with ever smaller frames, higher temperature rises, and consequent reduced amounts of active material. Fans also were subjected to the same thinking, and they were designed for lowest capital cost. This led to reduced diameters and increased blade widths with an ability, where direct driven, to run at 2 and 4 pole motor speeds.

In the late 1990's, it is suggested that this trend ought to be reversed. A re-emergence of larger fans, driven at slower speeds, and with a high efficiency across a greater portion of the characteristic, should be encouraged. The decrease in lifetime costs, where energy consumption and maintenance costs are added to the capital cost will be readily apparent. The matching of motors and fans, and the choice of control system concepts to permit optimisation, should become a design priority.

The table below emphasises how motor efficiency and power factor at full load may vary with speed for a typical manufacturer's 45kW motor of standard or 'energy efficient design'. The increase in fan efficiency with reduced speed could well be larger than that for the motor.

Frame size	Poles	rev/min	Full load eff %		Full load p.f.	
			Std	Energy Eff	Std	Energy Eff
225M	2	2955	90.0	93.9	0.85	0.91
225M	4	1470	92.0	93.8	0.86	0.87
250M	6	980	92.5	94.0	0.84	0.83
280S	8	735	92.0	94.0	0.79	0.75
280M	10	589	91.0	94.3	0.73	0.75
315S	12	490	90.5	94.8	0.73	0.75

Wherever continuous running is a requirement, the increased efficiency of both larger and slower units is apparent. Energy efficient designs can repay their increased first cost many times over. When allied with the power savings possible from high efficiency fans (often of a much higher magnitude - there are many fans with a 60% peak fan efficiency which could be replaced by 80% efficient units), the effects on overall electrical demand could be enormous.

Perhaps of greater importance, especially where a fan works for considerable periods at reduced load, is the 'flatter' motor efficiency curve (Figure 13).

d) Vee belt drives

The major fan manufacturers have families of designs covering a range of specific speeds and diameters. These ensure that all duties can be met by a fan running at a direct drive speed often with the impeller mounted on the motor shaft, thus eliminating the losses in vee belt drives. Greater effort should be made in the accurate determination of system resistance (how many enquiries for fans give an obviously rounded pressure of 500Pa?) Thus the need for pulley changing will be obviated, a worthwhile reduction in first cost result, and the transmission losses will be eliminated. Where belt drives are inevitable, greater thought should be given to their selection. It should here be noted that if a vee drive is either under or over engineered, efficiency will reduce as shown in Figure 14. Flat and toothed belts do not suffer to the same degree, whilst their peak efficiency is usually higher. With very small drives, the difference in power transmitted between say one and two vee belts or between two and three is obviously substantial. Figure 15 may be used to estimate the losses in a standard vee belt drive. Such losses will need to be added to the fan power to determine the power required for the motor. The use of larger single units is indicated, if transmission losses are to be reduced.

Especial attention is drawn to the increased losses in vee belt drives when used at lower than their maximum rating on variable speed units. Whilst the fan impeller power on a constant orifice or fixed system will vary as rotational speed ^{3}, the losses in the bearings and shaft seals will vary approximately as the speed. Drive efficiency will fall from about 94% at rated full load to around 70% at half load so that the full 'cubic' saving is not achieved where vee belts are fitted (Figure 16). Again the preference for direct drive fans with variable speed is indicated.

9 Considerations in Fixed Output Systems

a) Considerations in fixed output systems

Almost without exception, both fans and motors can be made more efficient in larger sizes. Where a number of units have to run continuously, consideration should be given to providing a manifold or plenum to couple up the duct systems and use one common fan/motor unit. Even allowing for the additional duct resistance, power savings can often be made. The tables below are typical but emphasise the improvements possible. They show how motor efficiency increases with increasing motor frame size whilst fan efficiency also increases with fan size due to 'scale' effects.

i) **Motors**

Frame size	Motor kW	Rev/min	Efficiency %	Power factor
63	0.25	2800	67	0.83
71	0.55	2800	71	0.85
80	1.1	2800	76	0.85
90L	2.2	2820	78	0.88
132S	5.5	1440	85	0.83
160M	11.0	1440	88	0.85
200L	22.0	970	90	0.80

ii) **Fans**

Impeller dia. mm	Rev/min	Peak fan static efficiency %	Relative tip speed factor
630	2880	73.8	1.000
800	2270	74	0.995
1000	1815	75	0.992
1250	1440	77	0.989
1600	1125	78	0.987

b) Fixed speed motors

These drives are almost invariably from three phase induction motors. They should be matched to the fan to absorb as near nameplate load as possible, for both efficiency and power factor are likely to fall at reduced load. Multi pole slow speed motors can often be used to eliminate mechanical reductions, lower power factor can be compensated with additional capacitance. Efficiency as distinct from power factor, does not fall significantly at higher pole numbers.

10 The Need for Flowrate Control

Every fan is selected and installed for a given flowrate and system pressure, but there will be many occasions when the demand will not be at this design maximum. Boiler induced draught units will have to cope with gas flows varying according to the amount of fuel being burned and therefore the boiler output. A fan on a grain drying installation will have to blow through more crop as the harvest progresses. On a ventilation plant there may be differences between winter and summer duties whilst on VAV (variable air volume) systems the fan capacity and system requirements must be continuously balanced.

For these and many other examples, the fan manufacturer needs to provide or advise on capacity control systems. Before considering specific cases, it is necessary to determine how the system demand may vary as on this will depend not only the best method of control to us, but also which type of fan is most suitable. To operate the fan at a higher rate than necessary is wasteful in energy. Whilst increasing the initial cost, fan control systems will usually more than pay for themselves over the life of the fan.

The manner in which fan demand may vary can be categorised as follows although combinations of these are also possible.

a) Constant orifice systems

In these the plant remains unchanged, but the air flowrate through it may need to vary. When there are fixed elements such as straight ducting, bends, takeoffs etc., and the flow is fully turbulent, then we may apply normal systems resistance 'laws'. Thus system pressure p_S (flowrate Q)^2. If the capacity control is to maintain its efficiency constant then as fan power

$$P \propto Q \times p$$

$$P \propto Q \times Q^2$$

or $$P \propto Q^3$$

As fan capacity $Q \propto N$ it will be seen that speed variation is the optimum solution provided that power source efficiency can also remain constant over the range required. With AC electric motors, good efficiencies are maintained down to about 50% power (i.e. 80% fan flowrate).

b) **Parallel path systems**

Here the airflow may vary, but pressure required remains virtually constant. Examples which come to mind are mechanical draught systems where one fan may cater for more than one boiler. As boilers are shut off or started up according to demand, so the gas flowrate will vary. Provided the common ductwork is short, however, the pressure drop through each boiler and therefore through the system will remain unchanged. If the capacity control is to maintain a constant efficiency then as

power $P \propto$ flowrate Q x fan pressure p_s

i.e. $P \propto Q$

Similar situations can arise in central extract systems where dampers in parallel ducting legs may be shut according to whether a machine is or is not operating and therefore emitting dust or fume. With an on-floor grain drying plant, the floor area to be ventilated will increase as the harvest progresses but at constant grain depth and drying rate the pressure demand would remain unchanged.

c) **Series path systems**

The airflow needs to remain fairly constant, but pressure required will vary. For a fan ventilating a tunnel during construction, the air requirements at the working face will remain constant, depending only on the number of men working and air required to cool or supply the machinery. The length of ducting taken to a fresh airsource will, however, increase as the work progresses. If the fan control is to maintain a constant efficiency then as

power $P \propto$ flowrate Q x fan pressure p_S

i.e. $P \propto p_s$

Similar situations can arise in drying plants with bottom ventilated bins where pressure will increase with the depth of bed.

d) **Variable air volume systems**

In a VAV system, as applied to the air conditioning of a building environment, the airflow rate to each separate room or occupied space is varied both individually and continuously. Thus the instantaneous cooling demands of a room may be satisfied. Such a system is shown in Figure 17 and consists of a central unit (1), ducting (2), flow variators (3) and supply air terminals (4). Each flow variator is controlled by a room thermostat (5) and demands a constant pressure in the ducting. This is maintained by the pressure transducer (6) which controls the fan flowrate by altering fan speed, inlet guide vane angle, disc throttle position, impeller pitch angle or such other method of flow variation as installed.

The system pressure required may be divided into three main parts:

p_a : Pressure drop in the air handling unit, which varies generally as the square of the fan air flow (any filters p_f may be $\propto Q$) $p_a \propto Q^2$

p_b : Frictional pressure drop in the ducts, which varies as the square of $p_b \propto Q^2$ the air flow.

p_c : Constant pressure drop across the flow variator. This can amount to between 10% and 50% of the total pressure drop in the system. $p_c = c$

Reference to Figure 18 shows that the resulting system curve of "orifice" is far from the usual square law relationship where $ps \propto Q^2$. When assessing the suitability of the fan we must, therefore, consider that the resultant $p_s = (p_a + p_b + p_c) \propto Q^2 + c$.

It must be emphasised that no type of fan flowrate control is applicable to all installations. The type selected will depend on the turndown ratio required, how the system resistance varies and the presence of contaminants or high temperatures. Some of the features and advantages/disadvantages of the various designs are detailed below.

11 Damper Control

Radial inlet vanes are considered later, for whilst they also act as a damper, their major intention is to promote pre-swirl into the fan inlet. Thus, they materially affect the fan geometry and an additional power saving results. A comparison of the flowrate control for various types of damper is shown in Figure 19.

12 Variable Speed Control

Where high efficiency fans, such as centrifugal units fitted with backward inclined, backward curved or aerofoil impellers or premium efficiency downstream guidevane axial flow fans are installed on constant orifice or series path systems, then reduction in flowrate by varying the speed is preferred. In this way full advantage can be taken of the fan characteristic without sacrificing the inherent low energy demand. It should be noted that speed variation is not usually suitable for parallel path systems due to the reduction in pressure developed ($P_s \propto Q^2 \propto N^2$) with decreasing flow.

Whether speed variation can be used on VAV systems will depend on the fixed element of system resistance due to the flow variator. Where this is 10% of the total fan pressure at maximum duty the it is acceptable, but at 50% the variation in flowrate possible will probably be unacceptable.

Suitable prime movers for variable speed include:

i) AC electric induction motors with inverter drive.
ii) Slip ring and commutator type AC electric motors.
iii) DC electric motors.
iv) Variable vee rope drives with AC electric motors.
v) Steam turbine and reciprocating motors.

Multi-speed dual wound or pole changing electric motors can be used when the operating requirements are clearly defined. For example there may only be specific winter and summer, or continuous and overload, duties to be met. In conjunction with damper control, a wider duty variation is possible, and this combination is often a very simple solution to the problem.

Where continuously variable control down to about 50% design flowrate is required, the economy achieved by slip couplings of the eddy current, scoop control fluid, or powder type may be indicated. There is the additional advantage of improved starting by gradually 'letting in' the fan inertia. In all such cases close consultation between system designer, fan manufacturer, and coupling manufacturer, is necessary to achieve the best results in energy saving.

A steam turbine drive with a gearbox to give optimum matching of fan and turbine speeds is usually the most efficient. It is only considered for industrial applications, however, where a suitable steam supply is available.

The table below shows approximately drive efficiencies for a 15kW input at ¼, ½, ¾ and full speed.

Prime mover and control	Range	Usage	Approx. efficiency ¼ N	½ N	¾ N	N
DC motor with AC input through thyristor control	Speed and torque adjustable over range	Speed reduction required at full torque	70	86	88	89
AC motor with variable vee rope drive	Speed adjustable down to $^1/_6$ N. Torque increases with decreasing speed	Infrequent requirements for speed reduction	70	80	83	85
AC motor with inverter voltage and frequency control in rotor circuit	Speed adjustable down to $^1/_{12}$ N.	Torque reduces substantially with speed	50	60	77	85
AC slip-ring motor with resistance control in rotor circuit	Full speed control only possible down to ½ N.	Minor speed adjustment or easy staring as control losses substantial	22	45	67	89
AC motor with slip coupling	Adjustable over whole range with electronic equipment and tachometer generator	Limited speed reduction and easy staring	20	41	62	82
Steam turbine and gearbox with variable supply	Adjustable over whole range but requires suitable steam supply	Good speed reduction but requires gearbox to match optimum turbine speeds	70	86	88	90

It is not always realised that centrifugal impellers of backward bladed design, whilst shown on performance data as having a smooth continuous characteristic of pressure against flow, often have a small order discontinuity close to their peak pressure point. This discontinuity usually increases with impeller width and is the result of a rotating stall 'cell' between adjacent blades. Manufacturers try to obtain the maximum airflow from a given casing size by incorporating wide impellers. This results in the performance being obtained with the smallest space envelope.

For a given resultant pressure rise there is a relationship between the blade inlet and outlet radii. The inlet cone throat diameter is dictated by the blade inlet diameter. Thus there is an optimum width of impeller for the correct inlet throat area/impeller inlet blade area ratio. An increase in this value will result in the impeller being susceptible to inlet disturbance and the resultant discharge airflow may contain disturbing pulsations. These can be difficult to deal with, and the downstream ducting may become 'live' to low frequency vibrations.

The area of instability is shown in Figure 20 which also indicates a typical VAV system curve. If speed control is used as a means of modulation then entry into this unstable area is inevitable. This is often overlooked with the availability of low cost (but lower efficiency) prime movers.

13 Variable Geometry Fans

The possibilities for varying the fan geometry are limitless. Many exotic methods have been tried on both centrifugal and axial flow fans. In all systems, the intention is to vary the inlet/outlet velocity flow triangles. At inlet, pre or contra swirl of varying amounts may be induced by the use of variable angle radial vane inlet controls. They have been used extensively for over forty years with backward inclined or aerofoil bladed centrifugal fans where they have proved particularly successful, and also with axial flow fans where the simplicity of a non-rotating control has been desired. The operating range at high efficiency with axials is, however, somewhat narrow. Mixed flow fans are becoming more popular and again this method of control is widely used.

For axial flow fans, the alteration of impeller blade pitch angle at rest has been available for many years but over the last two decades the means of varying the pitch angle in motion has extended from the high technology mine ventilation and mechanical draught installations into the more humble HVAC plant. This is now seen to be an extremely efficient and versatile form of control, rivalling the inverter drive on constant orifice systems. It also gives useful power savings on variable flow/constant pressure and constant flow/variable pressure systems.

Other less popular methods of centrifugal fan control have consisted of variable angle impeller tips and a rotating plate attached to the impeller backplate which can vary its axial position and, therefore, the impeller blade width. A cylindrical drum moving axially over the impeller periphery to achieve the same result has been more extensively used in North America.

The Radial Vane Inlet Control (RVIC) for centrifugal fans, and the Variable Pitch-in-motion Axial fan have, to date, been the most widely used and are now described in more detail, along with the Disc Throttle:

i) Radial vane inlet control

The full pressure development of a fan is achieved only when the air enters the impeller eye axially and without swirl. If the air or gas entering the fan is already spinning in the direction of impeller rotation, the fan will develop less pressure.

Both flowrate and power absorbed will thus be reduced. It is the purpose of this control to induce pre-rotation. In effect, it alters the design pressure/flow characteristic whilst largely maintaining the fan's efficiency. Thus the power consumption can be considerably reduced with lowering fan capacity.

Radial vanes are most effective with backward bladed high flowrate fans where the pressure curve rises considerably above the duty condition, the power is non-overloading, and the impeller inlet velocity vectors are of such a magnitude that they can be materially affected. With other types, especially the forward curved, power savings are not nearly so great and often only marginal.

The relationship between control arm movement and flowrate reduction is intermediate between the two previous types. Such dampers should never be used on direct pneumatic conveying or high dust burden extract systems as they require many parts within the air/gas stream subject to erosion and/or corrosion.

A typical performance characteristic for a backward aerofoil centrifugal fan is shown in Figure 21. Super- imposed are the effects of the various types of system and thus the energy savings achieved. It should again be noted that a typical VAV system will have a system characteristic intermediate between the parallel path and constant orifice systems. When the fixed element of system resistance is a large proportion of the total, then the power savings will approach those for parallel paths, whilst if it is small, then the power saving will be similar to that for a constant orifice system.

The mechanical design of the vanes and particularly the mechanism can cause problems because of the need for continuous maintenance and greasing. This is due to the high friction and corresponding high operating torque required for the operation of the actuating mechanism. This mechanism usually comprises an external ring and a number of actuating levers, one lever for each vane (see Figure 22). The vanes are supported by a larger hollow collar at the centre to allow the fan shaft to pass through. The ring is normally supported by a number of rollers. The actuating levers are connected to the ring via double links to overcome the great differences between the paths of the levers and the external ring. It is such mechanical problems which have resulted in doubt as to their reliable operation for VAV systems, especially as the fans are often of double inlet design necessitating cross linkage between the two assemblies.

As with speed variation, when considering the use of RVICs as a means of control, then the resultant area of instability may lead to problems. In 'wider' impeller designs this area can be large, see Figure 23. This has lead some to claim that one should not consider their use if a flowrate of less than 50% of design is required. Below this ratio simple damper control would have to take over, with its resultant inefficiency. However, by correct impeller/RVIC design selection, the area of instability can be very small with modulation over the entire VAV system curve totally stable. Normally a turndown to 20% can be achieved with a single speed drive motor and this is generally sufficient for VAV system use. By using a two speed fan, operation down to 10% of design is feasible.

It should be noted that due to the relatively large clearances necessary at the centre support, zero flow is impossible and even with complete closure there will be a leakage of up to 8%. As well as inducing pre-swirl, the RVIC imposes an additional and increasing resistance as the vanes approach full closure. This is the explanation for the corresponding reduction in efficiency, as this loss of energy is then attributed to the fan/RVIC combination. RVICs are very expensive and the price for two fitted to a double inlet fan can even exceed the price of the bare fan itself.

ii) Variable pitch-in-motion axial flow fans

One of the most important parameters in the design of any turbo machine is the angle which the outer edges of the blades make with the tangent of the peripheral motion. As this angle is increased, so the volume flowrate will also increase, and this applies to axial, mixed flow or centrifugal fans. At the same time the pressure, which is a function of the swirl, remains substantially constant.

It will, therefore, be seen that if the pitch of the blades of an axial flow fan could be altered in motion, then an effective method of volume control would be available. The technology to do this already existed with the aircraft propeller, albeit where the number of duty hours was considerably less than the humble ventilating fan. Nevertheless, over the last few years, the systems necessary have been simplified to enable a sufficiently reliable fan to become available for normal HVAC applications.

As previously stated, only variable pitch axial fans can adequately meet the needs of constant orifice, constant flowrate or constant pressure systems. The energy savings made have been amongst the highest achieved, and a reasonably good efficiency is maintained over a turn-down of 4:1. The aerodynamic performance of such fans is, of course, similar to normal adjustable pitch-at-rest axials and a typical characteristic is shown in Figure 24. The centrifugal force on an individual fan blade can be considerable and is a function of the blade weight and its rotational speed. For a typical application, this force can be as much as 600 times the dead weight. These forces are usually resisted by anti-friction bearings

of the ball or roller type. Such bearings have a lower capacity under the virtually static conditions prevailing, and in the early days failure was not uncommon. With increasing experience, however, the problems have been overcome.
Levers at the base of each blade convert the equal pitch angle adjustment into axial movement of a sliding member within the impeller hub. This may be controlled in a number of ways:

a) By movement of pneumatic bellows against a spring as shown in Figure 25. The bellows are expanded by compressed air through a rotary air seal onto a shaft extension.

b) By an actuator (either pneumatic or electric) giving axial movement through levers to the stationary race of a ball thrust bearing, the revolving race being coupled to the sliding actuator within the hub.

In all cases, when the fan is running, a force must be applied to each blade to maintain the required pitch angle or it would rotate to a position near zero pitch angle where the centrifugal forces on it were in balance. Weights are sometimes attached to the blade root, at right angles to the blade pitch, to produce a counterbalancing moment and thus reduce the actuating force necessary. In the event of compressed air supply failure, the flowrate will, of course, revert to minimum unless some alternative is available.

iii) Disc throttle (UK Patent No. 2,119,440B)
The unit comprises a profiled circular plate supported co-axially within a centrifugal impeller. It is necessary for the inner edges of the blades to be parallel to the impeller axis so that a close clearance can be maintained with the periphery of this disc throughout its movement. The plate is carried by an axially extending shaft which projects outwards through the inlet venturi and is moved axially by means of an actuator of any convenient kind. The actuator is supported from the fan casing by suitable brackets or rods and where the travel is particularly long, an additional sliding bearing may be incorporated to support the shaft. A cross-section of the arrangement is shown in Figure 26.

Movement of the rod alters the position of the disc axially with respect to the impeller's blades and this effectively controls the flowrate by varying the active width of the blades. The disc does not rotate and it will be seen that there are, therefore, a minimum of moving parts. This produces an inexpensive device, and a high efficiency is maintained for a considerable turn-down. A soft rubber ring can be attached to the outer edge of the disc so that when the damper is withdrawn up to the venturi, the inlet flowrate is almost zero. Conversely, with the plate close to the impeller backplate, the flow is at a maximum and almost the same as that for a fan without a Disc Throttle.

Stability of the fan increases as the active width of the blades is reduced, as shown in Figure 27. This, therefore, permits the control to be used with very wide impellers to achieve the maximum flowrate from a given space envelope, without the risk of entering the stall range. Its availability and effectiveness has encouraged the development of a special range of impellers having dimensions calculated to make the best possible use of the Disc Throttle. The control offers a substantial energy reduction compared with conventional dampers. There is also an additional power saving compared to Radial Vane Inlet Controls. With the damper plate acting on width, operation is unaffected by blade shape and these may, therefore, take many of the forms commonly used in centrifugal fans, such as backward inclined, backward curved, aerofoil, shrouded radial and radial tipped. Flowrate control is substantially linear over a wide range. Even forward curved bladed fans may be fitted when an additional power saving over normal dampers is made, albeit small, in contradistinction to the Radial Vane Inlet Control.

Again, with narrower width high pressure impellers, the power savings become less but the other advantages outlined remain. The Disc Throttle is a competitive solution to all centrifugal fan flowrate control problems.

14 Conclusions

The advantages of a good fan efficiency are clear - low running costs which can lead to the additional capital cost being recouped in a very short period of time - often less than two years. A high efficiency impeller may not necessarily be more expensive as, with a reduction in internal losses, the fan may even be reduced in size for a specific duty.

No form of flowrate control is applicable to all types of system and care must be taken to distinguish between different types of system. Speed control by the use of inverters with induction motors is not a universal panacea.

Graphs of the type shown in Figure 28 are common, but attention is drawn to some of the assumptions made and to the fact that they are only applicable to constant orifice systems, where $P \propto Q^3 \propto N^3$. It must be appreciated that they are approximate and that they refer to specific items of equipment. The general conclusions are, however, valid. In the analysis, the backward bladed fan has an assumed static efficiency of 80%, whilst for the forward curved and variable pitch axial, this is 60% both at the design flowrate. The differences would be smaller if both axials and centrifugals were selected on a total pressure basis as recommended in the fan test standards ISO5801/2. Special attention is drawn to the use of wide backward bladed centrifugal fans with 2 speed (dual wound 4/6

pole shown) motors and disc throttle dampers. This is a relatively cheap installation rivalling more sophisticated methods in its control efficiency.
DC motors with thyristor control probably surpass all others, but AC motors with inverter drives are almost as efficient and much more reliable. Speed control, whilst the preferred method for constant orifice or fixed systems, and also usable in many constant flow systems, is not applicable to constant pressure systems.

You would expect a fan manufacturer to say it, but more care should be devoted to selection of appropriate equipment. Where comparisons are to be made on the basis of absorbed power, performance certification schemes such as AMCA become necessary and performance data needs to be independently validated.

Remember that:

$$P \text{ (Power input) kW} = \frac{Q \times p_r \times 1000}{\eta_f \times \eta_m \times \eta_t} \quad \text{where :}$$

Q	=	flowrate	m^3/s
p_f	=	fan (total) pressure	Pa
η_f	=	fan (total) efficiency	%
η_m	=	motor efficiency	%
η_t	=	transmission efficiency	%
P	=	input power	kW

The need to avoid unnecessary energy conversions is obvious, and direct drive fans should be considered wherever possible.

ETSU, BRECSU and others can take justifiable pride in the manner in which they have brought to public attention, the reduction in running costs by changing from normal to high efficiency motors, when a saving of perhaps 5% can be made. How much greater would be the savings if the many fans with impeller efficiencies of 50 to 60% were changed for units having efficiencies of greater than 75%.

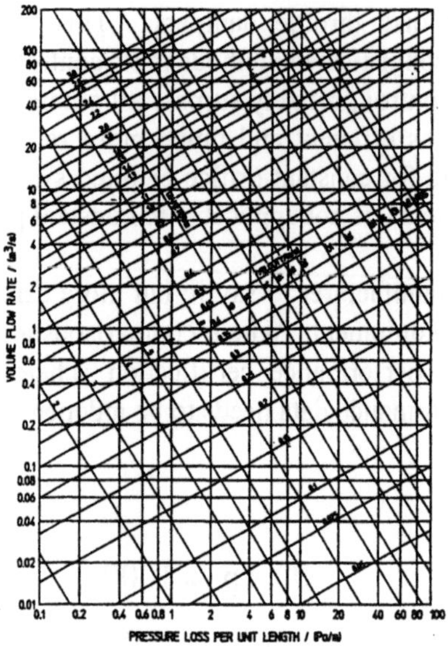

Fig 1 Flow of air in circular ducts

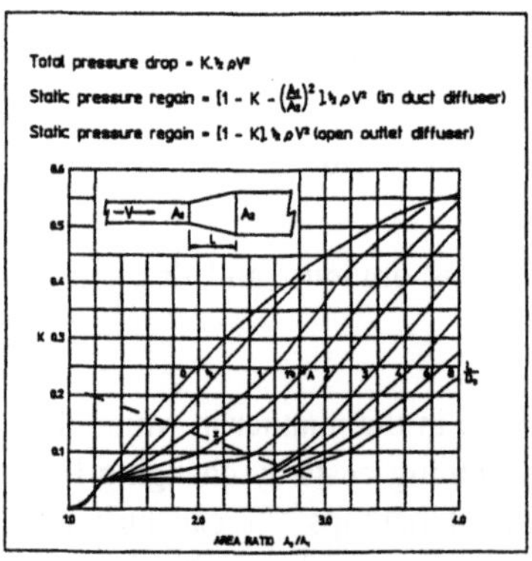

Fig 2 Losses of in-duct diffusers

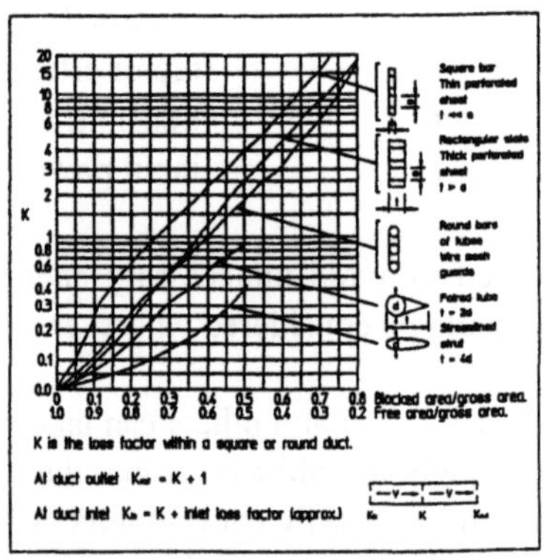

Fig 3 Losses at guards, screens and obstruction

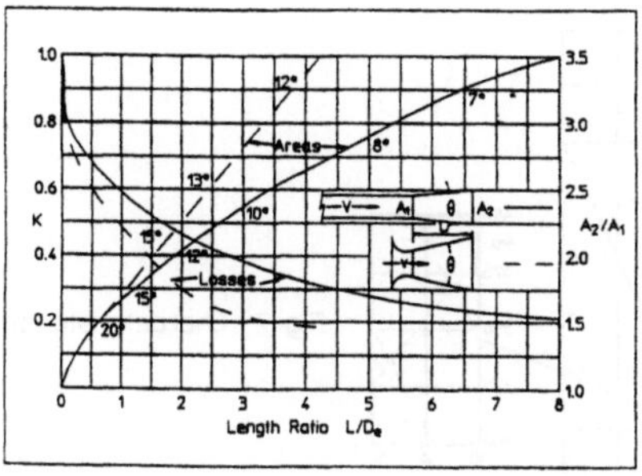

Fig 4 Losses of open outlet diffusers of optimum proportions. (Applies to circular/circular or rectangular/rectangular expansions.

Fig 5 Losses at right angle bends in ducts of circular section

K for bends turning through angles other than 90° can be expressed approximately as a percentage of the 90° value.

Angle of bend	15°	30°	45°	60°	75°	90°	120°	180°
Easy bend (R/D > 1)	17	33	50	67	83	100	115	125%
Sharp bend (e.g. mitre)	5	15	30	50	75	100	140	- %

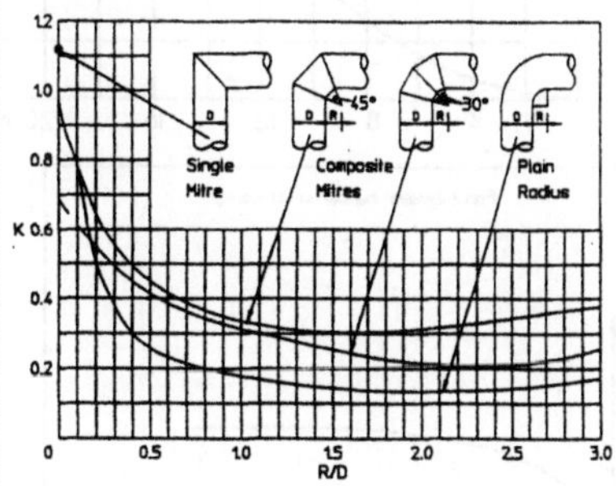

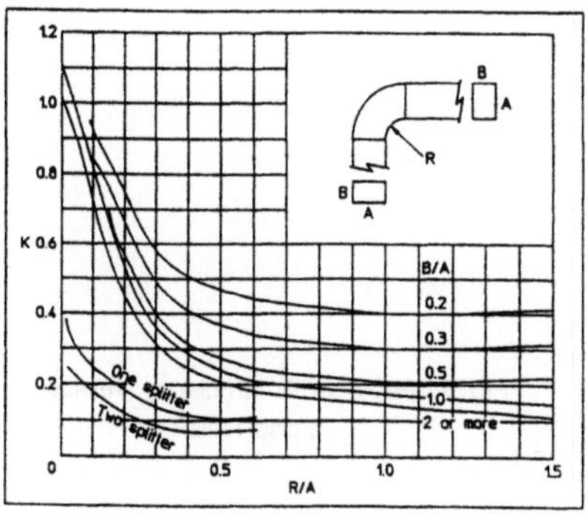

Fig 6 Losses at right-angle bends in rectangular ducts

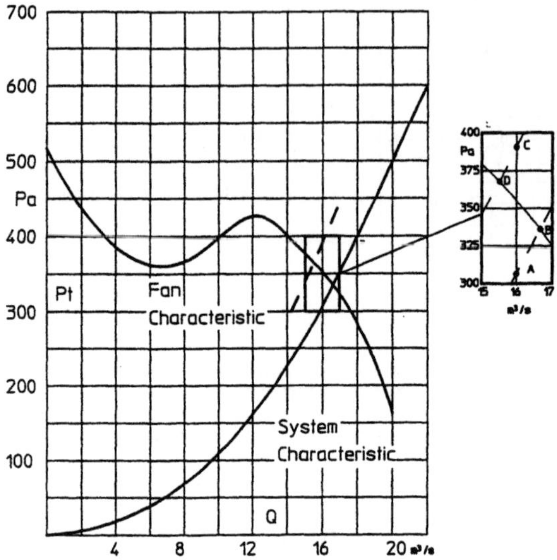

Fig 7 Characteristic curves

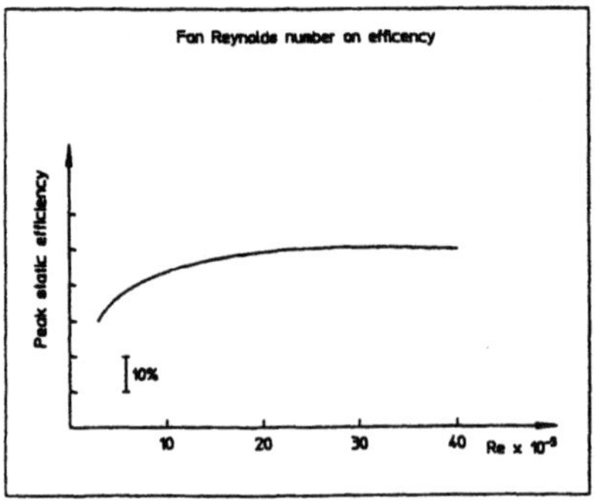

Fig 8 Effect of Fan Reynolds Number on efficency.

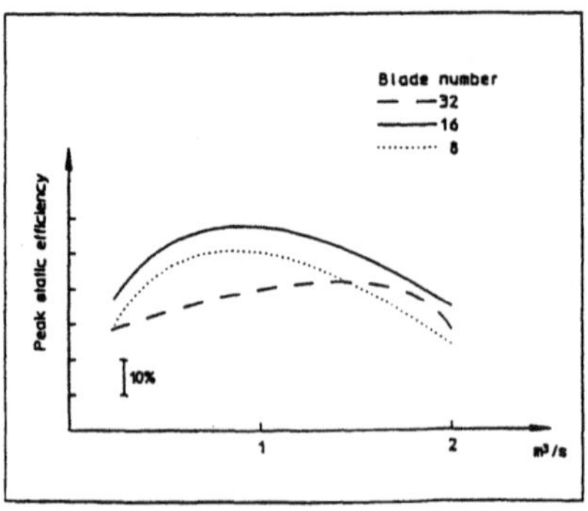

Fig 9 Peak static efficiency versus blade number.

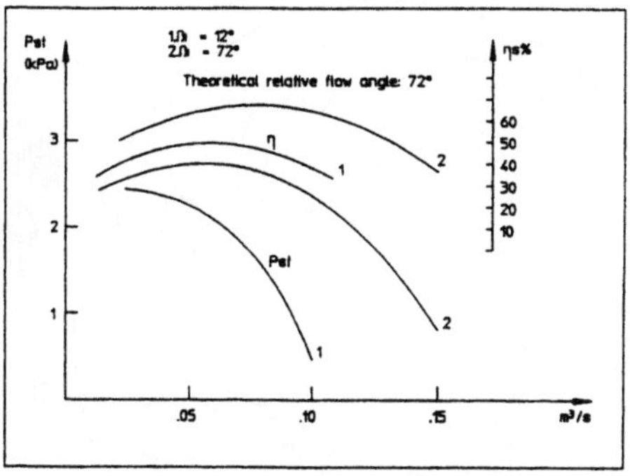

Fig 10 Effect of inlet blade angle on performance of a backward Curved Centrifugal Fan.

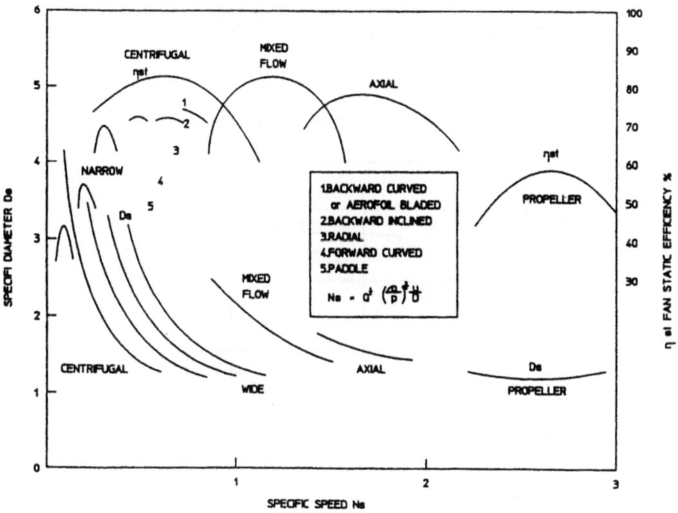

FLOW COEFFICIENT ø(phi) $= \frac{Q}{U D^2}$

PRESSURE COEFFICIENT Ψ(psi) $= \frac{P}{\rho U^2}$

POWER COEFFICIENT λ (lambda) $= \frac{\Psi \phi}{\eta / 100}$

SPECIFIC SPEED $N_s = \frac{\phi^{1/2}}{(\Psi_{st})^{3/4}} = \frac{N\, Q^{0.5}}{(P_s/\rho)^{0.75}}$

SPECIFIC DIAMETER $D_s = \frac{(\Psi_{st})^{1/4}}{\phi^{1/2}} = \frac{D (P_s/\rho)^{0.25}}{Q^{0.5}}$

Q - Volumetric flowrate m³/s
P - Fan pressure Pa
U - Impeller tip speed m/s = πDN
D - Impeller diameter m
N - Rotational speed rev/s
ρ - Inlet air density kg/m³ = (1.2 as std)

Subscripts
st - static
t - total

Fig 11 Specific Speed versus Specific Diameter and Efficiency

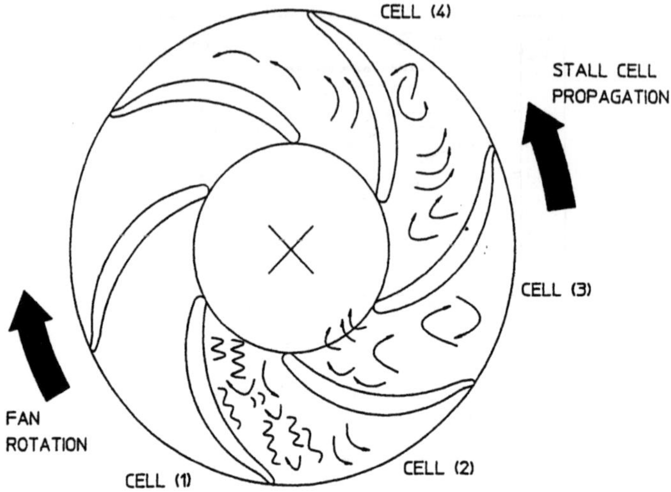

Fig 12 Rotating stall of Centrifugal impeller.

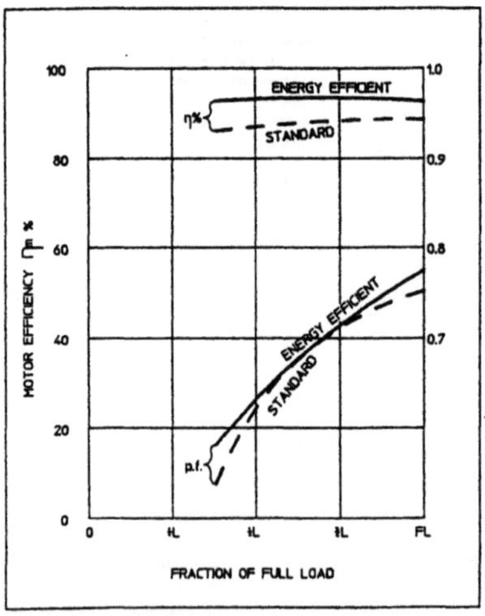

Fig 13 Typical efficiency and power values
for 37kW 4 pole motors.

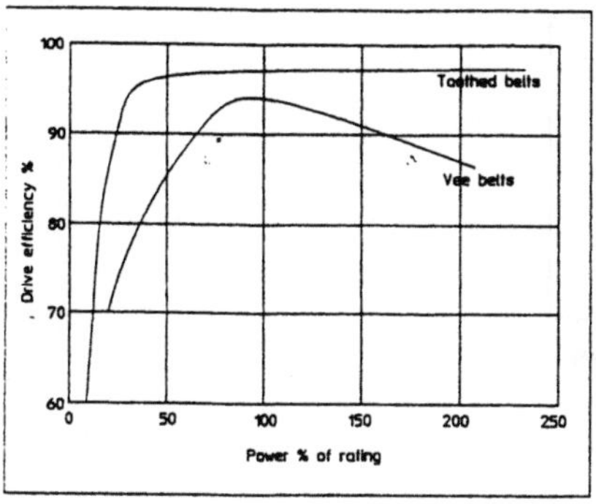

Fig 14 Efficiency of toothed and vee belt drives.

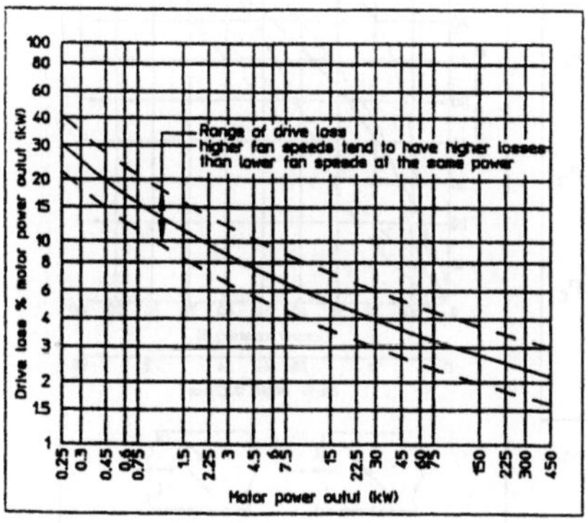

Fig 15 Estimated vee drive losses.

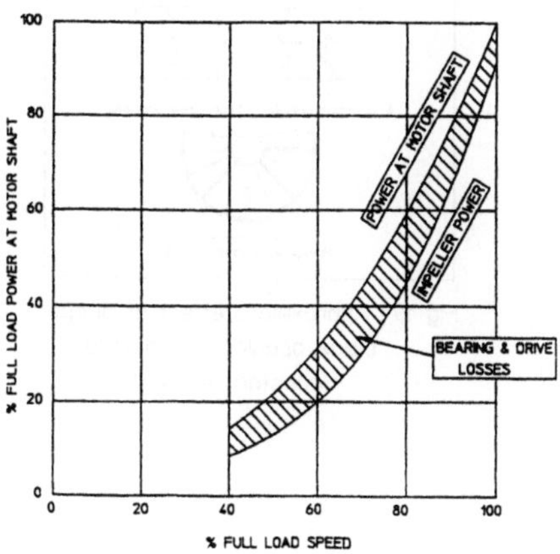

Fig 16 Power required at motor shaft by vee belt driven fan operating constant orifice system.

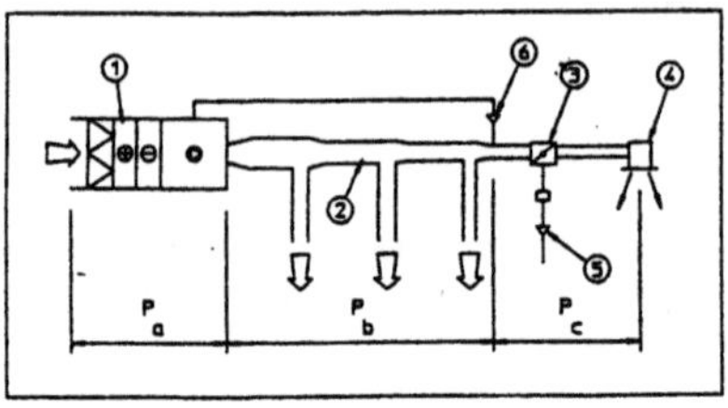

Fig 17 Variable air volume (VAV) system

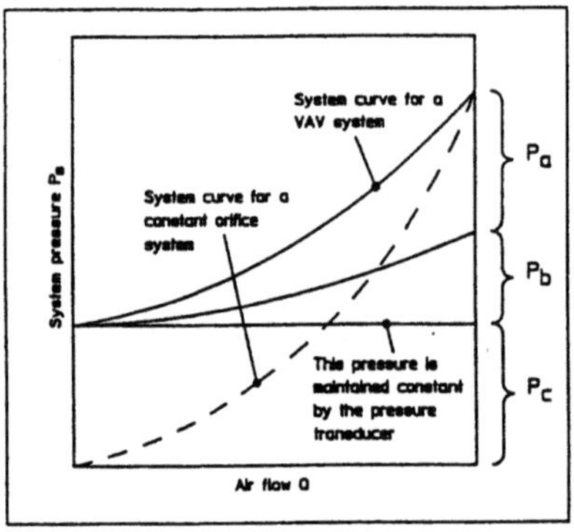

Fig 18 System pressure in a VAV system

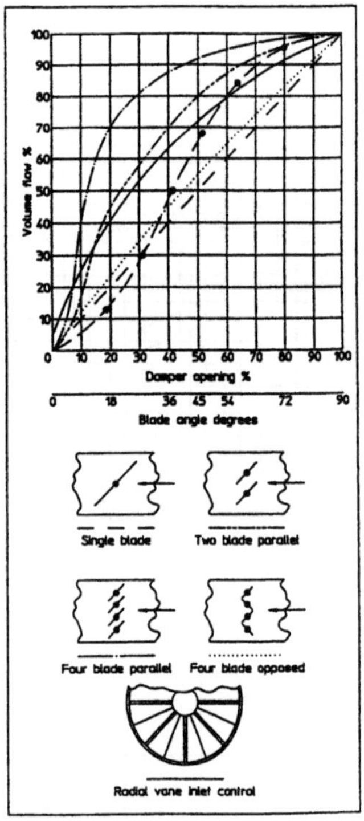

Fig 19 Approximate effect of damper blade opening on flowrate (constant system)

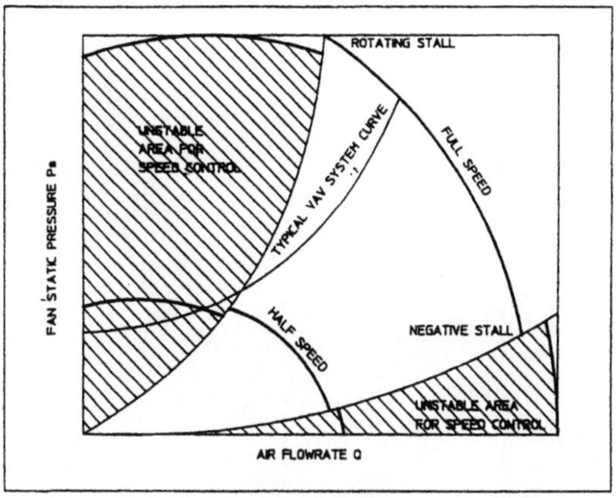

Fig 20 Instability with speed control of wide Backward Bladed Centrifugal Fans.

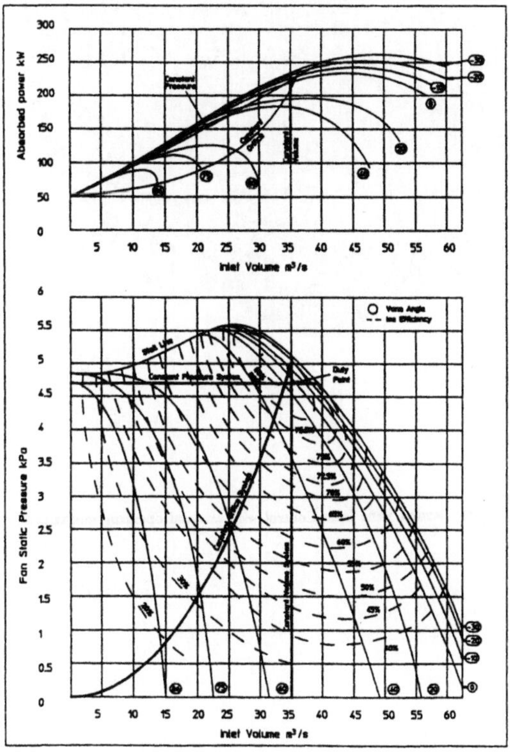

Fig 21 Performance characteristics of 1000mm Series 388 Type R50 Aerofoil Bladed S.I.S.W. Centrifugal fan running at 1910 rev/min fitted with RIVC control and handling

Fig 22 Radial Vane Inlet Control with external operator mechanism

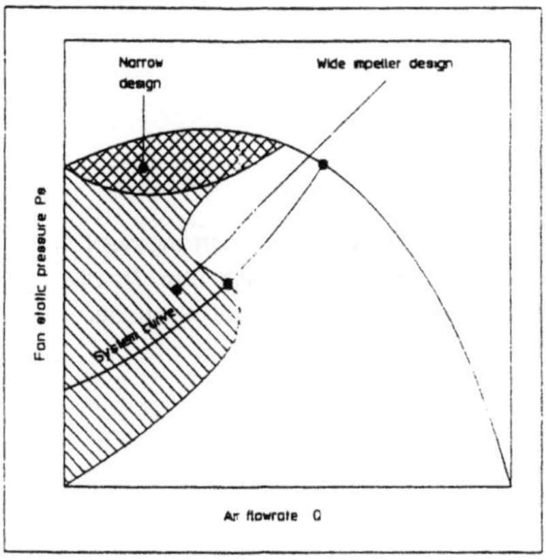

Fig 23 Instability with Radial Vane Inlet Control of Backward Bladed Centrifugal Fans.

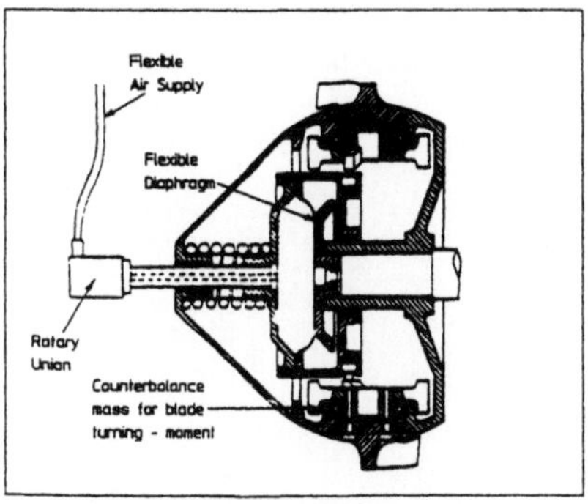

Fig 25 Cross-sectional arrangement of hub mechanism for variable pitch-in-motion axial flow fan (compressed air operation).

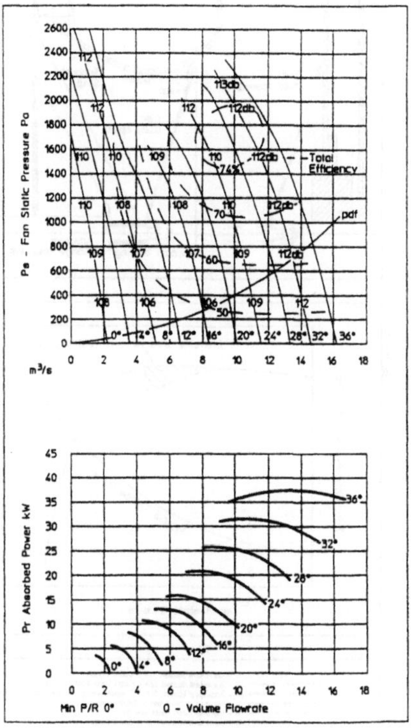

Fig 24 Characteristic curves for 710mm diameter variable pitch-in-motion axial flow fan at 2950 rev/min and handling standard air.

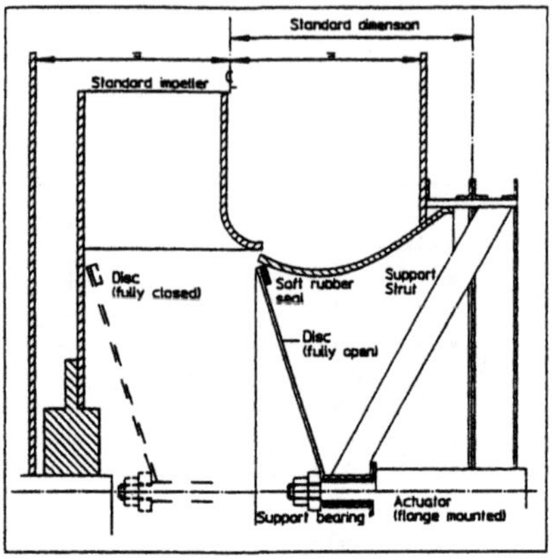

Fig 26 Cross-sectional arrangement of series 28 fan fitted with Disc Throttle arranged for pneumatic actuator.

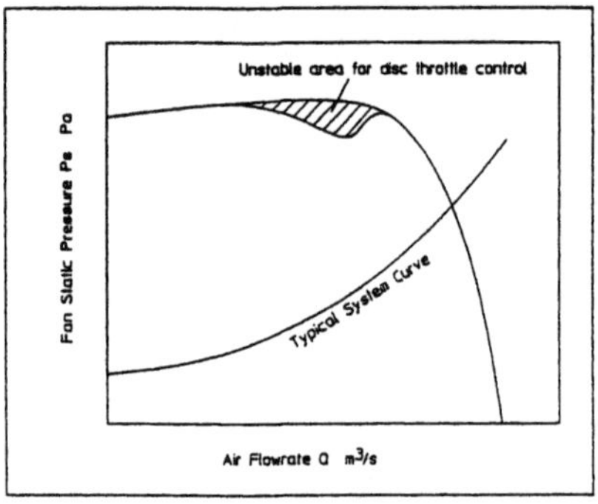

Fig 27 Instability with Disc Throttle Control of very wide Centrifugal fans.

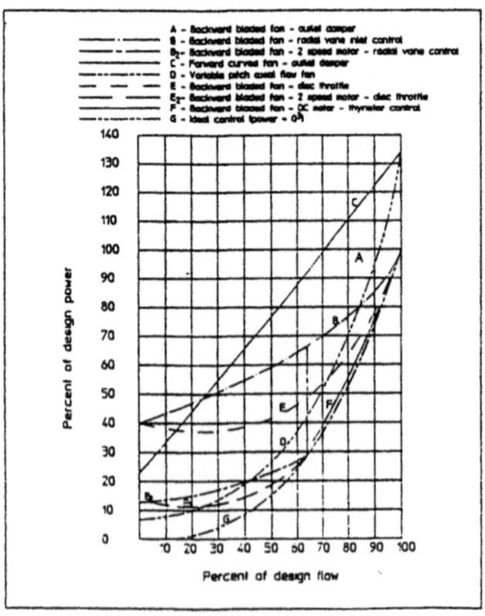

Fig 28 Power variation for various forms of fan and control on constant orific systems.

Energy Optimized Drive Systems Design

H. Auinger

Dr. techn. Herbert Auinger
Siemens Automation & Drives, Standard Drives Division (A&D SD),
Chairman of Technical Commission of German Electrical and Electronic Manufacturers Association
Electric Drive Systems (ZVEI TK 1/2), CEMEP Technical Board

Abstract. Electric drives represent about 70 % of the industrial power consumption. The total saving potential in Germany by consistentally using energy optimized drive systems has been estimated to be 20 THh. This corresponds to the equivalent of
8 fossil-fuel power station blocks. High efficiency motors alone represents an energy-saving potential which is a factor of 7...8 smaller.

Technical measures for the purpose of energy saving have been implemented in, the standard components available today. No fundamental new developments are required.

Essential for practical application is :

- <u>Move away from conventional fixed-speed drives</u>
  with downstream loss-intensive adaptation to the process requirement, whereby the excess energy available is dissipated, i.e. is converted into heat

- <u>Move towards low-loss variable speed electric drives</u> (VSD)
  with optimum process control, whereby only the required energy is made available

- Further more, it should be ensured that low-loss system components are consequentially used; e.g. no worm gears and no throttle- or bypass controls.

Electronically-controlled VSD's allow energy-saving operations in all operating phases, especially at partial loads, partial speeds as well as when accelerating- and braking. The capital investment costs are generally amortized in a relatively short time from the significantly lower operating costs.

1 Introduction, General Economic Significance

When it comes to energy saving, electric drives play a key role. Electric drive systems represent about 70 % of the industrial power consumption, i.e. power which is not directly consumed, but essentially converted into mechanical energy.

In a ZVEI study [1], the total saving potential in Germany by consitentally using energy-optimized electric drives has been estimated to be 20 TWh. This corresponds to the equivalent of 8 fossil-fuel power station blocks, or an annual energy cost saving of about 3 billion DM.

According to investigations made by the ZVEI and the German Statistics Office, this involves a product business with approximately 12 billion DM annual turnover in Germany. The production unit quantities lie in the order of magnitude of 5.4 million low-voltage motors, 100 million small motors and 700.000 drive converters.

Starting from the typical structure of electric drive systems, the central measures to optimize energy usage will be presented in the following explanations and discussed in more details using specific examples.

2 Structure of Electric Drive Systems

The typical structogram of a variable-speed electric drive system is shown in Figure 1. The electric motor, as electromagnetic/mechanical energy converter, is the central function element. The motor is either fed from the line supply directly or via an AC drive converter. Gearboxes (as torque-speed converter) are frequently used on the mechanical side as well as various mechanical transmission elements (couplings, jointed shafts, chain- and belt drives ...) and for adaptation to the power requirement.

There is also a series of electric drive components with combined functionality, which can be universally used in the industry and the trades. These include, for example, gear motors, or as a new product which is quickly establishing itself in the market, so-called compact drives. In the latter case, AC converter and electric motor are combined to form a single mechanical unit. Compact drives such as these form a drive system which can be immediately connected-up. As "intelligent drive", it can execute process-related open- and closed-loop control operations, protection and monitoring functions also with regard to minimum energy usage. [2]

Today, when designing and engineering variable-speed drive systems, using well-proven standard components, special significance is frequently given to criteria such as functional scope, procurement costs and reliability, but not to energy use in operation. In the industrial automation environment, variable-speed electric drives are generally used due to their advantages relating to improved processes and simpler process control. Typical reasons include, for example, shortening process times, higher production volumes, improved and uniform high quality and a higher productivity.

The performance features of the drive are characterized by mechanical quantities such as torque, speed and their speed-time relationship, the dynamic behavior when accelerating and braking, the ability to operate in 1-, 2- or 4-torque/speed-quadrant operation as well as a series of additional technical features, e.g. smooth torque, low noise etc.

In the following presentation, the focus is on electric drives for the trades and industry in the output range from several 1/10 kW up to several 100 kW. They consist of standard components which can be universally used and produced in high unit quantities (standard motors, AC drive converters, gear motors ...).

3 Energy-Saving Measures

Where is the potential for energy saving when using variable-speed electric drives?

3.1 Improved Efficiency of the Electric System Components

If the energy flow in an electric drive system is analyzed (Figure 2), then it becomes clear that the main potential for saving is not to improve the efficiency of the electric system components. Typically, drive converters and transformers have efficiencies of ≥ 97 %. Even electric motors have, intrinsically, a relatively good efficiency in the order of magnitude of 90 % at 10 kW. The efficiencies of larger motors are higher than the smaller motors, refer to Figure 3. In order to achieve noticeable improvements in the motor efficiency, it is necessary to use significantly more and/or better quality materials (copper, aluminum, sheet steel). Presently, this strategy is being followed within the scope of activities to increase efficiency of three-phase motors in the US and Europe [3, 4]. The efficiency definitions for the 3 efficiency classes for 2- and 4-pole squirrel-cage motors, voluntarily agreed between the EC and the European Manufacturer's Association CEMEP, is shown in Figure 3.

For drive converter-fed variable-speed drives, other motor principles can be used, e.g. permanent-magnet synchronous motors, which, according to their principle, have no rotor- and excitation losses, and therefore have higher efficiencies [5].

Only about 1/8 of the total potential, mentioned at the beginning of this presentation, can be saved by consequentially using higher-efficiency motors, i.e. this represents a fossil-fuel power station in Germany [1]. However, it is still sufficiently interesting, especially when one considers the relative short implementation phase. Concrete measures to support this include, for example, the EPACT legislation in the US [3] and the European "voluntary agreement" between the EC and CEMEP [4] to classify efficiencies and a drastic reduction of the market share of EFF3 motors. The increased costs of motors with higher efficiency, Classes EFF1 and EFF2 are generally amortized within a relatively short operating time in the range of several months up to max. 1-2 years. Referred to the total operating costs over the lifetime of an electric motor, the capital investment costs only represent a few percent; however, the accumulated energy costs mostly more than 95% [6].

The drive process represents an energy-saving potential which is a factor of 7...8 higher than for high efficiency motors. Therefore, drive concepts, optimized for minimum energy consumption and a process control using variable-speed electric drives are required.

3.2 Optimizing the Drive Concept

Mechanical transmission elements frequently have a significant impact on the losses, e.g. worm gears for slow drives, whose efficiencies can be extremely poor with ≥ 50 %.

The energy balance of a belt drive using standard components for a beverage filling station is shown as a typical example in Figure 4a. With the new drive concept, (Figure 4b) energy-savings of 44 % - from 1,67 kWh to 0,94 kWh – are achieved. This mainly results from the fact that only 4 % of the electric energy fed into the drive is required for the mechanical transmission when using a spur gear with belt drive. On the other hand, for a conventional drive with worm gear and chain drive (Figure 4a) it would be 32 %. This is also supplemented by the use of a smaller sized high efficiency motor with significantly reduced losses.

A drive retrofit is amortized, through lower power consumption, within 2 years. These relationships are characteristic for numerous, similar drive tasks, which were frequently designed according to criteria of functionality, capital investment costs and reliability, however not energy consumption in operation.

3.3 Avoiding of Loss Intensive Control Techniques

When operating pumps, fans, extruders or mills, flows, pressures or material quantities must be changed according to the requirements of the various processes involved. For conventional drives, the motor runs with a constant speed. The quantity reductions, demanded by the process, are implemented using conventional control techniques (throttles, bypass systems or vanes), which in some cases can result in extremely high energy losses.

However, these losses can be significantly reduced by using state-of-the-art drive systems, i.e. drives with electronic closed-loop speed control.

In this case, pump drives play a significant role. Pumps represent more than 50 % of the industrial power consumption. Many pumps are still operated by using the conventional techniques mentioned above. If the speed is controlled, pressure reduction using a control throttle - valve is eliminated. The pump pressure is always adapted via the speed. Further, the high spoiling losses are reduced to a minimum, as these are significantly dependent on the speed. As Figure 5 shows, overall, significant energy savings can be achieved.

3.4 Dynamic Drive Tasks with Accelerating and Braking

In addition to the drive tasks which have been investigated up to now, with quasi steady-state operating conditions, for many variable-speed drives, a good dynamic performance is required.

Typical motion control tasks include the transport and processing of materials, e.g. machine tools, rolling mills, elevator drives, centrifuges etc. In this case, the special capabilities and possibilities of variable-speed electric drives are becoming increasingly important, such as

- minimized energy consumption at partial load and partial speed, and
- low-loss acceleration, and
- ability to regenerate when braking.

While accelerating, kinetic energy (E_{kin}) is stored in the various masses being moved. For drives involving induction motors, which are directly fed from the line supply, additional rotor losses, with the same order of magnitude, must be assumed when accelerating. These are caused by the difference between the rotor speed and the line frequency (Figure 6a). Inversely, when braking, the kinetic energy, stored in the drive must be reduced or dissipated, i.e. converted into heat.

These losses can be completely eliminated when using an AC drive converter, if the frequency continually changes with the rotor speed and when braking, the mechanical energy is fed back into the line supply, refer to Figure 6 c. This allows a significant energy-saving potential for all motion control applications with periodic acceleration- and braking operations. Furthermore, the motor thermal stressing is lower which has a positive impact on its lifetime.

A comparison of the energy-saving which can be achieved with various measures is shown in Figure 7 using an elevator drive as an example. It involves a cable-type elevator with 1000 kg load capability, 17 m high and with 5 stopping positions.

Starting from a conventional drive with a pole-changeable motor (= 100 %), when using a smaller motor, fed from an AC drive converter, a loss-reduced gearbox and a regenerative feedback unit, energy savings of up to 81 % can be achieved.

With 1400 movements daily, the increased capital investment costs are amortized for the individual alternatives in 1.5...1.7 years.

4 Résumé

4.1 Sensitizing People to the Issues

Barriers, which have to be overcome, result, especially from

- old established procedures when engineering drives

  energy consumption and consciousness of the environment have no or only a low priority

- typical sales situations

  for standard products, > 80 % are sold through OEMs and plant constructors

- purchasing behavior of customers

  capital investment costs are often the decisive factor, and not the ongoing operating costs (lifestyle costs).

In order to permanently change the situation, a series of measures must be taken at various levels (politics, associations, manufacturers, engi-neers ...) [7], which should be systematically approached as follow-on activities from the EEMODS

Conference. The energy-saving potential and the associated economic impact are considerable.

In this case, in addition to the energy use balance, which can be directly assigned to the electric drive system, the following should also be considered:

⇒ the losses in the feeder cables and in the power supply network

⇒ the power losses in the drive system increase the ambient temperature and additionally load the air-conditioning of the plant.

For the total loss savings which can be achieved, a multiplication effect is obtained. The efficiency of power generation in fossil-fuel power stations is approximately 40 %; this means, that energy saving at the end user increases, referred to the primary energy requirement and therefore for environmental stressing, by an additional factor of 2.5.

4.2 Technology

Technical measures for the purpose of energy saving have already been implemented in the standard components available today. No fundamental new developments are required. Some clear statements can be made for practical applications:

⇒ Move away from conventional (Figure 8a)
fixed-speed drives
with downstream loss-intensive adaptation
to the process requirement, whereby the
excess energy available is dissipated,
i.e. is converted into heat

⇒ Move towards low-loss variable speed (Figure 8 b)
electric drives (VSD)
with optimum process control, whereby only the
required energy is made available

Electronically-controlled variable-speed drives allow energy-saving operations in all operating phases, especially at partial loads, partial speeds as well as when accelerating- and braking.

This is especially significant for all "motion control drive functions" such as: transport- and machining operations, e.g. cranes, material loading, centrifuges, presses, rolling mills, machine tools.

⇒ Furthermore, it should be ensured that low-loss system components are consequentially used:

- if possible, no worm gears and no hydraulic- or eddy current couplings
- avoid throttle-, choke- and bypass controls
- use EFF1 motors, especially for continuous operation applications.

The standard components of electric drive systems, available today, combined with sensitive sensor systems can be used as "intelligent drives" for the widest range of distributed automation tasks. In addition to their technological advantages, they especially offer economic and ecological benefits. The capital investment costs are generally amortized in a relatively short time from the significantly lower operating costs.

Combinations of standard drive components should be seen in conjunction with this, e.g.:

Gear motors, which are more effective at high torques and relatively low speeds than a direct drive with an appropriately larger electric motor designed for the full torque.

Compact drives, where the electric motor and associated AC converter form a single mechanical unit. This is a ready-to-connect variable-speed drive with the same mounting dimensions as a standard motor.

5 Literature

[1] ZVEI FV1 Brochure „Energiesparen mit elektrischen Antrieben" August (1999)

[2] Siemens-Brochure : "Combi Master – Variable Speed Drives 120 W to 7,5 kW" A&D SD (1999)

[3] EPACT Legislation in USA

[4] EC-CEMEP volontary agreement (1999)

[5] A. Binder, A. Storath, U. Kaumann „Moderne Antriebstechnik spart Energie – ein Vergleich energieeffizienter Drehstrommotoren" „Elektrie 52 (1998) pp. 47...55

[6] Siemens Brochure: "Our energy saving motors - making a profit for you" A&D SD (1999)

[7] „Maßnahmen zur Förderung der rationellen Energienutzung bei elektrischen Antrieben", Research-Report of Fraunhofer Institut Systemtechnik und Innovationsforschung (March 1999)

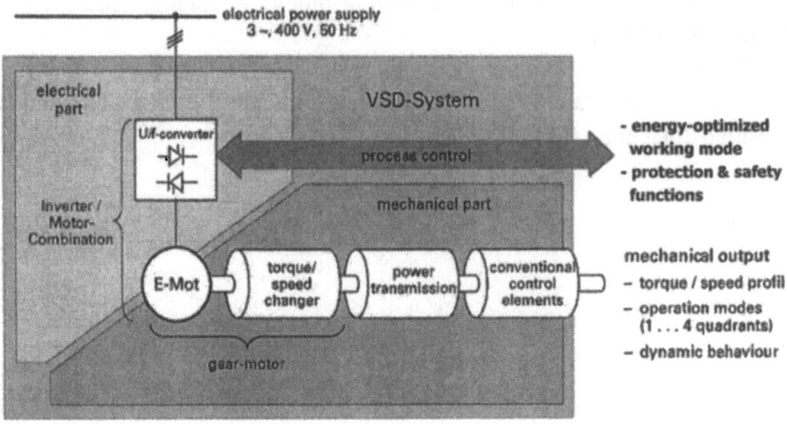

Fig 1: Typical functional structure of VSD-Systems

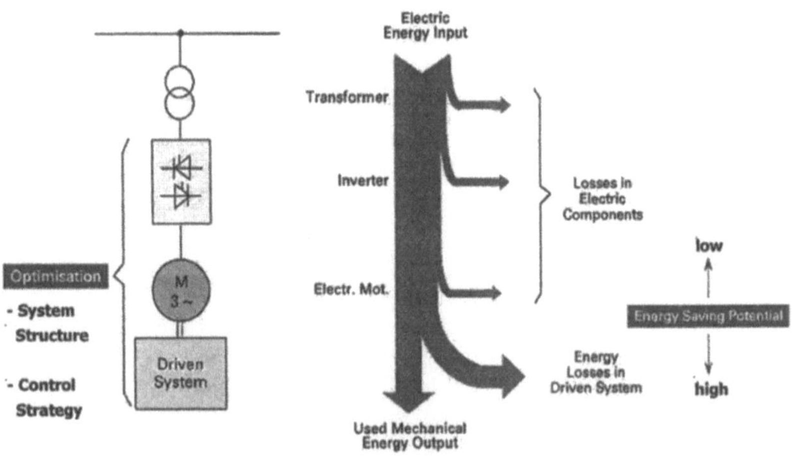

Fig 2: Typical Energy Consumption in Electric Drive Systems

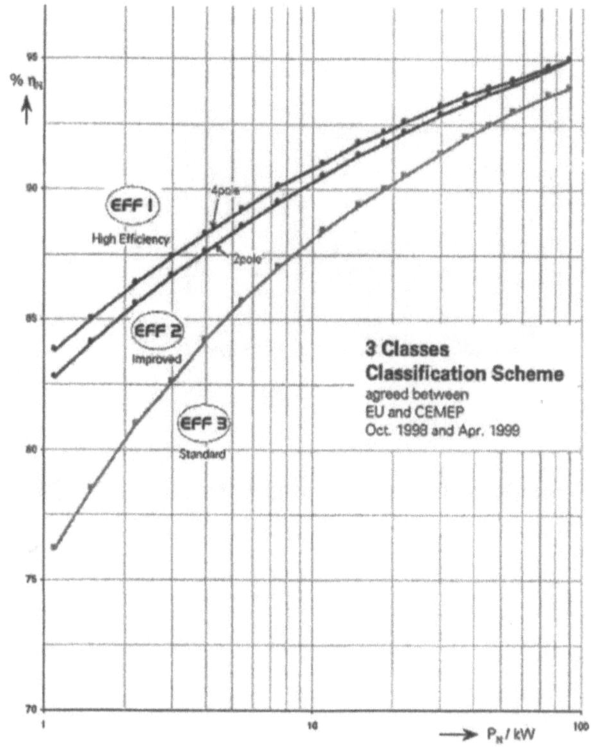

Fig 3: European Efficiency Classification for 2- and 4-pole Induction Motors

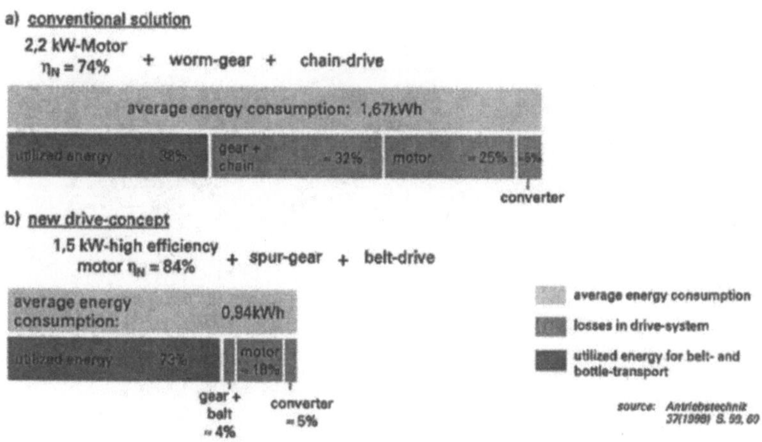

Fig 4: Typical example: Belt drive system für a beverage filling station

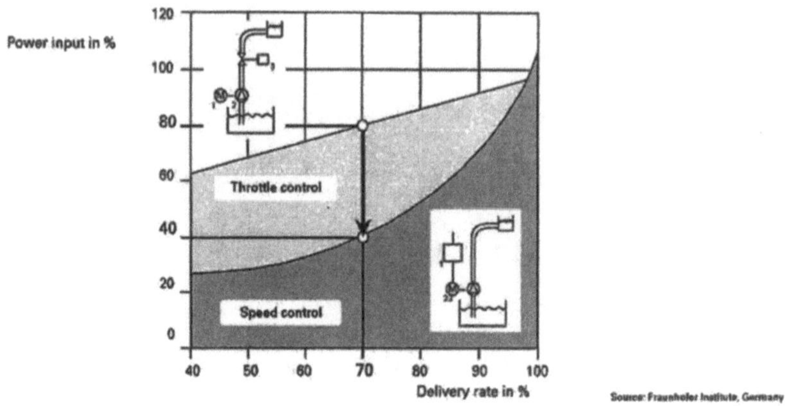

Fig 5: Power input of a pump with throttle control vs. one with speed control

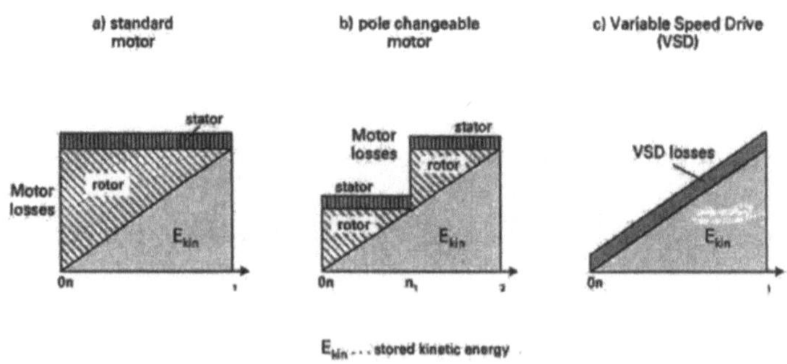

Fig 6: Energy-Consumption for an Acceleration Period

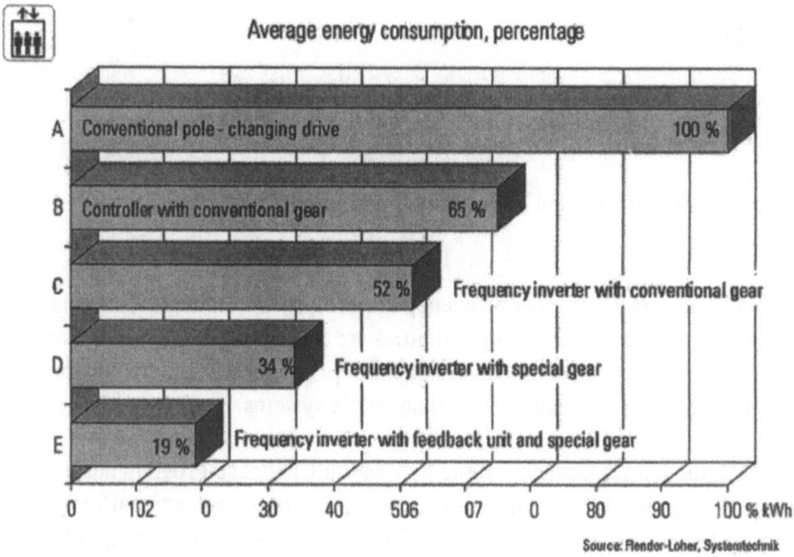

Fig 7: Energy - comsumption of an elevator drive system

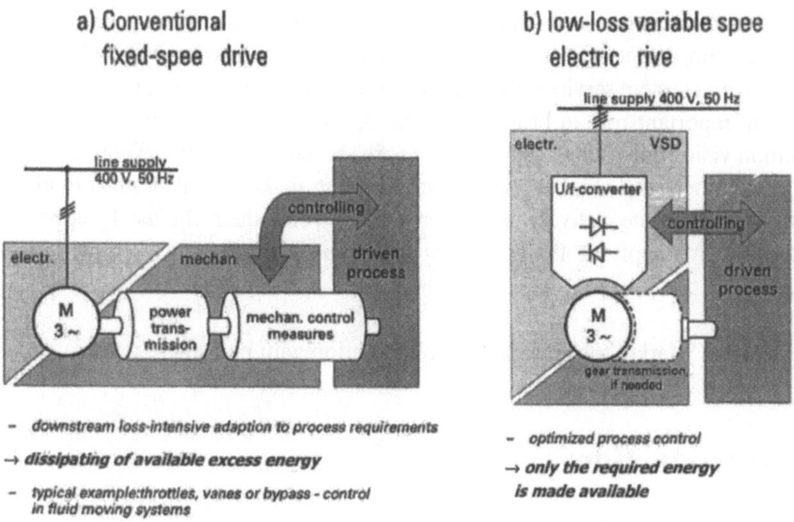

Fig 8: Advantageous application of variable speed electric drives

Using Collaboration to Achieve Industrial Market Change

Aimee T. Mckane

Lawrence Berkeley National Laboratory, Washington, DC

Abstract. The greatest energy efficiency opportunities in industrial motors can be found in the systems, yet these opportunities are frequently overlooked because the business interactions between industrial suppliers and the customers that they serve typically focus on components rather than whole systems.

Since 1993, the US Department of Energy's Motor Challenge Program has developed an array of program offerings that designed to encourage companies who use motor systems to focus on a systems approach. The collaborative approach applied to the industrial compressed air systems described in this paper, the Compressed Air Challenge®, was an outgrowth of that effort. This approach seeks to effect institutional and behavioral change, rather than technological change. The purpose of the collaboration is to strengthen and promote a "best practices" approach to compressed air systems to achieve greater efficiency. The goal is to replace the existing components focus in the customer/supplier relationship with one that emphasizes a system services perspective. Government and public-interest facilitators have an important role in bringing key market stakeholders together to develop a common vision for change. Project costs are shared among all stakeholders, as are the project benefits. Stakeholders share decision-making control, thus motivating them to participate actively and contribute beyond their financial support. A framework for applying the collaborative approach in other markets will also be discussed.

Keywords. Market transformation, collaboration with industry

1 Introduction

Many large opportunities for improving energy efficiency involve entire systems, yet they are often overlooked because they require a shift in thinking by both system users and the businesses that serve them. There are significant competitive advantages to implementing a systems or "best practices" approach as part of a company's overall total quality management effort. Motor systems are ubiquitous in US industry, representing about 23% of all electricity sold. Despite this, motor system efficiency is typically not on the "radar screen" as a cost-cutting and productivity improvement for most US companies because it is viewed as a supporting rather than line activity[1]

Industrial compressed air systems are found in nearly every industrial facility in the US, consuming nearly 90 billion kWh annually, but the per plant cost of operating these systems is rarely known, and therefore is not managed effectively. The efficiency of these systems could be improved by 20-50% through the application of best practices that rely primarily on changes in operation and maintenance procedures[2] Despite widespread recognition among industry experts, these best practices are not in common use. Anecdotal evidence from colleagues in other countries indicates that similar problems exist on a global basis.

This paper describes the results of a four-year effort to develop a framework for promoting change in the industrial compressed air systems market by addressing some of the behavior-based barriers to achieving greater system efficiency. Government can be an effective facilitator for this type of change, but it requires a different approach than is typical of policies such as directed actions (e.g., regulations), and persuasive actions (e.g., tax incentives).

The approach described here is an alternative type of persuasive market intervention that relies primarily on establishing an environment that encourages the participants in an existing market structure to interact in new ways. It is my contention that these interactions, by creating new business opportunities, will result in a permanent transformation of a market.

This type of intervention effects an *institutional and behavioral change*, rather than a *technological change* more typical of energy efficiency market interventions. It is assumed that the structural shifts resulting from institutional or behavioral change will create an environment for further technological innovation. Since it engages many aspects of a market, this approach is described as *collaborative intervention.*

Collaborative intervention places government in the role of a broker or facilitator, responsible for setting out general goals. Market participants are invited to be champions of these goals. In exchange, government acting as the broker can

---

1 Consumption figures are from the USDOE *United States Industrial Motor System Market Assessment,* December 1998. Statements about the position of motor energy efficiency in a company's total quality management approach is based on numerous interactions between Motor Challenge staff and industrial energy managers. This paper includes some material previously presented in papers prepared for the ECEEE 1999 Summer Study and the ACEEE 1998 Summer Study.

2 I started with the estimate of 20-30% published in USDOE's 1996 National Market Transformation Strategies for Industrial Electric Motor Systems and revised it upward based on numerous discussions with experts in the compressed air equipment and consulting businesses. This estimate has been consistently supported by compressed air system specialists.

recognize them for the risks they assume, and help them identify paths for reaching these goals. This approach seeks to exploit the different, and potentially complementary, roles and competencies of the public, private, and not-for-profit sectors. The four key elements of this strategy are:

- broadly defining the goals with no predetermined way of achieving them,
- creating an atmosphere of mutual respect,
- acknowledgment and acceptance that the participants will act in ways that are consistent with their economic and political self-interest, and
- establishing a high tolerance for the ambiguity and tension involved in forming coalitions across typical market structures.

The Compressed Air Challenge® (Challenge), described in this paper, is an application of this strategy to the industrial compressed air market. It builds on a similar approach used by the US Department of Energy's Office of Industrial Technology's (DOE/OIT) Industries of the Future Program to promote industrial research and development.. In addition, this market-based approach is a direct outgrowth of program development experiences in DOE/OIT's Motor Challenge Program. The goal of the collaboration is to replace the existing components focus in the customer/supplier relationship with one that emphasizes a system services approach. The need to develop additional activities directed toward building management awareness will be discussed.

This industry is suited for a collaborative intervention because of characteristics, including: market structures, market size, pressure to change, system improvement opportunities, and barriers to achieving those opportunities. Whether this market is capable of transforming itself without intervention will be addressed.

Finally, some guidelines for using the policy framework developed for the Compressed Air Challenge® in other market interventions are provided. This will include identification of the key elements to consider in selecting a potential target market

2 A Challenge to Change

In a market-based economy, products and services are constantly being created to meet perceived customer needs. Whether new products and services are successful and whether they displace existing products and services is driven primarily by how well they find their target market. This process of finding a good "fit" for new products and services is a function of both product availability and customer interest, yet most policy instruments to promote energy efficiency are directed at either the customer (demand side) or the producer (supply side), but rarely both. Many persuasive actions (rebates, tax incentives) are directed toward changing the demand side. While it is true that the supply side will react to these changes in the demand side, it is my contention that change will happen more quickly and last longer if the supply side can clearly identify a business opportunity independent of any temporary incentive.

This paper describes a role for the U.S. government to work with existing market structures while acting as a catalyst to achieve changes that a market may be unable to accomplish on its own. The Compressed Air Challenge is one example of how voluntary participation can be gained by recognizing and engaging private interests to create permanent changes (or transform) markets for the common good.

A key element of working within market structures is acknowledgment and acceptance that participants will act in ways that are consist with their economic and political self-interests. For example, not all suppliers will approve of a collaborative intervention. Suppliers who self-select to participate will contribute the suppliers' point of view, helping to ensure success by becoming "early-adopters". Engaging suppliers in a collaborative intervention recognizes both the durability of existing relationships between the supply and demand side as well as the potential for enlisting progressive suppliers as effective champions for change. Suppliers (manufacturers, distributors) have unique technical and market knowledge and the ability to support (or resist) change in their interactions with customers.

The success of these voluntary efforts depends on flexibility and the ability to align the market interests of the various stakeholders to a sufficient degree that a common arena for action can be identified. The *challenge* is to translate the public goals into private interests so that a market can be transformed through the actions of the market stakeholders.

3 Why Select Compressed Air?

A combination of *opportunity* and *possibility* are the reasons why the U.S. government selected the compressed air industry for a collaborative intervention. The *opportunity* is in the energy savings coupled with corresponding improvements in quality control and reliability of production.

Optimization of compressed air systems represents one of the largest non-process, industrial energy efficiency opportunities, with improvements of 20-50% readily achievable through the introduction of a best practices approach. Lack of information has been a primary barrier to realizing substantial improvements in the efficiency, reliability, and productivity of industrial compressed air systems. Compressed air systems in U.S. manufacturing account for $4.5 billion U.S. ($4.2 billion Euros) per year of energy costs and 21 MMTCE of total U.S. carbon emissions (1-2 percent of total U.S. emissions). The Compressed Air Challenge® seeks to save $150 million U.S. ($139 million Euros) in annual energy costs by the year 2010 (USDOE, 1998 Compressed Air Challenge Kickoff)[3].

Compressed air is industry's "fourth utility"; it is central to production for many industries, including: chemicals, textiles, general manufacturing, plastics, mining, glass, pulp and paper, shipbuilding, furniture, automobile and aircraft manufacturing, iron and steel-making, and petroleum refining. Compressed air is used extensively as a source of power for tools and equipment as well as in industrial processes for pressurizing, atomizing, agitating, and mixing applications.

Compressed air is the most expensive utility; a single compressor delivering 500 standard cubic feet per minute (scfm) 24-hrs per day can cost $100,000 U.S. ($93,000 Euros) year to operate. If the company using the compressor has a 5% net profit ratio, the cost of operating the compressor will be the equivalent of $2 million U.S. ($1.85 million Euros) in production (Foss 1994, 71). In spite of its cost, this level of waste in a manufacturing facility is not uncommon, due to poor system operation coupled with a perception by production staff that compressed air is "free". This is largely because compressed air is already present in distribution piping in the plant and the user is neither knowledgeable of nor responsible for the electricity bill associated with its use.

Compressed air systems are modified over time and frequently suffer from some or all of the following:

- improperly installed and/or leaking distribution lines;
- outdated or inadequate controls;
- poor maintenance of filters and other accessories;
- mismatch of compressors to load;
- excess compressor capacity;
- higher than necessary system pressure, and
- inappropriate applications.

Although the cost of operating compressed air systems is high, it is a small fraction of the overall cost of production and receives little attention, as long as the air supply is adequate. Compressed air systems are not well understood by the majority of plant operations staff; modifying a system, no matter how poorly it is operating, is perceived as a risk to production[4].

---

[3] USDOE *United States Industrial Motor System Market Assessment*, December 1998.

[4] The information in the preceding two paragraphs is the result of numerous conversations between the author and manufacturers, distributors, consultants, energy experts, and end users of compressed air systems over a period from 4/95-10/97. This material was used in the prospectus and presentation materials developed by the author and others for the Compressed Air Challenge.

The *possibility for change* arises from current market structures combined with internal and external pressures for change. The supply side of the compressed air market is highly specialized and relatively compact, with a limited number of participants- a feature that makes it a feasible target market. It is possible to seat representatives from all major equipment manufacturers around an average size conference table. Although there are thousands of equipment distributors, 600 of the most influential distributors are represented by a handful of associations. Highly skilled compressed air system consultants are so few in number that they can participate in a conference call. This allows for negotiation, consensus-building, and the rapid exchange of information to effect change.

The supply side of the compressed air system market is under both external and internal pressure to change. Internal pressures include low margins of profitability and the drive for greater economy of production. Externally, the market is being pressured from several sources: globalization, utilities entering the business of providing industrial customers with compressed air, the increased use of improved electric tools instead of air tools, and customer dissatisfaction with existing services and equipment performance.

4 The Industrial Compressed Air System Market

In a paper, Wayne Perry of Quincy Compressor described the existing compressed air equipment market and the relationship between equipment buyers and sellers. The major points made in this paper have been subsequently reaffirmed through several other sources and will be referenced several times in this section.

This is a market in which equipment distributors are the primary source of information for small to medium size companies. Distributors operate in an intensively competitive market in which customers typically buy substantial pieces of equipment (purchase prices can range from $50,000 to a million or more for a major new system) on a lowest first cost basis, margins for equipment sales are extremely tight, and long-term equipment service contracts are essential to economic survival. The focus on lowest first cost persists despite the fact that many large compressors cost more to operate in the first year than initial purchase price (Kemp 1998). Most companies administer separate budgets for capital projects and operating expenses, so economy in one area may have no value for the other.

Perry indicates that lack of complete information from manufacturers on equipment performance adds to purchaser confusion. This is further complicated by the fact that a compressed air system is dynamic and "most manufacturers cannot be of much help when predicting the behavior of their products in dynamic systems" - mainly due to lack of available training and experience (Perry 1998, 72). One equipment distributor described the situation as "providing pre-packaged solutions to an undefined problem on a lowest first-cost basis" (McMorrow 1998).

Customer frequently make purchasing decisions based on in-house advice plus distributor input. If it is a large company or a large purchase is anticipated, a compressed air consultant may used to evaluate the system. Often, significant savings can be obtained from relatively low-cost changes in the way that compressed air is used, stored, or supplied (Howe and Scales 1995) (Van Ormer 1997)(Foss 1997). However, since capital budgets and operating budgets are separate, there is little incentive to spend any extra capital on either efficient equipment or a comprehensive system approach to reduce operating expense. Purchasing a solution without first understanding the problem may result in a $50,000.00 solution to a $500.00 problem (Perry, 71).

The result is systems that are not understood, continue to operate poorly, waste energy, and cut into profitability. Many times, it is only when a new facilities person takes over or management begins to question overall costs that cycles like this can be broken.

This is a chicken or egg dilemma. Until end users ask for a different approach, manufacturers and distributors cannot afford to concentrate on one or they will go out of business before they have a market. Lacking information, end users do not know what they need or how to manage any perceived increase in risk from taking a non-traditional approach.

It should also be noted that inefficient industrial compressed air systems are not just an American phenomenon. From discussions and materials shared with my colleagues in Canada, the United Kingdom, the Netherlands, and India, industrial compressed air systems seem to be oversized or poorly operated on a global basis. There is also some anecdotal evidence that customers in countries where production is being constrained by the electricity supply have a greater interest in energy efficient components, but it does not necessarily follow that their systems are more efficient overall.

4.1 Why Doesn't the Market "Transform" Itself?

The average U.S. corporate customer is ignorant of the possible benefits from improving the operation of their compressed air system. A deep lack of trust among the stakeholders has made it extremely difficult for manufacturers and distributors to change the way that they interact with customers for fear of being undercut. It is an atmosphere where an outside intervention is helpful so that no single stakeholder has to be the one who blinks first. Changing the situation requires the creation of a new market, a distributor/manufacturer focus on system opportunities rather than equipment solutions. It is a high-risk change that requires a complete re-evaluation of what constitutes a successful customer relationship coupled with an intensive reeducation of consumers on the value of this approach. Providing an independent source of consumer information is an essential element of this change.

5 Creating a Collaborative Model: the Compressed Air Challenge®

The Compressed Air Challenge® is the result of work on Industry Partnerships for the USDOE Motor Challenge Program. The themes which ultimately led to this project were first identified in the April 1995 Roundtable on Market Transformation Strategies for Industrial Motor Systems breakout session on compressed air systems. A general point of consensus established at the Roundtable was that the major improvement opportunities are in the compressed air system, not the individual components.

Subsequent to the Roundtable, work began with CAGI to identify cooperative projects that addressed some of these needs. Over the next twelve months, CAGI proposed two major activities:

- develop data sheets standardized reporting of performance for rotary screw compressors and two types of compressed air dryers; and
- develop a training and certification program on compressed air system best practices (Correspondence with CAGI Energy Awareness Committee 1996).

Work on a standardized format for reporting equipment performance is well underway. The idea of a training and certification program on compressed air system best practices led to the Compressed Air Challenge®.

An initial proposal for a training and certification program was developed by myself, Neal Elliott of ACEEE, Mark Hanson and later Ron Wroblewski of the Energy Center of Wisconsin. The proposal was refined and reshaped over eight months through a series of meetings with CAGI, state and utility representatives, and other interested parties.

From its inception, the project model was to solicit multiple sponsors from market stakeholders with the objective of pooling funds to complete project deliverables. This approach was taken for two reasons: cost and developing ownership for implementation. The project was too extensive and costly for a single sponsor (or CAGI and DOE together) to undertake. Given the level of conflict and distrust in the industry, it was also critically important to build ownership and a new network of relationships during the development phase of the project. These networks of relationships are at least as important an outcome as the materials and training produced by the Challenge. Since the ultimate goal of the Challenge is to change market interactions and stakeholder behavior, those stakeholders (manufacturers, consultants, end users, distributors, state organizations, utilities) needed to become part of the process early on to develop ownership.

A business prospectus approach was used because we were seeking a combination of private, public, and not-for-profit sponsorship. In May 1997, stakeholders met to review the draft prospectus and try to reach consensus on a fundable project. Attendees included: representatives from equipment manufacturers and distributors, compressed air system consultants, state and federal government, utilities, energy efficiency organizations, state research and development organizations, and a utility ESCO. Many attendees brought opposing agendas to the meeting and, in some cases, substantial distrust based on previous skirmishes in this highly competitive market. End user companies were purposely excluded from the formation meeting so that the other stakeholders would not be distracted by key customer relationships.

The group reached a consensus on the formation of a national collaborative to bring together the best information on compressed air system design, operation and maintenance, and assessment to:

- $ increase the reliability and quality of industrial production processes,
- $ reduce plant operating costs,
- $ expand the market for high quality compressed air system services, and
- $ save energy- projecting annual savings of more than 3 billion kWh of electricity nationwide

The collaboration has three primary elements:

- $ a *customer awareness campaign* on the benefits of effective and efficient industrial compressed air systems;
- $ a nationally recognized *professional development program* to train plant operating personnel on compressed air system best practices, and
- $ a *certification program* for plant operating personnel who apply these best practices.

The group also decided that the project sponsors would each be asked to contribute $30,000 for each of the two project years and would comprise an Advisory Board with final decisions for the project. Another body, the Project Development Committee, represents a cross-section of stakeholders, whether or not they were sponsors. This permits participation and critical technical input by key stakeholders, such as the compressed air system consultants, who would be unlikely to commit $30,000 of their own funds. The Committee is responsible for the overall operation of the project, in cooperation with the Project Manager. Working Groups develop information and educational materials.

My work with the compressed air industry has been as a representative of DOE/OIT. I began the project as the project facilitator and now serve as the chair of the Project

Development Committee. The material that follows is from that perspective

5.1 Accomplishments

For a project of this type, progress has been rapid. Since the first Advisory Board meeting in September 1997, the Compressed Air Challenge® has accomplished the following:

$ in cooperation with USDOE, prepared, published, and sold 2448 copies of *Improving Compressed Air System Performance: A Sourcebook for Industry;*

$ developed and tested a one-day training program for plant engineers *Fundamentals of Compressed Air Systems;*

$ selected and trained the first group of 13 instructors to give the training;

$ developed and delivered a "train the trainer" program to qualify an additional 12 instructors;

$ coordinated delivery of 42 one-day training sessions during the spring of 1999;

$ created a national advertising campaign with full-page ads in *Plant Engineering* and *Plant Services* magazines and a national mailing to 5,000 plant engineers;

$ prepared a special insert for the USDOE *Energy Matters* newsletter;

$ created a web site *http://www.knowpressure.org;*

$ begun work on a 2-1/2 day advanced training session to be offered in early 2000; and

$ established a trademark to protect the Compressed Air Challenge name and logo.

5.2 Potential Barriers

The potential barriers to the continued success of the Challenge are significant and should not be underestimated. They include:

$ tension during project development between the need to assemble small groups empowered to draft materials so that the project can proceed and the need to keep the process open so that everyone who wants to participate gets the opportunity;

$ the importance of to protecting end users and consultants (who often bring in their most valued end use customers) from unwanted solicitations for products and services as the result of contributing to project development;

$ the program message concerning oversized systems. If the Challenge message successfully reaches its target audience, there will be a shift away from large equipment purchases and toward smaller equipment coupled with comprehensive services. This will require some major changes in the way that manufacturers and distributors interact with their customers;

$ volunteerism exhaustion before the project has reached its stated objectives; and

$ working the difficult politics of conflicting agendas.

5.3 Program Effectiveness

The initial funding raising goal was \$600,000 over two years, based on sponsor contributions of \$30,000 apiece per year; this goal was exceeded- 14 sponsors committed \$840,000. The organizational structure was also successfully developed and initiated with an active Advisory Board, a Project Development Committee, and several Working Groups.

A baseline for evaluating the program is being developed based on customer and expert surveys. Also contributing: a 1998 market research study conducted of approximately 300 plant operating and supervising technical staff conducted by the Energy Center of Wisconsin on behalf of the Compressed Air Challenge® and an extensive study of industrial motor systems applications conducted by XENERGY and ORNL for Motor Challenge. Strategies for assessing program impact include: training evaluation forms, post-training follow-up surveys, and feedback from consultants, distributors, and end users.

Events at a recent meeting of an equipment distributors' association clearly illustrated that the market changes that the Challenge hopes to promote have already begun. At this meeting, several leading distributors spent time outlining the business advantage of using a systems approach to compressed air systems. The message to their fellow distributors was "entering this new market takes planning and preparation, but the financial returns are substantial."

5.4 Key Success Factors

As work progresses and the potential for change becomes evident, individuals with differing points of view have become more engaged and active. Simultaneously, mutual respect has developed. Participants are willing to compromise, even over hotly-contested issues and closely-guarded positions of genuine disagreement, in the interest of project progress. The collective commitment of so many important market shareholders has created its own forward momentum. A sense of shared purpose has superseded personal agendas, however temporarily. The biggest challenge will continue to be maintaining a balance between developing an independent body of information for the public good and the commercial interests of those involved in its development and delivery. Government can be an effective moderator.

Factors contributing to the success of the Challenge include:

- $ Market influencers such as the investor-owned utilities and the compressed air controls industry are looking for business opportunities and threatening the status quo. The Challenge offers an opportunity for equipment manufacturers and distributors to look like good corporate citizens while keeping abreast of and influencing new market developments;
- $ utilities and utility consortia are looking for market transformation projects of manageable length and investment. The Challenge gives them a quality product for their critically important industrial customers at a highly leveraged bargain rate.
- $ the Challenge structure that allows the sponsors to share rights, form partnerships of their choice to deliver the resulting products and materials, and take credit for sponsorship;
- $ non-governmental sponsors have indicated that association with the DOE is a critically important public relations and marketing factor;
- $ compressed air users are able to participate without an overbearing time commitment. The presence of an influential corporate energy manager on the Project Development Committee has helped keep the project oriented toward end user needs;
- $ a strong base of quality technical information from which to draw; it requires consolidation and re-packaging; and
- $ working through representative associations allows input from the universe of interested parties to be managed.

5.5 The Importance of Champions

In the compressed air industry, which is small and highly specialized, many of the equipment manufacturers' personnel and consultants have worked for more than one company. As a result, today's protégé or mentor may become tomorrow's fiercest competitor or critic. For an outside party to be an effective facilitator in such a charged environment, they must first be accepted by the industry. There is no substitute for personal interaction in building that trust- logging the miles, visiting plants, appearing at association meetings, and the like.

A major result of this work was the identification of champions - representatives from each stakeholder group who were really willing to take risks to support and persuade others to support the Challenge. These representatives included: manufacturers, consultants, distributors, directors of state R&D organizations and a facilities engineering association, and representatives from national energy efficiency organizations. Without them and the influence that they wield among their peers, the project would not have come to fruition. Many of them have also invested huge amounts of their personal time and company resources in attending meetings, reviewing information, and drafting materials for the Challenge.

5.6 Next Steps

The Challenge sponsors agreed to a two year commitment to launch the Challenge. That time period ends in October 1999. Since sufficient funding exists, the Advisory Board has decided to extend the project in its current configuration the end of calendar 1999. Meanwhile, the Board has undertaken a long-term planning exercise to determine the most effective future for the Challenge. There is a consensus that the network established through the creation of the Challenge needs to be preserved and strengthened in order to complete the process of transforming the compressed air market. The sponsors would like to focus on effective use of the training products, rather than undertaking any large new development efforts. Future activities will include a major emphasis on marketing the training- both *Fundamentals* and the advanced training now under development. The Challenge will also be developing other informational products to build awareness among those who can't go to training and reinforce changes in behavior for those already trained. These will include: a toll-free technical assistance line, an enhanced website, and brief presentations that can be given at meetings and conferences.

Concurrently, USDOE's Office of Industrial Technologies has begun integrating and streamlining its industrial energy outreach programs into the Industries of the Future (IOF) BestPractices. Since many industrial facilities with compressed air systems also have other motor-driven and steam systems, there is an immediate opportunity to leverage the extensive network of participants (end users, suppliers, consultants, utilities, etc) developed through the Compressed Air Challenge to bring the more comprehensive BestPractices message and package of services to a large group already committed to energy improvements.

Participants in the Compressed Air Challenge have also identified a need to create an effective message for management concerning competitiveness, productivity, and reliability benefits to be realized from system improvements. This reinforces previous findings from market research conducted by the Motor Challenge Program. Creation of this "management message" is closely aligned with the goals of the IOF BestPractices and can be applied equally well to any motor-driven and steam systems. Moreover, it should be packaged as part of a company's overall continuous improvement effort. For this reason, it is likely that this work will be developed collaboratively with IOF BestPractices.

6 Applying a Collaborative Approach

The essential elements of the Collaborative Approach are: a market under pressure for change, a broad range of stakeholders, a market of manageable size, a public benefit opportunity (e.g.-energy savings, environmental benefit), a business opportunity, the presence of champions, and a facilitator. It is important that the potential facilitator takes the time to determine to answer to the question "why hasn't this situation already changed?" If current market information is insufficient to answer this question, more research is needed before proceeding with any planned intervention.

Since this activity is voluntary, a successful collaboration depends on analysis of each participating sector for the following:

- $ what is there potential contribution to the collaboration (why are they desirable participants)?
- $ what is their initial motivation to join the collaboration?
- $ what are their primary drivers? what do they hope to gain from their participation?

An analysis of the compressed air systems market participants is provided in Chart 1. Taking the time to conduct this analysis during project formation can really help structure a collaboration that is appealing to a broad base of stakeholders. This example, developed during the formation of the Compressed Air Challenge®, was based on trial and error. If this framework had been available during the earliest discussions with potential champions, the time required for project formation may have been reduced. By identifying the initial motivation and primary drivers, a facilitator can be far more effective in soliciting and recognizing the varying interests and agendas of the participants.

It is not necessary to select a market that is initially receptive to partnership or already known to the government in some way. It would have been difficult to identify a market that was initially less known to, or perceived as more resistant to partnership with government than the compressed air industry. The industry is a tightly-held, highly competitive industry with virtually no experience working with the public sector except for the Internal Revenue Service and the occasional antitrust threat.

A compelling characteristic of the market was pervasive mistrust among all parties including manufacturers, distributors, consultants, and the customers that they all served. It is important to recognize that initial conflict among participants often means an opportunity for transformation exists when something of value is at stake.

The Challenge began from a position of mutual suspicion, but with considerable curiosity. After an initial period of building trust, the lack of experience became an asset for the collaborative intervention. The market had few preconceived notions of what was possible, so the ideas and solutions were more forthcoming and creative than they might otherwise have been. The supply side has already begun to effect a market change in anticipation of the impact from the plant engineer training program.

Creating an Effective Government Role

The role of the U.S. government has been critically important to the Challenge's progress to date. The US government is in a unique position to encourage champions for change across the entire stakeholder spectrum, thus accelerating this transition while still working within market structures. Government can publicly recognize suppliers and end users as forward-thinking, empower would-be champions to build their own position within their organizations, and offer the market the prospect of avoiding future regulation. To realize this opportunity, a governmental entity must take the following into consideration:

$ First, ambiguity and shared control. An essential ingredient of a successful collaborative is the ability to listen and attempt to accommodate the desires, interests, ideas, and agendas of the other market stakeholders. Acknowledging publicly the skills, knowledge, and talent of people who have spent their entire careers in a specialized field is critically important. Ideally, the government facilitator will undertake a mutual exploration of the market with the stakeholders, each getting to know the other at different levels, and stakeholders educating the government facilitator on how the market works.

$ Second, need to empower a representative to facilitate. This is an activity that requires the facilitator to work "outside the box" to develop an understanding of the other stakeholders and to build their trust. Failure to do this will result in project failure as participants learn that the facilitator lacks the authority to negotiate.

$ Third, time. For the Compressed Air Challenge®, it was nearly four years from the first exploratory discussions to launch of the Challenge training program. While this type of time line is acceptable for funding of technical research, it is not typical of "applied program" activities in the energy efficiency field. A single budget cycle is not enough time to develop a meaningful collaborative intervention. The government facilitator must spend at least a year becoming known and trusted with the market(s) that he/she is attempting to influence before any tangible product can even be planned.

While the investment of time and money will decline in the outlying years, the continued presence of the U.S. government in some type of facilitating role will likely be needed for several more years before market transformation can be assured. Permanent change on this scale takes time

Table I: Compressed Air Systems Market - Market Analysis

Participant	Potential Contribution	Initial Motivation	Primary Drivers
Equipment Manufacturers & Distributors	Detailed technical & market information Customer influence	Brand enhancement Perceived threat to business	Sales Customer retention Increased profits
Specialized Consultants	System engineering Broad knowledge of applications/ problems	Increased visibility Legacy	Increased billable hours Recognition
End Users of Compressed Air Systems	Knowledge of specific applications & organizational dynamics	Alignment w/ green image May fix problem Concern about competitiveness	Reliability Profitability Efficiency Recognition
Influencers: Utilities Energy-efficiency organizations Government	Market credibility Outreach mechanisms Perceived neutrality	Meets mission Highly leveraged Politically beneficial	Customer education Customer retention Energy savings Recognition

7 Conclusion

Collaborative intervention is difficult but rewarding; it can be an attractive alternative to conventional prescriptive market intervention approaches. Each collaboration will have its own characteristics but the framework described in this paper can be applied to any market that meets the basic criteria for size, existing pressure to change, business opportunity, and public benefit opportunity. If these conditions are met and the facilitator is able to engage the support of champions, a collaborative intervention should be possible.

Our latest effort is the early stages of a collaboration in the pumping industry. The initial focus is on promoting a life-cycle cost approach to the procurement of pumping systems. The Hydraulic Institute (a trade association of U.S. pump manufacturers) is the primary participant in this activity, in cooperation with their European counterparts, EUROPUMP. There is a general recognition from the US participants that the American outreach effort will need to engage end use industrial customers in a meaningful way if implementation is to be successful. It is likely that this will be accomplished through work with key end use associations, including agencies responsible for U.S. government purchasing.

Collaboration requires skills and expectations different from more conventional technology deployment programs. These projects require a substantially longer development time and greater flexibility to accommodate the goals of market stakeholders. Because collaboration depends on building consensus, good communication skills are needed.

The rewards are several: a highly leveraged project, widespread and lasting market change, and a built-in exit strategy. If the collaboration is successful, a new market structure will be created which will perpetuate the project goals, even after government participation has ceased.

8 References

[1] [CAGI] Compressed Air and Gas Institute 1996. Personal communications to author. New York, N.Y.

[2] Compressed Air Challenge® 9/97-7/99. Advisory Board and Project Development Committee policy statements and meeting summaries. Madison, WI: Energy Center of Wisconsin.

[3] [DOE] U.S. Department of Energy 1996. National Market Transformation Strategies for Industrial Electric Motor Systems: Vols I & II Washington, D.C. : Office Industrial Technologies DOE/PO-0044.

[4] [DOE] U.S. Department of Energy 1998a. Fact sheet entitled "Compressed Air Challenge: A New Model for Industrial Partnerships" Washington, D.C.: Office of Industrial Technologies .

[5] [DOE] U.S. Department of Energy 1998b. *Improving Compressed Air System Performance: A Sourcebook for Industry.* Washington, D.C.: Office of Industrial Technologies.

[6] FOSS, R.S. 1997. *Compressed Air Systems Solutions Series.* Charlotte, N.C.: Plant Air Technology.

[7] HOWE, B. and B. Scales 1995. "Assessing Processes for Compressed Air Efficiency" *E-Source Tech Update*, November.

[8] KEMP, Henry. 1998. Personal Communication. Tampa, FL: Strategic Air Concepts.

[9] MCKANE, A.T., R.N. Elliott, and R. Wroblewski 1997. "Prospectus for a Compressed Air System Optimizations Training and Certification Program," September 19. Madison, Wisc.: Energy Center of Wisconsin.

[10] MCKANE, A.T., R.N. Elliott, J. Reese, and V.C. Tutterow. "Collaborative Intervention: Change from the Inside Out," in Proceedings: American Council for an Energy Efficient Economy 1998 Summer Study for Buildings, Asilomar, CA August 1998.

[11] PERRY, W. 1998. The State of the Industrial Compressor Market; in the 20th National Industrial Energy Technology Conference Proceedings, Houston, TX, April 22-23, 1998

[12] VAN ORMER, H.P. 1997. Personal communications to author. Pickerington, Oh.: Air Power, USA.

Pump System Effectiveness

Gunnar Hovstadius, Ph.D., Director, Technology

ITT Fluid Technology Corporation
35 Nutmeg Drive
Trumbull, CT 06611

When comparing different solutions for a pump system it is helpful to be able to easily identify the system effectiveness and to compare different solutions. In order to make an intelligent choice, some basic facts will need to be established. The first is: What are the process demands?

The Process Demands will generally steer you in a certain direction, i.e. do you need to be able to vary the flow rate and, if so, should it be continuously variable or can you vary your flow rate in steps? Can you use on-off batch pumping? What is your peak flow rate and how is the flow rate distributed over time?

The answers to these questions will determine if, and how, to regulate the flow. It will also give some guidance regarding the design of your pumping system.

An easy way of showing the flow demand is to use a duration diagram. A duration diagram (see Figure .1.1) in its simplest form shows how many hours during a year that you need a given flow rate - the dashed curve. The solid curve in the same diagram is interpreted differently. Each point on the solid curve tells how many hours during a year the flow rate exceeds the value on the y-axis.

This diagram is instrumental in understanding the pumping needs. The system must be able to deliver the peak flow but, from an economic point of view, it is also important to know at what flow rates the system is going to operate most of the time.

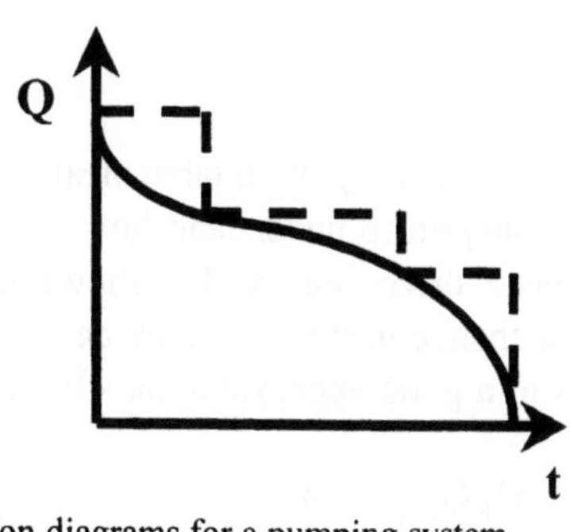

Figure 1.1: Duration diagrams for a pumping system

Given this information, the piping system can be designed. If, for example, the maximum flow rate occurs only for short periods of time, it may not pay to install a large diameter pipe. On the other hand, if the system is operated at peak flow rates for extended periods of time, this fact should be taken into account when specifying the pipe diameter. When the piping system is designed, it is time to calculate the system curve.

The system curve (see Figure 1.2) tells how much head, or pressure, is needed from each pump to push a given flow rate through the pipe system. The head has two components: the static head and the dynamic head. The static head indicates how high the fluid must be lifted or, if pumping into a pressurized vessel, the static pressure differential. The static head is independent of the flow rate.

The dynamic head is the pressure needed to overcome the frictional losses in the pipes and pipe fittings from moving the liquid at the desired flow rate. The frictional losses increase with the square of a change in flow rates in the turbulent flow region. So, if the flow rate is doubled, the dynamic losses increase by a factor of four. Now the duration curve can be used to determine where on the system curve the pump will operate and for how long. If a considerable length of time is spent high up on the system curve, a larger pipe diameter should be considered.

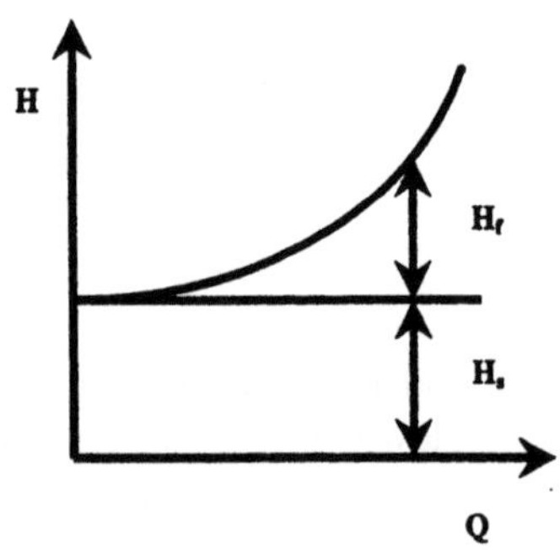

Figure 1.2: System Curve.

The pump, or pumps, can now be specified, plus suitable means for regulating the flow - if needed. Also, now it is helpful to understand how a pump performs in conjunction with a variable speed drive. Figure 1.3 shows how the pump's performance curve responds to a change in the frequency delivered to the motor. New curves can be calculated - to a good approximation - by using the Affinity Laws,

$$Q_1/Q_2 = n_1/n_2$$

$$H_1/H_2 = (n_1/n_2)^2$$

where n is the speed of the impeller in rpm, Q is the flow rate, and H is the developed head. The broken lines show how a particular point on the original curve moves to form a new curve. Pump efficiency remains essentially constant along these lines.

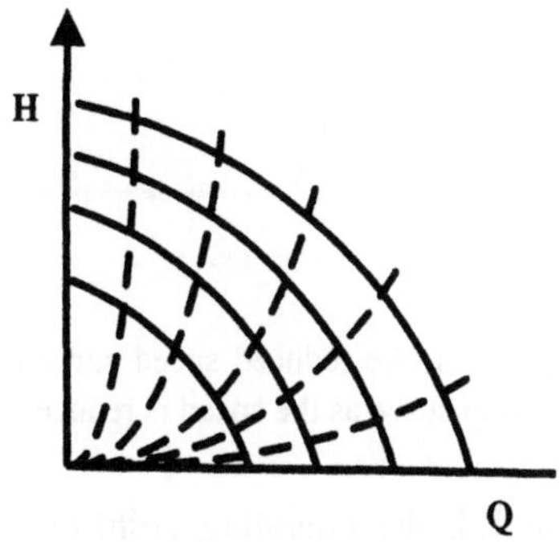

Figure 1.3: Lines of constant efficiency (broken) superimposed over speed-regulated pump curves (solid).

A common mistake is to use the Affinity Laws to calculate energy savings. Although this may be done as a good approximation sometimes, it can also lead to major errors. To find the operating point of a pump in a given system, the pump curves from Figure 1.3 must be superimposed on the system curve from Figure 1.2. It is important to note that the operating point is the intersection of the pump curve and the system curve. In systems without static head, the system curve follows the lines of constant efficiency in Figure 1.3, and pump efficiency remains essentially constant as the pump speed changes.

However, in systems with static head, the system curve does not start from the origin but at some non-zero value on the y-axis corresponding to the static head. Hence, the system curve does not follow the curves of constant efficiency in Figure 1.3. Instead, it intersects them. In turn, this means that the pump efficiency changes when the speed of the pump changes. This is quite the contrary to a pump system without static head, in which the pump efficiency remains fairly constant with changing speed.

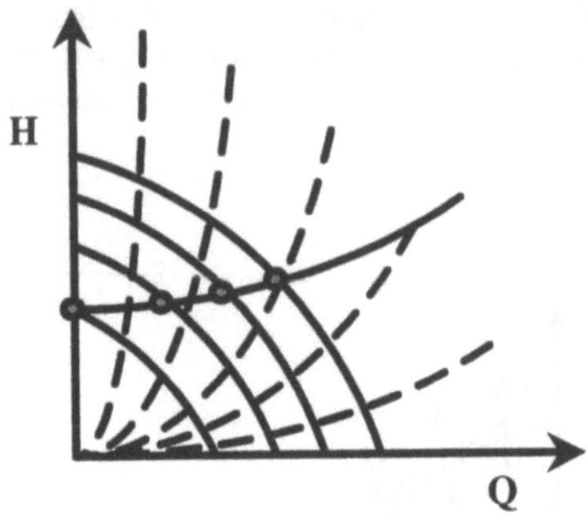

Figure 1.4: The operating point on the reduced speed curve moves relatively higher and higher on the pump curve as the speed is reduced.

As can be seen in Figure 1.4, the operating point on the reduced speed curve moves relatively higher and higher on the pump curve as the speed is reduced. Eventually, the pump runs at shut-off head if the speed is reduced too greatly.

There are two problems connected with this situation. First, the pump efficiency goes towards zero when the operating point moves towards the shut-off head. The second problem is that the pump then operates in a flow regime where it should not be operated. Operating the pump close to shut-off head for extended periods of time could severely damage the pump - even if this takes place at a reduced rate of speed.

The more static head there is in a system, the more dramatic these effects will be. At high static head, a relatively minor decrease in speed moves the pump operating point into an area where the pump should not be operated continuously. Decreasing the speed even more brings the operating point to the shut-off point. Then the motor is running and consuming energy, although generally less than it does at full speed. The flow rate is, however, equal to zero and the energy provided to the pump goes into heating the fluid through friction.

1 Specific Energy

A pump system is built to move a certain volume of fluid from one point to another (in circulating systems they are the same). A useful measure for calculating the cost of pumping is the specific energy E_s. Specific energy is measured as Wh/g, where W = power in watts, h = time in hours, and g = volume in gallons. See equation below:

$$\frac{Energy \cdot used}{Pumped \cdot Volume} = Specific \cdot Energy \qquad (1)$$

$$E_s = \frac{P_{in} \cdot Time}{V} = \frac{P_{in}}{Q} \qquad (2)$$

P_{in}= input power to the driver

E_s is a function of flow rate (Q). It is, therefore, necessary to evaluate this dependence. (This is especially important when a variable speed drive is contemplated in order to avoid pitfalls.)

It is useful to separate systems with and without static head since they have different characteristics.

1.1 Systems *without* Static Head or Circulating Systems

The specific energy here is dependent on the frictional head loss which, in turn, is determined by the losses in the pipe system (including throttling valves), and by the combined drive - motor - pump efficiency.

The combined drive - motor - pump efficiency has to be evaluated for each duty point. It is to be noted that the pump efficiency remains approximately the same in a system of this type when the speed is changed, whereas the drive-motor efficiency can drop drastically as the load is reduced. If on the other hand, the system curve is changed by changing the setting of a valve, this will change the duty point of the pump and, hence, its efficiency.

1.2 Systems *with* Static Head

In systems with static head the energy usage can be calculated in a slightly different way. The head needed from the pump can be separated into static and dynamic (friction losses). Substituting H_s+H_f for the total head in the expression for specific energy will generate the following expressions:

$$P_{in} = \frac{Q \cdot (H_s + H_f) \cdot \rho \cdot g}{\eta_{motor} \cdot \eta_{pump}} \qquad (3)$$

where.

ρ = density
g = gravitational const.
η = efficiency

$$E_s = \frac{H_s + H_f}{H_s} \cdot \frac{\rho \cdot g \cdot H_s}{\eta_{motor} \cdot \eta_{pump}} \qquad (4)$$

if,

$$\frac{H_s}{H_s + H_f} = f_{HS} \qquad (5)$$

then,

$$E_s = \frac{H_s \cdot \rho \cdot g}{\eta_{motor} \cdot \eta_{pump} \cdot f_{HS}} \qquad (6)$$

The "Hydraulic System factor", f_{HS}, indicates the relative amount of static head in the system. It is obvious that E_s now has a minimum value: $H_s \cdot \rho \cdot g$ would occur if all efficiencies were equal to 100% and there were no friction losses.

The different factors are all functions of the flow rate and will vary with the duty point. If a variable speed drive is used, they will vary with speed as the duty point moves along the system curve.

Motor efficiency will generally decrease as the speed is lowered and the motor goes below 75% of full load. The drop in motor – drive efficiency can be substantial if the motor load drops below 50% of full load.

The Hydraulic System factor will increase when the friction losses go towards zero, which happens when the duty point approaches the shut off head. The specific energy will always increase drastically as the duty point moves towards shut off head in systems with static head due to reduced pump, motor and drive efficiencies. In systems with high static head, this can happen even at a relatively moderate decrease in speed.

In such systems the area of usefulness of a variable speed drive can be somewhat improved by making sure that the system curve and the full speed pump curve intersect to the right of the pumps best efficiency point.

To calculate the cost of pumping the specific energy has to be calculated along the system curve. Combining this information with the information in the flow duration diagram the cost of pumping can be determined.

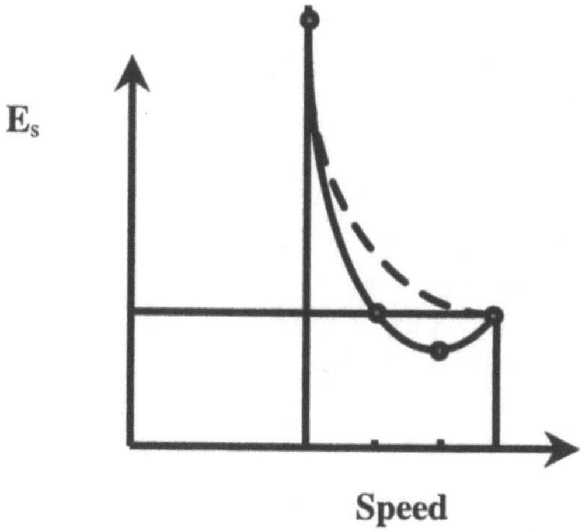

Figure 1.5: Specific energy as function of speed

2 Flow Regulated by Throttling

When the flow is regulated using a throttling valve, the system curve is changed. The specific energy can be calculated for each operating point by dividing the input power to the motor by the flow rate. E_s usually increases rapidly as the flow is reduced - typically like the dashed curve in Figure 1.5.

3 Parallel Pumps, Common Header

It is important to understand that parallel pumps discharging into a common header exhibit a phenomenon that is completely independent of any static head that may be present. The first pump discharging to the header pressurizes the header. The second and subsequent pumps that come on line must then pump into a pressurized header. The net effect is the same as pumping against a static head. The more pumps that are running the higher the pressure in the header, which limits the chances of saving energy by using a variable speed drive. See Figure 3.1 for a graphical explanation.

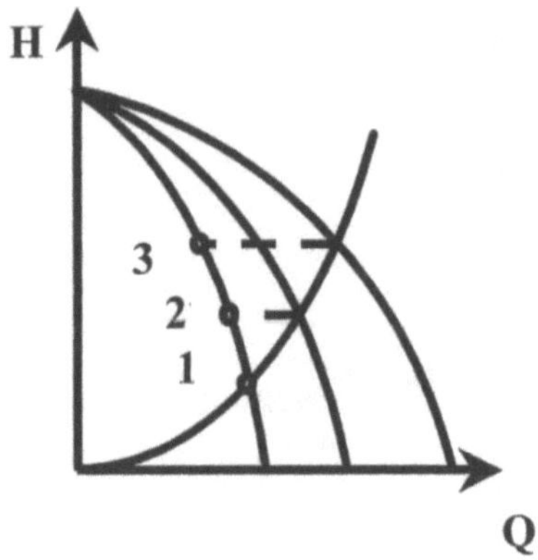

Figure 3.1: Parallel pump operation.

Running parallel pumps into a header increases the pressure in the header. With three pumps running, each operates at point 3. Thus, a fourth pump must deliver at least this pressure before it produces any useable flow. This situation is, therefore, identical to pumping against a static head.

4 Summary

To understand a pumping system it must be realized that all of its components are interdependent. Sub-optimization on the component level can easily be deceptive. When examining or designing a pump system, the process demands must first be established. For example, does the flow rate have to be regulated continuously or in steps? Can on-off batch pumping be used? What are the flow rates needed and how are they distributed in time? A duration diagram encapsulates the answers to these questions.

The true cost of pumping can be calculated by using the information from Figure 1.1 and Equation (6) where E_s has to be evaluated for all flow rates. In most instances, pumps are sized to deliver the peak flow rate with some margin upwards. With duration curves looking like the continuous curve in Figure 1.1, it is, therefore, common that the *normal* flow rate is around, or less, than 50 percent of the design flow.

If the system curve exhibits a fair amount of static head and the pump is oversized for most of the pumping needs, then problems can occur. It is not uncommon in such systems to find that the cost of pumping is considerably increased, compared to on-off pumping at full speed, when using a variable speed drive. That the combined efficiency of a variable speed drive-motor package can drop considerably as the load is decreased does not make the situation better. Throttling the flow, however, is still worse!

Variable speed drives can give good control of the flow rate. They generally reduce the operating cost in systems when compared to throttling valves, and in systems with little or no static head without throttling valves. In systems with high relative static head, extra care has to be taken when using variable speed drives to avoid the pitfalls of low pumping efficiency and operation in harmful flow regimes. The allowable speed range thus becomes restricted both from an operational and economical point of view.

Driver Selection Matched to Demand: A Consideration for Maximising Lifecycle Costs by Using the Correct Driver

Ian Metcalf

Brook Hansen, St. Thomas' Road, Huddersfield. HD1 3LJ.UK

Abstract. This paper will consider the various driver options available to the pump designer, consultant or engineer. Its aim is to be a practical guide to maximising lifecycle costs by using the correct driver.

It will highlight the importance of efficiency and reliability within the choice of driver, whilst taking into consideration the need to examine the system to obtain lowest life cycle costs. Given that the majority of pumps are driven by a cage induction motor, discussion will be centred on this type of driver. Mention will be made to the new motor labelling scheme and its likely impact on pump system design.

Examples of some of the cost benefits from using the best driver will be given, highlighting the economic impact that the driver has within the pumping system. Ultimately it will demonstrate the economic and environmental sense of using a pumping system which is both energy efficient and designed for optimum performance.

1 Introduction

This paper is designed to give practical assistance to all involved in the design and use of pumps within industry today. Emphasis is made on life cycle costs and maximising the pump and driver to achieve the best efficiency point.
As an example, a cooling system may provide a water supply to water-cooled elements manufactured by others. The manufacturer will specify the water demand for the elements based upon physical and theoretical calculations for the system plus an inherent safety margin.
The water supplied is then taken as a minimum so the system is designed to supply the specified water demand continuously.

As we now can see the whole system is becoming over designed which will mean higher capital expenditure and higher energy costs due to system inefficiency from over sizing.

Attention should be paid to the different types of electric motor available to the pump user. Although the energy cost will be the major factor in the Life Cycle Cost (LCC) of the unit, other costs (including purchase and maintenance cost) should be taken into consideration when selecting the type of motor and pump for a given application. The motor specification should be defined to minimise life cycle cost considering issues such as enclosure, starting torque, service factor, duty cycle and environmental conditions.

Although this document is principally about pumps, it should be mentioned that on a typical industrial site motors account for approximately two thirds of the electricity costs. The principles outlined will apply to all motors on a given site; not just those used as pump drivers.

2 Drivers

The driver must produce sufficient torque to start the pump, accelerate it to speed and run at a constant speed, without overheating or disrupting the electrical supply.

Throughout industry the environmental conditions vary considerably from hot and dry to wet and humid, clean to extremely dirty, and the driver has to be capable of continuous running whatever the conditions. As with most applications we expect reliable equipment together with low maintenance and trouble free running.

While there are different types of drivers available (such as diesel engines, steam engines, d.c. motors, permanent magnet motors, synchronous reluctance motors and wound rotor motors) the majority of pumps are driven by an a.c. induction motor. Hence this paper deals mainly with this type of driver.

2.1 Standard Induction Motors

The induction motor is by far the most commonly used driver in the industry. The variations in enclosure, (totally enclosed, drip proof, hazardous area) and mounting (horizontal or vertical, foot or flange mounted) are wide ranging. IEC and NEMA standards define the physical, as well as, electrical characteristics of these motors.

Typically a pump using a motor requires a base plate, pedestal or, in the case of vertically mounted a shaft seal, a flexible coupling, a coupling guard and labour to align the units.

High efficiency electric motors are easily adaptable for variable speed and variable torque applications, when operating against a low static head, without effecting the driver size. It is important to consult with the driver supplier in order to ensure compatibility is assured.

It must be noted that in cases of failure or upgrading, the driver can be easily changed thus reducing maintenance costs due to reduced downtime anu capital expediture.

2.2 Submersible Motors

Two types of submersible motor are available, the 'dry stator' version where the stator is isolated from the liquid being pumped by an oil chamber, and the 'wet stator' version where the stator is in contact with the liquid being pumped.

For many types of pump it is practical to use the wet stator type where the motor operates submerged in a liquid and thereby avoids the need for mechanical shaft sealing. This can be less efficient because of the friction losses in the liquid in the rotor gap and because the physical dimensions do not allow optimum design. In principle the same rules for motor design are valid for these motor types. However, the installation cost of a pump using this type of motor is generally less than a pump using a conventional motor.

Maintenance costs can also be reduced as the submersible motor is not fitted with a mechanical shaft seal.

2.3 Canned Motors

This design is similar to the submersible design in that the motor is part of the pump unit. This type of motor is typically used for hazardous conditions where leakage into the atmosphere cannot be tolerated, as the liquids pumped are often toxic or corrosive. Thus the stator and rotor are shielded from the fluid by a thin metallic shield or can. Although the canning material is generally a highly magnetically permeable material, such as Hastelloy, eddy current losses in the can and friction losses of the liquid in the rotor air gap reduce the efficiency of the unit.

Typically canned motors will be 1% to 10% less efficient than the standard induction motor.

Comparative installation cost, minimal space requirements, lower maintenance costs and environmental benefits are where most of the saving will be achieved over the lifetime of the pump driven by a canned motor due to its integral design.

3 Variable Speed Drives for Low Static Head Systems

When considering the running efficiency of any system, the use of a variable speed drive offers significant benefits by allowing the speed to be accurately set to the minimum allowable for a given process, hence allowing the opportunity for decreased energy input.

When considering utilising a variable speed drive with an existing motor, care should be taken to match the electrical characteristics of the motor and drive. It should be noted that a 10 % reduction in speed would give a 30 % reduction in energy consumption.

In general a variable speed drive will consume approximately 2 to 6% of the inlet power (94-98% efficiency). The motor itself may also be less efficient due to the influence of distorted waveforms, known as harmonics, in the power output from the drive. These efficiency penalties should be considered in any analysis of energy saving using variable speed drives.

The efficiency drop will be minimal but may vary from one motor manufacturer to another. It is also dependent on the type of inverter drive used, however there is no international standard for inverters. The following characteristics vary from one inverter to another and will affect motor performance:

- Total harmonic distortion
- Peak voltage
- Maximum dv/dt
- Maximum carrier switching frequency
- Maximum cable length between the motor and inverter

There is also the question of motor de-rating, particularly on constant torque characteristic applications, to compensate for both reduced cooling when the motor external fan is not generating the normal airflow at low speeds and the increased heating created by the harmonics in the inverter waveform.

Each motor manufacturer must answer the question of how much de-rating is required. There are no definitive rules that can be applied to all motors. However, in general, little or no de-rating is required on most centrifugal type pump drives. On other types of pumps, some de-rating may be required.

It may be that special windings/insulation/bearings are required depending upon the application, dv/dt rate and the voltage supply. Advice must be sought from both the motor and the drive manufacturer as variable speed technology is constantly changing and needs to be understood.

4 Efficiency and Energy Costs

Electricity consumed by British Industry for pumping fluids is estimated at more than £1,000M per year. The driver of the pump, which is predominantly an a.c electric motor, accounts for over 50% of electricity consumption within industry today.

The cost of electricity used by the average electric motor during its lifetime is approximately one hundred times its purchase price. This illustrates the importance of giving close attention to efficiency criteria. The current legislative situation within both Europe and North America demonstrates the wide range of efficiencies available to the pump user.

It should be noted that the efficiency stated on a motor nameplate is the full load efficiency. For a more accurate assessment of energy costs, the efficiency at the anticipated duty point should be obtained from motor manufacturers. It is also important that efficiency comparisons are made using results from a common testing method as testing standards vary by country and may give different results. An example of the energy cost calculation is given below to demonstrate the effect on running costs with a difference in motor efficiency.

Consider what 3% energy saving on a driver would mean.

Example 30kW 4 pole fixed speed pump.

Duty point	24 kW.
Running hours per year	6000 Hrs
Cost per kWh	0.07 Euro

Manufacturer A.	
Full load efficiency	92.5%
¾ load efficiency	92.5%
Energy cost per annum	**10897 Euro**
Manufacturer B.	
Full load efficiency	90.0%
¾ load efficiency	89.0%
Energy cost per annum	**11326 Euro**
Saving per annum	**429 Euro**

The duty point or loading of the motor can also affect its efficiency. In most cases the motor will be operating at a duty point below its rated output. As can be seen from the performance figures shown in Eurodeem and Motor Master + software that the difference in efficiency at full load between the standard and high efficiency motor vary by as 2 to 3%, but at half load the difference is considerably higher.

For most high efficiency motors the optimum running output is at ¾ load as this is where the motor has been designed to run. This is taking into account a certain amount of system loss.

When using the motor performance software data it can been seen that after the ¾ load efficiency has been reached, the efficiency curve drops off rapidly down to ½ load and any system working on ½ load and below will be running on low efficiency and low power factor which will give rise to high current consumption of the system.

Whichever type of motor is chosen, the efficiency will fall at light loadings. If a motor is constantly loaded below 40% there is one technique to improve the efficiency of the motor.

This is to run a star delta design of motor permanently in star. This reduces the flux in the machine by approximately 40%, thereby reducing the iron losses and increasing the effective motor efficiency.

The best way to avoid this problem is to ensure that the motor is sized as close to the pump duty point as possible.

5 Consider the Facts

From the information above we should consider all applications even if they are operating a trouble free system. It should not be assumed that just because the system runs trouble free that it is running efficiently.

With over 50% of industrial electricity consumption being accounted to electric motors any saving on efficiency will have a major impact on the overall electricity bill and will have an even further reaching environmental impact in the reduction of electricity demand leading to a reduction in greenhouse gasses.

What we need to consider is the awareness of pumping costs on each and every process in our factories.

6 Summary

Consider the application. Certain instances will require a motor that is built in to the pump in order to maintain reliability, however always try to use a high efficiency design.

The driver is an integral part of the pumping system and is one of the easiest things to change in order to improve the systems efficiency.

Energy efficient motors are now readily available from manufacturers with no added cost premium giving a logical argument for high efficiency motor specification to the equipment manufacturers.

Equipment fitted with high efficiency drivers will give lower running lifetime costs, ensure that equipment is compliant with the latest European guidelines and ultimately help preserve the environment due to reduced smoke stack emissions as less electricity is used.

Environmental Impact Assessment (LCA) Energy and Re-cycling for a Circulator

Senior Engineer Nils Thorup

Grundfos A/S, Research department, DK-8850 Bjerringbro

Abstract. The use of LCA in the product development process gives two clear arrears for improvement of the environmental impact of a Circulator. Most of the energy used in the whole life cycle, from cradle to grave, is used in the use phase, with the end user of the product. Typical 99% of the whole energy are consumed in this phase. All other phases like raw material productions-, transport- and the disposal phase consume 1%. The main environmental improvement will therefore be development of high efficient pumps, and secure the right dimensions of the pump to the system were in it has to operate. An other important environmental improvement of the pumps environmental performance, is to make sure, that the resources, bounded in the can be recycled under disposal. The most important material in a typical Circulator is the amount of Copper. Therefor the Circulator should be constructed in a way, sow the copper can be recycled in the disposal stage.

The conclusion based on the LCA is general for almost all types of pumps, who run daily under the use phase. Only "special" pumps like "pumps in fire fighting equipment" have a different environmental impact grouping

At Grundfos we use the Danish LCA-tool EDIP (Environmental Design of Industrial Products) to secure that the environmental performances are documented in the product development process.

1 What is a LCA?

Life Cycle Assessment is an environmental tool to assess the environmental impacts of a product. The impacts are mapped from “cradle to grave”: There are several different tools to use. Grundfos use the Danish toll called EDIP (Environmental Design of Industrial Products)

1.1 When to Make a LCA (purpose)

A LCA can bee used for different purpose. The Object of the LCA is properly the most important thing to do. Grundfos use only LCA for product development.

1.2 LCA for a Smaller Pump

Figure 1: the old circulator

Figure 2: the new circulator

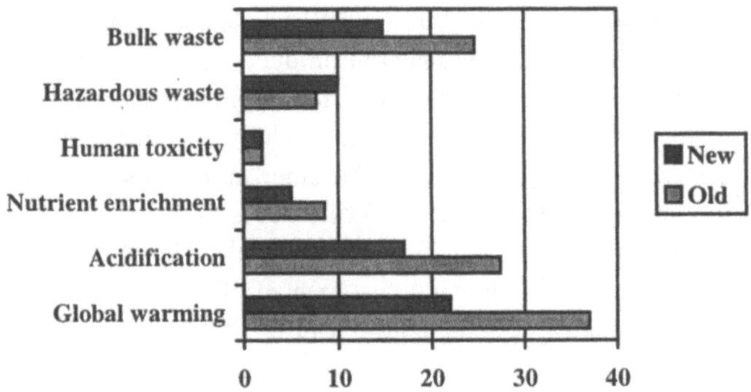

Figure 3: A typical LCA for a circulator used in the product design phase to visualise the environmental impact of the circulators. The reason why the impact category "Hazardous waste" is higher for the new circulator is only, that all electronic is categorised as "Hazardous waste" and will only be a potential risk if the circulator is disposed wrongly

1.3 Analysis of the Environmental Impacts

The analysis of the LCA show, that the main contributions to the environmental impacts are caused by the electricity use in the use phase, when the pump is operating.App. 99% of all energy consumed in the pumps entire life cycle is used in the use phase.An other important impact of the environment is the use of copper in the pump. This impact is comment from both products in this LCA.

2 Conclusion

Energy efficiency is the most important environmental parameter for a pump. A pump must be designed for disassembling/ recycling to ensure recovering of resources.

Promoting the Purchase of More Efficient Pumps A SAVE Study

David T Reeves[1], Hugh Falkner[2]

[1] Pump Engineering Consultant, UK
[2] ETSU, AEA Technology, UK

Synopsis. A successful SAVE study on improving electric motor efficiency in the EU led to the initiation of a similar study on pumps. It is generally acknowledged that the greatest opportunities for savings in pump life cycle costs lie in the field of application. These aspects are being pursued elsewhere. This study therefore concentrates on improving the efficiencies of the pumps themselves operating within the EU. Potential energy savings are well in excess of 50 MECUpa. The pumps initially chosen for evaluation are of the double entry split case and end suction types working within the duty ranges of maximum popularity. The analysis of efficiencies as published by makers shows considerable scatter. 'Higher efficiency' pumps will be defined to give makers and users the opportunity to produce and to buy pumps which will give lower life cycle costs. Methods of avoiding unnecessary deterioration in efficiency during pump lifetimes will also be publicised. Five partners chosen from across Europe are taking part in the study with assistance from the makers and users.

1 Introduction

A SAVE study on improving the efficiency of motors in the EU was initiated in 1995. As a result, the voluntary use of agreed efficiency classes for motors will be implemented this year. This is expected to lead to a general improvement in motor efficiencies with consequent savings in energy consumption in the future.

In view of this successful outcome, a SAVE study on improving the efficiency of pumps was initiated at the beginning of 1999. This paper summarises the current status of the project.

2 What's it All About?

We all know that most of the energy wastage associated with pumping is due to misapplication. This covers such things as over-estimating system losses, oversizing pumps, inappropriate pipe fittings, flow control by throttling or bypass, etc.. These aspects of energy efficiency are already under review elsewhere.

The purpose of this study is to concentrate on the pump itself. The basic objective is to encourage the use of more efficient pumps within the EU.

Pumps use more electricity than any other single type of machine in EU industry. It is estimated that the electricity used to run the pumps within the EU is around 12,000 MECUpa. Thus if, say, the pumps representing just 30% of the total energy consumption were on average 1.5% more efficient, the energy savings would be worth around 54 MECUpa.

3 Who's Involved?

The partners working on the project are:

ETSU (AEA Technology), UK (Project Co-ordinator)
CETIM, France
Darmstadt Technical University, Germany
David T Reeves, Pump Engineering Consultant
NESA – Energy Technology, Denmark

These parties have been chosen to represent a broad spectrum of technical, market and geographical interests. Pump makers and users are also involved as observers and advisers. All data exists, no new R & D is involved.

4 Which Pumps Are Covered?

To deal with every type of pump used in the EU would obviously be impractical. We therefore decided to concentrate initially on centrifugal pumps operating on 'clean' cool water. This includes duties such as water abstraction, water supply and cooling water. It excludes 'dirty' duties such as sewage and drainage which require special construction, materials, clearances, seals, etc..

Next we had to decide on the pump type. An examination of the market showed that the majority of purchases by value were single stage pumps. These were mainly of the double entry split case and end suction types, the latter being either close coupled with the impeller mounted on an extension of the motor shaft or long coupled with their own bearings. We have yet to decide whether to include at this stage pumps with in-line branches which are used mainly for the circulation of hot or chilled water but are also used for cold water pressure boosting.

Finally, we needed to specify an operating range. Again, to keep the study and the implementation within reasonable bounds, we looked at the areas of maximum purchase value. This resulted in limits for sizes, flows, heads and powers. The 50 Hz speeds of two pole and four pole motors only were selected. The resulting data is shown in Table 1. It can be seen that the shaft power range is quite wide, from 1.2 to 680 kW.

Table 1: Pump size limits selected

		2 Pole				4 Pole			
Pump Type	Size Limit	Outlet Size (mm	Equiv. Flow (m^3/h) approx	Head (m) probable max or min	Shaft kW @ max or min head	Outlet Size (mm)	Equiv. Flow (m^3/h) approx	Head (m) probable max or min	Shaft kW @ max or min head
Double Entry Split Case	Max	150	400	150	200	300	1500	140	680
	Min	80	130	50	23	100	150	15	7.4
End Suction own bearings	Max	100	200	130	96	150	400	50	67
	Min	50	30	20	2.5	50	30	10	1.2
End Suction close coupled	Max	100	200	130	96	100	100	50	20
	Min	32	15	20	1.4	50	30	10	1.2

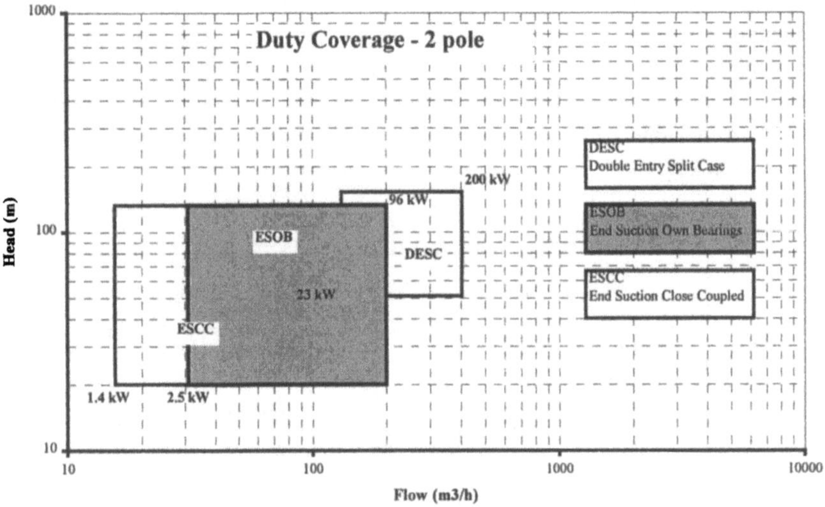

Figure 1: Range coverage at best-efficiency duties (2 pole)

To fully define the areas of interest we then produced Figures 1 & 2, which indicate the exact ranges of best-efficiency flows and heads which we will consider. The two pole coverage (Figure 1) shows the simple rectangular boxes of head and flow as defined by Table 1. However, at four pole speed (Figure 2) rectangular boxes would have produced some abnormally low or high specific speeds at those corners of the boxes which are not directly defined by Table 1. It was therefore necessary to change the rectangles to trapezia to avoid those duties where pumps in parallel or series (or even multistage) would be much more efficient.

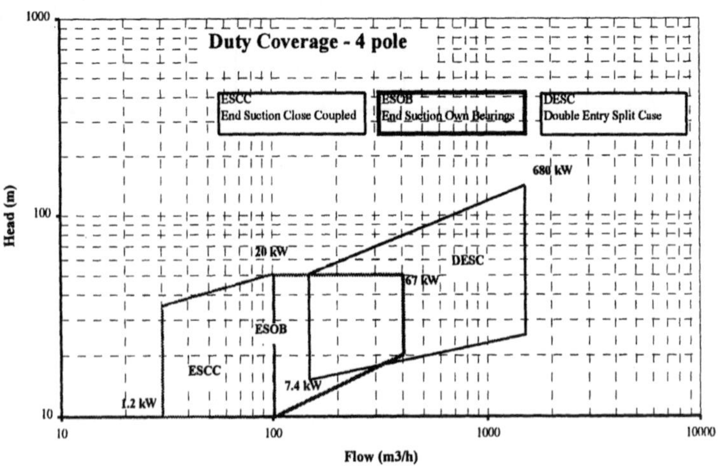

Figure 2: Range coverage at best-efficiency duties (4 pole)

A more detailed analysis of the market is on-going. This will hopefully confirm our initial choice of duty coverage and will assess the potential for energy savings and the cost-effectiveness of the various options we will consider.

5 What Performance Will we Analyse?

Firstly, we must collect data from pump makers in the form of hydraulic performance curves showing head, flow, efficiency and NPSH at maximum impeller diameter. The NPSH figure is needed since we will ignore any pump requiring more than 10m NPSH as 'abnormal'. This data is readily available and a large amount has already been collected. We must emphasise that no data will be published in a form that makes it identifiable with any particular maker. It is no part of the project to point a finger at anyone specific!

Secondly, we must analyse the pump efficiencies. The question therefore arises of which efficiency to analyse. The most simple is obviously the efficiency at best efficiency flow. However, this fails to take into account two relevant facts:

(a) On average, pumps operate at flows below their best efficiency flow.

(b) Pumps usually operate over a range of flows. Thus, an efficiency curve which has a high peak value but falls off rapidly away from the peak is not ideal.

After much discussion, in which we included makers and users, we have decided on a 'weighted efficiency' of the form:

$$\text{Weighted efficiency} = \frac{(60\%Q\text{x}1)+(80\%Q\text{x}2)+(100\%Q\text{x}2)+(120\%Q\text{x}1)}{6}$$

(where 60%Q means efficiency at 60% of best-efficiency flow)

6 How Do we Do the Analysis?

There are many ways of producing curves of pump efficiency against the mutual variables of head, flow and speed. One of the most traditional is to plot efficiency against specific speed (N_s).

$N_s = N \times Q^{1/2} / H^{3/4}$ (where N = speed, Q = flow, H = head)

However, this must be done for a discrete number of flows, since large pumps are more efficient than small pumps. This means that analysis by plotting efficiency points for many pumps on a curve of this type is impractical.

The method we have chosen is to plot efficiency against flow for each type of pump and running speed. This is also the method used by the Hydraulic Institute of the USA amongst others. The disadvantage is that head variation must be allowed for, since at a given flow pumps with high or low heads will have lower efficiencies than pumps with 'optimum' heads even though the pumps are of equal 'quality'.

The variation of head is accommodated by making a correction for specific speed. For example, Figure 3 shows a plot of peak efficiency against flow for several makes of end suction pumps having their own bearings at 2900 rev/min. A specific speed correction curve is also shown. Thus, at a given flow a pump of 2500 specific speed would have its efficiency point plotted directly, whereas a pump of 800 specific speed would have 8 points added to its efficiency before plotting. In this way, efficiency scatter should be a reflection of pump 'quality' only. Indeed, a check on those efficiencies which are relatively high or low on the plot indicates that the specific speed correction as used is about right and other factors are responsible for the scatter. Nevertheless, the precise form of efficiency correction is still under review.

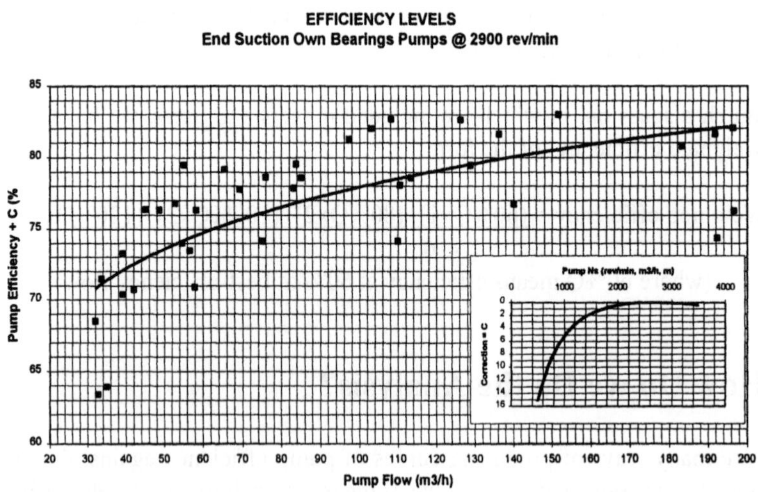

Figure 3: Sample plot of efficiency against flow

Although relatively few pumps are shown on Figure 3, the scatter of efficiencies lies roughly between four points above and seven points below the average. It is important that the extent of efficiency variation is brought to the notice of all involved, to emphasise the scope for choosing pumps of better efficiencies.

7 What Do we Do with the Analysis?

Figure 3 includes a line of average efficiency derived from the points plotted. Those points above the line represent pumps of 'above average efficiency'. A differently defined dividing line could be chosen, e.g. above average. Those points above the line could represent, say, 'Grade 1' (or 'Eff1' to mirror the SAVE motor classification). A decision on the various possibilities for grading, labelling, etc. will be taken towards the end of the project. The feeling is that a buyer should have some means of easily identifying 'higher efficiency' pumps.

One major problem with pumps when compared with motors is that accurate testing is extremely difficult and expensive. Initial thoughts are that any grading should apply directly to pumps which have been tested by the maker to Class B of ISO 9906, which should be in force before the end of 1999. This will give a test tolerance on efficiency of +/- 3%. Pumps which have only been tested to Class C (tolerance of +/- 5%) could then have 2% of their efficiency deducted before comparing with the Class B figures.

8 Can Efficiencies Be Improved?

It is not the intention of the project to tell makers how to make pumps! Work has been done on theoretically achievable efficiencies and we intend to compare these figures with the actual efficiencies we plot.

Hydraulic, mechanical and volumetric losses will be analysed and compared with achieved efficiencies for a number of pump sizes and specific speeds. Any improvements which appear to be possible will be evaluated for cost effectiveness. These will include surface roughness and wear ring clearances and geometry.

9 Are High Efficiencies Maintainable?

There is little point in encouraging the use of higher efficiency pumps if their efficiencies fall off rapidly in service. A small pump on clean water supply has been known to lose over 20% of its new efficiency in less than two years of operation. Although this is an extreme case, it has been shown that larger clean water pumps can lose, on average, around 5% of their new efficiency in the first five years of operation.

Part of this loss in efficiency is due to wear at the impeller/casing wear rings. The rate of wear will tend to be much greater in pumps which operate at flows away from the best-efficiency flow due to increased shaft deflection.

Most of the loss in efficiency will normally be due to a build-up of corrosion products in cast iron casings. This occurs with practically all pumps working with clean cool water. The rate of corrosion is affected by the catalytic effects of bacteria which are usually present even in treated drinking water.

Several methods of maintaining pump efficiencies at high levels will be evaluated to determine their cost-effectiveness. Obvious amongst these are regular monitoring and maintenance, using genuine spare parts and competent repair companies, avoidance of low-flow operation, improved wear ring materials, double-volutes in larger pumps to minimise radial thrusts and coating of flow passage surfaces.

The project will encourage users to think carefully about how they can avoid costly loss of efficiency over an extended period of time.

10 How Will this Affect the Maker and User?

It is the intention of the project to encourage the purchase of higher efficiency pumps. If this is successful, sellers of less efficient pumps will wish to improve their efficiencies. This will cost money. In co-operation with pump makers we will make evaluations of extra costs involved in such items as patterns, tools, materials and labour.

We will also be looking at the effects on buyers of improving pump efficiencies. The desired balance between capital and revenue costs varies considerably in different markets and according to who does the buying, e.g. whether a contractor or an end user.

Whoever is buying, it has been demonstrated many times that the cost of a pump is insignificant in comparison with its running costs. Most users are now aware of the importance of life cycle costs and this is being endorsed by the Europump Enersave project. The SAVE pumps project will underline the fact that paying a relatively small premium for a better pump will prove to be a very sound investment.

11 Conclusions

The efficiency levels at which pumps operate in the EU will be improved as a result of the SAVE pumps project. The project will propose policy options which will then be discussed with makers and users to arrive at agreed methods of implementation.

The end result may be the grading/labelling of individual pumps in an agreed manner so that sales and purchases of better products can be encouraged.

On the other hand, the end result may be a series of curves which enable purchasers to assess whether the pumps they are considering buying meet or exceed target efficiencies.

There again, it could be a combination of these options. The important outcome is to provide yardsticks which are easy to use and which promote the sale and purchase of more efficient pumps, to the benefit of users, manufacturers and the world we live in.

On Site Testing

Ilkka Sintonen

Marketing Manager, Oy E. Sarlin Ab, Finland

1 Pump Inspection Project

Existing systems are a larger opportunity for energy savings, through use of LCC methods, than new systems, for two reasons:

- There are 20 times as many pumps systems in the installed base as are purchased each year
- More existing systems have pumps or controls that are less than optimum since the pumping task can change over time.

Some studies have shown that over 30% of the energy consumed by existing pump systems could be saved through equipment or control system changes.

In Finland annually some 9 TWh of electricity is used in pumping. In paper and chemical industries pumping forms a half of all energy use. Electricity costs are over 95 % of the life cycle costs of pump units.

MOTIVA, governmental Information Center for Energy Efficiency in Finland, started in 1995 a programme, Pump Inspection Service Project, to promote energy savings in pumping. The program was focused on industrial pump applications, especially in heavy process industry.

The target was to create an advanced and complete package of services to analyse the energy efficiency of the pump operation. The service must be easy to buy and to apply by any industrial customer. The packaging of the services took place as competition, in which the inspection and maintenance service companies as well as pump manufacturers participated.

Three companies were accepted by the jury and are now Government Certified Inspection Service companies. In order to get the industry interested in using the services, the government has covered 15 % of the inspection costs during the first two years. The cost of review per pump has been app. 1700 EUR.

2 Findings

Some 26 companies ordered this Inspection. The finished reports included 79 analysed pump units, from 30 to 1260 kW each, totally 13 773 kW.
The efficiencies varied from 8 % to 92 %. Improvement suggestions were given to 39 units, 49 % of the analysed ones.

Sizing of a pump is regularly based on insufficient and inaccurate data. Because of the consequences of under-sizing, most pumps tend to be oversized.
Duration curves of the inspected pumps varied, as well. Typical cases are illustrated in the Figure 1. Due to these reasons many pumps operated far from their most efficient duty point.

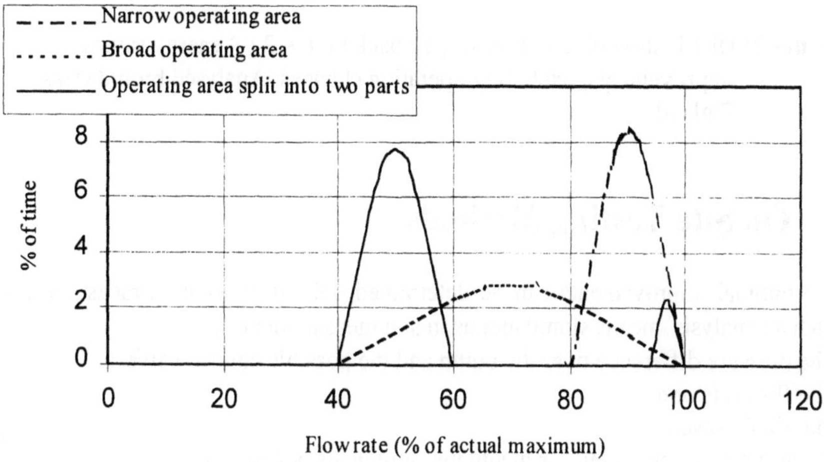

Figure 1: Typical duty cycles of the inspected pumps

The most common suggestions related to the changes in operation.
Impeller repairs or changes provided shortest pay-back times.
Pump replacements in some cases also gave pay-back times less than two years.
The suggested investments totalled app. 300 000 EUR.
The average pay-back time of all improvements was about nine months (cost of electricity was 30 - 47 EUR/MWh).

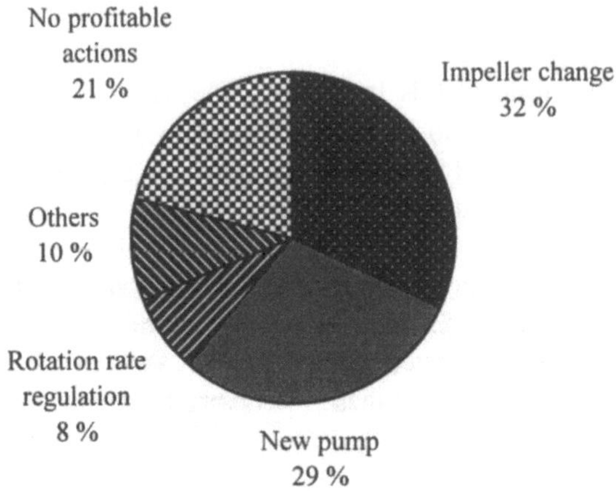

Figure 2: Distribution of short-term (pay-back time < 2 y) energy saving improvements, excluding operation changes. Analysed by IndMeas, Finland

3 On Site Testing Methods

The optimal improvements can be determined only after accurate measurements. Such an analysis means simultaneous in situ measurements of
- the pressure difference over the pump and the possible control valve,
- the flow rate and
- the shaft power.

The accuracy requirement for in situ measurements is 2 to 3 %.

Pressure measurements are performed by calibrated pressure probes. Joints need to be made for the measurements in most cases.

The shaft power can be calculated, when the motor intake power is measured by three-phase Watt meter method. The motor losses, that are subtracted from the intake power, are received from the supplier's efficiency charts with sufficient accuracy.

To measure the flow with 2 to 3 % accuracy is much more demanding in closed pipeline systems.
For example clamp-on ultrasonic measurement gives only 10 to 20 % accuracy.
The existing flow meters of the process are typically too inaccurate due to the effects of installation position, unless they have been calibrated in site.

Radiotracer methods can produce a momentary accurate reference value for the flow rate. The most common application is the transit time method (ISO 2975/VII and Nordtest NT VVS 082), where a pulse of radioactive substance is injected into the suction side of the pump. The speed of the pulse is then measured by two gamma detectors after the pump in a straight part of pipework. When the distance between detectors, diameter and thickness of the pipe wall are known, the flow rate can be calculated. If the transit time method is not applicable, it can be replaced by dilution method (ISO 2975/III).

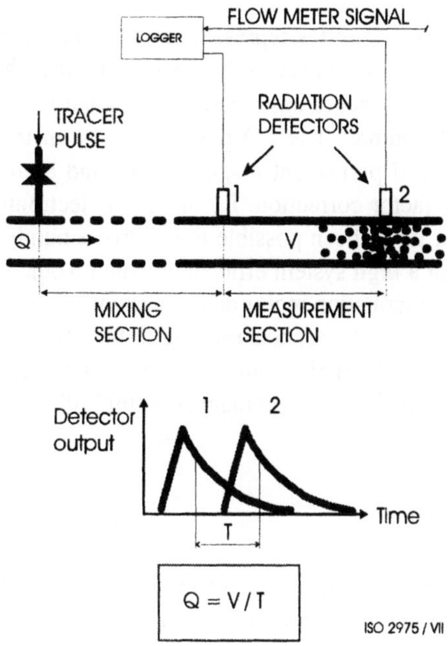

In pumping systems, where an open well or basin precedes the pump, the flow can be measured by continuous level measurement. The method presumes, that the volume between the predetermined levels can be calculated and the incoming flow during pumping is properly taken into account.

4 References

[1] Härkönen, Heikki (Ed.) (1997): Memorandum of Pump Inspection Seminarium on 11 February 1997. MOTIVA, Information Center for Energy Efficiency in Finland

[2] Ravila, Pekka (1998): Energy saving in industrial pumping process, Presentation. IndMeas Oy, Finland

SQ - Submersible Pump with Integrated Permanent Magnet Motor Drive

Henrik Ørskov Pedersen

Grundfos A/S, Motor Engineering, DK-8850 Bjerringbro, Denmark

Abstract The Grundfos SQ series represents a major breakthrough in design and performance of high-quality compact submersible pumps. The SQ series with an outside diameter of only 3 inches has a performance that is comparable to conventional uncontrolled 4" pumps. The SQ pump features an integrated motor drive system with a high-speed permanent magnet motor and a complete built-in drive unit including power factor correction. By using an electronically controlled permanent magnet motor it has been possible to achieve a number of unique features while still maintaining a high system efficiency which would not have been possible with a conventional induction motor driven system.
In the presentation some of the aspects relating to the efficiency of the motor system will be discussed. Additionally some of the benefits and features arising from the use of a fully controlled permanent magnet motor will be covered.

1 Introduction

Submersible pumps are typically installed in bore holes where they are used to pump water to the surface. The power range is from a few hundred watts up to several tens of kilowatts. The smaller ones are typically used for domestic water supply, small community waterworks and irrigation, but can also be used for tank applications and various environmental uses.

Due to the installation in bore holes the motors are relatively long with a small diameter. There are several advantages associated with a smaller diameter of motor and pump. The bore hole can be drilled with smaller diameter which contributes considerable savings and it will be easier to replace pumps in old bore holes.

In the following some of the aspects relating to the efficiency of the motor system will be discussed. Additionally some of the benefits and features arising from the use of a fully controlled permanent magnet motor will be covered.

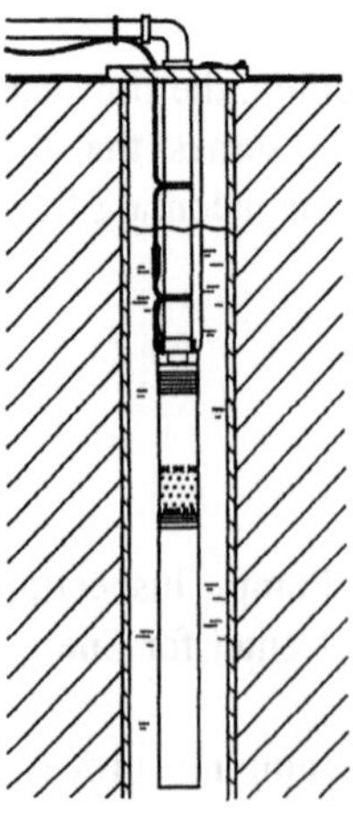

Figure 1 Typical installation of submersible pump in a borhole or well

2 Design Objectives and Constraints

The SQ pump has been developed to meet the following design objectives and constraints:

- Compact design - less than 3" overall outside diameter compared to conventional solutions requiring 4"
- Built-in electronic protection features
- Variable speed
- Motor efficiency high enough to overcome additional losses in the control unit and additional losses in a 3" hydraulic system compared to 4" hydraulics
- Extension of product range in terms of output power (1730 W compared to 800 W with existing induction motor solution)
- Robust, reliable and long life time

3 Why a High-Speed Permanent Magnet Motor?

A comparison between the an induction motor and a permanent magnet motor reveals a number of advantages and drawbacks for the permanent magnet motor.
Identical for both motor principles is that the dimensional restrictions call for a high-speed solution, which requires a complete built-in power-electronic drive-unit. This would also satisfy the demand for variable speed and simplify the implementation of a number of the desired protection features.
The permanent magnet motor has significantly higher efficiency than an induction motor of comparable size. Therefore in order to compensate for the lower efficiency in the 3" hydraulics and the losses in the power electronics the obvious choice would be a permanent magnet motor.
The lower losses in a permanent magnet motor leads to better thermal conditions in the motor. For a given cooling capability it is therefore possible to extract more power from a permanent magnet motor than from an induction motor and thereby extend the power range.
The implementation of various electronic protection features would be simplified using a permanent magnet motor due to the linear relationship between load and current.

On the negative side was that no prior experience with the robustness of permanent magnet motors existed for submersible applications. Therefore the introduction of a permanent magnet motor driven pump would require extensive testing in order to build this experience.

3.1 Motor Construction

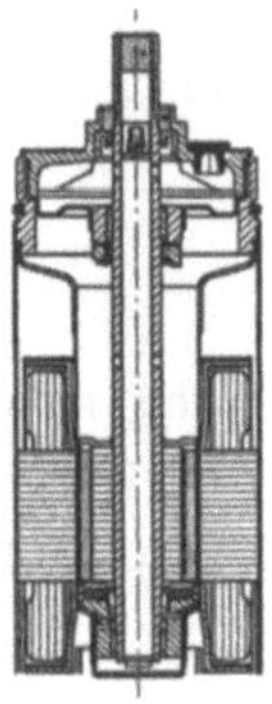

The motor is constructed as a canned motor, where the rotor and stator are separated by a stainless steel tube in the airgap in order to keep the electrical part of the motor dry. This also means that the rotor runs in a liquid-filled rotor chamber. This construction imposes a number of loss components that will not be found in a conventional motor - eddy-currents will be induced in the separating can and having the rotor chamber filled with water means additional friction losses on the rotor surface. Especially for a high speed motor these losses will be considerable.

Figure 2 Mechanical construction of motor with canned rotor

3.2 Efficiency Considerations

In order to obtain the best overall efficiency of the system it has been necessary to consider the different loss components in the motor and their dependencies on both motor design parameters and more overall system parameters like speed, size limitations, load requirements (torque) and thermal properties. Table I lists the motor loss components and their main dependencies.

Table I: Loss dependencies

Loss component	Motor design parameters	System parameters
Stator iron losses	Pole number Lamination geometry Lamination quality Airgap induction (magnet grade)	Speed Torque
Stator copper losses	Slot design Winding temperature (cooling) Airgap induction (magnet grade)	Torque Cooling
Eddy current losses in the airgap tube	Tube thickness Tube resistivity Pole number Airgap induction (magnet grade)	Speed
Rotor viscous friction	Airgap length Rotor diameter Viscosity of rotor fluid	Speed
Bearing friction	Radial forces	Speed

Eventually this led to a series of compromises in both motor design and in selection of operating speed and torque. As a combination of the restrictions on physical size of the motor and considerations regarding efficiency of both motor and hydraulic system a maximum speed of 10.800 rpm was chosen.

Compared to a conventional motor special attention had to be given to the losses in the airgap region with the eddy current losses in the tube and the viscous friction losses. As an example reduction of the stator copper losses would call for a high airgap flux density and thereby use of very strong magnets in the rotor, whereas the eddy current losses in the airgap tube will increase considerably with increasing airgap flux density and thereby call for use of lower grade magnets.

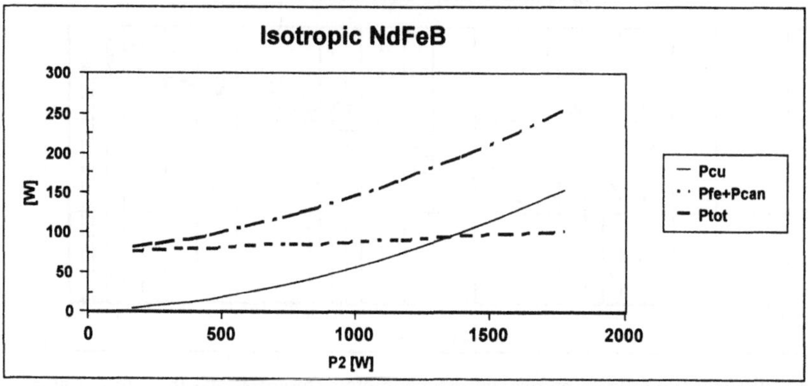

Figure 3a) Electric/magnetic losses as function of shaft power for isotropic NdFeB magnets at 10.800 RPM.

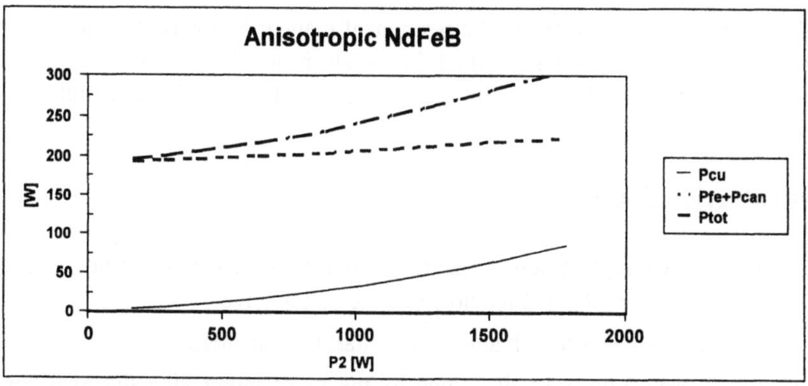

Figure 3b) Electric/magnetic losses as function of shaft power for anisotropic NdFeB magnets at 10.800 RPM.

The trade-off on this aspect is illustrated in Figure 3a) and 3b) in terms of the loss dependency on the shaft power and resulted in use of a 4-pole isotropic polymer bonded NdFeB magnet with the alternatives being the more expensive anisotropic NdFeB or the cheaper Ferrite magnets (results not shown).

An efficiency of >80% has been obtained for the 1.7 kW motor which is not especially high for a permanent magnet motor in general, but for a 3" submersible motor this is very high indeed since the losses in the airgap tube and the friction accounts for more than 30% of the total losses at rated load and speed.
Compared to solutions with a 4" induction motor the efficiency of the motor itself has been improved by ~10% as illustrated in Figure 4. Even the drive efficiency (including the power electronics) has been improved compared to existing solutions without power electronics.

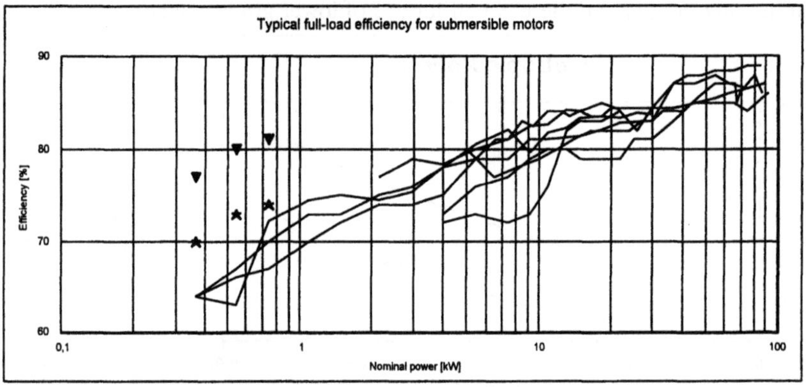

Figure 4 Efficiency of the 3" permanent magnet motors (motor & driver , motor alone - 0.63, 1.05 and 1.7kW) compared with efficiencies of typical 4" motors on the market.

Another very important aspect of the permanent magnet motor is the wide operating range with high efficiency, which has made it possible to cover the entire range of pumps with only three motor sizes while still maintaining a high efficiency at reduced loads.

3.3 Protective Features

The inclusion of a complete power electronic control unit has made it possible to integrate a number of protective features into the motor drive.

- Dry running protection with automatic restart is implemented.
- Overload protection is implemented by means of monitoring the load current. An uncontrolled pump will draw excessive current from the supply if the load is increased, but the overload protection automatically reduces the speed of the pump to compensate for the increased load and thereby maintaining a constant load current.
- A blocked rotor is sensed by the control unit and automatically switches off power to the motor.

- Over and under voltage protection. If the voltage exceeds 280 V or drops below 150 V the drive shuts down. When normal voltage is restored the drive automatically restarts.
- Overheating protection. The permanent magnet motor produces less heat due to higher efficiency than conventional induction motors. This leads to optimal running conditions for the motor. Additionally the control includes a thermometer and in case the temperature exceeds a certain level the motor is switched off until the temperature regains normal working level.
- Soft start system. This system limits the starting current and minimises water hammering.

3.4 Reliability Testing

The SQ pump introduces a number of completely new solutions to the concept of submersible pumps - one of them being a permanent magnet motor with integrated electronic control. This poses a lot of emphasis on the need for thorough reliability testing during the development phase in order to make sure the pump and in this case especially the motor will survive the harsh environment in a borehole.

A very critical component is the permanent magnets. Neodymium-Iron-Boron magnets will corrode within hours if they are subjected to the water in a well. Therefor means for preventing contact to the water had to be applied. Different coatings were analysed, but none were found that could meet the requirements with respect to media (water, chemicals), temperature and time. Ultimately it was chosen to hermetically seal the magnets by means of a laser welded stainless steel sheath which effectively prevents contact between magnets and water. In order to test for stability against temperature a high number of magnets were exposed to high temperature and thermal cycling over more than 1000 hours with a magnetic loading comparable to a worst-case situation in the motor. This test effectively showed that stability of the magnets is not a problem at temperatures of the magnets below 120°C - in the motor the temperature of the magnets will not exceed 85°C.

The complete pump as well as the individual components have been subjected to accelerated life tests in terms of thermal conditions (both continuously elevated temperatures and rapid cycling from below zero to maximum temperature), mechanical chocks and vibrations, electrical supply conditions beyond the specifications and excessive loadings.

All this to ensure reliable operation of the pump within the specified supply, cooling and load conditions as well as the reliability of the built-in protection of the pump against conditions exceeding the specifications.

4 Conclusion

By using a high-speed permanent magnet motor with a purpose designed built-in power electronic control unit it has been possible to develop a submersible pump with previously unseen features. By focusing on the performance of the complete system in the development of the individual components it has been possible to integrate an expensive permanent magnet motor and a micro frequency converter into a high quality product and still keeping the end price at an attractive level.
The compact solution has been achieved without suffering loss of efficiency compared to bulky pumps using more conventional technology.
The capability to control the speed of the pump according to the pumping needs combined with a high system efficiency over a wide operating range ensures low energy consumption.

Induction Motor Efficiency Measurement

Prof. S. Williamson, Dr. H. P. Sambath

Brook Hansen
St Thomas' Road
Huddersfield HD1 3LJ

1 Introduction

There can be no doubt that the most accurate methods for determining the efficiency of an induction motor are based on calorimetry. Such methods seek to determine losses directly be measuring the heat generated by the appropriately loaded motor. Unfortunately, however, calorimetric testing is expensive and time-consuming and as such is completely unsuitable for routine testing in industry. As a result, test methods that rely on assumptions and approximations of varying accuracy have been developed and enshrined in testing standards. This is a pragmatic approach to efficiency measurement, balancing the time and cost of testing against the accuracy of the result.

As energy efficiency has advanced up the agenda of motor development, manufacturers have sought to improve testing methods. This is a matter of much debate both nationally and internationally. The purpose of this paper is to compare the efficiencies of a group of 4 pole 400-3-50Hz motors determined in accordance with several current and proposed standards, with that determined from a calorimetric test.

2 Methodology

The first test batch of motors consisted of five 7.5kW 4 pole 400-3-50Hz machines, bought off-the-shelf from five leading manufacturers. These motors are identified as A, B, C, D and E in this paper, in order to maintain commercial confidentiality.

All five motors were industrially tested according to three current standard test methods and two proposed test methods. These five test methods are listed in Table 1, along with their distinguishing features

Table 1: Test methods for determining efficiency of cage induction motors

Method	I	II	III	IV	V
International Standard	IEC 61972	CSA C390	IEC 34-2	IEC 61972	IEC 34-2
Status	Proposed	Current	Current	Proposed	Current
Classification	Input/Output			Summation of losses	
Insulation class (Class B or Class F)	any	any	any	any	B
Direct measurement of electrical input power and mechanical output power	✓	✓	✓	✖	✖
Segregation of losses	✓	✓	✖	✓	✓
Core losses with voltage drop compensation	✓	✖	✖	✓	✖
Stray load losses using regression analysis	✓	✓	✖	✖	✖
Stray load losses from assigned values	✖	✖	✖	✓ 1	✓ 2
Stator resistance values with thermal probes in stator winding	✖	✓	✖	✖	✖
Stator resistance obtained from mean values before and after load tests	✓	✖	✖	✓	✖
Temperature corrected winding (I^2R) losses (both stator and rotor)	✓	✓	✖	✓	✖
Thermal equilibrium at rated load	✓	✓	✖ 3	✓	✖
Stabilisation of no-load losses	✓	✓	✖	✓	✖
Recording of frequency	✓	✖	✖	✓	✖
Dynamometer torque correction	✓	✓	✖	✖	✖

Notes:

1. IEC 34-2 assumes that stray load loss is 0.5% of input power at rated load (independent of the motor rated output power), and varies with the square of the input current at other loads.
2. IEC 61972 assumes that stray load loss depends on motor rated output power and allocates a percentage value at rated load. At all other loads, stray loss is assumed to vary with the square of input current minus the square of no-load current at other loads.
3. Motor temperature to be as near as possible to that attained in operation.

The principal difference between them is the way in which they deal with stray load losses, being classified as either 'input/output' or 'summation of losses'. The summation of losses methods are the lowest cost methods, allowing the assumption to be made that stray-load losses have a fixed value which is independent of motor design.
The minimum accuracy classification of the instruments used for making measurements in these tests is given in Table 2. In practice, the accuracy class of the measurement devices is often better than that specified by the relevant standard. Brook Hansen, for example, routinely use digital instruments and power analysers in their testing.

The motors were also tested in the calorimetric test facility in the Electrical Engineering Department at Nottingham University. The test facility and test procedure are the subject of a separate paper and will not be further discussed here. Suffice it to say that the Nottingham results will be used in this paper as the basis for comparison with the industrial methods for determining efficiency.

Table 2: Instrumentation accuracy (+/- percentage of full scale)

International Standard	IEC 61972	CSA C390	IEC 34-2
Electrical	0.2	0.5	1.0
Instrument transformer	0.2	0.5	1.0
Frequency	0.1	0.5	1.0
Speed	0.1	0.5	1.0
Torque	0.2	0.5	1.0
Resistance	0.2	0.5	0.5

3 Results and Observations

Table 3 gives the full-load efficiencies of all five motors, as determined by each of the five test methods given in Table 1. Table 3 also includes the efficiencies of the same motors calculated using direct loss measurements made in the calorimeter. The calorimeter results show that the least efficient motor (E) has losses that are 20.5% greater than the most efficient (A).

Table 3: Efficiency (%) of motors at full rated load - 7.5kW motors

	Motor				
Method	A	B	C	D	E
I	87.04	86.07	84.67	85.77	84.79
II	87.02	86.03	84.49	85.70	84.77
III	86.72	86.46	84.71	84.76	85.24
IV	86.02	85.50	83.51	84.60	83.92
V	88.19	87.88	85.58	86.93	86.57
Calorimetry	87.05	86.19	84.60	86.36	84.39

The data of Table 3 has been used to generate Table 4, which for each motor shows the difference between the efficiency calculated using the industrial test methods and that determined from the calorimeter test. The final three columns in Table 4 give the mean difference for each test method (average across all five motors) along with the RMS of the difference and its standard deviation.

Table 4: Difference between efficiency from industrial test methods and calorimetric test - 7.5kW motors

	Motor							
Method	A	B	C	D	E	Mean	RMS	STD
I	-0.01	-0.12	0.07	-0.59	0.40	-0.05	0.32	0.36
II	-0.03	-0.16	-0.11	-0.66	0.38	-0.12	0.35	0.37
III	-0.33	0.27	0.11	-1.60	0.85	-0.14	0.83	0.92
IV	-1.03	-0.69	-1.09	-1.76	-0.47	-1.01	1.10	0.49
V	1.14	1.69	0.98	0.57	2.18	1.31	1.43	0.63

The results presented in Table 4 indicate the following:

- The input/output methods of testing (i.e. I – III) give superior accuracy to the summation of losses methods, with Method I and Method II being the most accurate.

- Method IV, which assumes a full-load stray loss of 2.43%, for a 7.5kW motor, consistently underestimates efficiency by approximately 1.0%.

- Method V, which assumes a full-load stray loss of 0.5% of input power, results in an over-estimate of efficiency by approximately 1.3%.

4 Stray Load Losses

The last two observations suggest that much of the difference between the methods derives from the way in which stray load losses are allowed for. To investigate this proposition Table 5 has been constructed, showing the stray load loss assumed or calculated by each method, and that determined from the calorimeter tests. No value is given for method III because stray losses are not determined as part of that method. The data given in Table 5 is a percentage of the input power during the relevant test.

Table 5: Stray load losses (% of input power) at rated load - 7.5kW motors

	Motor							
Method	A	B	C	D	E	Mean	RMS	STD
I	1.42	1.86	1.27	1.26	1.56	1.47	1.49	0.248
II	1.26	1.64	0.85	1.09	1.36	1.24	1.27	0.296
III	not applicable - no segregation of losses							
IV	2.43	2.43	2.43	2.43	2.43	2.43		
V	0.5	0.5	0.5	0.5	0.5	0.5		
Calorim	1.15	1.34	1.20	1.43	0.84	1.19	1.21	0.226

The results show that Method I and Method II give values for stray-load losses that are in good agreement with the calorimeter results, once again reinforcing the view that these two methods are superior. Method V underestimates these losses and Method IV overestimates them, as expected.

A similar series of tests are also in the process of being performed on a group of five 0.75kW 4 pole 50Hz motors, from the same five manufacturers as the 7.5kW group. The programme is not yet completed, but sufficient tests have been carried out to enable the stray load losses for these motors at full load to be determined using Method I. The results are presented in Table 6, which also includes the losses assigned according to Methods IV and V.

Table 6: Stray load losses (% of input power) at rated load - 0.75kW motors

	Motor							
Method	A	B	C	D	E	Mean	RMS	STD
I	2.39	2.14	2.00	1.77	2.06	2.07	2.08	0.225
IV	3.00	3.00	3.00	3.00	3.00	3.00		
V	0.5	0.5	0.5	0.5	0.5	0.5		

Examination of the data presented in Table 5 supports the observation that the 0.5% assigned to stray load loss in Method V is a considerable underestimate whereas the value assigned according to Method IV is an overestimate. Comparison between the Method I data in Tables 5 and 6 also confirms that stray load losses are proportionately greater in the lower power motor, as assumed in Method IV.

5 Manufacturing Tolerances

The results that have been presented so far refer to equally rated motors produced by five different manufacturers. Rules governing the declaration of efficiency also permit error bands that allow for measurement tolerances and manufacturing variations. In order to investigate the effect of manufacturing variations, four nominally identical 7.5kW 4 pole 50Hz motors made by one manufacturer were tested in the calorimetric test rig. This group included Motor A used in the first part of this study, plus three more taken from the production line at random. The mean full-load efficiency of these four motors was found to be 86.97%, with a standard deviation of 0.083%. The difference between the losses in the most efficiency and least efficient of these four motors was found to be 0.17% of the input power. The maximum permissible difference between measured and declared efficiency according to the proposed Standard IEC 61972 equates to 10% of the total loss, which with a mean efficiency of 86.97% is 1.3% of rated input power. The implication of this is that the variations due to manufacture, based on this admittedly small sample of motors, are easily contained within the permitted tolerance band.

6 Conclusions

The results presented in this paper relating specifically to 7.5kW 4 pole 400-3-50Hz motors, indicate that efficiency measurement methods based on input/output tests are more accurate than those employing assigned values for stray load loss. The measurements made using the most accurate input/output method clearly show that the fixed allocation of 0.5% stray load losses is a considerable underestimate for both sizes of motor examined, independent of manufacturer. Further tests are needed on a range of motor sizes before this conclusion can be generalised, but the authors believe that they have demonstrated a procedure that will enable such a conclusion to be reached.

Experience Factors When Testing for Efficiency and Correlating Results with Design

Joseph A. Kline Sr.

U.S. Electrical Motor Division of Emerson Electric, USA

Abstract. Since the enactment of federal regulations in 1992, the U.S. and other North American countries have required that motors meet minimum efficiency standards. During this transition to date, test verifications have evolved to provide methodology and procedures to assure efficiency requirements. Experience in testing during this period have given insight into factors which have been shown to provide consistent test results. Correlation of results with designed prediction have enabled adjustment of parameters to reflect actual product performance values.

Introduction

At this writing, the U.S. Department of Energy has not yet published the final Federal Code that will regulate efficiency of integral horsepower motors. Present prospects are that the codes will be published in October 1999. However, the mandatory efficiency requirements did commence on October 24, 1997. All defined products manufactured after that date were subject to the Federal Energy Act of 1992, commonly referred to as EPACT 92.

Both Canada and Mexico have enacted federal codes related to efficiency which utilize test standards IEEE 112 method B and CSA 390 method 1 which conform to the modified National Institute of Standards and Technology's (NIST) National Voluntary Lab Accreditation Program (NVLAP) test procedures.

The difference in the U.S. requirements for efficiency is that certification of a product's efficiency may be obtained by third party NVLAP approved testing labs or self-certified by the manufacturers own NVLAP accredited lab. The federal rules for these are forthcoming.

Typical Efficiency Test Requirements

Figure 1 shows a blocked diagram of a typical system built to meet the repeatability and accuracy requirements of NVLAP.

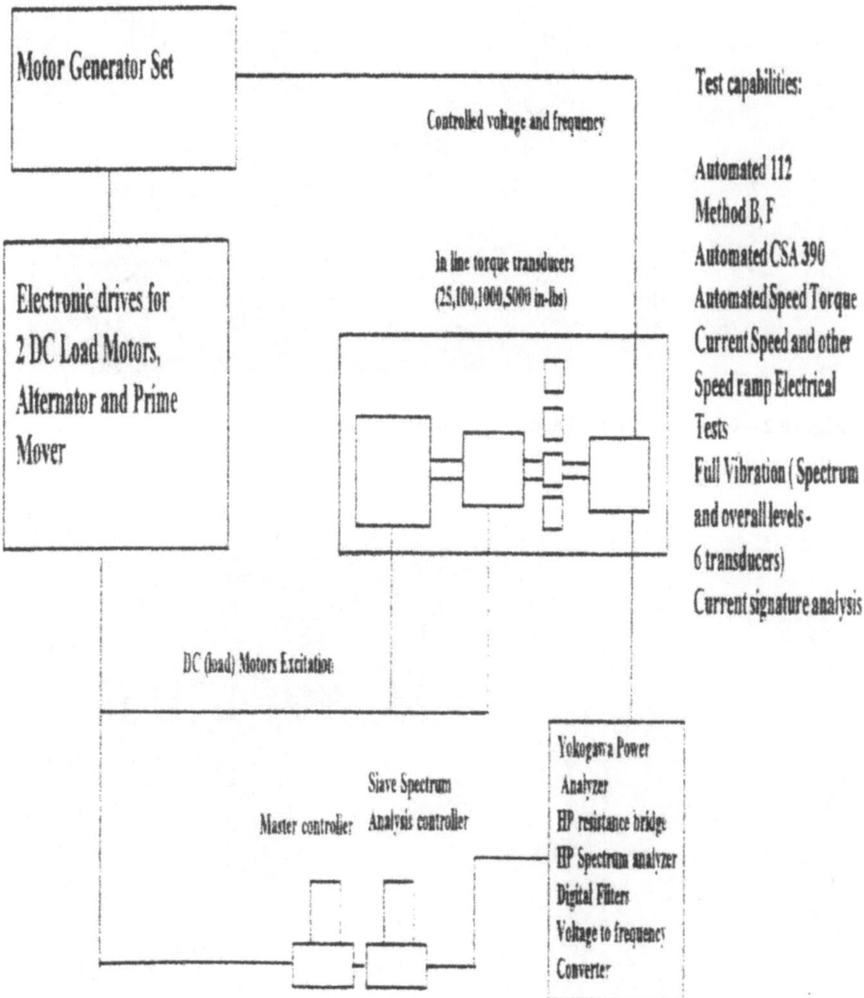

Figure 1: Typical Test System Block Program

Actual pictures of components of this system are shown in Figure 2 and 3.

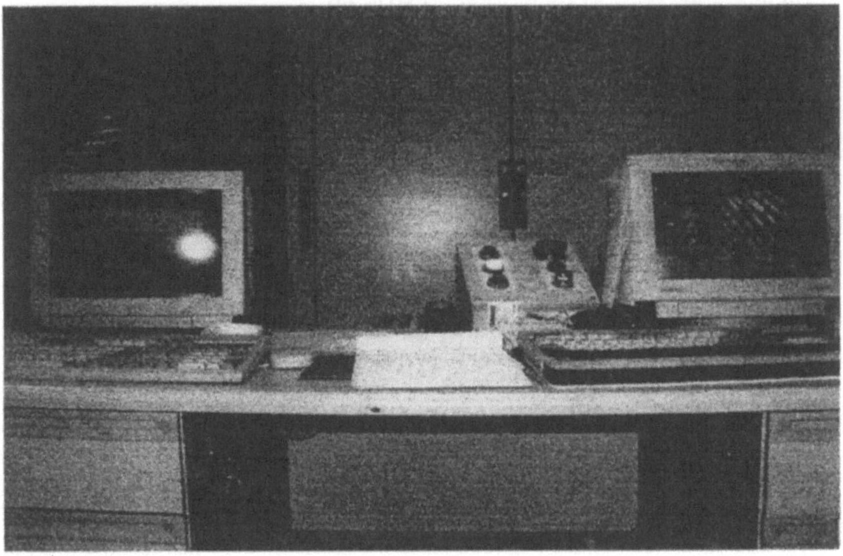

Figure 2 – Computer Control and Power Instrumentation

Figure 3 – Load Test Stand

Basic Instrument Accuracy Requirements Per IEEE 112 B

Following are the accuracy requirements per IEEE 112 B. Most instruments quote accuracy based on full scale.

Parameters	Accuracy
• Watts	± 0.20 %
• Current	± 0.20 %
• Voltage	± 0.20 %
• Frequency	± 0.10 %
• Torque	± 0.20 %
• Resistance	± 0.20 %
• Temperature °C	± 1.0
• RPM (slip)	± 1.0

Typical Firmware/Software System

Figure 4 depicts a typical automated test stand configuration for testing motor efficiency.

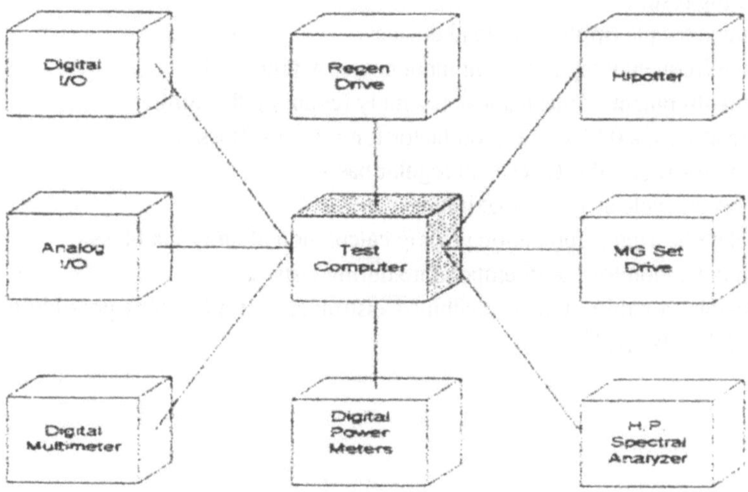

Figure 4 – Software/Firmware Block Diagram

System Functionality

- Digital I/O – Monitors control system relays, lights, sensors, and miscellaneous digital I/O.
- Analog I/O – Communication to PC to monitor thermocouples.
- Digital Power Meters – Measurements of volts, current, watts, torque, and speed.
- Digital Multi-meter – Used to measure winding resistance.
- Hipotter – Used to perform hipot prior to test (safety).
- MG Set Drives – Used to provide precise voltage regulation to motor under test.
- Regeneration Drives – Returns low power back to the source.

Experience Factors

The following are recommendations to assure repeatability and accuracy of efficiency testing:

- Sensitivity of torque instrument is adequate for motor under test.
- Load horsepower which is no more than three times the motor under test.
- Load cells are re-calibrated prior to every test.
- Capability for simultaneously reading input power and torque and speed.
- Stabilize readings by taking multiple data (eg.2030pts) at each load point and averaging results.
- Minimize torque ripple from load devices.
- Voltage frequency regulation on input power within 0.5%.
- Careful alignment of mechanical assembly (coupling friction).
- Assure at least a 0.98 correlation factor for stray load losses.
- Base motor repeatability tests on regular basis.
- Calibration cycle at 1 year maximum.
- Complete bearing stabilization prior to calculation of stray load loss.
- Continuous monitoring of temperature during tests.
- If possible, a calibration of combined instruments at all ranges possible (eg. CT's, PT's, W, A, V, etc.).

Correlation of Data

The obvious reason to do efficiency testing on motors is to provide correlation of test data to design and to assure compliance to requirements. As simplistic as this sounds it is a quite complex problem. It is impossible to test an infinite number of one rated model motor to obtain that „perfect" normal distribution. Rather, it is necessary to provide a statistical evaluation based upon base sample products to assure that the test is repeatable and consistent. This will monitor manufacturing variability to assure that the motor will be within nominal loss spreads.

The ideal automatic test system would continually update the performance design program to identify design changes and improved manufacturing processes or identify processes which need to be examined.

Conclusion

The testing of motors for efficiency is very complex and is variable based upon rating and speed. Only good procedures, good high accuracy instruments, good practices, as well as automated systems will give the best results.

References

1. Joseph A. Kline Sr. – Energy Efficient Motors U.S. Manufacturers Achievements, „Energy Efficiency Improvements in Electric Motor and Drives Conference" – Lisbon, Portugal – October 1996, Editors A. de Almeida, P. Bertoldi and W. Leonhard, Springer, Berlin, 1997.
2. Robert G. Bartheld and Joseph A. Kline Sr. – Comparison Testing for Efficiency in Accordance with IEC 34-2 and IEEE 112 Procedures . , „Energy Efficiency Improvements in Electric Motor and Drives Conference" – Lisbon, Portugal – October 1996, Editors A. de Almeida, P. Bertoldi and W. Leonhard, Springer, Berlin, 1997.

Approaching the Truth in Induction Motor Efficiency

Haataja J., Pyrhönen J.

Lappeenranta University of Technology, Finland

Abstract. The importance of motor efficiency in the motor market is increasing. Therefore the accuracy of the efficiency value given by the manufacturer is significant. The accuracy of the efficiency determination depends on the test method used and the precision of the loss determination by the test method. Determining methods can be grouped into two broad categories: direct measurement methods and indirect methods (loss segregation methods). In the following the effect of different methods in the given motor efficiency has been investigated in more detail. While testing low voltage, polyphase squirrel cage induction motors IEC publication 34-2 allows both a direct and an indirect method to be used, but according to NEMA the IEEE-112 test method B - Input-Output - with loss segregation is to be used. European manufacturers prefer to use the "summation of losses" according to the IEC 34-2. When comparing IEC 34-2 to IEEE 112 it can be noticed that the latter one defines stabilised operating temperatures for the measurements, more accurate measuring instruments and more precise procedures to determine all losses than the IEC 34-2. The IEC 34-2 assumes an operating temperature, assumes a value for stray load loss and permits measurements at any motor operating condition with less accurate instruments. On the other hand the IEC 34-2 tolerances for motor losses are smaller than in the IEEE 112. However, it is important to note that IEEE states that besides the nominal efficiency value also a guaranteed minimum efficiency shall be shown on the nameplate. Depending on the manufacturing tolerances this minimum value can be better than the worst value that the tolerance allows. The test results shown below were measured for a high efficiency 690/400 V, four pole, general purpose 5.5 kW motor using both the IEEE 112 method B and the IEC 34-2 indirect (summation of losses) and direct methods.

Keywords. Energy efficiency, induction motor

1 Motor Measurement

In IEC 34-2 and in IEEE method B a dynamometer is used to load the motor. Both direct methods use load cell to measure motor torque. The dynamometer has electrical control for setting the load torque and holding it steady while readings are being taken.

The measurement consists of several steps, rated load temperature test, test under load, and no-load test. The apparent total loss (input minus output) is segregated into various components. In direct methods stray-load losses or additional losses are defined as the difference between the apparent total loss and the sum of the conventional losses. In IEEE the value of stray-load loss thus determined is plotted vs. torque square, and a linear regression is used to reduce the effect of random errors in the test measurements. The smoothed stray-load loss data are used to calculate the final value of total loss and the efficiency.

1.1 Rated Load Temperature Test

The motor is coupled to a dynamometer and operated at rated load until it is thermally stable. If the motor has a high efficiency it is important that the actual operating temperature is monitored and used for the evaluation. Otherwise too high a temperature for the windings must be used for the determining.

In this test the temperature measurements were based on the stator winding resistance measurements. The measurement of the temperature was done after a rapid shutdown. The stator resistance behaviour as a function of time was measured. These readings and a fitted curve (with a mathematical formula) have been plotted in figure 1.

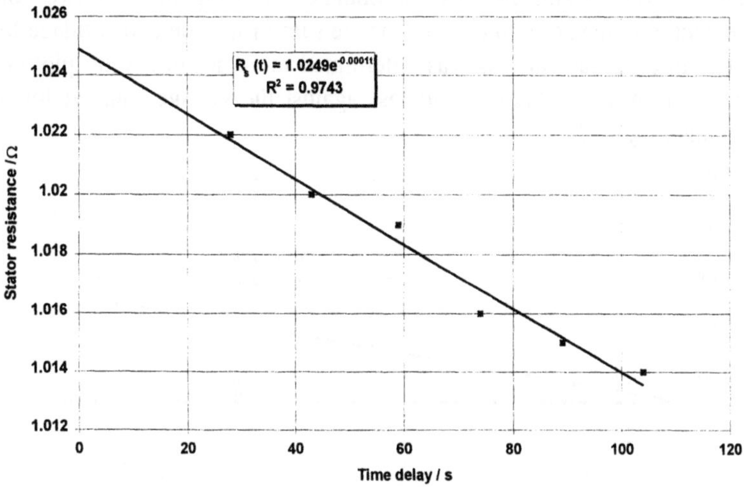

Figure 1. The resistance values measured after a heating test

1.2 No-Load Test

The test was made by running the machine as a motor at rated voltage and frequency without load. The machine was operated at no-load until both the temperature and input power had stabilised before making the measurement.

The reading of the input power is equal to the sum of the no-load motor losses. Subtracting the stator ohmic loss (at the temperature of this test) from the input gives the sum of the friction, windage, and core losses. The average values of the load test are shown in table I.

Table I: No-load test measured values of a high efficiency 690/400 V, four pole, general purpose 5.5 kW motor

Voltage / V	Current / A	Input Power / W	Speed / min^{-1}
430	6.2	271.1	1500
400	5.5	238.4	1500
297	3.7	164.8	1500
256	3.1	138.4	1500
202	2.5	114.8	1500
150	1.9	95.4	1500
89	1.3	78.9	1500

Separation of the core loss from the friction and windage loss was made by reading the voltage, the current, and the power input at the rated frequency and at voltage ratings from 125 % of rated voltage down to a point where further voltage reduction started to increase the current. Power input minus the stator ohmic loss is plotted vs. voltage, and the curve so obtained is extrapolated to zero voltage. The intercept with the zero voltage axis is the sum of friction and windage losses. In this test the mechanical loss was determined more accurately by plotting the input power minus the stator ohmic loss against the voltage squared for lower voltage range, figure 2.

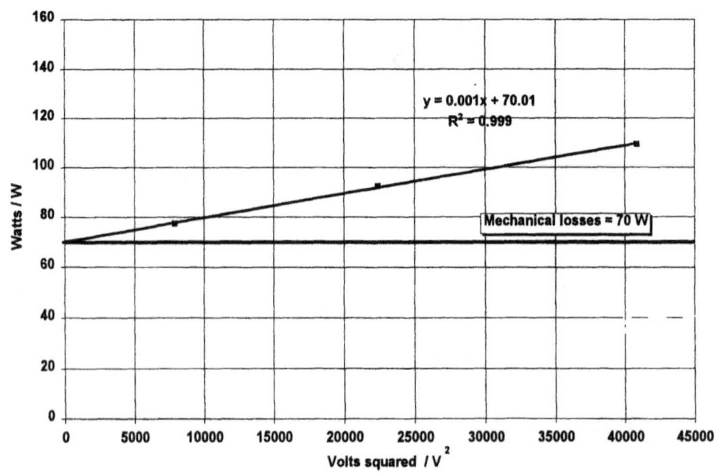

Figure 2: Core and mechanical loss curves plotted against the voltage squared

The core loss at no load and at the rated voltage is obtained by subtracting the mechanical and stator ohmic losses from the total input power at no-load.

1.3 Tests Under Load

The motor was loaded by a dynamometer. The load tests were made to determine the efficiency, the power factor, the speed, the current, and the temperature rise of the motor. The average values of the load test are shown in table II.

Table II: The measured load test values

Average Voltage / V	Average Current / A	Power in / W	Speed / min^{-1}	Torque / Nm
401	6.9	2552.6	1484	14.48
400	9.0	4380.8	1479	25.44
401	10.2	5233.0	1474	30.66
401	11.3	6042.7	1470	35.52
400	13.3	7384.3	1462	43.47
400	14.6	8171.0	1457	47.79

1.3.1 IEEE-112 Ambient Temperature Correction of the Stator and the Rotor Ohmic Losses

A corrected stator ohmic loss for each of the six load points is calculated by using the common value of the stator resistance obtained from the rated load temperature test corrected to an ambient of 25 °C. In equation

$$R_s = R_t \frac{t_s + k}{t_t + k}, \qquad (1)$$

the resistance R_t is the measured resistance at the conclusion of the rated load temperature test; t_t is the hottest winding temperature measured during the temperature test, t_s is equal to t_t corrected to an ambient of 25 °C and k is 234,5 for IACS conductivity copper.

A corrected rotor ohmic loss for each of the six load points is calculated using the value of the slip for each of the points corrected to an ambient temperature of 25 °C and using the ambient corrected value of the stator I^2R loss for each load point. In equation

$$s_s = s_t \frac{t_s + k}{t_t + k}, \qquad (2)$$

the measured slip s_t is the measured slip for the particular load point.

1.4 Stray-Load Loss

In IEEE-112 the input power and output power are measured for several load points. The apparent total loss is segregated into various components with stray load loss defined as the difference between the apparent total loss and the sum of the conventional losses. The value of the stray load loss thus determined is plotted vs. torque squared and a linear regression is used to reduce the effects of random errors in the test measurements. The stray load loss should be zero when the motor torque is zero. This is shown graphically in figure 3. The smoothed stray load loss data are used to calculate the final value of the total loss and the efficiency.

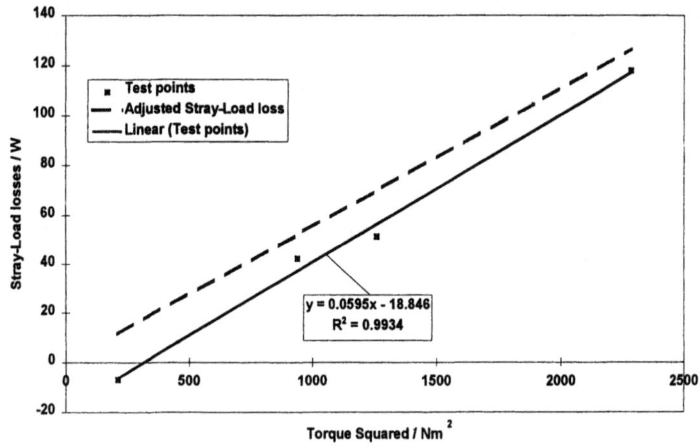

Figure 3: Plot of Stray-Load loss vs. Torque Squared

2 Motor Losses and Efficiencies

The effect of different methods of segregating the total losses into various components with the stray load loss on the motor efficiency has been shown in figure 4. To clarify the illustration only ohmic and stray load losses and efficiencies at different partial loads are shown. All the other losses (mechanical and core losses) are the same for all different standards and also load independent.

In general the efficiency determined by using the IEC 34-2 gives a higher value than by using the IEEE 112. The efficiency variation in this test was 0.7 %-units at the rated power. This is mainly because in the IEEE a higher reference temperature is used to calculate stator and rotor ohmic losses. In the motor size investigated in this paper the actual stray load losses according to IEEE seem to be bigger than 0,5 % of the input power. The stray-load losses at the rated power according to IEEE are 0.8 % of the input power.

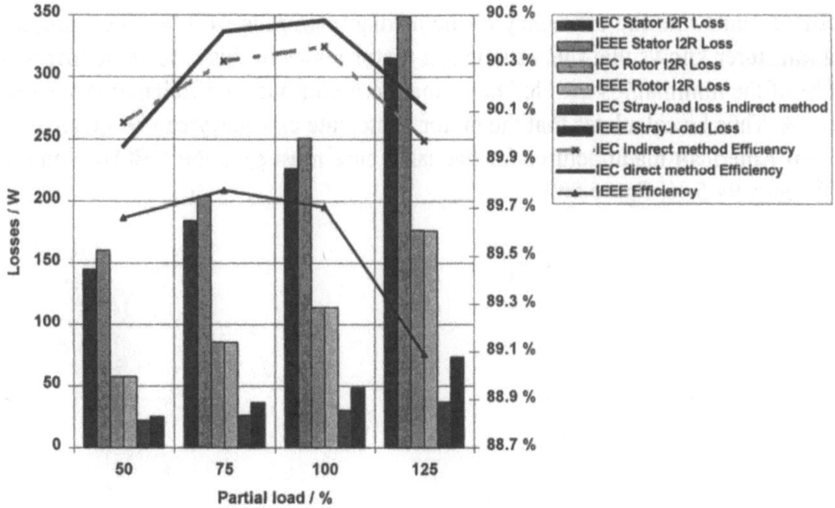

Figure 4: The test motor's partial load losses and efficiencies using IEC 34-2 indirect and direct method and IEEE 112 Method B

3 Tolerances

For both standards let us assume that measurements for a large population of motors have been made and the median motors have the efficiency η_{sample}, Fig. 5a. IEC 34 accepts a 10 ... 15 % loss increase in a motor compared to the value calculated from the nominal efficiency given for a specific motor design, $\Delta\eta = -0.10 \ldots 0.15(1 - \eta_{nom})$. If the manufacturer relies on his manufacturing process it is possible to give an efficiency value for the motor that is larger than the efficiency of the median motors, η_{IEC34}, figure 5a. When comparing the efficiencies of induction motors according to NEMA and IEC 34 it must be remembered that according to NEMA the motor efficiency shall be identified on the nameplate by the caption "NEMA Nominal Efficiency" and this efficiency must have a standard value given in NEMA nominal efficiency table. For example values ... 94.5 %, 95.0 %, 95.4 %, 95.8 %, 96.2 % ... are given in the table. Thus if the median motors have the efficiency η_{sample} = 94.7 % the nominal efficiency of the machine must be given as NEMA Nominal Efficiency 94.5 %. Thus there will be quite a large difference between the given efficiencies for the same motor.

Let us assume that the efficiency of the median motors is 94.9 %. The European manufacturer knows that the variation in motor losses in this case is smaller than 10 % of the nominal losses. He knows that no motor has an efficiency lower than 94.5 %. Thus he calculates that the motor nameplate efficiency can be given as 95 %. An American manufacturer for the same case must give the NEMA Nominal Efficiency 94.5 %, figure 5b.

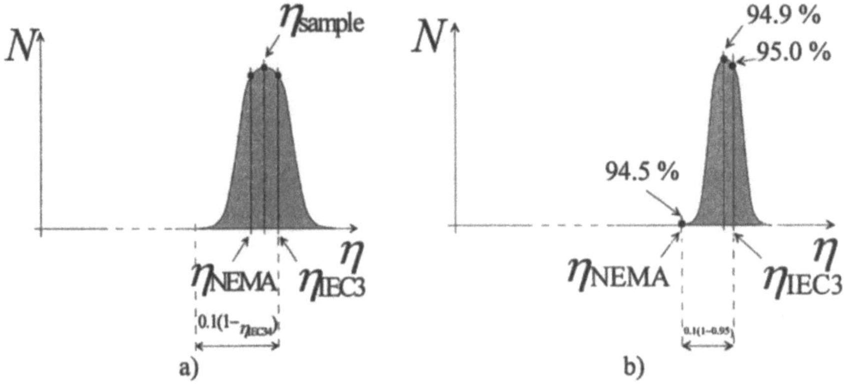

Figure 5: a) Comparison of NEMA and IEC 34 efficiencies, b) an example

According to IEC 34 the motor efficiency is given as 95 % and the same motor is going to be 0.5 %-units less efficient according to NEMA. In practice the NEMA value in this case is the minimum efficiency for this motor type and the IEC 34 value is a value that is not reached by - say - 60 % of the motors!

4 Conclusion

In this paper the methods defining 3-phase low voltage induction motor's efficiency has been investigated utilising the IEC 34-2 and the IEEE 112 B methods. The IEEE method is more accurate and contains less assumptions than the IEC-method. The efficiency values utilising the IEEE-method are lower than those given by the IEC method mainly because of the higher and fixed reference temperature (25°C) and the actual stray-load losses are higher than 0.5 % of the input power. In the high efficiency motor investigated here the stray load losses are not notable so that the difference between the IEC summation of losses compared to the IEC direct method is not large. The efficiency difference at the rated power is only 0.11 %-unit.

To make the market embrace motor efficiency more seriously the manufacturers′ efficiency values presented in the catalogues should be comparable. Thus a common reference temperature and a stability criterion as in the IEEE method should be also included into the IEC 34-2 standard.

5 References

[1] IEEE Standard 112-1996, Test Procedure for Polyphase Induction Motors and Generators

[2] IEC 34-2: 1972, Methods for determining losses and efficiency of rotating electrical machinery from tests (excluding machines for traction vehicles)

[3] IEC 34-1: 1994, Rotating electrical machines-Rating and performance

[4] NEMA Standards Publication No. MG 1-1993, Motors and Generators

Precise Efficiency Data for Electric Motors - an Illusion?

Herbert Auinger

Dr. techn. Herbert Auinger
Siemens AG Automation and Drives, A&D SD
ZVEI FV1 – Technical Commission

Abstract. Significant differences can exist between

- test results from various determination methods
- the designated nominal efficiency η_N, and
- the actual efficiency under practical operation

Pragmatic, clear and easy to handle rules, covering the overall context, are needed.

The expenditure for a "precise" determination method appears only to be justified, if it is ensured that the improvements which are achieved are not lost within the scope of accuracy, reproducibility and existing tolerances. The full load efficiency of a 4 kW-Motor can change by + 2 to - 4 percentage points due to the field operating conditions. Until now, no information has been provided on this issue. On the other hand, considerable effort is being made to come up with "more accurate" methods, whose results frequently only differ by less than 1 percentage point. In this case, the question must be asked whether the priorities have really been set correctly.

Committees of IEC and CENELEC are responsible in creating appropriate specifications. At the end of 1996, the EC gave CENELEC a clearly formulted mandate. Unfortunately the presented Draft IEC 61972 CDV is not sufficient and it has been rejected at the voting procedure. Therefore the introduction of the European high efficiency motors will be realized on the basis of the existing IEC 60034-2.

1 Introduction

Significant differences can exist between

(1) test results from various determination methods
(2) the nominal efficiency η_N, specified in Catalogs and
(3) the actual efficiency under practical operating conditions.

The following explanations are intended to create an awareness of the various inter-relationships.

2 Standards Situation

2.1 IEC 60034-2

Worldwide the most established method of determining motor efficiency is the "summation of losses"-method according to IEC 60034-2. [1] This is the preferred method for squirrel-cage induction motors. Specific advantages include: simple to handle, no principle-related restrictions regarding applicability, and extremely good reproducible results, even under field conditions. The measurement uncertainty only needs a relately low proportion of the tolerance permitted by IEC 60034-1.(Figure 1)

In practice, this has often led to the fact that "optimistic nominal values" of η_N were designated, i.e. values were been rounded-off upwards. An emphatic discussed main point of criticism is the fact that the load-dependent stray load losses are taken into account as a fixed percentage, i.e. 0.5 % P_1.

2.2 American Philosophy

For the North American market, the stray load losses must be measured in accordance with IEEE 112B. [2] The method of determining individual losses, (similar to that specified in IEC 60034-2) is extended with a series of input-, output- measurements in quarter-load stages between 0.25...1.5 P_N. The stray-load losses are calculated from these six measuring points according to their reference dependency $V_{SL} = K \cdot T^2$ using a linear regression analysis, which reduces the influence of the measuring errors.

As for every direct efficiency measurement, the unavoidable errors when measuring the input- and output power have, with increasing efficiency, an increased impact on the accuracy and reproducibility of the results. Various sources state, independent of one another, that in spite of using extremely highly accurate measuring equipment and even when taking extreme care, a typical uncertainty range of ± 0,5 percentage points can be expected [3, 4]. When referred to the total losses in the range η_N > 90 %, there is a progressively increasing uncertainty range as shown in Figure 1, with e.g. ± 10 percent for η_N = 94.5 %. As the stray-load losses represent approximately 10...15 % of the total losses, the inaccuracy when determining them becomes a factor of between 10...6 higher, and is, for example, ±(100...60)% for $\eta_N = 94.5$ %.
When this method is practically applied, there are other prerequisites which differ from those IEC 60034. The most important differences are:

- The necessity to have approved testing laboratories
- Statistical documentation of several motors of the same design; a sample test result alone is not sufficient
- η_N must correspond to a discrete numerical value from a specified selection table.

2.3 New Standards Work for IEC SC 2G

Against the background that High Efficiency Motors have started to be introduced in Europe and the fact that in future the motors will be sub-divided into three efficiency classes at the end of 1996, the EC gave CENELEC the mandate to draw-up an improved Standard to determine and designate the efficiency. [5]

The Draft Standard IEC 61972 CDV was been created within the framework of this mandate, [6]. However, this only handles the specific aspect of the determination methods. The other tasks, especially the designation topic, e.g. information which must be made accessible to users, and the required secondary conditions, have not been addressed until now.

Two determination methods are presented in IEC 61972 CDV. Method 1 essentially corresponds to the procedure specified in IEEE 112B - and its applicability reaches principle-related limits of the tolerances, defined in IEC 60034-1, in the range $\eta_N > 93$ % (refer to Figure 1). Method 2 corresponds to the present summation of losses method. However the fixed percentage values for the stray-load losses are unreasonably high for high-efficiency motors. The real values for European motors of today's production are considerably lower.

Unfortunately the Draft IEC 61972 CDV is not sufficient and it has been rejected at the voting procedure. The introduction of the European high-efficiency motors will be realized, as a last resort, on the basis of the existing IEC 60034-2.

3 Practical Example

Concrete values for the full-load efficiency of a 4-pole 4kW motor (Figure 2) will clearly show the present situation.

3.1 Test Results

The efficiencies, determined according to various methods, are entered in the lefthand part of the diagram, including the associated areas of uncertainty which typically exist. The differences are essentially caused by the stray-load losses, which were determined according to IEEE 112B with 1.3 %. According to IEC 61972 CDV, method 1, there are only slight differences due to the load-dependent correction of the iron losses.

3.2 Designated Nominal Value η_N

All 3 test results justify a nominal value of η_N = 85.5 %.

The lower limit of the tolerance range, permitted in accordance with IEC 60034-1, is η_{min} = 1.15 η_N - 15 % = 83.325 %. Based on IEC 60034-2, it would be easily possible to designate η_N higher, e.g. with η_N = 86 %.
NEMA MG1 only permits selected specific numerical values in the specified steps. The sample test result lies precisely in the center between 84 % and 85.5 %. The decision whether to round-off up or down depends on the statistical average from a higher number of test results on similar motors. A second condition is that for η_N = 85.5 % the minimum efficiency of all motors must be $\eta_{\min} \geq$ 82.5 % (this corresponds to + 20 % tolerance in the losses).

3.3 Efficiency in Field Operation

As shown in the righthand section of Figure 2, the electrical supply conditions, existing on the public power supply network, according to EN 50160, regarding voltage amplitude, symmetry and harmonic content can have a significantly higher impact on the efficiency than the previously investigated differences when determining and designating. [11] This is especially true also for the efficiencies at partial loads. In this particular example - the amplitude of the supply voltage alone - can change the full-load efficiency by up to + 1 to – 3 percentage points. In conjunction with this, IEC 60034-1 and NEMA MG1-30 [1, 2] provide clear warning information regarding the increased temperature rise due to increased losses (20 K for 0.9 V_N), and the required restrictions (no continuous operation, de-rating).

The motor operating temperature also has a significant impact on the efficiency. Both the ambient temperature as well as the temperature rise of the motor at any instant in time, are of special significance. Full-load operation where the motor temperature has stabilized, which for larger motors is only achieved after several hours of operation, is in practice, more the exception than the rule. [10, 11]

When it comes to the load-dependent stray-load losses, whose precise determination is a core issue [6, 8, 9], then one should be aware that the age of the motor and the winding configuration play a major role. Professor Oberretl determined 30 years ago [12], that the stray-load losses after 6 months operation are typically 40% lower than for a new motor. There can also be significant differences between star- and delta winding configurations.

4 Conclusion

If it is of interest to have the best possible method of determining efficiency, then this is realized best in the overall context. There must be clear goals and a clear understanding of where the benefits really lie. [10]

The designated nominal value η_N has increasing importance in the market as a quality criterion. Its significance must be seen in the sense of a guaranteed product feature which can be used to objectively evaluate motors by comparing the technical data and in turn selecting a particular product.
With regard to real applications, it must be simple to prove this value with a good reproducibility.

The increased expenditure for a "precise" determination method appear only to be justified, if it is ensured that the improvements which are achieved are not lost within the scope of the measuring accuracy, the reproducibility and the existing tolerances when designating η_N. The main focus should be on the practical benefits for the user. The full load efficiency can change by + 2 to - 4 percentage points due to the field operating conditions. Until now, no information has been provided on this issue. On the other hand, considerable effort is being made to come up with "more accurate" determination techniques, whose results frequently only differ by less than 1 percentage point.

In this case, the question must be asked whether the priorities of Standardisation activities have really been correctly set. Suggestions for continuing the Standards work are presented.

5 Literature

[1] DIN EN 60034-1 (VDE 0530, Teil 1) „Drehende elektrische Maschinen – Bemessung und Betriebsverhalten“, November 1995

[2] NEMA MG1 – 1993, Part 12 and Part 30: “Test and Performance” and „Application Considerations” for constant Speed AC Motors

[3] G. Gray, J. Marting : „Efficiency Testing of Medium Induction Motors, a Comment on IEEE Std.
112 – 1991 IEEE Transactions on Energy Conversion 11 (1996) pp. 495...499.
[4] US-Dept. of Energy Federal Register Vol. 63 No. 122 (1998)
[5] European Commission Mandate M/244 (Dec. 1996)
Mandat to CEN and CENELEC for the Elaboration and Adaption of Standards for Test Methods and for the Measurement, Calculation and Designation of Efficiency of General Purpose Low Voltage Tree Phase Induction Electric Motors
[6] IEC 61972, 2G/102/CDV, August 1998 „Method for determining losses and efficiency of three-phase, cage induction motors"
[7] A. de Almeida, P. Bertoldi, W. Leonhard „Energy Efficiency Improvements in Electric Motors and Drives" ISBN 3-540-63068-6, Springer Verlag Berlin, Heidelberg, New York 1997
[8] N. Glew „Efficiency Measurement Testing Standards – Stray Loss, the key to Efficiency Determination" included in [7] pp. 249...265
[9] N. Glew „Stray load losses in Induction Motors; a Challenge to Academia – Cambridge Conference Paper (Sept. 1997)
[10] H. Auinger, „Consideration about the Determination and Designation of the Efficiency of Electric Maschines", included in [7], pp. 284-304 in German : „Betrachtung zur Ermittlung und Kennzeichnung des Wirkungsrades elektrischer Maschinen", Elektrie 51(1997, pp. 2-9)
[11] H. Auinger, „Wirkungsgrad von Elektromotoren unter realen Einsatzbedingungen" Elektrie 53 (1999), pp. 74...77
[12] K. Oberretl „13 Regeln für minimale Zusatzverluste in Induktionsmotoren" Bulletin Oerlikon, Nr. 989/990 (1969) pp. 1...11

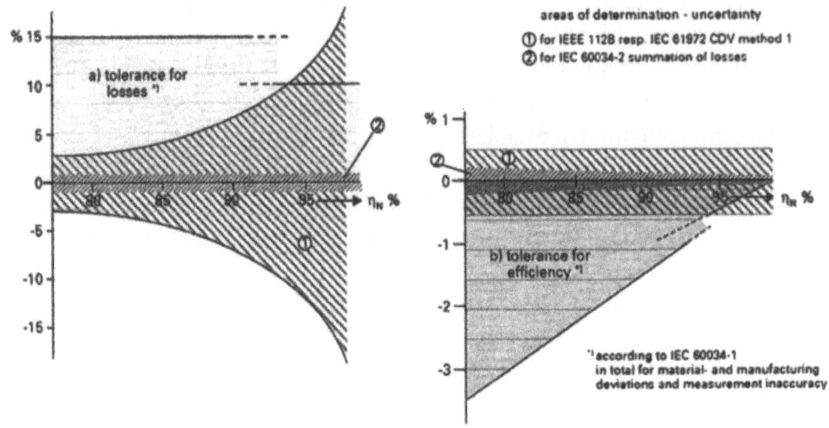

Fig 1: Permitted Tolerance compared with typical Determination-Accuracy (Reproducibility)

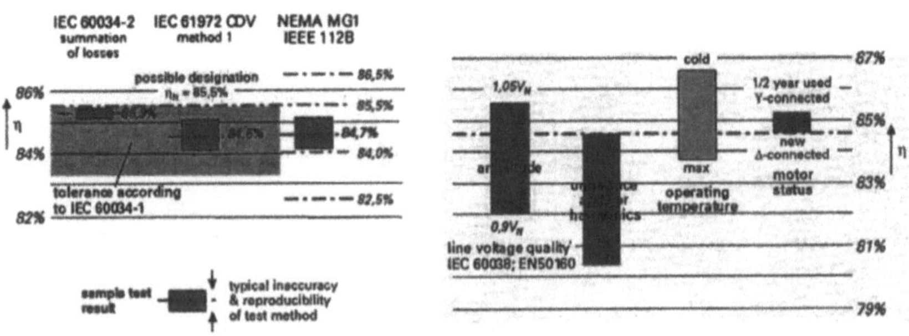

Fig 2: Full load Efficiency of a 4kW-Motor as Representative Example

Calorimetric Method to Measure Total Losses of the Induction Motor

Elzbieta Ogonowska – Schweitzer

Electrotechnical Institute, 04-703 Warsaw, 28 Pozaryskiego str., Poland

Abstract. The advantage of the calorimetric method is possibility of loss determination irrespective of value supply frequency and of the shape of voltage pattern. The main principle of the calorimetric method is measurement of the heat produced in the machine. The heat is transferred to the surrounding by conduction, convection and radiation.
In the paper the new, based on the principle of the calorimetric method, laboratory stand for measurements of total loss of the induction motor is described. The stand is designed to test medium power encapsulated induction motors. The measurement chamber is similar to well known Thomas' cylinder used in ordinary ventilation tests. Measurement of the temperature rise of the airflow through tunnel is made by means of the copper nets being resistance thermometers.
The measurement results of losses of the induction motors carried on by calorimetric method are close to value obtained by other methods. The calorimetric method fulfils requirements as a tool to determination of the value of motor efficiency fed from inverters.

1 Introduction

The main goal of the total loss of the electrical machine measurements is possibility of the calculation of its efficiency. There are few different ways of loss determina-tion described in international and national standards:

- indirect method i.e. losses segregation method based on the measurement of the electric parameters of the motor and calculations of the losses,
- direct method based on input and output measurements, applied to motors up to 1kW only, because of low accuracy of this method,
- calorimetric method, recommended to determination of the losses of large direct current generators.

Direct and indirect methods are suitable for determination of the values of the losses of the motors fed from mains, but they are not applicable yet to loss determination of the motors fed from inverters, because of well-known problem with power measuring.
The advantage of the calorimetric method is possibility of loss determination irrespective and of value supply frequency and of the shape of voltage pattern.

The calorimetric method of loss determination given in standard IEC 60034-2A has been devised mainly for large generators, but the principles used can also be applied to other electrical machines.
The based problem of losses determination in induction motors is to provide the accuracy of the calorimetric measurement during load test.

2 Measurement Method

2.1 The Laboratory Stand

General idea of the calorimetric method was realised in measurement stand which is schematically presented on Figure 1. The measurement chamber is similar to well known Thomas' cylinder used in ordinary ventilation tests. Here, the reverse Thomas cylinder is applied.

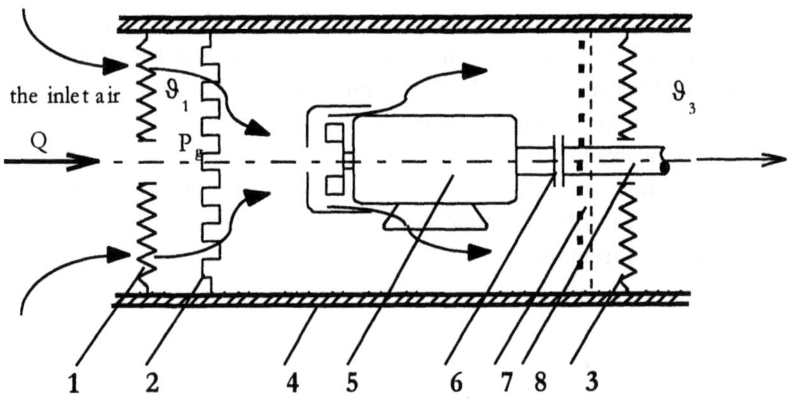

1 - the first measuring net,
2 - the heating net,
3 - the second measuring net,
4 - insulated wall of chamber,
5 - motor,
6 - the shaft coupling with heat insulation,
7 - equalize plates,
8 - cardan shaft.

Figure 1: Scheme of the measurement chamber

The measurement chamber is cuboid shape. The walls of this chamber were made with plywood board insulated with foamed polystyrene. The motor was set up on wood beams. Tested motor is positioning between two measurement nets.

Additional heating net is placed on the inlet of the tunnel, over the first measuring net. Measuring nets working as resistance thermometer are made from copper conductor placed uniformly in whole cross-section of the tunnel. The temperature rise to measure is low value and it can be up to few Kelvins, so differences of resistance of nets are low also and they can be anticipated up to 3%. Typical measuring bridges can't measure these low values of differences, so special measuring unit was designed and built.
The cardan's coupling is preserved against heat conducting by distance insulator.

2.2 Measurement Principle

The difference of temperature of the air measured on inlet and outlet of tunnel is proportional to the losses generated in the motor.
The heat balance equation, before switching on the heating net is given by

$$P_t = Q \cdot c_p(\vartheta_3 - \vartheta_1) = Q \cdot c_p \cdot \Delta\vartheta_t \tag{1}$$

After the switching on the heating net, the heat balance equation is given by

$$P_t + P_g = Q \cdot c_p(\vartheta_3 - \vartheta_1)_{tg} = Q \cdot c_p \cdot \Delta\vartheta_{tg} \tag{2}$$

where:

P_t -losses dissipated in motor, [W],
P_g -additional heating power, [W],
$\Delta\vartheta_{tg}$ -difference of temperature between nets 1 and 2 after added heating power, [K],
$\Delta\vartheta_t$ -temperature rise of the air flow through tunnel caused by losses of motor, [K],
Q -volume rate of flow of cooling medium, [kg/s] ,
c_p -specific heat capacity of cooling medium, [W· s/kg· K].

$$P_t = P_g \frac{1}{\frac{\Delta\vartheta_{tg}}{\Delta\vartheta_t} - 1} \tag{3}$$

where:

P_t -losses dissipated in motor, [W],
P_g -additional heating power, [W],
$\Delta\vartheta_{tg}$ -difference of temperature between nets 1 and 2 after added heating power, [K],
$\Delta\vartheta_t$ -temperature rise of the air flow through tunnel caused by losses of motor, [K].

The value of total loss of the inspected motor obtained by calorimetric method was always less then the values calculated according to means of indirect and direct methods. The reason of such difference was probably irregularity of distribution of velocity, direction and temperature of the air flow through outlet measuring net. The resistance of this nèt is strongly dependent of above parameters. To eliminate influence of this measurement calibration of chamber was necessary. The calibration was done during unload run of the motor, when whole energy consumed by the motor is transformed to the heat. During the calibration process the coupling device was connected to the motor journal. There were developed few different measurement where different values of unload losses were taken into account. For each known value of unload losses of the motor the steady thermal state was obtained. Then the measurement of the losses were done. It was assumed that the losses obtained by calorimetric method are linear function of real value of the motor losses. The correction coefficient can be calculated by linear regression method. In unload state of the motor:

$$P_k = a \cdot P_1 + b \qquad (4)$$

where:

P_k - total loss of the motor obtained by calorimetric method, [W],
P_1 - real value of the total loss, [W],
a - the slope of the line,
b - the y intercept of the line.

If there were $a > 0$ and correlation coefficient $c \geq 0,9$, then the result of the regression is satisfied. Knowing value of a (the correction coefficient), the value of total loss may be obtained from the formula:

$$P_t = \frac{1}{a} \cdot P_{tk} \qquad (5)$$

where:

P_t – the value of total loss, [W],
P_{tk} – the value of losses calculated from formula (3), [W].

3 Measurement Results

Two motors of high shaft 132mm were tested:

- high efficiency motor P_N = 15HP, U = 460V, f = 60Hz, n = 1755rpm, tested under supply from mains (380V, 50Hz), and supplied from inverter for frequencies f = 30, 40, 60 Hz,
- general purpose motor P_N = 5,5kW, U = 380V, f = 50Hz, n = 1450rpm.

Table 1 shows calibration results and calculations of the regression coefficient for inspected motors. For each motor five different measurements were carried on.

Table 1: Results of calibration.

P_1 [W]	Motor no 1		Motor no 2	
	P_k [W]	Regression coefficients	P_k [W]	Regression coefficients
200	215		170	
300	258	a = 0,832	266	a = 0,829
400	347		373	
500	441	c = 0,983	430	c = 0,987
600	540		503	

It can be observed that high value of c and very similar value of a obtained. During calorimetric tests of motors the values of the total loss were obtained. In the same time the measurements and calculations of the losses according to other methods were carried on. There were: indirect method, B - method according to IEEE standard and direct method.

Tables 2 and 3 show the results of measurements and calculations losses and efficiency. Table 2 is for motor no 1, table 3 is for motor no 2 (1 - calorimetric method, 2 - indirect method, 3 - B - method, 4 - direct method). Table 4 consist results of measured and calculated values of losses and efficiency for high efficiency motor fed from inverter. The tests were carried on for 30, 40 and 60Hz with constant value of rated load.

Table 2: High efficiency motor no 1, P_N = 15 HP, U = 460 V, f = 60 Hz tested under supply from mains U = 380 V, f = 50 Hz.

P [W]	Losses P_t [W]				Efficiency [%]			
	1	2	3	4	1	2	3	4
2500	385	367	339	440	86,8	87,5	88,2	82,6
4600	536	507	487	569	89,5	90,2	90,6	88,1
6900	785	768	754	771	89,8	90,0	90,2	89,8
9200	1203	1198	1198	1205	88,3	88,5	88,0	88,9
11500	2365	2087	2642	2427	83,0	85,0	81,0	85,4

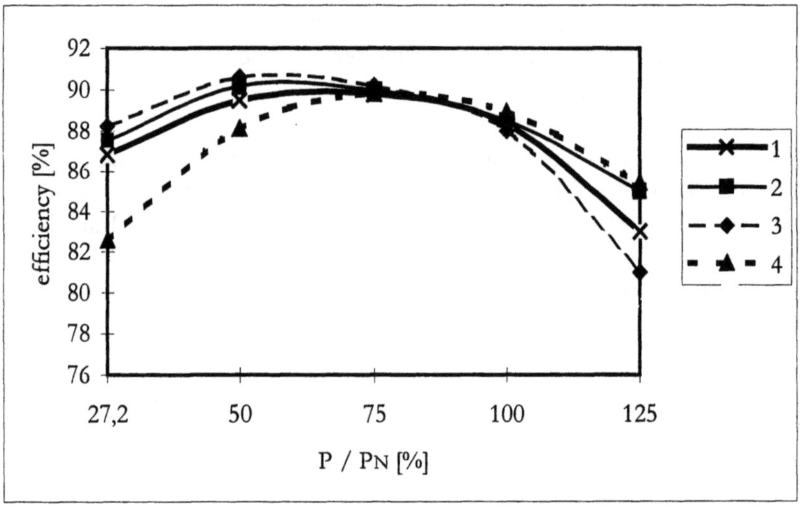

Figure 2: Efficiency of motor no 1.

Table 3: Motor no 2, P_N = 5,5 kW, U = 380 V, f = 50 Hz.

P [W]	Losses P_t [W]				Efficiency [%]			
	1	2	3	4	1	2	3	4
1375	385	412	385	428	77,1	78,7	78,8	76,0
2750	536	499	479	505	84,7	85,3	85,5	84,4
4125	785	660	634	667	86,2	86,8	86,7	86,1
5500	1203	1035	911	1079	84,1	86,2	85,4	83,6
6875	2365	1503	1305	1530	82,1	84,4	82,3	81,8

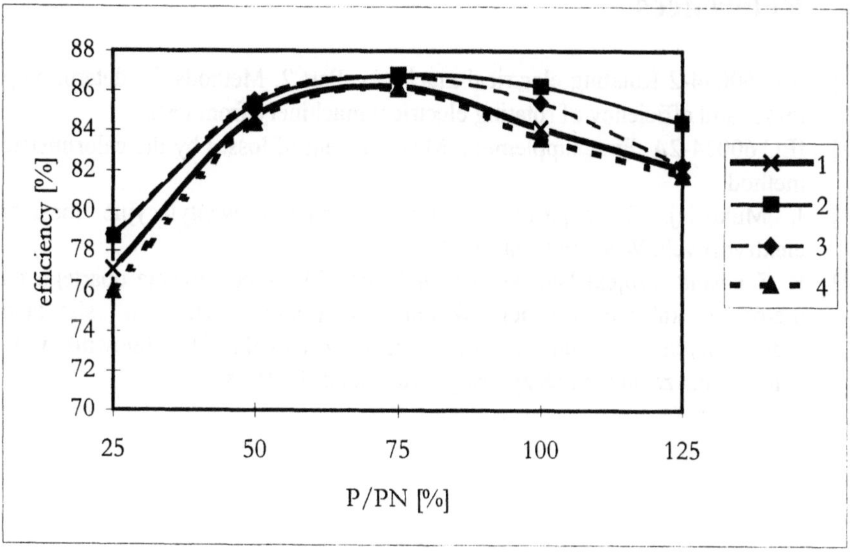

Figure 3: Efficiency of motor no 2.

Table 4: Motor no 2 fed from inverter.

U [V]	f [Hz]	P_1 [W]	Losses P_t [W] calorimetric method	Losses P_t [W] direct method	Efficiency, [%] calorimetric method	Efficiency, [%] direct met hod
460	60	12387	1618	1541	86,9	87,6
340	40	8496	1455	1426	82,9	83,2
295	30	6792	1744	1769	74,3	74,0

4 Conclusions

The measurements results of losses of the induction motor carried on by calorimetric method are close to value obtained by other methods. The calorimetric method fulfil requirements as a tool to determination of the value of motor efficiency fed from inverters.
The main drawback of the method is that it is slow and difficult to realise.
It is necessary to provide more tests for different types of induction motors.

5 References

[1] IEC 60034-2 Rotating electrical machines. Part 2. Methods for determining losses and efficiency of rotating electrical machinery from tests.

[2] IEC 60034-2A. First supplement. Measurement of losses by the calorimetric method

[3] J. Mukosiej, R. Zapasnik. Badania cieplne i wentylacyjne maszyn elektrycznych. WNT Warszawa 1964

[4] R. Zapasnik. Projekt badawczy nr 8T10A01609. Opracowanie opartego na metodzie kalorymetrycznej ukladu do pomiaru strat w silnikach indukcyjnych szczególnie energooszczednych i zasilanych z falowników w celu wyznaczenia ich rzeczywistej sprawnosci, 12/1998.

Precision Calorimetry for Stray Load Loss Determination

K. J. Bradley[1], A. Ferrah[1], R. Magill[1], P. Mcleod[2], J. C. Clare[1], P. Wheeler[1], P. Sewell[1]

[1] The University of Nottingham
[2] Dyson Research Centre

Abstract. The paper explains how the use of calorimetric techniques can lead to a dramatic increase in the accuracy of measurement of both total and stray load losses in induction motors. Analysis of the errors involved in determining stray load loss from total and component losses is. The need for a separate generated supply is explained. An explanation of the calorimetric technique is given together with details of a particular design. Accuracy is ±5.6 W on the 3.2kW range. The repeatability of the loading system, and calorimeter was developed to be ±8W when testing 7.5 kW motors. A comparison is presented for a 30 kW motor of carefully obtained C390 results and Calorimeter results for stray load loss.

1 Introduction

The precision measurement of stray load loss in induction motors has been a problem of long standing. Current techniques lack precision. Only when accurate results are available will it be possible to evaluate effectively methods of stray load loss prediction. Improved motor design and manufacturing quality control may then follow in a bid to improve overall motor efficiency.

1.1 Current and Calorimetric Methods for Measuring Stray Load Loss

The technique used to measure stray load loss in induction machines is defined in current and draft standards [1..3]. It is based upon the derivation of the total loss from the input and output powers. This is potentially inaccurate with high efficiency machines. The Figure 1 shows the error involved in measurement of total loss for different machine efficiencies and different accuracy of measurement. Figure 1 also shows the errors involved when using a calorimeter with 0.5% error to measure total power loss directly. Several researchers have built calorimeters [4..8]. The one described here and in [9] is believed to be the most accurate.

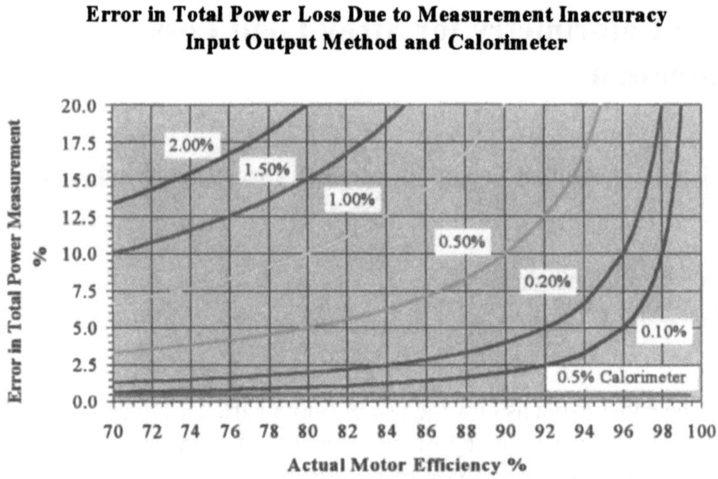

Figure 1

An excellent accuracy for power measurement is 0.2%. This requires 0.1% full scale accuracy instruments. To obtain this accuracy the instrument will need to average the readings over at least one slip cycle. Figure 1 shows clearly the advantage of calorimetry particularly for high efficiency machines.

Table 1 shows a typical full load breakdown of losses for a 7.5 kW motor.

Table 1: Breakdown of full load loss for a typical 7.5 kW, 4-pole, induction-motor

Source of Component Loss	Percentage of total loss
Stator conductor loss	45
Rotor conductor loss	25
Iron loss	15
Windage and friction	5
Stray load loss	10

The error in stray load loss estimation for such a 7.5kW machine caused by errors in total loss measurement is shown in Figure 2. For the purposes of calculation all other component losses are assumed to have no error.

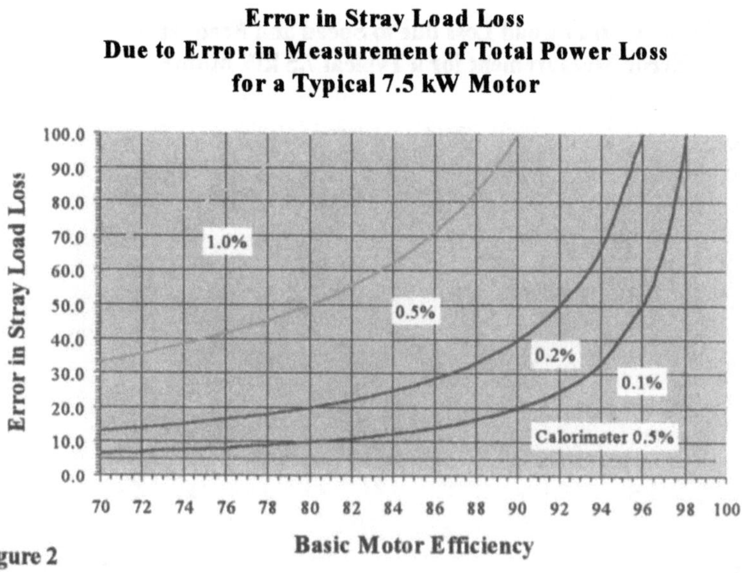

Figure 2

Quite clearly, even with excellent measurement accuracy of 0.2% the stray load loss error for a typically 86% efficient 7.5 kW motor is approximately 30%. The calorimeter is giving rise to only 5% error, which is much more precise.

The input output test method is improved by taking measurements at a number of different load points. By fitting a curve to the variation of residual loss (or stray load loss) with load it is possible to compute, new and more accurate values for power loss at each load point. However, stray loss has the same pattern of variation of loss with load as stator and conductor loss.

Rotor conductor loss in small machines is evaluated using slip which is derived from the shaft speed and the supply frequency. The Figure 3 shows the error in stray load loss for the typical 7.5 kW, 4-pole, motor due to measurement inaccuracy in frequency and speed components. At a typical 3% full load slip the error in stray load loss from using excellent accuracy 0.01% instruments giving about 0.2% measurement error is about 8%. Anyone used to testing induction machines on the normal mains supply will realise that 0.015-rpm (0.01%) speed stability is uncommon. With high power, low slip machines it is usual to measure slip directly. There is therefore a possible 40% or so error in stray load loss estimation already. For the standard tests the results for reduced loads will obviously have lower slip and greater potential error.

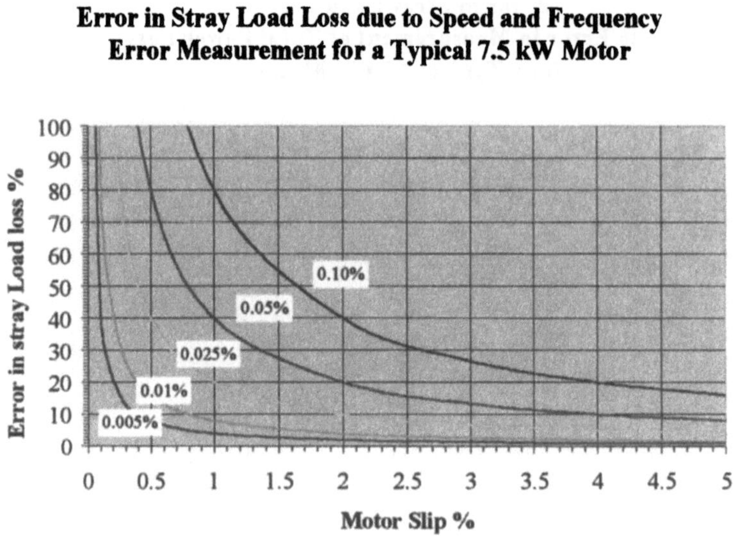

Figure 3

It is not unusual for part load tests sometimes to yield negative values for stray load loss. There are many potential sources of error in addition to that described above. The evaluation of the correct iron loss from no load tests can be particularly difficult since it is necessary to measure the stator resistance precisely at the instant of power measurement in order to subtract the conductor loss from the total no load loss. Windage and friction losses tend not to be repeatable if standard grease lubrication is used, oil seals are fitted and bearing pre-load applied by wavy washers where double ball bearings are fitted. The method is influenced substantially by the accuracy of the evaluation of these component losses.

With care the standard procedures can be made to give good results that reflect the particular model of stray load loss assumed. Calorimetry, with it greater accuracy in total loss measurement offers greater precision in stray load loss estimation but it has to be combined with the best techniques for evaluating component losses.

1.2 Calorimetry Based Method for Measuring Stray Load Loss

The calorimeter measures the power loss in the motor directly. Ideally, the calorimeter operates by placing the motor inside a perfect, thermally insulated enclosure or box while the machine is under load. The air inside the box is heated by the power loss from the motor. Heat energy is extracted from the air in the box by a coolant and heat exchanger located in the box. In steady state the air in the box is at a constant temperature and the power lost by the motor is equal to the measured directly from the mass flow of coolant, its temperature rise and its thermal capacity.

Of course, the simple description above is more difficult to achieve in practice but the principle can be applied giving errors in the measurement of power loss below 0.5% independent of the efficiency of the machine being tested. Sources of error in a practical calorimeter are.

1. Heat leakage paths as shown in the non-ideal calorimeter of Figure 4.
2. In air-cooled calorimeters the difficulty in maintaining a set mass of coolant. Most flow meters measure volume flow of air and the density of air depends upon temperature and pressure.
3. In air-cooled calorimeters the changing specific heat of air with its humidity.
4. The temperature gradients in the coolant in inlet and outlet pipes which have a large cross-section in air-cooled calorimeters. This may exceed the precision with which the mean temperature of the coolant needs to be measured.

The solutions to these problems are described below.

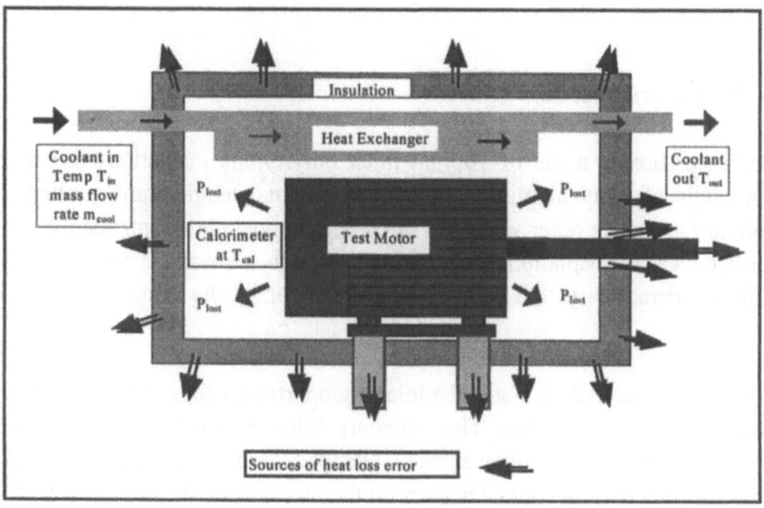

Figure 4

2 Experimental Equipment for Calorimetric Measurement of Total and Stray Load Loss

2.1 Description of Calorimeter

Figure 5 shows a schematic diagram of the calorimeter and test motor supply and loading system that is explained below.

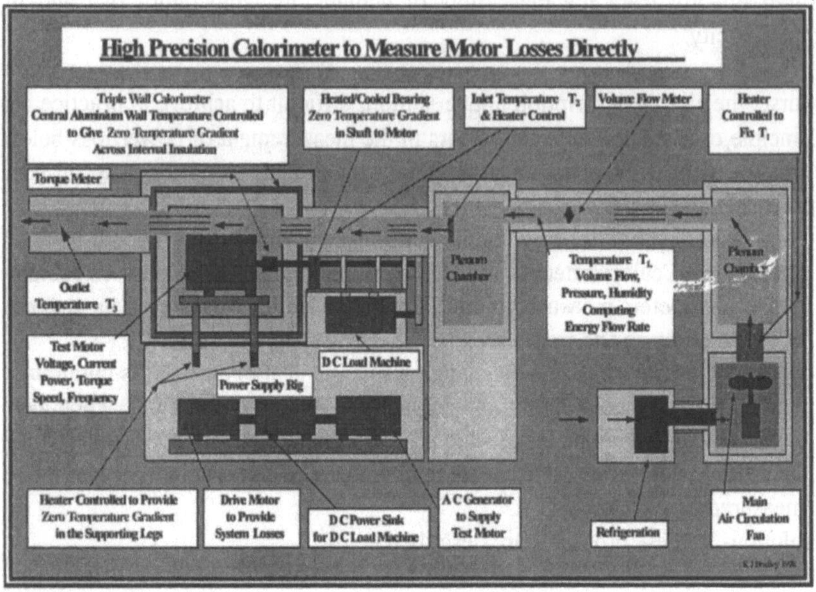

Figure 5

2.1.1 The Calorimeter Box

The calorimeter box is made of 100mm thick polystyrene insulation with inner and outer walls of 6mm aluminium to ensure minimal temperature gradient in those wall surfaces. The outer walls are fitted with temperature-controlled heaters and a further layer of insulation. Heat leakage through the walls is minimised by forcing the temperatures of the outer walls to be virtually the same as the inside walls.

The test motor is mounted on a steel bedplate, supported by four, 50mm diameter, 2mm wall thickness, steel tubes. The tubes are filled with insulation to prevent convection currents. Temperature sensors clamped to the tubs just inside and outside the calorimeter box control heaters on the legs outside the box. The heaters minimise the temperature gradient along the leg so reducing heat loss. A similar approach is used for the tubular drive shaft with a heated and cooled housing for the bearing located just outside of the calorimeter.

2.1.2 Description of Coolant Control

Air mass flow is controlled to give the same thermal capacity as a set amount of dry air. This is done by computing the thermal capacity of the cooling air from measurements of volume flow rate, pressure, humidity and temperature at one point in the inlet duct. The air temperature at this point is controlled to be constant within 0.1C. The result is converted to the equivalent mass of dry air. A PI controller is used to control the inlet fan speed to force this equivalent mass to be equal to the set point required for the particular range.

A full flow turbine mounted in an inlet duct designed to BS6169 measures the air volume flow rate. The repeatability of this transducer is quoted as 0.05%.

The air is cooled to 6C at entry to the main coolant fan. This ensures the inlet temperature to the calorimeter is close to ambient temperature while allowing sufficient control range for the two heaters in the inlet duct.

2.1.3 Description of Temperature Measurement

Temperature measurement has a resolution of 0.001C. Temperature gradients can be as high as 2C across inlet/outlet ducts at their entry/exit from the calorimeter box. To obtain a good approximation to the average temperature of the mass flow, temperatures is measured half way along the 2m, 150mm diameter, heavily insulated, inlet and outlet ducts connected to the calorimeter box. The outlet duct is also fed from an air to air heat exchanger. Mean temperatures are determined using a single PT100 probe and a small temperature averaging heat exchanger in the duct.

The measured inlet duct temperature is controlled to be constant to within 0.002C by a second heater located in the second plenum chamber. This level of control ensures low noise on the calculated power derived by the calorimeter.

2.1.4 Calorimeter Calibration

The calorimeter is calibrated once with a resistive heater supplied with dc. Power is computed from precision measurement of direct voltages and currents. Linearity is to 0.5W on the 3.2 kW range.

2.2 Precision Control of Test Conditions

2.2.1 Test Motor Supply and Loading

A separate ac generator is run at constant speed is used to supply the test motor. Voltage stability and control is 0.1V and frequency stability is to 0.01Hz. A torque controlled loading rig is used to apply the load. Figure 6 shows speed of a 2.2 kW test motor maintained within ±0.15 rpm and input power maintained within ±5W over a 10 hour test period following warm up of the machine.

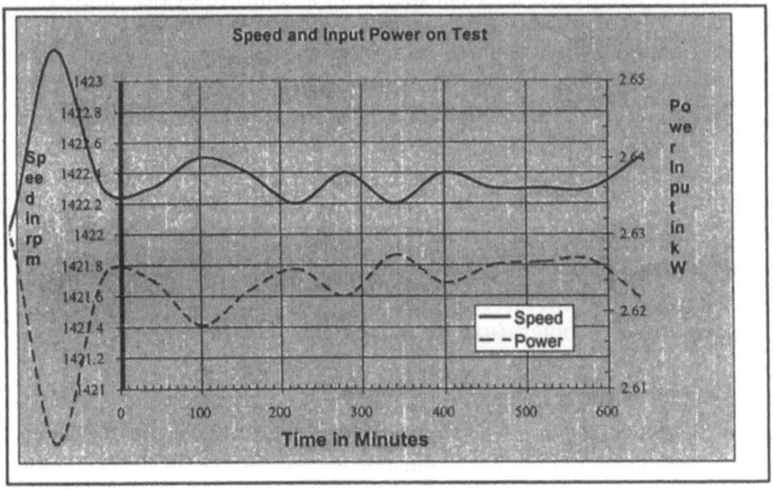

Figure 6

2.2.2 No Load Test Rig

The no load test rig consisted of a brake with very little windage and friction loss. The power supply for no load tests was as described above and immediately following measurement of the no load power, the supply to the test motor was removed and the test motor stopped. Readings of stator resistance were taken at 10 second intervals for the first minute after disconnection of the motor from the supply. These were used to extrapolate the winding resistance back to the switch off time. Accurate iron loss values could then be determined after compensation for stator winding resistance drop both on no load and on load.

3 Repeatability of Results and Overall Accuracy

The overall accuracy of the calorimeter on the 3.2 kW does not exceed ±5.6 W. This is made up of 1W calibration uncertainty error, 0.5 W linearity error, 2W repeatability (of the calorimeter), 2 W air circulation error and 0.1W of noise. It has been found that the calorimeter has a slightly lower error at lower power levels. Tests using the calorimeter to measure the power dissipated in the 30 W stirring fan and comparing the result with direct measurement showed no more than 1W overall error in 30W.

The repeatability of results for test motors measures not only the performance of the calorimeter but also the loading system and supply system performance. It was found that repeatability to within 8 W was possible for 7.5 kW motors when the bearings of these test motors were modified. That is, the grease was removed where possible and replaced with light oil, oil seals removed and pre load wavy washers removed.

4 Test Results

The Figure 7 shows a comparison for a 30 kW test motor of calorimetric and C390 results. The C390 test method was refined over several months and instrument accuracy continually monitored. The C390 test had some 11-load points. The test results were taken in the calorimeter, which maintained a constant ambient, at all load points. Winding resistance was by direct measurement after each test but thermocouples imbedded in the windings were used to correct for resistance at the time of measurement using cooling curves. Very precise calibration of the torque meter was performed and repeated several times.

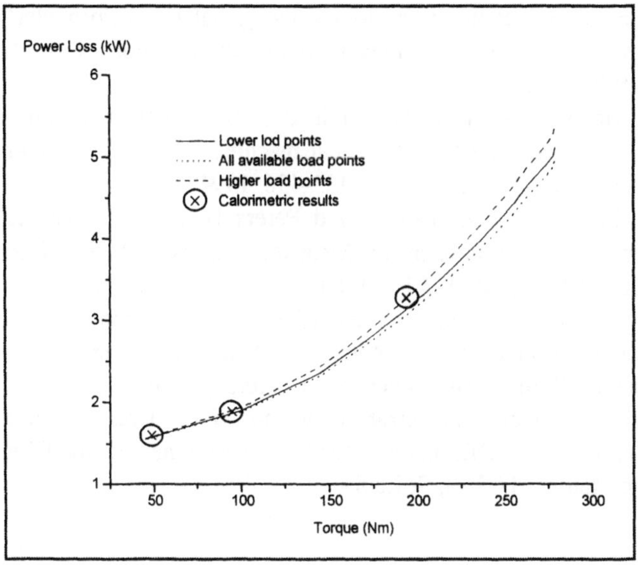

Figure 7

Even with all this care, it can be seen that stray loss predicted by the C390 is a function of the choice of load points. Some non-linearity in the torque transducer results in the greatest agreement with the calorimeter being achieved when only those load points close to full load are considered. Full load corresponds to the calibration point of the torque transducer. Including low load points in the tests increased the error since they introduced problems in the curve fitting of the stray load loss. Considering all points and discarding those that were furthest from the least squares fit also produced increased error. This was because the curve fit was biased by the inclusion of the low load points.

5 Conclusions

It is clear that even with extreme care, the results obtained by the standard test methods are not as reliable or as accurate as those of the calorimeter. The calorimetric method of determining stray load loss is therefore superior to the standard test methods and shows much promise as a tool to help in improving induction machine efficiency.

6 References

[1] IEEE Standard 112-1991

[2] Canadian Standard C390 M1985

[3] IEC16972 (Draft)

[4] Turner D R, Binns K J, Shamsadeen B N, and Warne D F, 'Accurate measurement of induction motor losses using balance calorimeter', IEE Proceedings, Electrical Power Applications, Vol. 138, No 5, September 1991

[5] Baholo B, Mellor P H, Howe D, Birch T S, 'An automated calorimetric method of loss measurement in electrical machines', Journal of Magnetism and Magnetic Materials, Vol. 133, PP 433-436, 1994

[6] Jalilian A, Gosbell V J, Cooper P, and Perera B S P, 'Double-Chamber Calorimeter, a New Approach to Measure Induction Motor Harmonic Losses', IEEE, 1997 pp MB1-7.1 to MB1-7.3

[7] Aglen O, ' Calorimetric measurement of losses in air cooled and water cooled asynchronous motors' ICEM 96, Vigo, Spain ,Vol 3, pp 256-262

[8] P D Malliband, MPhil Thesis, Cambridge University,1998

[9] P MacLeod, K J Bradley, A Ferrah, R Magill, J C Clare, P Wheeler, P Sewell,'High Precision Calorimetery for the Measurement of the Efficiency of Induction Motors', IAS 98, St Louis.

A Test and Measurement System for Test of Electrical Drive Systems

P. Sandholdt, E. Ritchie, J.K. Pedersen

Aalborg University, Institute of Energy Technology, Department of Electrical Energy Conversion, Pontoppidanstraede 101, DK-9220 Aalborg East, Denmark

Abstract. A general test and measurement system for electrical drive systems, is presented. This paper describes the test and measurement system, data acquisition, the load emulation capabilities. Experimental results from load torque emulation and sample tests are given.

Keywords. Drive test, measurements, energy efficiency, torque emulation.

1 Introduction

Electrical drive systems are widely used in industry applications. The main advantages of drive systems are that it is possible to control the electrical machine speed and torque. Many other facilities are included in modern drive systems making them attractive for applications in highly automated systems for manufacturing, process control, transport, etc.
Dynamometers have been used to perform steady state load tests of electrical machines [1], [2] for many years. Mainly performance is evaluated by measuring the torque-speed characteristic, power losses, efficiency, etc.
A versatile electrical drive test system was designed around a dynamometer and implemented for testing drives from 250 [W] to 7.5 [kW] shaft output power. The automated measurement and test system for testing of electrical machines and drive systems is described.
Emulation of different torque-speed load characteristics is provided. The test system is capable of emulating loads like pumps, fans, conveyors, etc. Testing under flexible conditions is possible and time efficient using this automated measurement and test system.

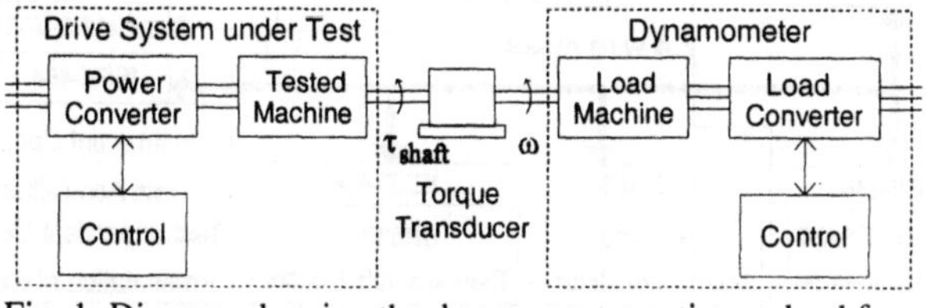

Fig. 1: Diagram showing the dynamometer acting as load for the drive under test.

2 Configuration of Test and Measurement System

The dynamometer system comprising the drive under test and a torque transducer, the load system controlled is shown in Fig. 1. The controllable load machine and converter comprises a servo drive system capable, of bidirectional power flow.
A DSP (ADSP-21062) is used in the test system to generate steady state and dynamic torque reference values for the dynamometer. The DSP is installed on a EZ-LAB board providing 33 [MHz] clocking.
The main purpose of this system is to generate suitable reference values for the load drive system and control the load drive system during steady state and dynamic situations. The DSP is equipped with A/D, D/A and decoder boards to provide adequate input and output facilities.
A Pentium computer is used to control the test system. The control and data handling is performed by a software package called Visual Engineering Environment (HP VEE). This facilitates direct control of instruments and equipment, using serial communication or the IEEE-488 bus. Control of an input/output data acquisition board can also be performed.
The DSP board is plugged into the PC bus and a C compiler is used to generate the DSP code. To communicate between the HP VEE software on the PC and the DSP program a DLL file is provided with the HP VEE program. The DLL file is created using a Visual C program. The Visual C program is then linked with the DSP Host Interface Library providing full bidirectional and online communication between the HP VEE program and the DSP program. HP VEE requires a header (*.h) file to be

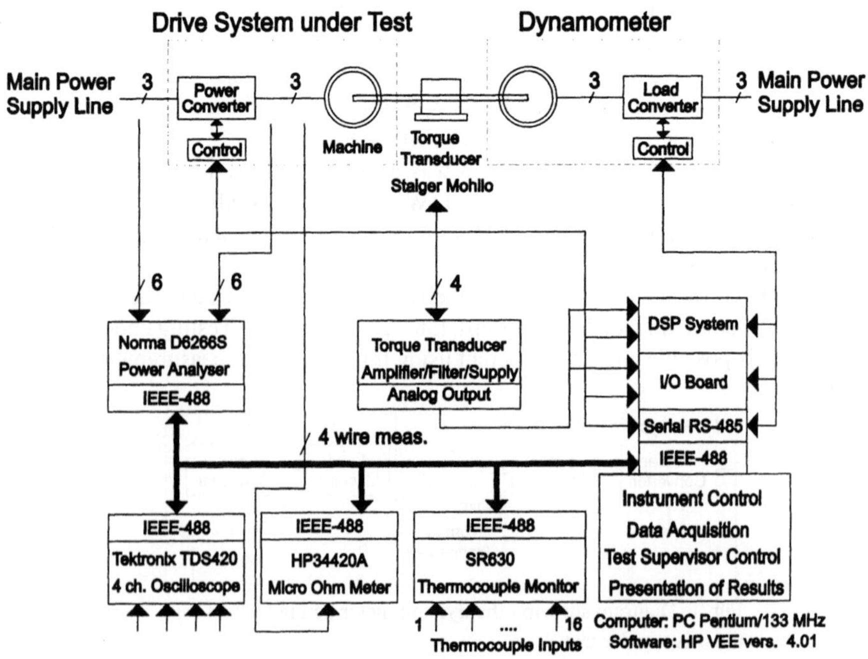

Fig. 2: Diagram showing the test and measurement system for drives.

included in the HP VEE program, declaring the functions implemented in the DLL. The experimental arrangement is shown in Fig. 2.

2.1 Data Acquisition

The Drive Under Test input and output voltages and currents are measured using a Norma D6266S power analyser, equipped with 6 current and 6 voltage channels, enabling simultaneous measurements of power converter input and output. The basic measurement accuracy is as low as 0.035 [%]. The power analyser is provided with computing facilities to perform calculations such as harmonic analysis, and filtering. In the test system, the converter input is measured using a two Watt-meter configuration, whereas the converter output is measured using the three Watt-meter method.

The torque transducer utilizes strain gauges and has a measurement accuracy of 0.1 [%] of full scale. The torque transducer is provided with several exchangeable measurement elements of ranges 5, 10, 20, 50, and 100 [Nm], minimising the measurement errors for different torque ratings.

A 4 channel, Tektronix TDS420, oscilloscope is used for dynamic measurement recordings.

Measurements of stator resistance was done automatically within 10 seconds, by an ohm meter. The ambient temperature is measured at three points using a thermocouple monitor. Thermocouples fitted to the machine may be measured too.

2.2 Data Handling

Often measurements of each line current, voltage, power etc. will be saved for each measurement, to detect asymmetrical effects.

For a complete drive system, the number of measurement values may be: 5 line currents, 5 phase-neutral voltages, 3 line-line voltages, 5 values for active power, reactive power, apparent power, power factor, phase angle, etc., plus mean and total values for the drive input and output. With torque, speed, temperature and resistance measurement values more than 140 values are stored for each measurement point.

For ease of data handling and graphing, measured values are transferred from the HP VEE program via Dynamic Data Exchange (DDE) to a spreadsheet file. In the spreadsheet file default graphs are provided, such that for instance the torque vs. speed graph may be plotted by simply activating the graph. This means that the time spent on data handling after the test is performed is reduced to a minimum.

3 Emulation of Load Torque

Many loads such as pumps, and fans have a torque-speed relationship which can be written as Eq. (1), if discontinuous friction contributions are neglected. Therefore the desired load torque that the load drive system should emulate may be in this form:

$$\tau_{ref} = \tau_0 \operatorname{sign}(\omega) + J p \omega + B_1 \omega + B_2 \omega^2 + B_3 \omega^3 \quad (1)$$

where
τ_{ref} is emulated reference load torque
τ_0 is a constant load torque
ω is the angular mechanical speed
$sign(\omega)$ is a sign function ($\operatorname{sign}(\omega \geq 0)=1$; $\operatorname{sign}(\omega<0)=-1$).
J is the mass moment of inertia
p is the differential operator
B_1 is the viscous friction constant
B_2 is the parameter for quadratic speed-dependent loads
B_3 is the parameter for triple speed-dependent loads.

The data input to the load system for torque emulation thus becomes the parameters τ_0, J, B_1, B_2 and B_3, making the computation of the desired load torque possible. To emulate the torque as function of the speed, the drive under test is speed controlled by setting the fundamental frequency f_{fund} and the load drive system is torque controlled according to Eq.(1) which is shown in Fig. 3.

4 Experimental Results

Tests was performed on a 2.2 [kW], 4 pole induction machine supplied with sinusoidal voltages and fed by a commercial frequency converter. The data for the tested induction machine and the frequency converter is listed in the appendix.

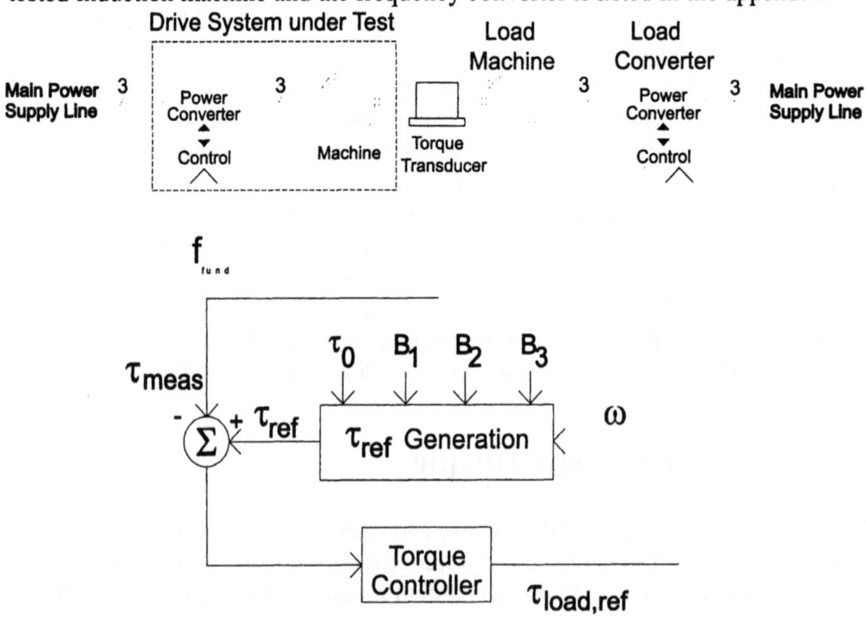

Fig. 3: Torque control schematic for the dynamometer with measurement of the load torque.

The results shown here were divided into tests performed on a line supplied induction machine and tests performed on the complete drive system.

4.1 Tests on Line Supplied Induction Machine

In this section results from a rated load test are shown. The determination of the induction machine torque-speed characteristic is in [3].

4.1.1 Rated Load Test

To evaluate the temperature rise of an induction machine a steady state load test was performed with a constant load torque. Other performance parameters such as efficiency, power factor etc. can also be evaluated. To emulate a constant load torque τ_0 in Eq.(1) was set to the desired value. The other parameters in Eq.(1) were then set to zero.

Determination of induction machine temperature was performed by loading the machine and measuring the stator winding resistance. This is usually performed at rated conditions. By comparing the hot resistance value with the cold value, measured prior to the test at a known temperature, the temperature rise can be determined. The calculation of the hot temperature is performed in Eq.(2).

$$T_2 = T_1 + \Delta T = T_1 + \frac{k_1 (R_2 - R_1)}{R_1} \tag{2}$$

where ΔT is temperature rise [°C]
R_2 is the "hot" resistance value [Ω] at temperature T_2 [°C]
R_1 is the "cold" resistance value [Ω] at temperature T_1 [°C]
k_1 is reciprocal of the temperature coefficient [K] at temperature T_1 [°C].

To compare the temperature obtained using resistance measurement a measurement of the machine winding and core temperature by inserted thermocouples was also performed.

The measurement of the temperature rise was performed as follows:

1. Measurement of the cold resistance value and the ambient temperature
2. Applying rated voltage and rated load to the machine
3. Every 10 minutes the load and the voltage has to be removed and the stator resistance and the thermocouples are to be measured

Step 2 and 3 are to be repeated until a stable temperature is measured. The time between two successive measurements may be adjusted, but 10 [min] was chosen here as a compromise between the number of measurement points and a long enough time to prevent cooling of the machine. The time required for the resistance measurement was approximately 30 [s] for each point making the off time equal to 5 [%]. The measured temperatures are shown in Fig. 4. A stable temperature is defined as a temperature which do not vary more than 1 [°C] withing 30 minutes. A stable winding temperature of 94 [°C] was obtained after approximately 180 [min]. The final temperatures reached were for the resistance measurement 94 [°C], while

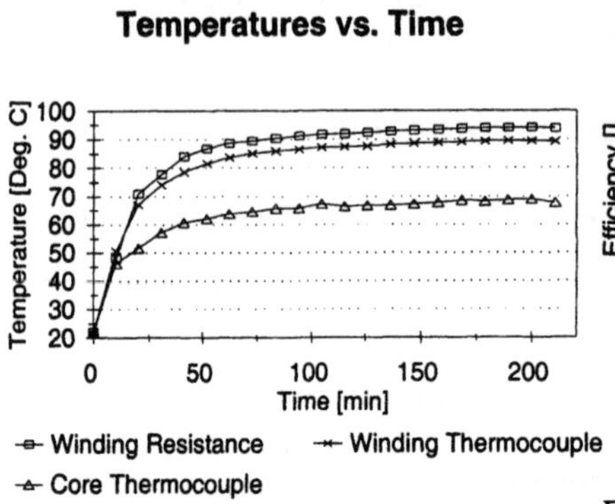

Fig. 4: Measured temperatures versus time for temperature rise test of sinusoidal supplied induction machine.

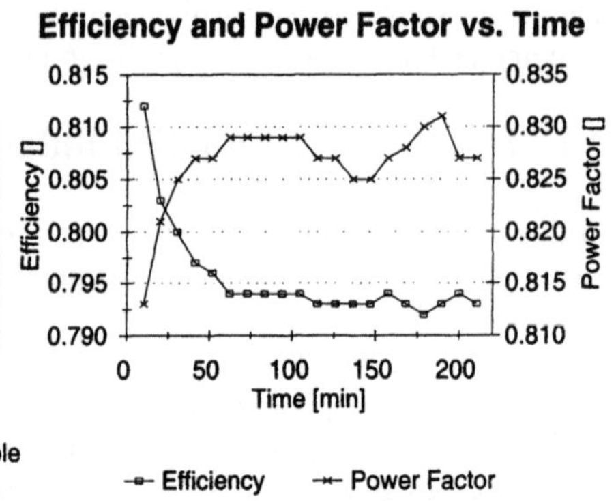

Fig. 5: Measured energy efficiency and power factor versus time for temperature rise test of sinusoidal supplied induction machine.

90 [°C] was obtained by the thermocouple measurement. The last temperature were measured immediately after de-energising the machine. In the future measurement of temperatures will continue after de-energising until the maximum temperature is reached.

For the same test the electrical quantities, such as machine line currents and voltages, and mechanical quantities, like torque and speed, are measured too. The results calculation of the machine energy efficiency is shown in Fig. 5.

As observed from Fig. 5 the efficiency stabilises after 60 [min] to a value around 0.79. The efficiency value after 10 [min], which was the first one measured here, is almost 2 percent points higher than the "hot" value. This makes it clear that efficiency measurements must be performed under specified conditions. Fig. 5 also shows the power factor, which increases with time because the stator resistance increases.

4.2 Tests on Drive System

In this section results arising from a rated load test, performed on the drive system including the speed dependent torque emulation.

4.2.1 Rated Load Test

A rated load test similar to the one in 4.1.1 was performed on the drive system. The frequency converter was set to a switching frequency of 4.5 [kHz] and programmed for a constant load torque application. The drive was driven with a fundamental frequency of 50 [Hz]. The results are shown in Fig. 6 and Fig. 7.

The final temperatures reached in this case are 99 [°C] by resistance and 94 [°C] was obtained by the thermocouple measurement. The higher machine temperature measured here with the converter supply was due to the extra harmonic losses associated with PWM voltage waveform.

This also affects the measured efficiencies as shown in Fig. 7. For the last 10 measured efficiency points, the averaged was 0.961, 0.783 and 0.753 for the power

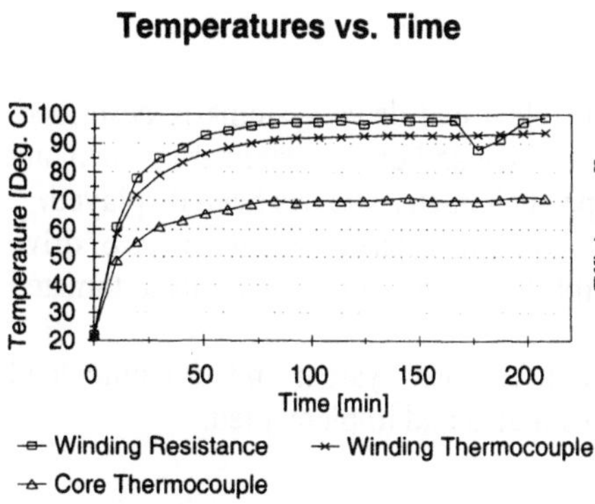

Fig. 6: Measured temperatures versus time for rated load test of drive system.

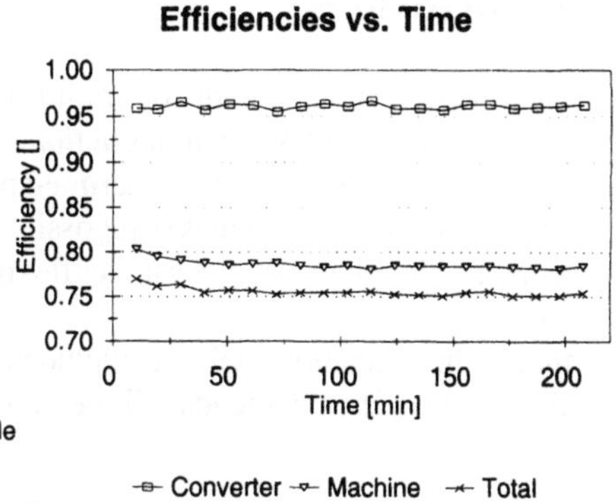

Fig. 7: Measured energy efficiencies versus time for rated load test of drive system.

converter, the machine and the total energy efficiency, respectively.

4.2.2 Emulation of Speed Dependent Load Torque

Here emulation of Eq.(1) with $B_2=300\cdot10^{-6}$ [Nm/(rad·s)^2] and the other parameters are zero was performed. The mechanical load corresponding to that may be a pump or a fan. The result is shown in Fig. 8. As seen, the torque is emulated very well. This can also be observed in Fig. 9, where the absolute and relative errors are shown as function of the speed for the same test as Fig. 8. As could be expected the relative error at a low torque value is high, because even a small absolute error will cause a high relative error. Apart from the point around 0.3 [Nm] the relative torque error is less than 1.0 [%]. From Fig. 9, the absolute error is observed to be less than 0.05 [Nm].

The torque was here measured with a torque element of 50 [Nm]. If an element with a lower value were to be used instead, e.g. 20 [Nm], even lower errors could be expected.

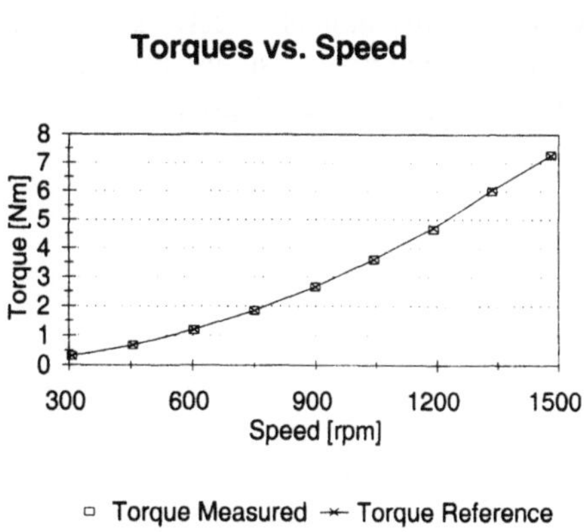

Fig. 8: Emulation of torque proportional to the speed squared with $\tau_{ref}=300\cdot10^{-6}\cdot\omega^2$.

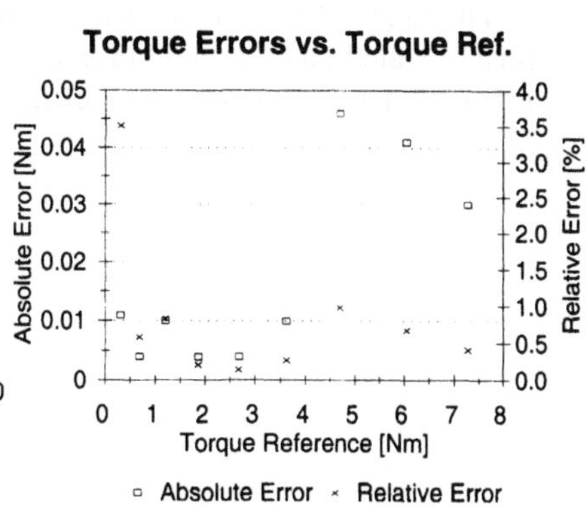

Fig. 9: Absolute and relative torque errors for emulation of the torque curve shown in Fig. 8.

5 Conclusions

An automatic testing system for testing of electrical drive systems is described. Results of rated load test on an induction machine and drive system are shown and a sample result of emulation of a torque-speed characteristic has been displayed. The developed test system makes it possible to make test and measurements of drive systems and electrical machines under flexible load conditions in a highly automated system.
In future, new features will be included in the test system and dynamic load emulation of mechanical loads will be investigated and implemented.

6 Appendix

Data for tested machine:
Motor 3~, 2.2 [kW], 50 [Hz], 380-400 [V] (Y), 4.9 [A], 1430 [rpm], cos $\phi = 0.81$.

Data for tested power converter:
Danfoss VLT3004: Input: 3·380-415 [V], 50/60 [Hz], 7 [A]
Output: 3·0-380/400/415 [V], 1-500 [Hz], 5.6 [A], 4[kVA]

7 References

[1] Collins, E.R., Huang, Y. (1994): A Programmable Dynamometer for Testing Rotating Machinery Using a Three-Phase Induction Machine. IEEE Transactions on Energy Conversion, vol. 9, no. 3 (September), 521-527

[2] Stone, D.A., Birch, T.S., Leonard, K.N. (1994): A Versatile Computer Controlled Dynamometer Test System. Proceed. of Speedam, 353-358

[3] Sandholdt, P., Ritchie, E., Pedersen, J.K. (1998): An Automatic Test System with Steady State Load Emulation for Test of Electrical Drive Systems. Proceed. of International Conference on Electrical Machines, vol. 3, 2071-2076

Induction Motor Efficiency Standards

P. Van Roy, B. Slaets, K. Hameyer, R. Belmans

Katholieke Universiteit Leuven, E.E. Dept., Div. ESAT/ELEN
Kardinaal Mercierlaan 94, B-3000 Leuven, Belgium.

Abstract. In 1997 18 motors of 11, 55 and 75 kW have been tested in the laboratory of the Electrical Engineering department at the KULeuven. Special attention was given to the various standards, mainly IEC 60034-2 and IEEE 112 Method B. The aim was to compare the various motors, and to rank them according to their efficiency. Based on these measurements, it was determined that the present IEC standard is not reliable, and does not offer the possibility to the consumers to make an instructed choice. The reason for this is the fact that the additional load losses, formerly refered to as stray load losses, are not properly taken into account. Comments on the new proposed IEC 61972 standard are included.

1 Introduction

Nowadays, with the growing emphasis on energy conservation and the increasing energy prices, the efficiency value has become very important and even dominant for applications in industry. The efficiency data given by the manufacturer are measured or calculated according to certain standards. These standards use different ways to incorporate the additional load losses. In the IEC 60034-2 standard, these losses are estimated at 0.5 % of the input power. The Japanese JEC 37 standard simply neglects this kind of losses. In the IEEE 112-B, they are deduced by measuring the input and output power; the losses not covered by the 4 other loss terms - the stator and rotor copper losses, the iron losses and the friction and windage losses - are then supposed to be the additional load losses. Efficiency values obtained from different testing standards can differ by several percent.

At this instant, a new IEC standard is been worked out. The standard gives two possibilities for the assessment of the additional load losses. The first one is a determination by means of the measured output power, the second one attributes a fixed amount to every machine of the same rated power.

2 Measurements

In 1997, 18 induction motors of 11, 55 and 75 kW were tested, in a program in collaboration with Laborelec, the Belgian national laboratory for electricity. The motor manufacturers were ABB, ACEC, Brook-Hansen, Leroy-Somer, Samco, Siemens and WEG.

2.1 Additional Load Losses

Measured additional load losses vary from 1.5 to 2.3 % of input power for the 11 kW motors tested (7 motors), from 0.4 to 3.0% for the 55 kW motors (6 motors) and from 0.9 to 2.7% for the 75 kW motors (5 motors). Values in the same range were found for some other motors tested in other power ratings. Similar values can be found in [1] and other references.

Figure 1 shows the result of using a measured value for additional load losses. For the 11 kW (a), 55 kW (b) and 75 kW (c) motors. On the left are the IEC 60034-2 values, on the right the IEEE values for efficiency at full load. The vertical scale is 1% between the marks. The IEC overestimates the efficiency (with one exception), but this is not the most important point. More important is the fact that from one motor to another, the additional load losses differ significantly.

(a) (b) (c)

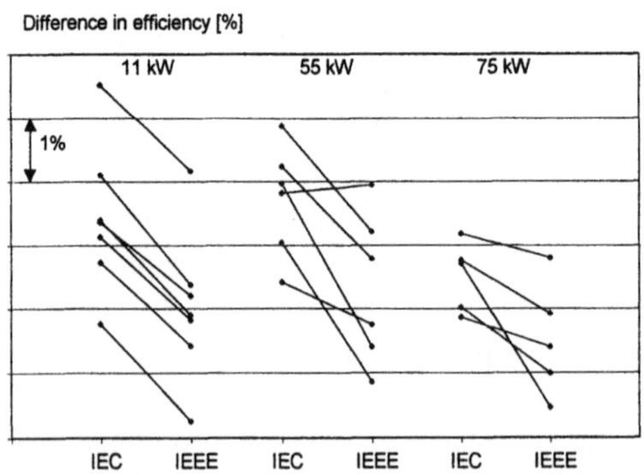

Figure 1: Ordering of motors based on IEC and IEEE full load efficiency

Based on the proposed new IEC standard, the additional load losses would be taken as 1.9% of input power for the 55 kW motors. This may be a good average value, but the efficiency of one motor of the tests would be overestimated by 1.1%, whereas the efficiency of another motor would be underestimated by 1.5%, with all possible values in between. Clearly this method is extremely unfair to the motor manufacturers, and to the customers who want reliable information on motor efficiency.

2.2 Accuracy

The accuracy of the efficiency was determined to be +/- 0.8 %, based on the accuracy of the measurement devices used.
From one motor manufacturer, four different motors of the same design were tested. The efficiency values are given in Table 1. The standard deviation between the measurements is 0.24%. This includes the deviation between the motors due to the manufacturing process.

Table 1: Efficiency at rated load for four motors of the same design

Partial load [%]	Motor A [%]	Motor B [%]	Motor C [%]	Motor D [%]	Average [%]
125	91.95	92.07	91.75	91.77	91.88
100	92.98	92.86	92.53	93.00	92.84
75	93.75	93.39	93.11	93.84	93.53
50	93.88	93.22	93.19	93.62	93.48

One motor was tested five times, by professional scientific personnel and by four groups of students. The difference with the average result is shown in figure 2. The standard deviation between the measurements is 0.12%.

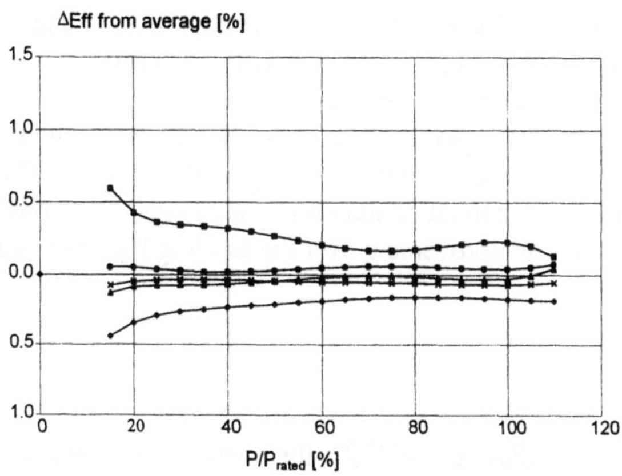

Figure 2: Deviation from average for 5 measurements on one motor

These results confirm the fact that a higher accuracy than +/- 0.5% is probably not achievable, even with the highest accuracy in measurement devices.

3 Additional Load Losses in the IEC 60034-2 and the New IEC 61972 Standard

In the new proposed IEC standard, the additional load losses are preferably determined by means of the measurement of the output power, like in the IEEE method [2]. This is the only relevant method. The alternative – with a fixed allowance – can not be defended. It is not important what average value would be used: it is the difference in additional load loss among motors of the same rating that is relevant.

For example: motor A may have a 93% efficiency according to the IEC standard. The real efficiency could be e.g. 91%, because the additional load losses are actually 2.5% for this motor. Motor B of the same power rating may have an IEC efficiency of 92%, and a real efficiency of also 92%, because the additional load losses happen to be 0.5%. The comparison of both motors according to any method using any assumed or average additional load losses is futile. This comparison would indicate motor A is the "best". In fact, motor B is clearly more efficient.

An argument to use a fixed amount of additional load losses, could be that these kind of losses are supposed to decrease during the first six months of operation. This assumption is debatable for several reasons: firstly, these changes are based on manufacturing techniques that were relevant 30 years ago. Nowadays manufacturing tolerances have become smaller.
Secondly, all motor manufacturers use similar designs and materials. For the user it is irrelevant with respect to the 'ab initio' comparison of the motor efficiencies.
Therefore, the use of a fixed amount of additional load losses in the proposed new IEC standard is therefore no improvement over the existing IEC standard.

4 Accuracy

In the American Energy Policy Act of 1992, the measurement error is taken into account. A round robin test involving 9 test facilities showed a measurement error of 0.7 to 0.9%. The variation in measured losses frequently exceeded 10% of the losses [3].

Motor efficiencies according to the NEMA nameplate labelling standard MG1-12.542 are determined based on the average value of a series of measurements on motors of the same design. Then the closest lower value in a standardised list is taken. This list contains the values 98.0 – 97.8 – 97.6 – 97.4 – 97.1 – 96.8 – 96.5 – 96.2 – 95.8 – 95.4 – 95.0 – 94.5 – 94.1 – 93.6 – 93.0 – 92.4 -.... Associated with this list is a second list of minimal efficiencies at rated load, voltage and frequency. Any motor of the same design must have at least this efficiency. For a 93.6% motor, the minimal efficiency is 92.4%. This constitutes a significant safety margin that may be larger than required. The method prevents users from assuming an undue accuracy in the efficiency determination. We would advise a similar list to be included in the IEC standard.

Measurement errors of 10% of the overall losses are perfectly possible. However, this 10% error for a 93.6% efficient motor, means an uncertainty of only 0.6% on the efficiency value, becoming 93.6 +/- 0.6%.

When comparing two motors, a difference of e.g. 0.2% is not relevant. Such a difference would also not be maintained after selecting a rated efficiency from the list. The difference between motors labelled at 93.6 and 93.0% may not be significant. However, when two motors are labelled as 93.6 and 92.4%, one can assume there is indeed a difference.

Temperature and non perfect power supplies are also important problems with respect to the efficiency and losses during operation. However, they are irrellevant with respect to the accurate assessment of the efficiency for comparison of motors by users. All motors will suffer from these problem in the same way.

5 Efficiency at Partial Load

Most motors are overdimensioned for safety reasons, and because of the standard power ratings. This means motors will usually be used at the range 50 – 75 – 100 % of rated power. It is essential that manufacturers mention the efficiency at 75 and 50% load. If we consider the two motors shown in figure 3, we notice their rated efficiency differs by 1 %. However, the difference in efficiency at the partial load is more important in this case. In other cases, the efficiency of a motor with a lower rated efficiency, has a higher efficiency at partial load. The rated efficiency does not give a good picture. The efficiency at partial load should be included in all manufacturers' information.

In the tests performed at the KULeuven, the motors were ordered based on the average weighted efficiency between 25 and 100% load, using the partial load as weighting factor. In this way, the ordering is based on the energy consumption.

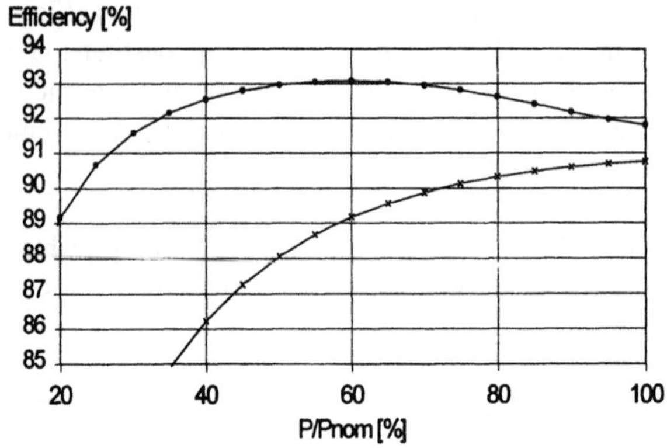

Figure 3: Efficiency of two motors

A possible "average weighted efficiency" could e.g. be defined as (1 x Eff_{100} + 0.75 x Eff_{75} + 0.5 x Eff_{50})/2.25, or if one wants to stress the efficiency at 75% load, (0.75 x Eff_{100} + 1 x Eff_{75} + 0.5 x Eff_{50})/2.25, or something similar. Perhaps motor efficiency labelling could be done, based on such an "average weighted efficiency", reflecting the energy consumption.

6 Practical Observations Regarding the New IEC Standard

6.1 Load Test

In order to measure the stator resistance, the motor has to be shut down. It is assumed that the motor is loaded at rated load, then shut down for stator resistance measurement before the highest load point, then loaded at 150, 125, 100, 75, 50 and 25% load. After this the motor is shut down for stator resistance measurement after the lowest load reading. The resistances are assumed to vary according to a straight line between the first and second measurement. However, will the motor temperature not increase slightly when loaded at 150% load? The exact evolution of motor temperature is not straightforward. It depends of shut-down time for resistance measurement, time at 150 and 125% load, size of the motor, ...

Certainly, the aimed at accuracy in terms of temperature seems to be very high. Is a single measurement of rated load temperature not sufficiently accurate? A difference in temperature of 10K leads to a difference in efficiency of only +/- 0.1%. Also, the larger the motor, the longer the thermal time constant becomes.

6.2 No-Load Test

Allowing the motor to reach a stable temperature at no-load is preferable from the theoretical point of view. The proposed method will however significantly increase the overall time needed for the various measurements. In practice, the no-load test can be performed immediately after the load test. The stator resistance measurement can be performed before and after the no-load test, and an average value can be used. This average resistance should be accurate enough, compared to the extra cost of the theoretically more accurate method.

7 Conclusion

The main conclusions from this study are:

- The efficiency determination according to the European IEC standard 60034-2 is not reliable.
- The additional load losses must be measured, and can in no way be replaced by any kind of fixed allowance, as the difference in additional load losses between motors of the same rating is too significant to be ignored; the difference from one motor to another can exceed 2% of input power, far exceeding the measurement error.
- The partial load efficiency is just as important as the full load efficiency with respect to energy consumption.
- The labelled efficiency value should preferably be taken from a standardised list, in order not to give the impression of a higher accuracy than actually achieved.

8 References

[1] C.N. Glew (1997): Efficiency Measurement Testing Standards Stray Losses, the Key to Efficiency determination. In: Anibal de Almeida: Energy efficiency Improvements in Electric Motors and Drives. Springer, Berlin, 249-265

[2] IEEE Std 112-1991: IEEE Standard Test Procedure for Polyphase Induction Motors and Generators. IEEE Power Eng. Society, New York, NY.

[3] The Bonneville Power Administration United States Department of Energy (1993): Energy-efficient Electric Motor Selection Handbook

[4] IEC Std 34-2 (1972): Rotating electrical Machines - Methods for determining losses and efficiency of rotating electrical machines from tests.

[5] IEC Std 61972:1998: Method for determining losses and efficiency of three-fase, cage induction motors.

[6] H. Auinger (1997): Considerations about the Determination and Designation of the Efficiency of Electric Machines. In: A. de Almeida: Energy efficiency Improvements in Electric Motors and Drives. Springer, Berlin, 284-293

[7] Anibal de Almeida (1997): Energy Efficient Motor Technologies. In: Anibal de Almeida: Energy efficiency Improvements in Electric Motors and Drives. Springer, Berlin, 1-17

[8] R.G. Bartheld and J.A. Kline (1997): Comparative Efficiency Measurements IEC 34-2 vs IEEE 112. In: A. de Almeida: Energy efficiency Improvements in Electric Motors and Drives. Springer, Berlin, 266-273

[9] Paul G.Cummings (1997): Efficiency Measurement Testing and Labeling. In: A. de Almeida: Energy efficiency Improvements in Electric Motors and Drives. Springer, Berlin, 274-283

The European Negotiated Agreement to Improve Motor Efficiency

Paolo Bertoldi[1], Gerald Kuehnemund[2]

[1] European Commission, Directorate General for Energy
[2] CEMEP

Abstract. The European Union is pursuing a number of measures to improve end-use equipment efficiency through a variety of policy instruments. One of the most effective policy options to achieve market transformation is minimum efficiency performance standards (MEPS). However, after the difficulties and controversy following the adoption of legislation for MEPS for domestic refrigerators/freezers, a new policy instrument, i.e. the negotiated agreements by manufacturers, has been used to achieve market transformation. Based on the positive experience made in two agreements, one for washing machines and one for TVs and VCRs, the European Commission and CEMEP, the European association of motor manufacturers, have negotiated a new agreement for electric motors.
This paper describes the proposed motor agreement, the expected energy savings of about 6 TWh, and the advantages and disadvantages of the motor agreements compared to mandatory MEPS, in the view of the Commission and the motor manufacturers.

The paper concludes that the negotiated agreement could be a viable policy option, which allow flexibility in the implementation of the efficiency targets and therefore the adoption of cost-effective solutions for manufacturers. The main question that remains to be answered is whether the negotiated agreement can deliver the results in the long term compared to what could be achieved through legislation.

1 Introduction

The European Commission under the SAVE, and the PACE programme, a European Union (EU) action programme to improve electricity end-use equipment (Bertoldi 1996), is pursuing several actions to improve energy efficiency of equipment in the domestic, commercial and industrial sectors. These actions include labelling and classification schemes, information and dissemantion tools (e.g. EURODEEM) MEPs, minimum efficiency standards and/or negotiated agreements, and technology procurement.

Since the beginning of the programme particular attention has been paid to motor systems (Bertoldi 1997), because they account for about 30% of total electricity demand in the EU. In addition, experts working for the Commission indicated that the traditional barriers to penetration of energy efficiency technologies were particularly difficult to overcome in the industrial sector.

The need for governmental action to transform the market is, at present, subject to increasingly long discussions among policy-makers in the European Union: several policy-makers claim that 'if energy efficiency is economically viable it should happen by itself' and insist on the need to reduce legislation and governmental intervention in economic affairs, and use market forces. This has largely affected energy efficiency policy and programmes and it has resulted in increased difficulty to introduce legislation for MEPS. This paper describes the experience gained in the EU in the use of negotiated agreements as a policy option alternative to MEPs, to transform the market and foster the penetration of more efficient domestic appliances. This paper presents the details of the motor agreement negotiated between the European commission and CEMEP.

It is widely accepted (Engleryd 1995; Geller 1994; Swisher 1994) that to enhance the average efficiency of equipment present on the market today, a range of different actions is needed in order to exploit the full energy efficiency potentials of the available technology. Market transformation in terms of energy performance is the goal of any energy efficiency action for equipment. The various instruments (minimum efficiency standards, labelling, incentives, procurement etc.) are intended to interact and influence the market, which in the case of energy efficiency does not function well enough on its own, because of the well known barriers to energy efficiency. Users information could be effective in shifting the whole energy efficiency distribution normal curve. Labelling actions focus customers' attention on operating costs and other environmental aspects of the equipment, thus enlarging the market share of efficient equipment. To eliminate the "bad" equipment from the market MEPS, setting the lowest acceptable performance level, are a well known and tried method. As it is described in this paper, negotiated agreements have been used in the EU as an alternative to MEPS established through legislation. Technology procurement (i.e. the process of aggregating purchase power in order to specify new high efficiency equipment) acts on the higher end of the market by accelerating and expanding the penetration of new and more efficient products into the market place. Technology procurement might also include incentives aimed at both customers and/or manufacturers. For every product group a new and specific application or combination of these policy instruments has to be chosen in order to transform the market for appliances in the domestic, commercial and industrial sectors (Bertoldi 1997).

2 Minimum Efficiency Standards

MEPS are a very powerful and effective tool in transforming the market: they act on the low-efficiency end of the market and are particularly effective when consumers are not influenced by information and labels in particular for appliances for the commercial and industrial sectors. In the European context the introduction of MEPS presents some additional aspects. One of the aims of the EU is to create an internal market for, inter alia, tradable goods: therefore EU Member States cannot introduce national legislation which may lead to barriers to trade unless justified on the basis of very limited criteria such as protection of human health, as indicated in article 36 of the EU Treaties. Energy efficiency is not considered an area sufficient to justify barriers to trade. Some Member States (e.g. Denmark, the Netherlands, etc.) which have a very ambitious energy efficiency policy have notified the European Commission (the "Commission"), as required by the European legislation, of their intention to introduce in their territory MEPS (for refrigerators in the case of the Netherlands; for washing machines, dishwashers and tumble dryers in the case of Denmark). The Commission has the power to stop the introduction of national legislation if it is established that it might create a potential barrier to trade among Member States, and it is not justified according to the criteria laid down in the EU Treaties. To solve the conflict between the internal market and energy efficiency policies of Member States, the solution normally adopted is to introduce common MEPS throughout the Union and the EEA area (Norway, Iceland and Liechtenstein).
The first appliances for which minimum efficiency requirements were introduced were refrigerators and freezers, the largest electricity consuming domestic appliance in the Union, with a total consumption of about 120 TWh per year (about 20% of total domestic electricity consumption). The European Directive introducing minimum efficiency requirements was adopted on 3 September 1996. The first level will come into force on 3 September 1999. The all process form the policy conception (1992) to the entry into force (1999) has been very long.
The lessons learned from this process were: 1) the extremely long time it took from the time of the policy definition to the entry into force of the Directive - more than seven years; 2) the strong opposition by manufacturers to the introduction of any legislation which might affect their market and the strong reaction that politician have when the proposed legislation could result in job losses.

3 The Negotiated Agreement Instrument

First of all it must be noted that often in literature the term "voluntary agreement" is used to describe this policy instrument, however the Commission decided to define this policy instrument as "negotiated agreement" to highlight the negotiating process between public authorities and industry; in addition, it was felt that the term "voluntary" was somehow misleading, because if the participation by industry is

voluntary, once the agreement has been concluded the delivery of the result is not any more voluntary, but governed by sanctions in case of non-compliance.
In recent years, the need to enlarge the mix of policy instruments to promote energy efficiency has resulted in a growing interest in policy measures based on collaboration and sharing of responsibility between public authorities and industry. Recently the Commission adopted a Communication on Environmental Agreements, which aims to promote and facilitate the use of effective and acceptable Environmental Agreements as part of the efforts to broaden the range of policy instruments; this Communication describes the essential characteristics of this instrument.
During the discussions in Council leading to the adoption of the "Refrigerator Directive" several Member States insisted on the use of more flexible instruments, in particular negotiated agreements with manufacturers to reach the energy efficiency improvement comparable to the one achieved with MEPS. It must be added that an EU Directive can regulate only tradable goods, i.e. each appliance shall meet the set efficiency requirement to be allowed to be placed on the EU market, and of course the requirement is the same throughout the EU. This rules out the possibility to have legislation including average or composite target values for energy efficiency (e.g. fleet undertakings or sale-weighted average to be reached by a certain date).
European motor manufacturers have also indicated their strong opposition to regulatory approaches, as products different in energy efficiency are sold in different Member States for different end use applications and drive systems. Manufacturers expressed their willingness to reach negotiated agreements to achieve energy savings where a target average efficiency is agreed. According to manufacturers, this would give them more freedom in reaching the target, by selling more high efficiency products, by having flexibility when to phase out low efficiency units and therefore reduce costs . In addition, manufacturers have highlighted the advantages of having a demand driven process, which would favour cost-effective solutions and which would allow manufacturers to have a pro-active role in setting quantified criteria, in implementing the measures and achieving the results. This is a clear advantage in a very competitive market.
In addition, negotiated agreements might present advantages for public authorities as negotiated agreements can be more flexible and quicker to implement than regulation. In addition, negotiated agreements are easier and quicker to upgrade than legislation allowing them to follow technological evolution and market changes. Since industry has the best knowledge of the production processes and the technologies required to improve efficiency, it is clear that co-operation between public authorities and industry can facilitate the establishment of ambitious energy efficiency targets. Moreover the extra flexibility can be used to identify cost-effective solutions, which might lead to higher efficiency improvements than a simple cut-off line as in the case of regulation. In addition, by co-operation on energy efficiency improvements, public authorities and industry can share responsibility in setting ambitious targets, which are at the same time realistic and achievable. The Commission considers that the negotiated agreement can be a valid alternative to the

introduction of legislation for mandatory MEPS, if they include the following three elements: i) commitments by manufacturers accounting for most of the appliances sold on the Union market (80% at least), ii) quantified commitments to significant improvements in the energy efficiencies of the appliances they produce over a reasonable time-scale; and iii) an effective and independent monitoring and reporting open to public scrutiny.

As indicated in the Commission Communication on Environmental Agreements, it is recommended that negotiated agreements are binding commitments of the parties and therefore shall take the form of a contract according to civil or public law, as contracts are binding for each of the parties and offer the possibilities to resolve case of non-compliance through court actions. However, since the EU treaties impede the Commission to sign contracts with the private sector to implement policy actions, negotiated agreements concluded between the Commission and manufacturers have to take the form of a unilateral commitment with the endorsement by the Commission. The endorsement takes the form of an official exemption by the Commission to the Community competition rules. However, to avoid non-compliance negotiated agreements must include additional and credible instruments to discourage non-compliance. These can include mechanisms to ensure loss of public image for the non-compliant company (e.g. press announcement by public authority) or the threat to introduce legislation by public authorities.

It is essential that negotiated agreements are in conformity with the rules of the EU treaties. In particular, the freedom of co-operation of the companies is limited by the requirement of preserving effective competition: in practice any form of co-operation has to be done in respect of Articles 85 and 86 of the EU Treaty, which rules out anti-competitive practices. To this end negotiated agreements have to be notified to the EU competition authorities to be cleared before they enter into force.

However, as it has been clearly indicated in the Communication on Environmental Agreements, besides the above indicated advantages of the negotiated agreements approach (and of environmental agreements in general), there are also certain risks associated with such agreements. In particular, there is the risk that the negotiated energy efficiency targets reflect little more than "Business as Usual" (BaU). Moreover the risk of "free-riding", which increases the more demanding the targets are, must be clearly assessed. In particular, the number of appliances imported into the EU must be clearly assessed, as usually only EU based manufacturers member of the EU trade association are likely to enter into the negotiated agreement process.

To evaluate the effectiveness of the negotiated agreement instrument in improving energy efficiency, it is important to determine the efficiency improvement that would in any case have taken place in the BaU scenario. It is worth noting that the definition of the BaU might be complex and sometimes controversial, and, in some cases, it might be more appropriate to compare the energy efficiency target (e.g. 20% improvement in 10 years) with the historical trend.

The main characteristic of the negotiated agreement is that is a target-based instrument. Any negotiated agreement shall promote energy efficiency beyond the BaU scenario and shall aim at least at the economic potential, which, because of existing market barriers, does not coincide with the BaU scenario.
During the process leading to the conclusion of a negotiated agreement, the Commission negotiates with the European Manufacturers Associations, which has a mandate from the associated manufacturers, as for some sector the number of manufacturers is too large (more than 100). In addition, manufacturers of the same type of equipment might have conflicting views on the efficiency improvements, this has to be settled by European Manufacturers Associations, which however tends to defend the weakest manufacturers. Another important element is that the negotiated agreement covers only products of participating manufacturers. As indicated before in order to comply with competition rules, each negotiated agreement must be open to the participation of manufacturers who are not members of the European Manufacturers Associations nor based in the Union. The technical background for the Commission to negotiate the agreement is the same as in the case of mandatory minimum efficiency standards, i.e. the negotiations are based on an exhaustive technical/economic analysis.

4 The Proposed Negotiated Agreements for Motors

CEMEP and the European Commission have developed a new system of efficiency designation. This system has been designed for a product range of totally enclosed fan ventilated (IP 54 or IP 55) three phase A.C. squirrel cage induction motors in the power range of 1.1 to 90 kW, with 2- or 4-poles, rated for 400 V-line, 50 Hz, S1, Duty Class, in standard design. These motors are according to various studies the most important power ranges, which have to be improved in their efficiency in order to reduce electrical energy consumption.

This power range has been subdivided into three bands of efficiency. The lowest band is called eff3 and covers round about two thirds of the motors currently sold in the EU. Therefore you may say that eff3 represents the current standard efficiency. The next class is called eff 2 and represents almost a third of the motors currently sold and finally the class eff1 which represents motors with an efficiency even higher than the efficiency asked for by the US EPACT legislation. This class of motors has a current market share of 1 to 2 percent.

In future the efficiency class of all the motors covered by this agreement has to be designated on the nameplates and in the catalogues. Additionally the efficiency at full load and at three-quarter load will be quoted in the catalogue. By this, efficiency will become an easy to handle quality criterion and OEMs will have the chance to use the different classes of motors for a cost effective efficiency optimisation of their applications. OEMs and the engineers of the project business are the people who have the best knowledge of operating conditions and saving potential. They may decide whether an eff2 motor is sufficient or whether an eff1 motor is the cost and energy efficient solution.

Round about 3.5 million standard motors that fit to the definition of the agreement are annually sold in the EU. The participants of the agreement cover round about 2.9 million units. Round about 70 percent of these motors belong to efficiency class eff3. The market share of these motors has to be reduced by 50 percent till the year 2003. When in the long run only the mentioned 50 percent of the current eff 3 motor stock will be replaced by eff 2 motors, the annual energy consumption will be reduced by 3 TWh. Each percentage point in favour of eff1 would in the beginning enable an additional reduction of additionally 400 GWh annually. The parties are optimistic that the total energy saving by the promotion in favour of eff2 and eff1 will in the long run accumulate to at least 6 TWh annually.

5 The Manufacturers´ View

CEMEP decided to participate in the development of a negotiated agreement as it has seen some major advantages in comparison to minimum efficiency standards. While some major reasons have already been outlined above there is also the issue of political effectiveness. Of course a minimum efficiency standard is the most effective instrument with regard to motor efficiency. But out of the ecological perspective not the efficiency of the motor but the efficiency of the electric drive system is the main issue. For this goal a negotiated agreement will enable the more efficient solution.

Manufacturers´ scepticism against a directive is mainly based on its' long terms impacts. Electric drive trains comprise several components, each of them with a certain impact upon the total efficiency. Although the saving potential of the drive system is the dominant one, the saving potential of motors can be significant too. However, this is the case for gears, belts, and other mechanical devices, too. This means that an efficiency policy that once has started to regulate components has to continue this process will in the next step regulate all other components, too. How, in an environment of regulated components with resultant cost increases for all applications, OEM's will still have the possibility to invest in system optimisation and in the use of power electronics is an open question. Shall systems design and power electronics be regulated, too?

6 Conclusions

Although the motor negotiated agreements are not yet cleared by the European competition authorities, it is expected that it will be implemented during year 2000. In principle all the major European producers have agreed to sign the agreements, therefore the free-riding risk is mainly limited to imported goods. The market share of the agreement participant is expected to be higher then the requested 80%. Therefore, it must be concluded that in the present situation and for the products considered so far, there is little risk that imported goods would be advantaged and could increase their market share.

The EU experience with negotiated agreement has shown the following positive elements (compared to equivalent legislation): much faster process from the start of the negotiations to the introduction of measures (entry into force of the negotiated agreement) compared to legislation; much lower administrative costs (for example no need to involve the Council and European Parliament, which is very expensive); much greater flexibility in the choice of the targets, which can be defined in terms of maximum consumption limit or fleet levels; the inclusion of soft targets, which might results in worthwhile additional savings; efficient monitoring mechanism (perhaps even more effective than in the case of legislation); greater co-operation by manufacturers and their proactive role in defining feasible and effective efficiency targets.

The main question that has yet to be answered is: will the negotiated agreement deliver the envisaged results? The European Commission believes that agreements constructed along the lines indicated in the previous sections will work, in any case the threat of introducing legislation will remain on industry in case of non-compliance. The Commission will closely survey the market on an annual basis, using the sales-weighted data to track any efficiency improvement and to be sure that the targets will be met. The results will be made public to inform the wide public of the results achieved by industry and to create an additional demand for more efficient products. The Commission will continue in parallel to prepare Directives for mandatory MEPS for equipment for which the market conditions or the manufacturing industry structure make a meaningful negotiated agreement impossible, or when industry will not agree to meaningful efficiency targets not with my name.

7 References

[1] Bertoldi, P. 1996. "European Union Efforts to Promote More Efficient Use of Electricity: the PACE Programme" *in Proceeding of the ACEEE 1996 Summer Study on Energy Efficiency in Buildings*, pp. 9-11.Washington, D.C.: American Council for an Energy-Efficient Economy

[2] Bertoldi, P. 1997. "European Union Efforts to Promote More Efficient Appliances" *in Proceeding of the ECEEE 1997 Summer Study on Energy Efficiency*, pp. 1-2-id24

[3] Engleryd, A. 1995. "Technology procurement as a policy instrument". Swedish National Board for Industrial and Technical Development, Stockholm

[4] Geller H., Nadel S. 1994. "Market transformation strategies to promote end-use efficiency". American Council for an Energy-Efficient Economy, Washington, DC.

[5] Nadel, S. 1994. "Minimum efficiency standards: options for federal and state action". American Council for an Energy-Efficient Economy, Washington, DC.

[6] Nilsson, H. 1992. "Market transformation by technology procurement and demonstration". Swedish National Board for Industrial and Technical Development, Stockholm

EPAct Legislation – The United States Experience of Minimum Efficiency Standards for Induction Motors

By Anthony Balducci

National Electrical Manufacturers Association

1 Background on the Department of Energy's Regulatory Energy Efficiency Program

Minimum energy efficiency standards for appliances have had a relatively long history in the United States. After the oil shocks of the early 1970s, the U.S. Congress passed the Energy Policy and Conservation Act of 1975 (EPCA), setting energy conservation targets for residential appliances. This legislation was strengthened by amendments to EPCA in 1978, which directed the Department of Energy (DOE or Department) to prescribe minimum efficiency standards for certain household appliances. The statute provided that within five years of prescribing an appliance efficiency standard, the Department had to determine whether the standard should be amended. The statute required that each proposed standard's maximum efficiency level must be "technologically feasible and economically justified." The statute also allowed states to set their own standards if they had a "significant interest" in doing so. Additionally, EPCA stipulated that DOE establish test procedures to be used in evaluating compliance with efficiency standards and in the labeling program administered by the U.S. Federal Trade Commission.

In the early 1980's, the Department of Energy issued a "no standard" standard. As a result, the Natural Resources Defense Council and other petitioners including the State of California, sued the Department for its failure to prescribe "substantive" standards. The U.S. Court of Appeals found for the plaintiffs and, in its decision, laid out an important framework for a stringent appliance standards program, which, to this day, provides important guidance for the Department of Energy.
It is against this backdrop that interested parties (commonly referred to as "stakeholders") — appliance manufacturers, energy efficiency and environmental groups, utilities, and states — began negotiations to develop a proposal for a Federal program which would eventually form the basis for the National Appliance Energy Conservation Act of 1987 (NAECA). NAECA set standards and established schedules for mandatory review of standards for each covered product while strengthening the Federal preemption of states in appliance standards matters.

Residential products covered under NAECA included refrigerators and freezers, room air conditioners, central air conditioners, clothes washers and dryers, and several other appliances. Some of these standards set minimum energy efficiency levels, while others were prescriptive (e.g., clothes washers were required to have a cold-rinse option). The Energy Policy Act of 1992 (EPAct) went on to establish minimum standards for commercial and industrial equipment, including commercial heating and air-conditioning equipment, water heaters, and electric motors.

The appliance standards program initially enjoyed relative stakeholder support. The program represented somewhat measurable energy savings which pleased environmental and energy efficiency groups. Many of the appliance manufacturers and their trade associations reluctantly supported Federal standards because they replaced disparate state standards that would have placed higher and higher costs on manufacturing. The proliferation of state standards would be a "nightmare" for manufacturers and consumers. The consumer as well as the manufacturer would lose due to a loss of efficiencies in production and distribution and would result in higher prices. The support from each of these sectors, diminished as the Department faced serious organizational, budgetary, and analytical problems, which slowed the issuance of standards.

In 1995, as a result of the Department's poor analyses and mismanagement of the program, the United States House of Representatives voted 260-165 to prohibit DOE from issuing new or amended standards, the United States Senate followed by voting unanimously to impose a moratorium on the Department's Office of Codes and Standards. Clearly, the will of the House and Senate was expressed in the strongest possible terms, which the President signed into law. During this time period, the Department revamped its process and published a "process improvement rule". This rule outlines actions that the Department is encouraged to follow when conducting its mandated review of energy efficiency standards. Since the passage of the rule, the Department has made their rulemakings more robust and transparent. The following year, appropriators added strong report language instructing the department to keep faith in using the Secretary of Energy's Appliance Standards Advisory Committee to ensure that the process improvement rule was adhered to. The Congress then lifted the moratorium. The Department has made great strides in improving its process for developing rules, but has a ways to go.

2 Energy Use of Electric Motors

The United States Industry spends approximately $30 billion annually on motor systems energy costs. In the US there are approximately 90 Million motors installed, with 40 million of these motors in industrial and manufacturing operations. The initial purchase price of an electric motor is only a small portion of the life cycle cost of the motor. As an example a typical 50 hp motor operating 8000 hours per year for ten years:

- Energy consumption makes up 91% of the total lifetime costs
- The actual purchase price represents only 2.4% of the total 10 year energy consumption

This example justifies an energy efficient motor for this application. The US market has several efficiency drivers for electric motors: (1) User Benefits - Cost savings due to energy savings, Less maintenance (energy efficient motors run cooler and lasts longer); (2) Utility Capacity - Energy crisis, avoidance of construction of New plants at a cost of $3-$4 billion, and rebate programs; and (3) Government Programs - Energy Policy Act of 1992.

3 EPAct Legislation and Electric Motors

In 1992, the United States Congress passed the Energy Policy Act of 1992 "EPAct". One section of EPAct, section 122, included a provision for energy efficiency standards and testing requirements for "electric motors". EPAct defined the term "electric motor" and provided for exemptions to the regulations. Motor efficiency is only one of many topics dealt with in EPAct - appliance standards, natural gas, clean coal technology, even global climate changes are involved. EPAct is difficult to follow and much of the material is written as amendments to earlier legislation. The Department of Energy was tasked by EPAct to publish a rule to codify the energy efficiency standards and test requirements. DOE also was given the authority to develop labeling requirements, certification requirements, and enforcement requirements.

3.1 Energy Efficiency Standards for Electric Motors

EPAct defines an electric motor as a "...any motor which is a general purpose T-frame, single speed, foot-mounting, polyphase squirrel-cage induction motor of the National Electrical Manufacturers Association, Design A and B, continuos rated, operating on 230/460 volts and constant 60 Hertz power as defined in NEMA Standards Publication MG1-1987."

The regulations require motors manufactured (alone or as a component of another piece of equipment) for sale in the United States meet certain energy efficiency levels, see Table 1. The regulations apply to imported motors as well. DOE has included IEC equivalent motors in the definition of covered motors. Members of NEMA put the affect of EPAct at 75% of the 1-200 HP motors sold in the United States. The effective date of energy efficiency standards for motors took place on October 24, 1997.

Table 1: Energy Efficiency Standards effective October 24, 1997

Number of Poles		Nominal Full Load Efficiency					
		Open			Enclosed		
		6	4	2	6	4	2
Motor Horsepower	Std KW Equivalent						
1	.75	80.0	82.5	---------	80.0	82.5	75.5
1.5	1.1	84.0	84.0	82.5	85.5	84.0	82.5
2	1.5	85.5	84.0	84.0	86.5	84.0	84.0
3	2.2	86.5	86.5	84.0	87.5	87.5	85.5
5	3.7	87.5	87.5	85.5	87.5	87.5	87.5
7.5	5.5	88.5	88.5	87.5	89.5	89.5	88.5
10	7.5	90.2	89.5	88.5	89.5	89.5	89.5
15	11	90.2	91.0	89.5	90.2	90.1	90.2
20	15	91.0	91.0	90.2	90.2	91.0	90.2
25	18.5	91.7	91.7	91.0	91.7	92.4	91.0
30	22	92.4	92.4	91.0	91.7	92.4	91.0
40	30	93.0	93.0	91.7	93.0	93.0	91.7
50	37	93.0	93.0	92.4	93.0	93.0	92.4
60	45	93.6	93.6	93.0	93.6	93.6	93.6
75	55	93.6	94.1	93.0	93.6	94.1	93.0
100	75	94.1	94.1	93.0	94.1	94.5	93.6
125	90	94.1	94.5	93.6	94.1	94.5	94.5
150	110	94.5	95.0	93.6	95.0	95.0	94.5
200	150	94.5	95.0	94.5	95.0	95.0	95.0

Covered are general-purpose motors as described in NEMA MG1, while definite and special purpose motors are exempt from the regulations. EPAct defines definite purpose motors as those designed for use under service conditions or a particular type of application other than usual and which cannot be used in most general purpose applications. Special purpose motors as defined as any motor which has special operating characteristics, special mechanical construction, or both. All state and local efficiency regulations are preempted by EPAct.

3.2 Test Standards for Electric Motors

EPAct requires that the test method to be used to determine energy efficiency be per NEMA standards or IEEE-112B. There is also a requirement that motor manufacturers test facilities be certified by third parties to test for motor efficiency. Motor manufacturers do have the option of determining motor efficiency with computer correlation programs that are supported by actual tests.

Once the manufacturer is "certified" to test motors, the Department of Energy will issue a certification number. This certification process could be an added burden to off shore manufacturers as obtaining certification of their methods will be more difficult.

3.3 Labeling Standards for Electric Motors

The Department of Energy was tasked by EPAct to develop labeling requirements for electric motors. These requirements include that

- the energy efficiency of the motor be on the permanent nameplate attached to the motor,
- the manufacturer prominently display the energy efficiency of the motor in equipment catalogs and other material used to market the equipment, and
- other markings as determined by the Department necessary solely to facilitate enforcement of the standards.

The Department proposed that the motor efficiency and that a "recognizable" logo that identifies the motor as meeting the requirements of EPAct be on the nameplate.

3.4 Enforcement and Prohibited Acts

It is unlawful for a manufacturer or private labeler to distribute in commerce any new covered product, which is not in conformity, and distribute in commerce an improperly labeled covered product. Any manufacturer or private labeler that knowing commits a prohibited act is subject to a $100 civil penalty per violation. Each item of a covered product gives rise to a separate violation.

If the Department gets a complaint in writing, they may require test records or correlation data to verify nameplate accuracy. The Department may also spot test motors for efficiency.

4 EPAct Summary

The efficiency standards for electric motors became effective on October 24, 1997. The products covered are 1-200 horsepower, 2, 4 and 6 pole general purpose, three phase induction motors. These motors are to be tested in accordance with NEMA standard MG1-1987 and IEEE-112B. All motors will be marked with a Department of Energy (DOE) energy mark, including a manufacturer's specific compliance number issued by the DOE. Each motor nameplate will include NEMA nominal efficiency. All state and local efficiency regulations are preempted by EPAct.

The Challenge to Define the New IEC Test Method (for the Measurement and Declaration of Efficiency of Induction Motors)

Neville Glew

ALSTOM Electrical Machines Ltd.
RUGBY England

PROLOGUE

The title of my presentation today was not selected by me, it was chosen by the conference Committee. In deciding to include the word "challenge" in the title, I wonder if the organisers were conscious of the di tionary definition of the word challenge and / or had a foreknowledge of the problems that were encountered throughout this project?

My dictionary defines the word CHALLENGE as:-

* A call to engage in a contest or fight!
* An act or statement likely to produce conflict or confrontation !
* A demand for an explanation or justification !
* The quality of requiring full use of ones abilities, energy and resources – difficult but stimulating !

All the above aspects are to be found to some degree in this project as I will try to demonstrate during this presentation. In fact, the forces, passions and influences that have been brought to bear on this subject since the Lisbon conference of October 1996 have turned what could have been a fairly dull and routine procedure into a series of events that have all the elements of politico commercial intrigue leading to a technological epic. To me, the project appears to have taken on the role of a mission to remedy the neglected of a generation.

It has:-

* Unexpectedly assumed the dimensions of a global agenda.
* Experienced the subtle pressures of national standards policies of a superpower.
* Been exposed to the time pressures of a bureaucratic programme.

* Come under the influence of the European political agenda which has adversely effected technical accuracy.
* Barely withstood the commercial pressures which have deliberately sought to impede progress.
* Opened up serious possibilities of reconciliation of the pan-Atlantic divide.
* Struggled to deal with national / commercial interests in committee seeking to ignore technical advancement. And finally,
* Suffered from the commercially encouraged strategic voting designed to delay progress

ACT ONE SETTING THE SCENE

Scene 1.1 The Fundamental Losses

The origin of the fundamental losses that occur in loaded induction machines is very well documented elsewhere and will not be further discussed here. The only aspect of the fundamental losses that has been debated in relation to the new standard is the accuracy of the instruments and the repeatability of the measurements necessary to determine the losses. This subject is not really contentious it being a function of proven procedures and the level of investment in the appropriate measuring instruments.

Scene 1.2 Stray Losses

These losses are in essence a whole family of losses. They are variously referred to as stray losses or additional losses and are largely the electromagnetic consequence of placing real stator and rotor windings in discrete slots. Their magnitude and application is further complicated by their dependence on the quality of the machine design and of its materials and manufacture.

The electromagnetic field produced by the windings has superimposed on it a whole series of harmonics identified as mmf harmonics, permeance harmonics and saturation harmonics which react with each other and the surrounding materials to produce a complex array of losses which are variously dependent on the applied voltage and / or the load current.

The magnitude and measurement of these " additional " losses is, and has been contentious for a very long time and the subject has continued to be the major debating point between the panel of experts preparing the new standard.

Scene 1.3 Historical

Perhaps it is because of the length of time I have been in this business, but I find increasingly, that a look backwards to the designs, procedures and practices of yesteryear is not only historically interesting but it is definitely beneficial.

On the subject of stray losses, a very early reference was made by Smith in a JIEE[1] paper of 1907 in which he said of the losses " ....*the exact nature of which is clearly not known* " Seventy four years ago in 1925 Rogowski, published a paper[2] in Germany in which he listed the "stray loss" as a % of input power for a range of 25 small induction motors with outputs up to about 15kW. His measured figures range from 0.7% to a fairly large 4.8%. There were a plethora of papers on the subject published in the 1960's and 1970's (see list in Ref [3]) all of which identified errors in the prevailing test methods and without exception discredited the use of 0.5% of input power as the assigned figure for stray load loss as the alternative to its measurement.

Prior to the issue of the third edition of IEC 34-2[4] in 1972, and until the issue of British Standard 4999 Part 33[5] in 1977, the applicable specification in the UK for the calculation and declaration of efficiency was British Standard 269[6] (1927) which made allowance for stray load loss by the expedient method of reducing the " nominal efficiency " derived by the summation of the (fundamental) losses method by 0.625 % at rated CMR (today's S1) load and 0.5 %, 0.375 % and 0.25 % at partial loads of 80%, 60 % and 40 % respectively. It is interesting to note that this arbitrary notional reduction in efficiency has been shown , in the light of current work, to be more accurate in terms of stray load loss magnitude than its replacement, the 0.5 % of input power contained in the now infamous Clause 9.1.3 in the current edition of IEC 34-2 and its derivatives.

The third edition of IEC 34-2 contained another area of inaccuracy in the figure used for the reference temperature for calculating the I^2R loss of the stator winding. Up to 1996 when amendment number 2[7] was issued , the standard stated that the reference temperature to be used for Class A, E and B temperature classes should be the low figure of 75°C. If one bears in mind that motors with Class A, E and B insulation systems are permitted to operate safely with full life expectancy at mean temperatures of 95°, 105° and 120° C respectively, then it can be clearly seen that the figure of 75° C is barely adequate for a Class A system let alone the more popular Class B. This defect was recognised and somewhat belatedly changed to the higher figure of 95° C for a Class B system by the issue of amendment 2 to IEC 34-2 in December 1996.

In specifying such a low figure for the reference temperature for a Class B system i.e. 95° C, (which is the Class A operating temperature limit), one could argue that it is still too low for those motors that are operated at full rated power and permitted class temperature.

ACT TWO THE ENERGY CRISIS

Scene 2.1 Origins

There is no doubt that what we now call the "energy crisis " has its roots firmly planted in the world oil crisis of the mid seventies. The fourfold increase in oil prices at that time caused all economies to pause and take a very deep breath and consider their vulnerability. If this global event is considered in parallel with the various nuclear generation accidents, such as Three Mile Island and Chernobyl, acid rain effects caused by conventional power generation, ozone depletion attributed to CFC's bringing the consequential global warming, then it comes as no surprise that world leaders in science and politics responded with various programmes of action.

In Europe, the Council of the European Union published in July 1989 its "PACE " programme on the promotion of energy efficiency in the Union countries. This was followed by another initiative , the SAVE programme (Specific Action for Vigorous Energy Efficiency) in 1991.

Many other nations outside Europe introduced similar programmes at or about the same time, some of which were based on technical / commercial persuasion with regard to the use of higher efficiency small induction motors. In the United States however, the government, after trying the persuasion strategy finally introduced legislation in the form of the Energy Policy Act of 1992 which made it legally binding that all general purpose motors, as defined in NEMA Specification MG1, should have a minimum efficiency as measured by a defined procedure. This act was implemented on October 24th 1997 and is now part of the law of the United States. (see ref [8]).

Scene 2.2 The Lisbon Conference

The reaction in Europe was somewhat slower, due perhaps to the requirement to achieve consensus between the many independent nations. However, following the Coimbra Report to the European Commission on " Actions to Promote Energy Efficient Motors in Europe", the Commission called an International Conference in Lisbon in October 1996 to hear from invited speakers from 37 countries the state-of-the-art position on energy efficient motors, drives and incentive programmes.

The conclusions of the conference were:-

(i) By far the highest saving in electrical power can be achieved by using variable speed drives to efficiently match the drive motor power with its demand wherever possible.

(ii) Substantial savings in electrical power can be achieved by utilising higher efficiency induction motors, especially in the smaller sizes.

(iii) That the current method of measuring the efficiency of induction motors, as defined in IEC 34-2 is universally discredited and as a consequence is inadequate for the proposed European programme and therefore requires revision.

{ Please note this third conclusion in the light of other actions that have been brought before this conference }

ACT THREE THE MANDATE FROM THE EUROPEAN COMMISSION

Scene 3.1 Definition

Resulting from the discussions that took place at The Lisbon Conference, the European Commission (DG XVII ENERGY) issued a mandate No. M 244[9] to CEN and CENELEC on the 13th December 1996. The mandate requests and requires of CEN and CENELEC (in fact only CENELEC), to accept the following tasks.

(i) **" To define test methods and standards for the measurement and calculation of the efficiency of general purpose low voltage 3 phase induction motors.**

And (ii) **To establish standards for the designation of efficiency of general purpose low voltage 3 phase induction motors fed by a sinusoidal network.**

As the term " designation " referred to in the second part of the mandate above relates to "identification " or "labelling" of the product, it was outside the scope of a "Standards" making authority such as CENELEC which therefore reduces the technical content of the mandate to :-

" To prepare a new standard for the measurement and declaration of the efficiency of small 3 phase induction motors "

Scene 3.2 Receipt, Proposals and Recommendations

The Technical Board (BT) of CENELEC noted the receipt of the mandate from The European Commission at its 91st meeting held on 11-13 March 1997 and at the same meeting asked the Chairman and Secretary of the appropriate Technical Committee (TC) , namely TC 2 for advice on the feasibility of executing the mandate. As Chairman of CENELEC (CLC) TC 2 Committee, I presented my report[10] to the CLC BT on the 24th May 1997 with the following proposals and recommendations.

Scene 3.2.1 Proposals.

(i) That in line with the October 1996 edition of the IEC-CENELEC Agreement[11] on Common Planning of New Work etc. (known as the " Dresden Agreement ") that CENELEC should offer the work required in the mandate to IEC (IEC Sub- Committee SC2G Working Group 2).

(ii) That the officers of CENELEC TC2 and IEC TC2 should co-operate to produce the required standard within two years from the date of approval to proceed. (subsequently 16th July 1997 - therefore completion by July 1999)

(iii) That as a consequence of the irrefutable evidence and conclusions with regard to basic loss measurement technology and possible improvements, that the BASIS of the new standard should be derived from the appropriate parts of IEEE 112 and revised and rewritten to the specific requirements of the IEC rules and the mandate.

(iv) That the European Commission be invited to finance research and development in Europe of alternative technologies for the measurement and declaration of the efficiency of induction motors to form the basis of a subsequent revision of the new standard.

Scene 3.2.2 Recommendation

That the CLC Technical Board accept the mandate M244 from the European Commission and execute it via the protocols of the Dresden Agreement subject to the formal acceptance of IEC TC2 and the agreement of the Commission to a revised completion date of July 1999.

Scene 3.2.3 Decisions

At its 92nd meeting held on 1-3 July 1997, the Technical Board of CENELEC accepted the report of the Chairman of CLC TC2 **EXCEPT** for clause 4.1.3 of the report which is summarised above as Scene 3.2.1 (iii) and states that the basis of the new standard should be IEEE 112. Although this exception was qualified as recorded below, it did represent the first attempt to bias the optimum technical result by commercial / political influence expressed through a (some) national permanent delegates to the CLC BT.

The qualification as advised to the chairman of CLC TC2 states :-

" The ".Name of Country.." Permanent Delegate (to the BT) stated that his experts considered the requirements of this standard (i.e. IEEE 112B) to be very time consuming and expensive and it was agreed that any draft should be on the basis of all technology available at European and International level. It will therefore be up to the members of the IEC Working Group to convince other National Committees of the technical superiority of the proposed IEEE draft "

The technical , political and commercial battle lines were thus drawn and throughout the subsequent two years vigorous argument and counter arguments were expressed.

Scene 3.3 The Way Forward

Considering the rejection by the CLC BT of the clause in the report by the Chairman of CLC TC2, it is pertinent and appropriate to present in précis that part of the report dealing with the possible and probable technology sources that were available to provide the optimum technical solution to fulfil the mandate.

Scene 3.3.1 Basic Technology Requirements (Being Clause 3.1 of the original report)

" Although there is now a near unanimous world wide consensus calling for a revision of IEC 34-2 in favour of more accurate methods, an extensive survey has been unable ,(as yet) to find any new basic technology upon which to base a radically new methodology. For a new standard to include new methods it would mean that the methods would have to have been declared and proven in use before the necessary acceptance and confidence could be gained.

This eagerly needed new technology does not exist and as such it may be appropriate for the Commission to sponsor further research into the subject in academia as a parallel activity through the Europe wide energy savings initiatives.

Accepting that no new technology exists, the only way forward that remains is to utilise IMPROVED versions of existing technology.

Scene 3.3.2 Improvements in Existing Technology (Being Clause 3.2 of the original report)

" The two basic methods that are used for the measurement and declaration of the efficiency of induction motors are :-

(i) The input output method.

And (ii) The summation of losses method with assigned value of stray load loss.

Scene 3.3.2.(i) Input Output Method.

Resulting from an extensive literature survey, it has been concluded that the latest and perhaps the most accurate application of input output methods have been developed in the United States in support of EPACT 92 and are contained in IEEE 112B and Canadian Standard C 390. There are no known and proven alternative sources available at the moment.

Scene 3.3.2.(ii) Summation of Losses Method with Assigned Value of Stray Load losses.

The major problem with this popular and widely applied method is the magnitude of the assigned value of stray load loss. The value of 0.5% of the power input specified in Clause 9.1.3 in IEC 34-2 is much too small for small induction motors of conventional manufacture. Many researchers have discovered that such a low figure is only likely to be achieved on well designed and carefully manufactured motors with outputs greater than (say) 1MW.

By plotting the results from these many sources, it can easily be shown that the % of stray load loss is on average a function of the motor output. Research in Europe and North America compiled by myself and others in the field have shown that a loss function which decreases continuously as a % of input power to be more realistic than either the current fixed % of IEC 34-2 or the step graded allowance of the IEEE document.

Scene 3.3.3 Instrument Accuracy.

It is considered probable that a substantial increase in accuracy of the foregoing methods would be achieved by using measuring instruments and their associated transformers of a higher accuracy class than that specified in IEC 34-2 Clause 3.

ACT FOUR IEC WORKING GROUP (SC2G WG 2)

IEC Technical Committee 2, to whom the European Commission Mandate was assigned, delegated the work to its subcommittee SC 2G who in turn delegated the work to one of its existing working groups WG2.

Within WG2, two related projects were already approved as work items, namely:-

(i) Test Methods for Induction Motors. Documents 2G/77NP and 2G/85/RVN.

(ii) Measurement of Efficiency of all Types of Machines Documents 2G/82/NP and 2G/xx/RVN.

The scope of the European Commission Mandate occupies a small part of the common ground contained within the above two projects. The working group agreed that although the mandate was specifically a European issue dealing with a small niche in the machines product portfolio, the issues and solutions had a global consequence and therefore in accepting the task to produce the restricted scope testing standard, it was aware that a solution to the current mandate would not only satisfy the European requirements but would also occupy an important and appropriate place in a world standard.

The representation in the working group is incredibly diverse. Four continents are represented by experts from fourteen countries, all of whom are active participants and our kept in order by efficient working group officers. It is worth noting that working group members although nominated by national committees are not national delegates. They are experts in their own right and may express opinions that are not necessarily the views of their own national committees. Never- the- less , it is inevitable that in such meetings both national and regional interests can and do intrude into the otherwise technically based forum but as all members are aware of this situation, due account can be taken where necessary. The aim of the convenor is to seek consensus and rarely to resort to the voting procedure.

Scene 4.1 The Technical Issues.

It became clear from the earliest meetings of the working group that the statement from the 92nd meeting of the CENELEC BT **NOT** to use IEEE as the basis of the new document did not find favour with the vast majority of the working group experts. A small minority pursued a commercially influenced national view that remained unconvincing to the other experts.

It would be tedious and inappropriate to attempt to relate the detailed negotiation that took place within the working group , suffice to say that the meetings were full of strong and lively debate , the experts concluding that the main issues were :-

(i) The definition of realistically achievable accuracy in the measurements leading to the declaration of efficiency.

(ii) That wherever possible to assign advantage to procedures that are based on testing the product.

(iii) To ensure that within the procedures that use an assigned value for stray load loss, that its magnitude relative to product rated output should follow the shape as determined by consensus based on test data submitted to the working group and be marginally greater than the test result envelope.

(iv) That the applicable output range of the motors should realistically span the outputs where the test technology and procedures are currently achievable.

Scene 4.2 National Comment and Vote on Draft IEC 61972

The working group issued a first committee draft for comment as 2G/100/CD[12] on 13th March 1998 and subsequently a revised committee draft for voting as 2G/102A/CDV[13] on 23rd October 1998 with a closing date for national votes on 31st March 1999.

The question put to the vote before national committees was :-

> "Does the national committee agree to the circulation of the committee draft as a FDIS ? (Final Draft International Standard)

The voting was a simple majority of P-members in favour i.e. 14 which equals 61% of the total voting , with 10 voting against, which is 38% of the total voting. This meant that the IEC rules for acceptance had **NOT** been met i.e. 68% in favour. With the exception of one country all the negative votes were by European countries. This is not surprising as there had been a concerted effort to procure a negative vote by a strong commercial lobby

Scene 4.3 Reprise.

At the last working group meeting in Paris in April 1999, the working group noted the result and pattern of the national committee voting and by a nearly unanimous vote in the working group, immediately agreed to proceed with a new committee draft with vote (i.e. CDV), taking full account of the comment that had been received from 15 of the 26 nations that had voted (not all P-members).

The new CDV(14) was circulated to Working Group members on 30th July 1999 and is expected to be circulated for vote during August 1999. On the basis of the absorption of the vast majority of the national comments submitted with the previous CDV, it is expected that the new pattern of voting will be sufficient to meet the IEC acceptance criteria. But who knows, the nations will decide!

ACT FIVE FINALE.

Scene 5.1 Australian / New Zealand Standard

Although Australia voted against 2G/102A/CDV because of " errors and lack of clarity " they have already produced a draft AS/NZ(15) Standard for comment which contains all the essential elements of Draft IEC 61972. (The Australian comments were taken into account during the production of the latest CDV in Paris.)

Australian / New Zealand Standards would appear to be keen to reap the benefit of the substance of Draft IEC 61972 ahead of Europe.

Scene 5.2 CEMEP / EU Agreement.

Earlier in this conference , you will have heard of the voluntary agreement on the levels of induction motor efficiency and product labelling made between CEMEP and the European Commission.

At the time of writing (early August 1999), I am led to believe, that despite the fact that the European Commission issued its mandate M 244, reflecting the facts that had emerged from its Lisbon Conference by requesting a new and more accurate method of testing for the declaration of the efficiency, the voluntary agreement specifies an unchanged IEC 34-2 as the agreed method ????

It is a matter of fact that the new standard was not approved by July 1999 as was promised, due, I would suggest, to a not insignificant level of tactical voting. Nevertheless, the new CDV is promised in August 1999 and all the portents suggest that early next year IEC 61972 will be published. When this occurs , will the CEMEP / EU Agreement be revised to call for the new standard or not ??? Even if against all reason, the latter prevails, then the mandate will not have been in vain for as the first vote proved, the majority of nations will approve and no doubt will use and specify the new standard.

Such a scenario makes the basis for efficiency declaration for those European motor manufacturers with a global business an interesting prospect!!

EPILOGUE.

The " Challenge " to produce a new standard for the measurement and declaration of efficiency of small induction motors has turned out to be an epic. But as in all classic epics , at the final curtain it is usual for right to prevail -- however, until the early part of next year we may have to be satisfied with this conference as an entr'acte.

To date, the project has survived:-

BARGAINING in BRUSSELS
DELEGATION via DRESDEN
FORMULATION in FRANKFURT
MACHINATION in MONTREAL
PERTURBATION in PARIS
and near **LACRIMATION in LONDON**

The next meeting of the working group will be in Switzerland in January 2000,

Will the project reach its

ZENITH in ZURICH

I hope so and then we will really be able to report that :-

ACCURACY has overcome EXPEDIENCY

ACKNOWLEDGEMENT.

Whilst the views and opinions expressed in this paper are those of the author, the support is acknowledged from colleagues at ALSTOM Electrical Machines Ltd, the Rotating Electrical Machines Association (REMA), Technical Committee 2 of CENELEC and last and by no means least, those friends and colleagues from around the World on Working Group 2 of IEC subcommittee SC 2G who have shared the " CHALLENGE "

REFERENCES.

(1) The Experimental Determination of Losses in Motors. Smith. JIEE 39 p437 1907.

(2) Zusatzliche Verluste in Kleinen Drehstrommotoren. Rogowski & Vieweg. Archiv fur Elektrotechnic Vol 14 p574 1925.

(3) Stray Load Losses in Induction Motors; a Challenge to Academia. C.N.Glew. IEE Power Engg. Journal Vol 12 No. 1 Feb 1998.

(4) IEC 32.2 Third Edition 1972.

(5) British Standard 4999 Part 33 1977.

(6) British Standard 269 1927.

(7) IEC 34-2 Amendment 2 Nov. 1996.

(8) An overview of how ac induction motor performance has been affected by the October 24 1997 implementation of the Energy Policy Act of 1992. Bonnett. IEEE PCIC paper 98-17.

(9) Mandate to CEN and CENELEC for the elaboration and adoption of standards for test methods and for the measurement , calculation and designation of the efficiency of general purpose low voltage three phase induction electric motors. European Commission DG XVII Energy Brussels 13 Dec. 1997.

(10) Report to the CENELEC TECHNICAL BOARD by C.N.Glew Chairman CLC TC2 May 1997.

(11) IEC - CENELEC Agreement of Common Planning of New Work and Parallel Voting Oct. 1996.

(12) IEC 2G/100/CD 13th March 1998.

(13) IEC 2G/102A/CDV 23rd October 1998.

(14) IEC 61972/CDV - revision of 2G/102/CDV - working group reference WG 2/34/WD issued to Working Group 30th July 1999.

(15) Draft AN / NZS 1359.102.3 dated 1st May 1999.

Evaluating the Impact of a Government-Sponsored Motor-Driven System Efficiency Program: The U.S. Experience

Paul E. Scheihing[1], Mitchell Rosenberg[2], Sue Weil[2]

[1] United States Department of Energy, Office of Industrial Technologies
[2] XENERGY Inc.

Abstract. This paper summarizes an independent evaluation of the impact of the United States Department of Energy's Motor Challenge program. The Motor Challenge was initiated in 1993. The program was developed and is managed by the Office of Industrial Technologies (OIT) in partnership with U.S. industry. The primary objective of Motor Challenge is to increase the energy efficiency of motor-driven systems used in the manufacturing sector. Motor Challenge pursues this objective by developing, in cooperation with industry associations and other interested groups, tools and information services that promote best practices in motor system design, purchase, and management. The program distributes these materials, decision tools and training through suppliers and others, to end users of motor systems to promote a systems approach in the way these systems are managed, maintained, upgraded and improved.

The paper will present an overview of the Program's objectives and offerings, an analysis of motor system energy use and savings opportunities taken from the recently released *U.S. Industrial Electric Motor Systems Market Opportunities Assessment (Market Assessment)*, and the objectives, methods, and key results of the evaluation. The program evaluation estimated that, as of September 1999, the Motor Challenge program has motivated industry to make changes in motor system purchase and management practices that result in $24.9 million in energy savings per year. These savings are net of those associated with changes in motor-related practices that customers and vendors would likely have made in the absence of the Program.

Also discussed, are plans to integrate the Motor Challenge activities within a broader, plant-wide energy efficiency technology delivery initiative called Industries of the Future BestPractices. The paper concludes with key conclusions and lessons-learned as to how any future motor system efficiency government-sponsored program could be designed to be more effective.

1 Introduction

This paper summarizes an independent evaluation of the impact of the United States Department of Energy's Motor Challenge program. The potential benefits of Motor Challenge in terms of reduced carbon emissions and increased industrial competitiveness are significant. Motor systems – pumps, fans, compressors, conveyors, and all manner of specialized production equipment – account for over 23 percent of all electricity used in the United States and for 63 percent of all electricity used in industry. A recent comprehensive study, *U.S. Industrial Electric Motor Systems Opportunities Assessment (Market Assessment)*[(1)], found that motor system energy use could be reduced an average of 15 percent in industrial facilities using mature, well-understood technologies. For individual systems, savings can range as high as 50 to 75 percent using thoughtful engineering techniques and a system approach. The full 15 percent energy efficiency improvement opportunity in motor systems by U.S. industry would translate into an annual energy cost savings of $4 billion and carbon emission reductions of 21 million metric tons of carbon equivalent (MMTCE) – equivalent to removing more than 4 million cars from the road.

The overall DOE strategy is directed at accelerating and assisting industry to move toward these potential savings by getting key leaders to demonstrate to the rest of industry that the savings are cost-effective and synergistic with core business objectives. The results of the independent evaluation summarized in this paper demonstrate that the motor system savings are beginning to more broadly be captured throughout the manufacturing sector. The evaluation also clearly demonstrates that the opportunities in energy efficiency have only begun to be tapped. Much more focus of system efficiency improvements, and the tools and methods to achieve them by both end users and their suppliers will be required to achieve the full potential benefits and to accelerate greater savings in the future.

2 Motor Challenge Overview: Objectives and Offerings

2.1 Objectives

Motor Challenge pursues an objective to develop best practice information and tools in cooperation with industry associations and energy efficiency organizations, and then to distribute technical products and training either directly by the Program, or through Allied Partners (vendors and others) to end users of motor systems to promote a systems approach in the way these systems are managed, maintained, upgraded and improved.

Motor Challenge was launched in October, 1993. From the start, the program faced the challenge of changing ingrained business and engineering practices among end-users and vendors without resort to providing them financial incentives. In the first three years of operation, the Program experimented with a variety of program activities, services, and delivery strategies. Within the past two to three years, the Program has settled on a set of core strategies and activities that focus on the development and dissemination of technical and business decision support tools and the development of partnerships with industry organizations that represent key segments of the motor systems market. Throughout the development of Motor Challenge, program managers have been guided by the following technical and program design principles:

- ***Promote a "systems" approach.*** Industrial engineers have long known that careful matching of the elements of a motor system -- motors, controls, couplings, and process machinery -- to the work to be performed yields far more savings than upgrading the efficiency of the individual components. The Market Assessment found that over 71 percent of total potential savings came from systems-level measures such as improving the configuration and control schemes in pump, fan, and compressor systems. The Market Assessment also discovered that very few plant operations staff are implementing management practices necessary to achieve the system efficiency opportunities.
- ***Harness the business motivations of end-users, manufacturers, and vendors in disseminating technical information and promoting energy efficiency.*** Throughout, Motor Challenge has emphasized not only the energy savings associated with improved motor system efficiency, but other benefits of efficiency improvements such as increased control over production processes, reduced waste, and improved production environment for workers. The Program has also sought to work with manufacturers and vendors to identify and exploit competitive advantages associated with promoting efficient motor systems and the benefits of life cycle costing.

2.2 Motor Challenge Offerings

The program's core technical activity has been to develop a set of technical decision tools to support end-users in their efforts to identify and implement cost-effective motor system efficiency measures, as well as vendors and others in their efforts to promote and sell efficiency-related products and services. To date, the most important of these tools have been as follows:

MotorMaster+ Motor Selection and Management Software. MotorMaster+ contains a database of efficiency, price, and other catalog information for more than 25,000 three phase, integral horsepower electric motors produced by major manufacturers. Using this database, the algorithms contained in the program, and information on motors currently in use, vendors and end-users can identify specific models which will provide the most cost-effective replacement for a failed motor, given its specific size and application.. The program can also be used to analyze the benefits of replacing versus repairing a failed motor. Other modules support motor inventory management. To date, over 23,000 copies of MotorMaster+ have been distributed to end-users, vendors, utilities, and government agencies. Utility and government program managers have also used the MotorMaster+ database to set product eligibility requirements and incentive levels for premium efficiency motor rebate programs.

Pump System Assessment Tool - PSAT. PSAT estimates pump system efficiency based on a limited number of on-site measurements. It can be used to assess the overall efficiency of a pump system relative to its optimal performance. It can be used to determine if further engineering analysis is justified to improve the pump itself and its system components and controls.

ASDMaster. ASDMaster is a software program that assesses the feasibility and cost-effectiveness of adding Adjustable Speed Drive controls to a motor system. The program was developed under the auspices of the Electric Power Research Institute (EPRI). Motor Challenge has licensed this software from EPRI and sponsors training in its use and distributes the software to trainees.

OIT is now undertaking a major tool re-engineering effort to design tools that are easier to use by both end user and suppliers, and which address the steps of profiling energy use within a plant, screening for opportunities, and implementing energy efficiency actions. One will be able to evaluate the full range of energy efficiency improvement opportunities, including motor systems, in a manufacturing facility.

Training Programs. Motor Challenge has conducted or co-sponsored technical training sessions in a wide range of motor system topics including: use of MotorMaster+ and motor selection; basic pump system efficiency topics and recently the use of PSAT; basic ASD operations and use of ASDMaster. 4,536 individuals representing an estimated 2,923 establishments have registered for these courses. Approximately 62 percent of these were commercial and industrial end-users; the remainder were vendors, consultants, utilities, and government agencies.

Showcase demonstrations. Showcase Demonstrations develop information on the field performance of efficient motor system technologies and design practices. In exchange for technical assistance from the Program's technical experts, customers arrange for monitoring and verification of energy savings associated with various motor system efficiency measures. Motor Challenge uses the documentation of the Showcases to develop case studies on advanced technologies which facility managers can use to assess the applicability of similar measures to their own facilities. To date, 13 technical motor system case studies have been developed.

Information Services. The Motor Challenge compiles and disseminates technical information on a wide variety of motor system topics through the following channels: a staffed information clearinghouse, a world-wide web site, a newsletter distributed to over 25,000 individuals, and through an Allied Partner network (see below) of 208 companies and organizations as of September, 1999.

Program delivery channels. Motor Challenge tools and services are disseminated to actors in the targeted markets through the following channels.

End-user direct channels. Initially, Motor Challenge employed a partnership approach to deliver program services. The Motor Challenge Partnership was begun at the inception of the Program in 1993. To become a Partner, an organization completed an application and signed an agreement committing to pursue the goals of the program. Partners were provided access to the information clearinghouse, the decision tools (any partner employee could register for and receive a copy of MotorMaster+ software), some training opportunities free of charge, the newsletter, and a variety of free publications. Motor Challenge Partners were under no obligation to implement specific practices, designs, or equipment purchases. Starting in 1997 as focus was shifted to other initiatives, end user companies and plant locations were no longer required to enter into a partnership agreement in order to access software or other available materials. However, individuals receiving products were requested to register themselves with the Program.

In addition to the partnership initiative, Motor Challenge has pursued a variety of strategies to deliver its services directly to end-users. These include broadcast methods such as newsletters, direct mail, and two national teleconferences, as well as provision of information on demand through the Internet and the information clearinghouse. Finally, Motor Challenge has developed partnership agreements with organizations such as the Technical Association of the Pulp and Paper Industry (TAPPI) and the Electrical Apparatus Service Association (EASA) to distribute materials to their members under the partner organization's auspices. Finally, the program deploys a number of account representatives to market the program directly to larger end-users.

Allied Partnerships: Vendor and Consultant Channels. Motor Challenge has recruited a number of organizations to help disseminate its materials and services to end-users. Among vendors, Allied Partners are primarily manufacturers, distributors, dealers, and consultants who perceive a value in providing their own customers with information on how to increase the energy efficiency of their facilities. Allied Partners agree to promote energy efficiency with their customers as well as within their own organization. Each Allied Partner is asked to complete an Action Plan outlining the activities they agree to undertake – product dissemination, training, etc. In exchange, Motor Challenge makes most of its resources available to Allied Partners in quantity at no cost or minimal cost. Allied Partners have access to a broad array of Motor Challenge publications and decision tools, which they can distribute to industrial end users in the course of their daily business or in conjunction with customer education meetings or workshops. At the end of FY1998, 104 vendors or consulting companies of various kinds were registered as Allied Partners. Based on survey results, these organizations have distributed approximately 1,800 copies of MotorMaster+. Some Allied Partners were able to achieve a high degree of leverage for the program. For example, one vendor trained over 2,700 individuals in the use of MotorMaster+.

Allied Partners: Governments, Utilities and Trade Associations. Motor Challenge also approached utilities and government agencies with an interest in industrial energy efficiency to become Allied Partners. These organizations used the technical resources available from Motor Challenge – in particular MotorMaster+ -- to help structure their own programs to promote the purchase of energy efficient motors. They also distributed tools and information to vendors and end-users who participated in their programs. Fifty-six electric utilities, 16 government agencies, and 24 trade and research organizations were registered as Allied Partners as of the end of 1998. Altogether, these organizations worked with over 12,000 industrial end-users using Motor Challenge tools and materials. Over 8,000 copies of MotorMaster+ were distributed through this channel. In some cases, such as EASA, one industry organization represented over 2,500 separate vendor establishments. Of these, 2,000 are in the US and it is estimated that 50 of the large establishments have independently enrolled as Allied Partners. The other members, although not counted as Allied Partners since they had not formally registered with the program are covered separately in the analysis.

Industry Partnerships. Over the past several years, Motor Challenge has developed partnerships with associations representing important groups of end-users, manufacturers, and vendors in the industrial motor system markets. The basic approach has been to harness OIT's technical resources to address motor system-related technical issues and opportunities that affect a broad range of the associations' members. Industry partnerships also serve as the main vehicle to develop new types of information, tools and training offerings. Examples of these partnerships include the following:

- *The Compressed Air Challenge.* In 1997, Motor Challenge developed an industry partnership with the Compressed Air & Gas Institute (CAGI), a trade organization of 45 manufacturers of compressed air system equipment. Nine sponsors, including CAGI and DOE, donated substantial funding to create a Compressed Air Efficiency Council in fourth quarter FY1997. This council was formed to increase customer awareness and develop national professional training and certification programs for plant engineers. In first quarter FY1998, the initiative was renamed "Compressed Air Challenge" and began developing marketing materials. The Compressed Air Challenge's first program initiative was to develop an introductory training program on compressed air system efficiency. The training sessions were rolled out in 1999 and over 1,400 people have been trained as of September, 1999.

- *Pump System Initiative.* The Hydraulic Institute (HI), a trade organization of approximately 70 pump manufacturers, has marketed a video training program entitled "Energy Reduction in Pumps and Pumping Systems" with student and instructor workbooks. HI has formed a "Life Cycle Costing" committee with DOE's facilitation assistance that will be developing products to assist end users address life cycle cost factors in managing and maintaining their pump systems.
- *Pulp and Paper Industry Initiative.* TAPPI became an Allied Partner in fourth quarter FY1997 and distributed over 400 copies of MotorMaster+ to pulp and paper mills across the country. TAPPI has 33,000 members and provides Motor Challenge tools and information to them mostly free of charge.

Table 1 below summarizes the volume of program activity and participation through September 1999:

Table 1: Volume of Program Activity and ParticipationCumulative from FY1994 through September 1999

Program Activity	Units of Service Delivered	Number End-User Facilities Served
DOE Direct Channels		
Publication Requests	11,082	N/A
Technical Assistance Cases	1,101	N/A
Bimonthly Newsletter	25,000	~7,900[1]
Copies of MotorMaster+ distributed	6,387	~1,900[2]
Motor Challenge Partners	N/A	943[3]
Technical Training Workshops	31 sessions	443[4]
Allied Partner: Vendors and Utilities		
Training Modules Ordered	182	N/A
Training Events Sponsored	98 sessions	1,282[4]
Copies of MotorMaster+ Distributed	17,150	~5,100[2]
Publications Distributed	120,500	N/A

1 Recent survey of *Energy Matters* newsletter readers found that less than half were end-users. XENERGY applied ratio of MotorMaster+ copies distributed to end users/unique facilities to estimate number of facilities receiving the newsletter.

2 XENERGY estimate based on ratio of copies shipped/separate establishments for registered MotorMaster+ users.

3 The Motor Challenge Partnership program component was suspended in 1997. These end-users were included in the MotorMaster+ analysis.

4 Estimate based on average number of attendees per session for program-sponsored trainings.

2.3 U.S. Industrial Electric Motor Systems Opportunities Assessment

OIT commissioned a study of industrial motor system energy usage and savings opportunities to serve as a blue print for the implementation of the Motor Challenge program. The study was conducted in 1997 and published in December 1998. The objectives of the Market Assessment were to:

- Develop a detailed profile of the current stock of motor-driven equipment in U.S. industrial facilities;
- Characterize and estimate the magnitude of opportunities to improve the energy efficiency of industrial motor systems;
- Develop a profile of current motor system purchase and maintenance practices;
- Develop and implement a procedure to update the detailed motor inventory profile on a regular basis using readily available market information; and
- Develop methods to estimate energy savings attributable to the Motor Challenge program.

Table 2 shows the distribution of motor system energy between manufacturing and non-manufacturing, and among major subdivisions within those categories.

Table 2: Motor System Energy Use by Major Industry Group

Industry Categories	Net Electric Demand* (million kWh)	Motor System Energy (million kWh)	Motor System Energy as % Total kWh
Manufacturing	**917,834**	**541,203**	**59%**
Process Industries	590,956	419,587	71%
Metal Production	152,740	46,093	30%
Non-metals Fabrication	106,107	50,031	47%
Metals Fabrication	68,031	25,492	37%
Non-Manufacturing	**167,563**	**137,902**	**82%**
Agricultural Production	32,970	13,452	41%
Mining	44,027	39,932	90%
Oil and Gas Extraction	33,038	29,866	90%
Water Supply, Sewage, Irrigation	57,528	54,652	95%
Total All Industrial	***1,085,397***	***679,105***	***62.6%***

* 'Net Demand for Electricity' is the sum of purchases, transfers in, and total onsite electricity generation, minus sales and transfers offsite.Sources: MECS 1994, Department of Agriculture 1992, Census of Mineral Industries 1992, Burton Environmental et al. 1993.

Table 3: Potential Systems-Level Motor Energy Savings by Manufacturing SIC and Application

SIC	Industry Category	Estimated Savings (GWh/Year) Fan System	Pump System	Compressed Air Systems	Other Proc. Systems	Motor Upgrade	Motor Downsizing	Improved Repair	All Systems	As % of Total Energy
20	Food and Kindred Prod.	157	1,250	494	517	1,376	585	295	4,674	12.4%
21	Tobacco Products									
22	Textile Mill Products	170	593	408	166	743	305	121	2,506	15.0%
23	Apparel & Other Textile Prod.	1	0	68	15	47	22	8	162	13.9%
24	Lumber and Wood Prod.	153	243	324	341	432	336	184	2,013	8.8%
25	Furniture and Fixtures	87	5	78	33	173	68	26	471	12.7%
26	Paper and Allied Prod.	1,082	6,293	773	881	3,197	845	870	13,942	14.0%
27	Printing and Publishing	52	17	74	90	305	153	39	731	12.3%
28	Chemicals and Allied Prods.	942	7,556	6,813	994	4,219	1,409	1,255	23,188	16.1%
29	Petroleum and Coal Prod.	271	6,159	1,352	169	1,736	459	453	10,599	20.4%
30	Rubber and Misc. Plastics Prod.	113	1,851	813	411	1,498	435	303	5,424	14.8%
31	Leather and Leather Prod.	27	0	0	0	22	6	3	58	11.8%
32	Stone, Clay, and Glass Products	31	18	96	20	117	45	14	343	15.4%
33	Primary Metal Industries	738	1,537	2,150	1,085	3,199	983	749	10,441	11.9%
34	Fabricated Metal Prod.	34	181	303	80	298	195	46	1,137	15.6%
35	Industrial Machinery and Equip.	28	195	200	94	368	208	44	1,138	15.4%
36	Electronic and Other Electric Equip.	18	1,554	513	43	609	222	93	3,053	23.1%
37	Transportation Equip.	353	1,109	941	242	1,195	340	235	4,415	14.9%
38	Instruments and Related Prod.	71	119	123	78	263	169	39	862	13.3%
39	Misc. Manufacturing Industries									
	All Industry Groups	*4,330*	*28,681*	*15,524*	*5,259*	*19,799*	*6,786*	*4,778*	*85,157*	14.8%

Table 3 shows potential motor system energy savings by application for each two-digit SIC (Industry) group.

Table 4 shows the concentration of the large savings opportunities (above 1 billion kWh/year) in certain applications for a targeted number of industry groups:

Table 4: Large Motor System Energy Savings Applications within Industry Sectors

Motor System Application	Industry Group with Greater Than 1 billion kWh per Year in Savings
Fan Systems	Paper
Pump Systems	Food, Paper, Chemicals, Petroleum, Rubber, Primary Metals, Electronics, Transportation
Compressed Air Systems	Chemicals, Petroleum, Primary Metals
Other Process Systems	Primary Metals
Motor Upgrade	Food, Paper, Chemicals, Petroleum, Rubber, Primary Metals, Transportation Equipment
Motor Downsizing	Chemicals
Improved Repair Practices	Chemicals

These 22 large energy savings areas account for 69 percent of all potential savings.

3 Evaluation Results

The evaluation summarized in this paper was independently conducted by XENERGY Inc., a company with extensive experience in energy product market assessment and program evaluation. The primary objective of the evaluation was to estimate the energy savings that can be attributed to Motor Challenge (MC) activities. In the course of gathering and analyzing the data needed to estimate energy savings, XENERGY also compiled information on the program's effects on the markets for industrial motor systems and on participants' response to the various program offerings. This paper focuses on the energy savings estimates and presents selected findings on market impacts and constituent response to the program.

The evaluation research and analysis activities were designed to answer three basic questions:

- ***How many industrial facilities and vendors used MC materials and tools to make changes in motor system purchase, management, and design practices?*** To address this question, XENERGY analyzed records maintained by Motor Challenge program administrators to establish the universe of establishments that had participated in the program. Surveys of various participant groups were then conducted to ascertain the extent to which they had used MC materials and tools to change motor system practices.

- ***What portion of reported changes in motor system practices were attributable to Motor Challenge?*** For research operations, this question is usually phrased: What portion of the changes would likely have been made in the absence of the program? Generally, responses from participant surveys were used to assess the extent to which MC materials and tools contributed to reported changes in motor system practices. XENERGY used information from other market research to supplement the survey analysis. These analyses resulted in the estimation of a "net" number of motor systems affected by the program.
- ***How much energy did those changes save?*** XENERGY used the results of the Market Assessment to estimate the base level energy usage and energy savings associated with the motor systems and the changes in motor system practice identified in the first two steps.

It is important to note that estimated savings associated with Motor Challenge participants upgrading the efficiency of replacement motors are incremental to savings realized as a result of the promulgation of regulations to implement new efficiency standards for some integral horsepower polyphase motors contained in the Energy Policy Act (EPAct) of 1992. In October 1997, federal regulations went into effect which required all general purpose integral horsepower motors up to 200 horsepower sold in the United States to meet efficiency standards developed by the National Electrical Manufacturers Association. In this evaluation, the program is only credited with savings associated with the purchase of new premium efficiency motors (higher efficiency than the EPAct standards), where the upgrade from EPAct standards can be attributed to the program. Also, the evaluation credits the Motor Challenge program influence on end users to replace a higher percentage of failed standard motors rather than repairing them relative to what the users would have done in the absence of the program.

3.1 Evaluation Methods

Motor Challenge is a multifaceted program that has been in operation for six years. Given constraints of time and budget, it was not possible to assess the effects of all the program's components. The primary objective of the evaluation was to estimate energy savings attributed to the Program. Therefore, it was decided to focus evaluation resources on program components that met the following criteria:

- Established energy program evaluation methods could be used to quantify energy savings; and,
- Established methods could be used to demonstrate a plausible link between program activities and actions taken by end users to save energy.

3.2 Key Findings

Below are the highlights of the evaluation:

Energy Savings

- The annual energy savings attributable to the program total 520 GWh per year. Table 5 shows the distribution of these energy savings attributable to different program components. The monetary value of these savings is estimated at $24.9 million per year as of September 1999.

Number and Characteristics of End-Users Served

- Between 4,900 and 8,500 end-user facilities received information materials, tools, and training directly from the program.
- Approximately 19,000 end-user facilities received services related to motor system efficiency from utilities, vendors, and consultants who used Motor Challenge materials to deliver those services. There may be some overlap between these customers and those who were served directly by the program.
- As of September 1999, there were 5,655 *registered* MotorMaster+ users representing 3,664 unique end-user facilities. These facilities and individuals figure prominently in the evaluation because they could be identified through program records and interviewed directly regarding the effect of the program on their motor system purchase and management practices. On average, the registered MotorMaster+ users are large industrial facilities. XENERGY estimated that they use roughly 20 times as much motor system energy as the average manufacturing plant and 5 times as much as a typical utility-sponsored motor program participant. Altogether, MotorMaster+ users consumed 165,120 GWh/year in electricity versus 1.1 million GWh/year for industrial users as a whole. Thus, even though registered MotorMaster+ users represent less than one percent of all industrial facilities, they account for 15 percent of total industrial electricity use and a comparable portion of motor system energy.

Table 5: Summary of Energy SavingsAttributable to the Motor Challenge Program (as of September 1999)

Component	# of Facilities Served	Number Acted due to MC	Net Program Energy Savings (MWh/year)	Net Program Energy Savings ($/year)
End-Users				
MM+	3,664	653	50,687	$2,432,971
ASD Training	370	87	22,475	$1,078,779
Pump Training	455	218	30,829	$1,479,797
Showcase	13	9	24,148	$1,105,600
Energy Matters	7,937	2,646	35,173	$1,688,305
Teleconference	318	318	4,227	$202,912
End-Users Subtotal	*12,757*	*3,931*	*167,539*	*$7,988,365*
Vendors & Partners	**# of Entities Served**	**Number Acted due to MC**	**Net Program Energy Savings (MWh/year)**	**Net Program Energy Savings ($/year)**
Allied Partners				
Vendor/Consultant	104	68	88,352	$4,240,872
Utility/Government	95	60	93,840	$4,504,340
EASA	1,948	487	23,325	$1,119,594
Assoc. Facil. Engineers			218	$10,460
ASD Training	371	223	131,431	$6,308,695
Pump Training	202	101	15,724	$754,744
Non End-Users Subtotal	*2,720*	*939*	*352,890*	*$16,938,705*
Total			***520,429***	***$24,927,070***

Effects of the Program on End Users

- ***MotorMaster+.*** Among registered MotorMaster+ users interviewed for the evaluation, 19 percent reported that they made changes to motor system design, purchase, and maintenance practices that they would not have been made in the absence of the program. These changes included upgrading efficiency of replacement motors; opting to replace rather than repair failed motors; improvements in motor system maintenance practices; and installation of ASDs.
- ***Training.*** Twenty-four percent of end-users who participated in the ASD training program and 48 percent of those who participated in the Pump System training program reported that they implemented improvements to the efficiency of their systems which they would not have made in the absence of the program.

- ***Energy Matters Newsletter.*** DOE conducted a survey of the readers of the *Energy Matters* newsletter independently of this evaluation. The survey found that one-third of end-users who received the newsletter reported that they had made changes in the way they purchased or managed motor systems as a result of reading the newsletter.

Number and Characteristics of Participating Vendors and Allied Partners

- ***Allied Partners.*** As of September 1999, 201 organizations were enrolled as Allied Partners of the Motor Challenge Program. These included 104 vendors and consultants and 95 utilities and government agencies that offered energy-efficiency services to their customers or constituents. The Electrical Apparatus Service Association, the industry association for motor repair shops, and the Association of Facility Engineers also joined the program as Allied Partners. The number of member firms for these organizations is shown in Table 5.
- ***Training Program Participants.*** 902 individuals representing 573 vendors, consulting firms,, utilities and government agencies participated in the ASD and Pump System training programs. Roughly half of the trainees were vendors or consultants.

Effects of the Program on Vendors and Allied Partners

- ***Vendors and Consultants.*** Among a sample of the 104 vendors and consultants participating in the program as Allied Partners, over 60 percent reported that they had used Motor Challenge materials to help convince customers to purchase energy efficient motor equipment. Many had also made changes to their practices in regard to motor system design, specification, and sales which resulted in increased energy efficiency for their customers. These changes included specification of energy efficient motors, specification of ASDs instead of mechanical devices for speed control, and application of more efficient overall design and control schemes for pump and compressed air systems.
  "The information provided by the Motor Challenge program (especially MotorMaster+ software) acts as a "proactive tool", allowing consulting firms to convince companies to think about buying and using efficient motors. Clients cannot deny the possibility of real savings when MotorMaster+ clearly highlights the potential for real savings. It's a very convincing decision tool." [quote from Energy Consulting Company]
- ***Utilities and Government Agencies.*** Ninety-five electric utilities and government agencies used MC materials and tools to structure or enhance their own programs to increase motor system efficiency for their customers and constituents. Typically, they distributed MM+ or used the database and analysis to structure incentives to customers for purchasing premium efficiency motors. Over 12,000 customers were reached through these programs.

"This isn't a "pie in the sky" program that has no influence on people's bottom lines. It's viewed as practical, useful, and is generally welcomed by most manufacturing, commercial, and industrial facilities. At least this is the feeling that I have gotten over the years." [quote from Electric Utility]

See the Final Report of the *Evaluation of the Motor Challenge Program*[(2)] for more details on the methodologies and assumptions made in assessing energy savings for the Program.

4 Evaluation Results in Context

Over the 6-year life of the program, Motor Challenge has established communication channels with technical and management decision-makers that represent a large portion of US motor system consumption. At a minimum, Motor Challenge has identified technical and management personnel (registered MotorMaster+ users) in 3,664 facilities that account for 15 percent of total industrial motor system energy use, or roughly 103,000 GWh per year. In addition to these end-users, the program has identified potential decision makers in 2,000 to 4,000 facilities through its Information Clearinghouse, and training activities. The customer identification records that support these operations are a key resource in advancing the mission of the program.

Capture of potential motor efficiency upgrade savings. The *Market Assessment* identifies and quantifies two basic groups of motor system efficiency improvements. The first consists of improvements to the inherent efficiency of motors themselves; the second consists of improving the way in which the components of a given motor system work together to accomplish the designated task. The first group of measures includes upgrading the efficiency of failed motors, replacing rather than rewinding failed motors, and improving rewind practices. The *Market Assessment* estimated the potential savings from such measures at 4.3 percent of total motor system energy use. In the following paragraphs we estimate how much of these potential savings were captured by participants in the program.

The evaluation found that Motor Challenge participants are already following good practices in efficient motor purchase decisions. Moreover, we found that Motor Challenge has played an important role in helping participating end-users to realize savings from motor efficiency upgrades.

Roughly 15 percent of the stock comes up for replacement each year. Thus, for the MotorMaster+ users interviewed for this study, potential savings from motor efficiency upgrades amount to:

$$\textit{Potential Savings} = \textit{15\% stock turnover} \times \textit{4.3\% potential savings} \times \textit{103,000 GWh/year}$$
$$= \textit{664 GWh/year}$$

MotorMaster+ users attributed actions that led to 51 - 60 GWh per year in motor efficiency upgrade savings to the influence of the program. These actions included purchasing more premium efficiency motors that they would have in the absence of the program, and rewinding 2.5 percent fewer motors that they would have in the absence of the program. According to this analysis, then, Motor Challenge participants captured 10 percent of potential annual savings available from motor efficiency upgrades due to actions influenced by the program.

These results can also be understood in the context of the industrial market for motors. MotorMaster+ users attributed the purchase of 10,303 premium efficient motors to the influence of the program (instead of purchasing EPACT qualifying motors). This is roughly 6 percent of the number of units of all premium efficient motors sold in 1998 to the industrial sector.

Capture of potential system improvement savings. Motor Challenge has barely scratched the surface in terms of helping end-users realize potential energy savings from system-level improvements. Between MotorMaster+ users, training session attendees, and users of various information services, Motor Challenge has reached 6,000 to 8,000 end-user facilities directly. Using the results of the *Market Assessment* and the evaluation surveys, we estimate that these facilities used approximately 200,000 GWh per year in motor system energy. The *Market Assessment* found that industrial facilities can save, on average, 10.5 percent of total motor system energy usage through system-level measures. Thus, for the population of end-users directly served by the program, potential savings from system-level measures can be estimated at 21,000 GWh per year. Our best estimate is that these facilities captured at most 300 GWh per year in savings, or 1.4 percent of the available potential.. Applying the information from the evaluation surveys to findings from the *Market Assessment* we estimate that facilities served directly by Motor Challenge used approximately 200,000 GWh per year in motor system energy. Further, we estimate that these facilities

5 New Directions: Industries of the Future BestPractices

Industries of the Future: Since the mid-1990s, OIT has been working within its industry-driven, customer-focused, "Industries of the Future" (IOF) strategic framework. Under IOF, OIT brings together stakeholders in energy-intensive industries—including companies that may otherwise be competitors—to create a strategic vision of the future of their industry and roadmaps to help them achieve their visions. The IOF concept enables OIT and other technology-supporting organizations to target industries' specific technology needs and get the most out of limited resources.

IOF BestPractices: To help industry make greater use of OIT's resources, OIT is integrating and streamlining its industrial energy outreach programs. Existing tools, resources and activities from OIT's Challenge programs—Motor, Steam, Compressed Air, Combined Heat and Power, and the Industrial Assessment Centers—are now being integrated into Industries of the Future (IOF) BestPractices, a single delivery vehicle facilitating easier, more comprehensive access to all OIT resources, products, and services.

BestPractices brings together the best available and emerging technologies and practices that are available today to assist industries in improving their competitive position. Through BestPractices, industry will be able to better access near-term and long-term technological solutions to improve energy efficiency, enhance environmental performance, and improve productivity for their total manufacturing plant operations.

BestPractices is OIT's technology delivery arm of the IOF framework. By promoting near-term energy saving opportunities as well as long-range R&D, OIT can have a greater impact on industrial energy efficiency. Targeting the most energy-intensive industries, BestPractices will focus on energy efficiency systems, equipment, and procedures that can be implemented immediaely for substantial bottom-line impacts.

OIT BestPractices encourages a perspective that views individual energy-consuming components as parts of a total system, focusing on the significant savings that can be found across an entire plant. With this approach, plants can realize near-term, cost-effective savings between 10 to 30 percent in three to five years. By applying the best technologies and practices available, companies can begin saving money, cutting waste, and increasing productivity immediately. Therefore, the tools, resources and activities that were deployed under Motor Challenge will now be used and integrated as part of a broader delivery of resources to promote total plant-wide efficiency. OIT's customers will now become aware that BestPractices is where to go to receive the types of products previously delivered through the Challenge programs – including Motor Challenge.

6 Lessons Learned and Conclusions

With 6 years of experience in developing the Motor Challenge program there have been many valuable lessons learned. Below are a few key lessons learned:

- Motor Challenge has already established extensive and effective channels to personnel in end-user facilities and through a large number of key Allied Partner vendors. However, the majority of the potential savings in end-user facilities have not been achieved. The next major effort must be to develop a set of tools and materials that will support end-users and vendors in achieving system-level savings.
- Program record-keeping must be enhanced to enable managers and implementation staff to better characterize establishments quickly as to function (end-user v. vendor v. utility or trade association), industry, and size. This will aid in program marketing, client relations management, and evaluation.
- Leveraging the market is essential to maximizing the effect of any national market transformation program, such as Motor Challenge and now BestPractices for DOE. Suppliers must see motivational factors to joining and promoting a energy efficiency program. This Win-Win situation needs to be developed and leveraged.
- Tools need to be made simple and developed for decision-making at various stages of project implementation: general plant profiling, screening for technology opportunities, and implementation of projects. Consideration of time to be spent, or not spent, by different participants in energy efficiency project implementation should be respected when developing tools.
- Programs should develop activities for not only awareness and promotion, but also for implementation in partnership with industry on a plant level. More extensive resources are needed to assist Allied Partners to more easily convince and assist end users in project implementation.
- Working on a plant-by-plant basis to demonstrate the leading plants in implementing best management practices for motor-driven systems (motor, pumps, compressed air systems) will go a long way in encouraging other companies to accelerate energy efficiency initiatives – industry has a tendency to follow leaders.

Acknowledgments

The efforts of many people are acknowledged in developing and shaping the Motor Challenge program; and for the results that are described in this paper. Special thanks go to the OIT management team, especially Denise Swink and Marsha Quinn for their support of the program. Exemplary performance and perseverance is recognized on the part of DOE employees Chris Cockrill, Julia

Oliver and Rich Been for their support of the program throughout its existence. Thanks to the staff of the Washington State University Energy Extension Office for their efforts in operating the Clearinghouse, and in developing MotorMaster+ software – especially Gil McCoy for his innovation – without this product the program would not have been as successful. For the tireless efforts of the staff from Oak Ridge National Lab (Showcase Demonstrations, Pump Training) and Lawrence Berkeley National Lab (Industry Partnerships) and National Renewable Energy Lab (Energy Matters). For the customer-friendly outreach conducted by the staff of Macro International, that amongst many things, helped to design and implement a successful Allied Partnership and two Teleconferences. For the thorough and comprehensive work done within the Market Assessment by XENERGY Inc. And, finally, the tremendous partnerships that have been created by thousands of DOE's industry partners in getting energy saving results.

7 References:

[1] XENERGY Inc. (1998): United States Industrial Motor Systems Market Opportunities Assessment. U. S. Department of Energy and Oak Ridge National Laboratories, Washington, D. C.

[2] XENERGY Inc. (1999): *Evaluation of the Motor Challenge Program: 1993 - 1998.* U. S. Department of Energy and Oak Ridge National Laboratories, Washington, D. C.

Oliver and Rich Beck for their support of this program throughout its existence. Thanks to the staff of the Washington State University Energy Extension Office for their efforts in organizing the Conferences, [illegible]

Canada's Energy Efficiency Regulations for Electric Motors: Four Years Later

Valerie Whelan

Equipment Standards Officer
Office of Energy Efficiency, Natural Resources Canada
580 Booth Street, Ottawa, Ontario K1A 0E4

Abstract. This paper provides an overview of the Canadian Energy Efficiency Regulatory Program for electric motors. The existing *Energy Efficiency Regulations* are reviewed. The Canadian experience with respect to implementation of the regulations for motors as well as various aspects related to compliance monitoring and enforcement of the regulations are discussed.

Keywords. Electric motors, energy efficiency regulations, compliance

1 Introduction

Canada's *Energy Efficiency Regulations* (the Regulations) form a key element of Natural Resources Canada's (NRCan's) Efficiency and Alternative Energy Program that consists of a mix of voluntary and non-voluntary initiatives designed to encourage the more efficient use of energy in Canada and to improve the efficiency of energy-using products available to Canadians.

Canada's *Energy Efficiency Act* was enacted in June 1992 and proclaimed January 1, 1993. This legislation is authorizing legislation that gives the government broad regulation making powers to prescribe energy-using equipment, to prescribe energy efficiency levels and labels for this equipment and specify compliance requirements.

The federal legislation applies only to energy-using products which are imported into Canada or traded interprovincially within Canada. There are five provinces in Canada (Ontario, British Columbia, Quebec, Nova Scotia and New Brunswick) that also have energy efficiency regulations. It is important to note that the federal legislation has no preemptive power over provincial legislation. This means that products subject to federal energy efficiency legislation may also have to comply with the regulations (if any) of the province in which the product is sold.

The first *Energy Efficiency Regulations* came into effect on February 3, 1995. These Regulations prescribed energy efficiency standards and labelling requirements for major residential appliances, water heaters, residential space conditioning equipment, fluorescent lamp ballasts and electric motors. Energy efficiency regulations for fluorescent lamps and incandescent reflector lamps came into effect on February 1 and April 1, 1996 followed by an amendment to the Regulations increasing the energy efficiency standards for electric motors on November 27, 1997. A fourth amendment specifying energy efficiency regulations for commercial space conditioning equipment (air conditioners, heat pumps, condensing units), oil-fired furnaces, gas and oil-fired boilers, ice makers, compact clothes dryers and dehumidifiers was implemented on December 31, 1998.

2 Background

As previously stated, the first federal *Energy Efficiency Regulations* for electric motors came into effect on February 3, 1995. These Regulations specified minimum energy efficiency standards for 1 to 200 hp polyphase electric induction motors as well as other requirements with respect to energy performance verification marking, reporting of information to NRCan and information to be provided on Customs documents upon importation. The motor energy efficiency standards specified at this time were the same as those that had been implemented in the provinces of Ontario and British Columbia in 1993. In 1995 and 1996, Ontario, British Columbia, Nova Scotia and New Brunswick implemented higher energy efficiency standards for electric motors which were the same standards as those that were to come into effect in the U.S. in October 1997. These more stringent energy efficiency standards for motors were implemented federally in Canada on November 27, 1997. These Regulations also revised the electric motor definition to more precisely identify regulated motors. The new definition included both IEC (metric) motors and other "definite purpose" motors which are excluded in the U.S. legislation. In addition, reporting requirements were revised to reduce the burden on dealers, and requirements pertaining to the import of noncompliant products that will subsequently be exported were changed.

3 Regulatory Experience

3.1 Implementation of Motor Regulations

3.1.1 Experience

Until the implementation of the federal motor energy efficiency amendment in 1997, the regulation of motors in Canada was a patchwork of differing minimum efficiency levels and effective dates in various provincial jurisdictions. The situation was further complicated by the U.S. Energy Policy Act motor regulations which were published in 1992 to come into effect in October 1997. The following provides an example of the differing requirements: The provinces of Ontario and British Columbia first implemented motor regulations in April 1993. The minimum efficiency levels specified at this time were not stringent and most standard efficiency motors met the standards. These same levels were implemented federally on February 3, 1995 and in Quebec on September 1, 1995. However, British Columbia and Nova Scotia had already moved to the higher minimum efficiency levels corresponding to the 1997 U.S. levels in January 1995 followed by New Brunswick in June 1995 and Ontario on January 1, 1995. This situation led to fairly widespread confusion with respect to regulatory requirements and made it difficult for motor dealers to comply with the regulations.

It became apparent shortly after implementation of the first federal motor regulations that the definition identifying regulated motors was too broad and it was therefore difficult to determine exactly what types of motors were covered. The fact that the motor definition in the U.S. legislation differed from that in the Canadian regulations also caused confusion. An attempt to harmonize Canadian and U.S. requirements for motors was undertaken during the development of the second federal motor regulations. The second federal regulations included a much more precise definition for the motors covered by the regulations as well as the more stringent motor energy efficiency standards that had been effected by the provinces and the U.S.

Communication of the regulatory requirements to affected parties also presented some challenges. Natural Resources Canada (NRCan) conducted extensive consultation activities during the development and implementation of the motor regulations. Activities included three workshops attended by 150 stakeholders, and three separate information mailouts to 600 stakeholders and numerous meetings with various other parties. Although communication with motor manufacturers and dealers regarding the regulations was generally good, it was very difficult to identify and contact original equipment manufacturers (OEMs) who would likely be affected by the regulations. It was also difficult to convince OEMs to attend the consultation workshops.

3.1.2 Observations

The regulation of a commercial/industrial product such as electric motors presents some unique difficulties as compared to other commonly regulated products. There are hundreds of thousands of constantly changing motor models, as compared to a discreet number of models for other products. This makes it very difficult to identify all motors which are subject to the regulations and forces the regulator to interpret or make changes to the regulations on a continuing basis in order to accommodate new motor models.

The importance of harmonizing motor regulatory requirements for motors and effective dates for regulation over different jurisdictions was clearly demonstrated by the difficulties encountered during implementation of the Canadian motor regulations. The majority of motors sold in Canada are manufactured in the U.S. by manufacturers who supply both the Canadian and U.S. markets. The requirement to meet different energy efficiency standards at different times within Canada and the U.S. placed a significant burden on motor dealers which had an impact on costs and their ability to comply with the regulations.

Adequate lead time for dealers to prepare for new regulations and widespread communication of new regulatory requirements during development and implementation of regulations to all stakeholders who will be affected by the regulations is imperative to ensure the smooth implementation of new requirements and obtain compliance. The lack of knowledge about the motor regulations, particularly on the part of OEMs, has led to uncertainties with respect to motors contained within equipment being imported into Canada.

3.2 Compliance Monitoring

3.2.1 Verification Mark Requirement

3.2.1.1 Experience

The Canadian *Energy Efficiency Regulations* require that regulated motors carry a verification mark indicating that the energy performance of the motor has been verified. The verification mark is the mark of a Standards Council of Canada accredited certification organization that administers an energy performance verification program for motors. This requirement forms a vital component of the regulatory compliance program. Through the verification program, a certification organization confirms that the motors meet the required energy efficiency standards and provides an ongoing monitoring function to ensure that motors manufactured after the initial verification continue to meet the standards. NRCan implemented this requirement because resources to monitor compliance to the Regulations were and are very limited. The costs of the energy performance verification programs are born by the motor dealers and the certification organizations will revoke the right of a motor dealer to use their verification mark if it is determined that the dealer's motors do not meet the standards.

Motor dealers have had some difficulties in understanding the verification mark requirement, often believing that having the motor tested at an accredited motor testing laboratory would comply with the regulations. This is not the case since accreditation as a certification organization is different from accreditation as a testing laboratory. Testing in a National Voluntary Laboratory Accreditation Program (NVLAP) accredited motor testing laboratory, as has been proposed for the U.S. regulations, has contributed to this misunderstanding.

3.2.1.2 Observations

The issue related to the harmonization of motor energy performance certification is difficult to address since the Canadian and U.S. regulatory requirements are different. Motor dealers are very concerned with the cost of motor testing to demonstrate compliance with the regulations and have requested that the U.S. Department of Energy recognize the certification organization motor energy performance verification programs as well as the NVLAP motor testing laboratory accreditation program as a means of certifying compliance to the U.S. regulations. The testing burden has been reduced by harmonization of the motor energy performance test procedure. There is also a Mutual Recognition Agreement between the Standards Council of Canada (SCC) and the National Institute of Science and Technology (NIST) in the U.S. that recognizes motor testing laboratory accreditation by SCC or NVLAP as equivalent so motor test data from either an SCC or NVLAP accredited motor testing laboratory is acceptable to demonstrate compliance.

3.2.2 Regulatory Reporting Requirements

3.2.2.1 Experience

The Canadian motor regulations require that an energy efficiency report containing motor energy performance information be submitted to NRCan prior to importation of regulated motors into Canada. This information is entered into NRCan's equipment database which is used in the regulatory compliance program.

There have been a number of problems regarding submission of motor energy efficiency reports. In the first Regulations, a report was required for every motor model. It was almost impossible for motor dealers to meet this requirement due to the large number of motor models and the fact that motor models are constantly changing. NRCan also experienced extreme difficulties in keeping the motor equipment database up-to-date. The second motor regulations contained a new method for reporting on motors that reduced the number of motor "models" to be reported on by assigning a unique motor identifiier (UMI) to represent a group of related motor models. The UMI consists of the name of the motor manufacturer and the motor rating. The new identifier was based on a proposal that the National Electrical Manufacturers Association (NEMA) had submitted to U.S. DOE for certifying motor efficiency. NEMA proposed that manufacturers submit the energy performance of the least efficient model of each motor rating (hp or kW, speed or no.of poles and the motor enclosure) they manufactured in order to certify compliance to the Regulations. NRCan required the name of the manufacturer as well as the motor rating since a report must be unique in order to identify the regulated motor and match information in the equipment database to information on Customs import documents. A motor rating alone is insufficient to do this.

The implementation of reporting by UMI has been very difficult due to the fact that motor dealers are totally unfamiliar with this concept. NRCan has initiated a number of specific communication activities in order to educate motor dealers on the requirement and has produced a separate brochure on motor reporting that was distributed to approximately 500 motor dealers. However, to date, NRCan has received only a small percentage of energy efficiency reports for motors.

3.2.2.2 Observations

It has become apparent that reporting on the basis of a UMI has not resolved the problems with reporting on regulated motors. The sheer numbers of motor types preclude easy determination of appropriate UMIs and the difficulties with matching equipment database information with information on Customs documentation have not improved. It had been NRCan's understanding that it would be relatively straightforward for motor dealers to report on the basis of UMI since motor ratings are common to the industry. It was also believed that motor rating information was already present on Customs documents so the requirement would reduce the regulatory burden on motor dealers. NRCan's experience has shown the neither has been the case.

There are other implications with respect to the fact that a single UMI may cover hundreds of motor models with a range of efficiencies. As originally conceived, the information provided in the motor energy efficiency reports would allow NRCan to create a comprehensive database on motor energy efficiency. This has been lost with the move to reporting by UMI. There is also a concern as to whether the UMI provides enough information to be useful in assessing compliance.

It now appears that an earlier suggestion by motor dealers for NRCan to issue a unique compliance number to each motor dealer/manufacturer that could be used as the motor "model" number for reporting and on Customs documents might be more viable as a means to match this information and ensure compliance. NRCan intends to explore this avenue in the future.

3.2.3 Regulatory Importing Requirements

3.2.3.1 Experience

The Regulations require that motor dealers must include certain information on the Customs release document. A copy of the Customs documentation is then forwarded to NRCan and the information on the Customs document is cross-checked against the information in the NRCan equipment database to ensure that the motors being imported comply with the Regulations.

With respect to motors, this requirement has been particularly problematic and NRCan does not believe that it has been receiving a high percentage of Customs documents related to motors. Efforts are currently underway with Revenue Canada Customs to increase the number of documents that are being forwarded to NRCan. NRCan has also been receiving Customs documents for non-regulated motors since it is difficult for Customs personnel to separate regulated versus non-regulated motors. There is also a major issue with motors being imported in other equipment. Regulated motors must comply with the Regulations regardless of whether they are being imported as standalone motors or incorporated within other equipment. Many equipment types contain regulated motors.

Also, it has proved to be extremely difficult to match information on the Customs documents with information in the equipment database because many documents do not identify a motor UMI, primarily as a result of motor dealer confusion with respect to the nature of a motor UMI.

3.2.3.2 Observations

NRCan is in the process of establishing a Memorandum of Understanding with Revenue Canada Customs regarding its monitoring requirements with respect to regulated motors.

The submission of Customs documents to NRCan in electronic format is being piloted since Customs is working toward an almost completely paperless environment in regard to importation of commercial goods. The situation with respect to border monitoring of motors regulated under the *Energy Efficiency Act* has been affected by the fact that there are no health and safety issues associated with the importation of non-compliant motors (or any other energy-using product for that matter). As a result, monitoring of these products may be less rigorous than products which have the potential to be dangerous to Canadian citizens. As has been discussed previously, NRCan intends to look at means other than a UMI to identify motor compliance with importation requirements.

3.2.4 Audit Testing

3.2.4.1 Experience

NRCan has recently completed a survey regarding testing electric motors for compliance with the *Energy Efficiency Regulations*. The survey involved contacting Canadian motor manufacturers and distributors who were asked questions pertaining to the development of an economical and effective audit protocol. The survey focused on obtaining information on the number of sizes and types of motors normally stocked, the locations of the warehouses, as well as feedback on the impact of the Regulations on the efficiency of regulated motors incorporated into other equipment. The survey was conducted by the Laboratoire des technologies électrochimiques et électrotechnologies d'Hydro-Québec.

3.2.4.2 Observations

The major recommendation from the survey indicated that NRCan should examine the feasibility of coordinating a motor audit testing program with certification organizations that administer energy performance verification programs for motors. It was felt that, in view of the large number of motor models in the market and the high cost of motor testing, this approach would be more viable. It was also recommended that a least one motor be selected and tested from each motor manufacturer/distributor and the testing should focus on the most popular motor types and sizes. NRCan intends to host a workshop with motor dealers and OEMs to receive comments on the survey and plan the logistics of a motor audit testing program.

4 Conclusions

NRCan's administration of the electric motor energy efficiency regulatory program during the past four years has demonstrated the importance of open and timely communication between regulators and regulatees in the development and implementation of regulations in order to effect compliance. Coordination of motor regulatory requirements with those in effect in other jurisdictions should also be given serious consideration. It is imperative that regulators ensure a level playing field for regulatees and quality assurance for motor purchasers in the administration of an effective motor regulatory program.

Motors: IEA Perspectives and Activities

Jean-Pierre Des Rosiers

Head, Energy Efficiency Policy Analysis Division
International Energy Agency, Paris

1 Context: The Need for *Effective* Efficiency Policy

Since its inception in the wake of the first oil market disruption, the International Energy Agency (IEA) has been concerned about energy conservation and efficiency. First for energy security reasons, and then later for economic and environmental sustainability as well. Today, the Kyoto climate commitments of IEA Member countries have given an added intensity and urgency to the agency's energy efficiency work.

As a reminder, at the December 1997 Kyoto meeting of the UN Climate Change Convention, industrialised countries -- most OECD countries and the transition economies in central and eastern Europe and the Commonwealth of Independent States -- committed to reduce their overall greenhouse gas emissions by an average of 5.2% from 1990 levels during 2008 to 2012. The EU pledged a 8% reduction, but the individual country commitments within the "bubble" range from a reduction of 28% (Luxembourg) to an increase of 27% (Portugal). The United States agreed to a 7% reduction; Canada and Japan both to 6% reductions. It should be remembered that though this is a rather ambitious target, it is generally viewed as but an early step in addressing the climate issue.

Due to the strong economic growth and changing lifestyles, energy-related CO_2 emissions in IEA countries have increased by 7 per cent between 1990 and 1996. Longer-term projections, including those undertaken by the IEA in its World Energy Outlook, suggest a constantly rising trend. The IEA business-as-usual scenario suggests that energy-related CO_2 emissions in OECD countries could be approximately 30 per cent above 1990 levels in the year 2010. Past trends suggest that closing the gap between the energy-related CO_2 projections and the commitments made in Kyoto is a formidable task (IEA, 1999). How will countries do this? After all the talk of emissions trading, AIJ, CDM, carbon taxes, etc., there are really only five methods: (1) enhancing end-use energy efficiency, (2) improving efficiency of the energy supply system, (3) increasing fuel switching toward low-carbon energy sources (renewables, hydropower, nuclear), (4) increasing agricultural and forestry sinks and (5) increasing carbon capture and disposal.

Given the short time to meet the targets, the option of enhanced **end-use energy efficiency** will need to deliver much of the Kyoto greenhouse gas reductions. The slow rate of capital stock turnover of the energy system and the undeveloped state of carbon sequestration methods mean that the other options can deliver more in the medium and longer terms.

The magnitude of the Kyoto goals, the desire to meet them with little or no disruption to the economy and the backdrop of increasingly liberalised energy markets, means that there is great need for ***effective*** energy efficiency policies. That is, policies that sufficiently reduce energy use and CO_2 emissions, entail low and equitable costs to consumers, businesses, and governments, and are robust with respect to fluctuating energy prices and the vagaries of public concern about energy.

The role of the IEA is to assist countries in monitoring and improving their present energy efficiency policies, and in identifying and exploiting new opportunities for improving energy efficiency. In other words, to promote and help design effective policies.

2 Technology and Market View

So what is ***effective*** energy efficiency policy for motors and systems? It is difficult to say precisely, but one thing is absolutely clear, to be effective policies take full account of the complex technical and market situation of motor systems. Some of the more important aspects of the situation are:

Electric motors systems are very large electricity users.

Motors account for over half of all electricity used in developed countries, typically 60-80% of industrial use of electricity and about 35% of commercial sector electricity use (Almeida & Fonseca, 1997). Motors use on the order of 400 TWh per annum in the EU industrial sector, and a similar amount in the US manufacturing sector (EC, 1996 and EIA, 1997).

In terms of first costs, there are strong economic signals to push for more efficient motors.

In industry, a motor typically consumes five times its purchase price in electricity each year of its 12 to 20 year life (Almeida & Fonseca, 1997).

There is a certain reluctance on the part of users to changing motors and systems,

because of the high costs of trouble shooting, especially if it involves process downtime.

There is significant potential for improved efficiency in the motors themselves.

Standard motors, when well maintained and operated near their design points, convert electricity input into mechanical output with average technical efficiencies of 77 to 94 percent depending on their size. Energy efficient motors operate at 84 to 96 percent technical efficiency.

Component oversizing is a common problem among motor drive systems.

Larger than needed equipment is often used to ensure that systems can meet the demands placed on them at every stage of the process. For example, motors are oversized: to handle starting electrical, mechanical, and thermal stresses, particularly with high-inertia loads; to provide a safety margin for the worst-case load over their lifetimes; and occasionally to handle plant expansion or be widely interchangeable within a plant. Motors and other components may be also oversized partly to compensate for difficulties in accurately predicting system flow rates and friction factors in advance, and to allow for the effect of the buildup of deposits on duct and pipe walls over time. The increased energy and capital costs of the larger equipment are perceived to be less than the risk of equipment failure (OTA, 1993).

The largest potential for efficiency gains lies in the systems aspect of motors systems.

The systems aspect compounds the energy losses caused by component inefficiency and oversizing. When each successive element of a system is sized to handle the load presented by the previous component plus a safety margin, oversizing can quickly become excessive. The losses of the entire system can be great. Although the motors can have an efficiency near or above 90%, the total system (including motor, shaft coupling, pump and throttle valve, for example) might deliver under 50% percent of the energy put into it. The efficiency of industrial compressed air systems could be improved by 20-50% through the application of best practices that rely primarily on changes in operation and maintenance procedures (McKane, 1999).

The systems aspect means there are many market players involved,

motor manufacturers, pump manufacturers, fan manufacturers, system designers, etc.

Motors systems are often "embedded" in products.

Product end users, those paying the electricity bills, often do not get a choice concerning the motor in the products they purchase. This coupled with the high sensitivity of original equipment manufacturers (OEMs) to motor purchase costs and relatively low concern for life cycle costs inhibits the use of high efficiency motors in such products.

Larger motors are often rewound, not replaced, when they malfunction.
Rewinding is initially less expensive -- less risky in terms of sizing -- than purchasing a new motor, but ultimately costs more because of degraded efficiency. The efficiency of a rewound motor is typically about 2 percentage points below that of a new standard motor. Work is needed to find methods of reducing core damage caused by the high temperatures that arise during motor rewinding (OTA, 1993).

3 Policy View

There are several general policies that are used (and/or contemplated) to improve energy efficiency. They tend to be price-based. Examples are energy taxes, carbon taxes and emission credit allocation and trading schemes. Today, however, I would like to focus on more targeted policies, such as equipment regulations, voluntary agreements, technology procurement, information and technical assistance advice, etc. Following are some key elements that help make energy efficiency policies effective:

Given the complexity of the market, there is great need to engage the market players. Successful motor efficiency initiatives require a great deal of stakeholder consultation and involvement. Government can encourage and facilitate efficiency measures, but cannot act alone in designing and implementing the measures.

The use of motor efficiency standards and/or voluntary targets can help transform the market and make it more efficient. This could be done at relatively low cost to the society if the standards and targets are implemented in an orderly fashion with enough lead-time for manufacturers to change products and production lines.

The complexity of the motor system situation calls for a comprehensive market transformation effort.The US Motor Challenge programme is a good example of such an effort. It involves the harnessing of market forces and commercial interests to promote motor system efficiency. It promotes the use of energy-efficient motors, drives and driven equipment as well as motor systems integration and optimisation. Its mechanisms include an Information Clearinghouse providing data on the practicality and profitability of electric motor system strategies; design-decision tools such as motor systems database software; Showcase Demonstration projects; and training, workshops and conferences. The programme works within existing market structures and makes heavy use of partnerships with commercial interests, such as equipment manufacturers, distributors, users and other companies.

Rigorous programme measurement and evaluation is essential. In light of the major investments that will likely be required of governments and industries in order to reduce greenhouse gas emissions, it is important that thorough measurement and evaluation processes be implemented. Governments need to be able to measure the real impact of their policies and be in a position on an on-going basis to change those policies that do not yield results beyond autonomous improvements of energy use.

Beyond national policies and programmes, there should be an effort to reduce the trade of substandard, low-efficiency products. As colour televisions have squeezed black and white televisions off the market on their own accord, we can expect high efficiency motors to become dominant as advanced countries and companies undertake measures to develop and promote improved products. But what of the substandard products that make their way to other parts of the world? Is there some sort of international government-industry initiative that might speed the phasing out of inefficient motors world-wide. The IEA might be good a forum for developing such an initiative. At the moment, we are doing something similar in trying to facilitate action to reduce standby losses from electronic equipment.

4 IEA Demand Side Management Implementing Agreement Activities

There has been a technology procurement for high-efficiency motors conducted by Annex III of the IEA Demand Side Management Implementing Agreement. The IEA Hi-Motors Competition was arranged to find electric motors with energy efficiency higher than in any motor before and get them to the marketplace. The competition goals were to establish electric motors with losses about 25-50% lower compared to average motors. The 3-phase, 4-pole, 1500 r/min, most common motor type on the market, was selected to be the competition product.

Two entries met the tough requirements of the competition, and were awarded with "IEA Award of Excellence". The smaller entry motor, 5.5 kW, from ABB Sweden represents the lower end the power range. The efficiency level at full load had to be at least 90.4% using IEC 34-2 direct method, and the tested entry motor reached the value 90.5%. The bigger entry motor, 75 kW, from ABB Finland met the efficiency value 96.0%, when the mandatory requirement was 95.8%. These new energy-efficient motors fulfilled all the mandatory requirements in prototype testing and will use 28-50% less energy and have pay-back times of 1-3 years.

5 IEA Secretariat Activities

As mentioned earlier, there is great need for *effective* energy efficiency policies. The IEA Secretariat seeks to help countries design effective policies by:

- conducting policy assessments,
- developing and enhancing analytical tools,
- compiling and disseminating focused and strategic information on energy efficiency programmes and
- facilitating international collaboration to enhance programme effectiveness and reduce costs.

Several examples:

Energy Efficiency Policy Profiles -- A series of publications describing the experiences, results and lessons learned from energy efficiency programmes. They emphasise the critical aspects of the policy framework and the programme planning, programme execution and measurement and evaluation activities, and highlight elements of good practice for each aspect. The first publication, "Towards an Effective Energy Efficiency Policy," will convey the general strategic elements of a good energy efficiency policy, such as:

Establish and maintain an effective market structure, through establishing real cost pricing; removing subsidies and cross-subsidies; using taxes and levies; securing clear rights and responsibilities of property ownership; etc.

Help market actors recognise their best interest and act on it, through analysing consumer behaviour; providing information and/or training on the energy efficiency aspects of products and behaviour; encouraging Energy Service Companies (ESCOs) and Third Party Financing; tapping special financing arrangements; etc.

Focus market interest on energy efficiency, through fostering voluntary agreements; establishing and enforcing building codes and minimum energy performance standards; integrating energy efficiency in procurement practices; using government purchasing to stimulate the market for advanced technology; etc.

Ensure access to good technology, through encouraging the development, adaptation and diffusion of energy efficient technology; improving district heating systems; expanding the use of combined heat and power; etc.

Develop and maintain a supportive institutional framework, through integrating energy efficiency in sectoral policies; ensuring the availability of impartial expertise; etc..

Act to ensure continuity, through establishing policy clarity; demonstrating leadership; implementing effective evaluation and monitoring techniques; strengthening international collaboration; etc.

The second publication in the series will focus on appliance and equipment labels and standards programmes.

Standby Power Initiative -- The Secretariat is facilitating information exchange and collaboration on in order to help governments and companies reduce standby energy losses of equipment (leaking electricity). Currently, there are three task forces working on definitions, CO_2 emissions estimates, and policy options.

6 Concluding Remarks

I believe there is a need for international work on motor and systems, and believe that an IEA Motors Initiative, similar to the one standby power losses, could be an excellent forum for such collaboration.

7 References

[1] Almeida, A. and P. Fonseca (1997) "Characterisation of the Electricity Use in European Union and the Savings Potential in 2010," *Energy Efficiency Improvements in Electric Motors and Drives*, eds. A. Almeida, P. Bertoldi and W. Leonhard.

[2] EC (1996) SAVE Motors Study Group, European Commission Directorate General for Energy (DG XVII), *Actions to Promote Energy Efficient Electric Motors*, October 1996.

[3] EIA (1997) Energy Information Administration, US Department of Energy, Manufacturing Consumption of Energy 1994, DOE/EIA-0512(94), December 1997.

[4] IEA (1999) "Energy and Climate Change: The Challenge," IEA Ministerial Document IEA/GB(99)27.

[5] OTA (1993) US Congress, Office of Technology Assessment, *Industrial Energy Efficiency*, OTA-E-560, August 1993.

[6] McKane, A. (1999) "Collaborative Intervention: Working With Market Forces to Effect Lasting Change," *Energy Efficiency and CO_2 Reduction: The Dimension of the Social Challenge*, Proceedings of the ECEEE Summer Study 1999.

Perspectives of Labelling and Standards of Electric Motors in Europe

Eberhard Jochem[1 2], Jürgen Reichert[1]

[1]Fraunhofer-Institute for Systems and Innovation Research (ISI), Breslauer Straße 48, D-76139 Karlsruhe, Phone +49 (0) 721 68 09-169, Fax + 49 (0) 721 68 09-270, e-mail ejo@isi.fhg.de,Internet: http:\\www.isi.fhg.de
[2]Centre for Energy Policy and Economics (CEPE), ETH Zentrum, WEC, Weinbergstr. 11, CH-8092 Zürich, Phone +41 (0)1 632 0650, Fax +41 (0)1 632 1050, e-mail: ejo@cepe.mavt.ethz.ch, Internet: http://www.cepe.ethz.ch

1 Introduction: Economic Potentials of Motor Electricity Savings in Europe

Since the last International Conference on "Energy Efficiency Improvements in Electric Motors and Drives" in 1996 (Almeida et al., 1997), the understanding of the innovation process in this broad field of energy use has substantially increased. More is known about the technical and economic potentials, about the barriers to implement existing economic opportunities and possible measures and policies of governments and manufacturers or motor using OEM as well as final users of the services of electrical motors and drives.

During the second half of the 1990s, major changes of the context of the use of electrical motors occurred in OECD countries, but also in Eastern Europe and the CIS countries. These changes are still under way and induce major challenges and opportunities for a more efficient use of electrical motors and drives (see Section 2). These changes are likely to influence energy labelling and technical standards as well as complementary measures that are essential to the final success of these two instruments (see Section 3 and 4).

Technical and profitable efficiency potentials in Europe

Recent estimates based on surveys and analytical case studies concluded that the economic potential of electricity savings of electrical motors and drives is around 60 TWh in the EU Member Countries in the industrial and commercial/public sector. The potential savings are relatively evenly distributed over the two sectors: slightly above 60% in industry and around 38% in the commercial/public sector (see Table 1). Assuming an average electricity price for these applications of 70Euro/MWh, the yearly cost savings are in the order of 400Mill Euros per year which may induce an investment of at least one billion Euro within the next decade. Given the relatively high specific CO_2 emission per unit of electricity, the potential represents some 30 Mill tonnes of CO_2 that could be avoided at no additional cost until the end of the first commitment period of the Kyoto Protocol. This possible reduction would contribute to the achievement of the EU's commitment that also means a change towards sustainable development in the long term.

The economic potentials have been thoroughly estimated at a sectoral basis in the industrial sector with the highest potentials in the chemical industry, in pulp, paper and printing, and in the iron and steel industry; these are all industries with high energy services for pumping, ventilation, compressors and mechanical processing. In the commercial/public sector, most attractive electricity savings have been identified in five major service areas: pumping, lifts and escalators, ventilation, air conditioning and cooling (see Table I).

Table I: Estimates of economic electricity savings potentials of electrical motors and drives, EU, 1996

EEM + VSD Section	NACE	Economic Potential[2] TWh/a	CO_2 Reduction Potential Mio t[1]
Mining	10-14		
Food, drink & tobacco	15, 16	3.32	1.7
Textiles and leather	17-19	1.50	0.7
Paper and printing	21, 22	5.19	2.6
Chemicals	24	10.80	5.4
Ceramic products	26	3.56	1.8
Iron & steel	27.1-27.3, 27.51, 27.52	4.87	2.4
Non-ferrous metals	27.4, 27.53, 27.54	2.05	1.0
Engineering	28-32, 34, 35	2.33	1.2
Other industries	20, 25, 33, 36, 37, 45	2.83	1.4
Tertiary sector	01-05, 50-93	23.10	11.5
- Pumping	-	6.47	3.2
- Lifts, escalators	-	5.77	2.9
- Ventilation	-	3.93	2.0
- Air condition	-	3.00	1.5
- Cooling	-	3.00	1.5
- Other uses	-	0.93	0.4
Total EU15		**~60.0**	**23**

1) assuming average specific CO_2 emission of electricity production of 385 kg/MWh
2) assuming possible extrapolation of surveys in some EU Member Countries

Source: SAVE-Proj. "Improving the penetration of energy efficiency motors and drives" (to be published in 1999)

These estimates of the economic efficiency potentials relate to 5% of total electricity demand in industry and to some 4% of that in the commercial/public sector. Of major interest for manufacturers and policy considerations, however, is the distribution of the savings potential over the power range of the motors, where little information on the existing motor and drive stock in the EU is available. First results for the German commercial/public sector suggest, that major efficiency potentials are the power range of 0.75 to 4 kW (42%), 4 to 10 kW (27%), and 10 to 30 kW (18%) (see Figure 1). This means that energy labelling is most important in this power range between 0.75 to 30 kW in the commercial/public sector. Regarding the industrial sector, the power range between 30 and 70 kW becomes much more important, as the average size of pumps, ventilators, compressors and drives for mechanical processing is higher.

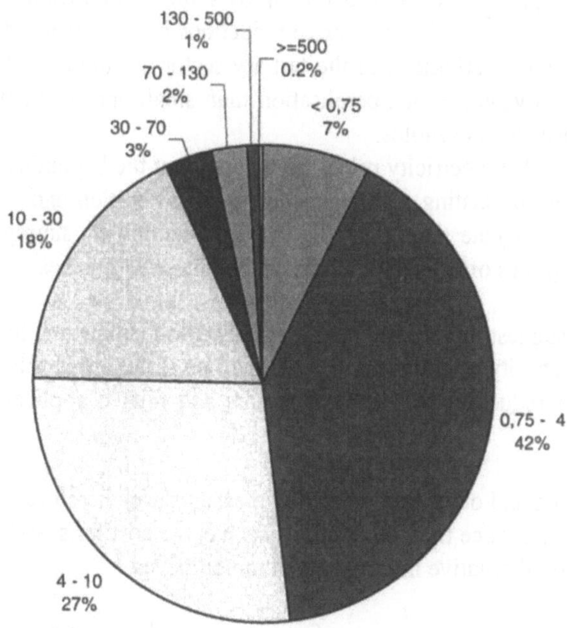

Figure 1: Distribution of Economic Efficiency Potentials of EEMs by Different Power Range (kW) in the Commercial/Public Sector, Germany, 1996

2 Contextual Conditions and Obstacles for Fast Dissemination

There is no doubt that the liberalisation of electricity supply in OECD countries, but also in Countries in Economic Transition (CIET) and higher income developing countries in the next decade, will have important implications on electricity efficiency in end-use sectors:

- Increasing productivity of electricity generation and distribution has been observed and will still occur, particularly in all early and fast liberalising OECD countries such as the USA, the United Kingdom, the Scandinavian Countries, Germany, and the Netherlands.
- As a consequence of the productivity increases and intensive competition, electricity prices have been declining by some 20 to 30% from its levels before liberalisation. Observed price reductions indicate that industry and large companies or institutions in the tertiary sector are more likely to be able to keep the advantages of liberalisation than small and medium sized companies or private households.
- It is likely that electricity prices drop sharply at the beginning of the liberalisation due to existing surplus capacities of generation and competition, but they may stagnate thereafter due to new oligopolistic supply structures, increasing prices of primary energies and higher energy taxes.

These trends suggest that the economic efficiency potentials that have been identified under electricity price levels of 1998 may be shrinking in the next few years, if similar cost reductions of efficient motors and related applications cannot be realised.

Energy labelling and other accompanying measures are, therefore, of major importance in order to reduce the transaction costs, i.e. the cost for search of information and planning of alternative investments of motor drives.

On the other hand, recent mergers of utilities as well as manufacturers of air conditioning systems, escalators, lifts, compressed air systems and cooling systems indicate a trend towards *energy service companies* offering service packages to energy consumers where they earn their profits from improved design of those systems (including efficient motors and drives) and, hence, saved electricity cost. This implies that after the liberalisation of electricity supply economic efficiency potentials still may be faster realised within the next decade than one could expect without these trends of liberalisation and energy service companies, particularly in the market of big electricity customers such as big companies, plants, buildings and institutions.

Obstacles to the use of efficient motors and drives

The major obstacles hindering a fast dissemination of high efficient motors or intelligent motor drives have been described in several publications that traditionally distinguish between general obstacles of energy efficiency and those specific to electric motors and drives (e.g. Jochem/Landwehr, 1997; Benerjee et al., 1997; Hall, 1997).

Lack of market knowledge, know-how and technical skills in several groups of actors along the innovation and decision chains (e.g. planners, OEMs, consultants, operating staff) reduce the speed of dissemination (see Figure 2). As about 60% of all electric motors are purchased directly by OEMs, their decisions on efficient motors and variable speed drives becomes very important. Specific qualifications like "intel inside" are not yet offered by them. Instead, the buyer is left with high transaction cost for searching and assessing available market information.

In addition, electric motors or drives have to be available in cases of re-investment and brake downs at extremely short lead times of only some hours. The huge variety in size, voltage and motor type makes it necessary for distributors or even end-users to store only a limited fraction of them. As storage cost of complete series would become very high, only often used motors and drives are stored for immediate delivery or use. As OEMs have to minimise their production cost and distributors their storage cost in order to remain competitive, new high efficient motors or variable speed drives at (initially) low market penetration are scarcely applied or stored, although the expert knows that the capital cost of motor systems may make up 20 to 30% of total life cycle cost and 65 to 75% the electricity cost.

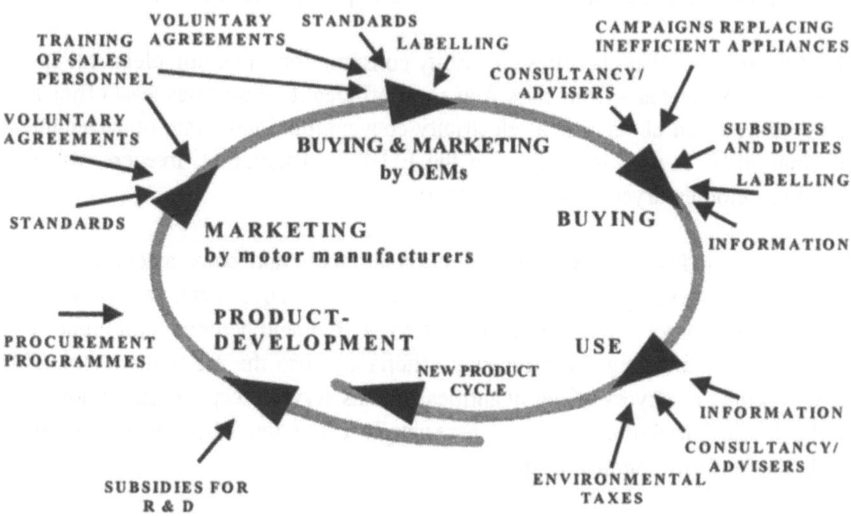

Figure 2: Policy measures to overcome existing barriers and to speed up dissemination of efficient motors and drives

One major conclusion of the analysis on obstacles and effective policy options is reflected in almost all analyses: due to several existing obstacles along the distribution chain from the manufacturer, OEM, distributor and vendor and purchasing end-user, only a *systemic bundle of policy measures* that is designed to overcome simultaneously the major obstacles is likely to increase substantially the dissemination of efficient electrical motors and drives.

3 Energy Labelling of Motors and Drives in the European Union

Transaction cost of the purchaser or investor of an efficient motor or drive may easily double or tripple the investment cost of the motor or drive itself in the case of small power sizes (see Bienek, this volume). This fact, of course, limits the efforts of an investor to search for new technical options and assess them. Instead, he will try to minimise his transaction cost by choosing exactly the same motor in case of re-investment or by deciding on those solutions he is familiar with or he gets recommended by the distributor or an external consultant.

A clear and simple information would be needed by the investor. This situation is somewhat similar as in the case of private households when they have to decide on investments of long-lasting electrical appliances such as refrigerators, freezers or washing machines. To accelerate the dissemination of more energy-efficient appliances, a framework energy labelling Directive was enacted in the European Union in 1992. The first energy label was for refrigerators that came into force in 1995. Since then, the average efficiency of the refrigeration equipment on the market in the EU has improved by 10 to 20%, depending on the Member Country.

This type of label that classifies electricity consumption of certain electrical appliances into six classes – from class A as most efficient to class G as least efficient – gives information about annual electricity consumption and some other technical information. It is now mandatory in the EU for refrigerators, freezers, washing machines, clothes-dryers, and dish washers.

To reduce standby electricity consumption of faxes and copy machines and to accelerate the development of energy-efficient office equipment, the Swiss Government sets target values in co-operation with manufacturers and importers. Manufacturers fulfilling the efficiency criteria can use the efficiency label. This label is aimed at buyers of big quantities such as service sector organisations and public bodies. Manufacturers are increasingly using the labels as part of their marketing and environmental strategy.

Both examples of energy labelling contribute to the role of an energy *labelling scheme for efficient motors* which may be introduced by the end of 1999 under the labelling Directive of the EU. This label will have three classes of different motor efficiencies where the numbers of minimum efficiency will depend on the power size and the type of the motor. The negotiation process of the definition of the three classes are linked with the expectation of a voluntary agreement of the European manufacturers that 50% of today's sales in the least-efficient class (eff 3-motors) will be shifted to the medium class (eff 2-motors) within a given time period. The labelling scheme will be limited to a power size between 1.1 kW and 75 kW where most of the efficiency gains is expected to be realised.

This form of labelling will open an easy way of making a real choice between motors with different efficiencies. It is expected to be very effective for smaller motors that are mass-produced. It seems to be less suited for big motors of several 100 kW power because of their more complex and individual specifications of application.

The labelling of energy-efficient motors may not be a favourite instrument for all manufacturers and importers as they will be obliged to certain performance standards which may not fit their present product strategies. Some importers may also question the new labelling scheme as a new non-tariff obstacle of foreign trade in this product area being in contradiction to the rules of GATT. These aspects will have to be discussed.

Finally, more theoretically based objections on the instrument of energy labelling or on efficiency standards are made on the argument that technical standards take away the incentives for further technical progress as the highest efficiency class freezes the technical standard. These arguments cannot be very valid given the experience with efficiency standards of boilers or new buildings which have been periodically adjusted to technical progress during the last 20 years. Efficiency awards like the Energy Star in the USA or the Blue Angel in Germany support the efforts of the manufacturers for further innovations. Periodical revisions of the present efficiency specifications of a labelling class and of the most recent best technical performance may lead to periodical adjustments of the specifications of energy labels or technical standards.

4 Complementary Measures Supporting the Wanted Success

Looking at the various obstacles along the product chain mentioned in section 2 (see Figure 2), it is necessary to stress the importance of a bundle of suitable measures in order to implement the economic potentials of efficient motors and drives. The voluntary agreement of the European motor manufacturers to shift 50% of the eff-3 motor sales to the eff-2 motor class already indicates that the energy labelling is not expected to speed up the dissemination of efficient motors just by this instrument. Energy labelling reduces the transaction cost and allows more transparency in the market of electrical motors. As electricity prices are falling and will fall in the future as a difficult contextual condition, accompanying and supporting measures become even more important:

- Information in the form of *market survey data* on a comparable basis, of a "source" book or a data bank like EURODEEM or OPAL is an important element. It will be easier to read with the classes of the energy labelling.
- In order to reduce the perceived risk if variable speed drives are introduced in production processes, *"show cases"* or *demonstration projects* that have been carefully selected may speed up the application in particular areas. The projects should be carefully selected to represent typical applications reflecting everyday circumstances. The results should be briefly documented to stimulate motivation and facilitate the planning and imitation of the project.
- The documentation of demonstration projects should be *published by neutral energy agencies* or other institutions such as *trade associations* of the user branch, as the end-users will have more confidence in the documentation material if it comes through these communication channels.
- *Procurement* will be facilitated by the classification of energy labels on electric motors. Procurement is a well-known measure of OEMs, but it could be much more exploited by end-users: large chains of hotels, banks, insurance companies and supermarkets may use it now more easily to get price reductions of their purchases and to set company-internal standards for certain applications of motors and drives. The procurement as a measure to enlarge the negotiating power of an end-use sector will also be supported by the experience in negotiating with electricity suppliers after the liberalisation of electricity supply. Once the communication channels among the end-users are formed they may also be used for procurement of efficient motors and drives.
- The development of *energy service companies* may also be useful to be supported. Electric utilities trying to bind their big customers by contracting of energy services may even take a more active role in order to compensate for lost sales of electricity by revenues from consulting, financing, professional operation and maintenance. But also equipment manufacturers experiencing saturation in many OEDC countries may compete in this new market of energy services. The label of energy efficient motors may help them to convince the end-user by the "intel inside"-effect and to reduce their own planning cost.
- *Professional training* of engineers, planners, installers and maintenance staff remains also of great importance, as the design, assessment, instalment, operation or maintenance of the motor applications demand high expertise and specific adaptation for individual investment project. Therefore, professional

training programmes with thoroughly prepared seminars like the Swiss RAVEL programme is still needed although the choice of electrical motors will be relatively easy after the introduction of energy labels.

5 Conclusions

Realising the economic efficiency potentials of electric motors and drives is a large challenge for energy policy and interested groups in the economy. The market of motors and drivers is very diverse and obstacles are manifold. Efficient motors are the first element that will be classified for energy labelling in the EU under the energy labelling Directive.

The electricity price decreases in the EU Member Countries as consequence of the liberalisation of the electricity supply may reduce economic potentials of efficient motors and variable speed drives and related applications. The energy labelling of efficient motors, the related voluntary agreement of European manufacturers to shift 50% of the sales of 3-eff motors to 2-eff motors, and recommended accompanying measures are likely to decrease the investment cost and transaction costs as well. This would result in an efficiency market volume in the EU of roughly 1 billion € in the next decade. Energy service companies are likely to take the lead in these activities.

6 Literature

[1] E. Jochem, M. Landwehr: Strategies to Promote Energy-Efficient Motor Systems, in: Energy Efficiency Improvements in Electric Motors and Drives, A. de Almeida, P. Bertoldi, W. Leonhard (Ed.), Springer 1997

[2] B. Benerjee, P. J. McCleer: Barriers Towards Successful Technology transfer and Commercialization of Energy Efficient Advanced Motors and Drives, in: Energy Efficiency Improvements in Electric Motors and Drives, A. de Almeida, P. Bertoldi, W. Leonhard (Ed.), Springer 1997

[3] S. Hall: Industrial Motor Standards in Latin America, in: Energy Efficiency Improvements in Electric Motors and Drives, A. de Almeida, P. Bertoldi, W. Leonhard (Ed.), Springer 1997

[4] J. Reichert, J. Schleich, G. Herzer, E. Tönsing, U. Rath, R. Hellmann: Measures for Rational Energy Uses of Electrical Drives (in German): Fraunhofer Institut Systemtechnik und Innovationsforschung, Endbericht zum Forschungsauftrag 1998 an das Bundesministerium für Wirtschaft, 1999

[5] International Energy Agency: Energy Efficiency Initiative, Energy Policy Analysis, Volume 1, Forth Ministerial Conference "Environment for Europe", Denmark 1998

Australia's National Greenhouse Strategy and Electric Motor Efficiency

Dr. Tony Marker, Manager, Energy Efficiency

Sustainable Energy Group
Australian Greenhouse Office

Abstract. Under the Australian domestic greenhouse strategy, all Governments have agreed to a program for improving the efficiency of certain equipment including electric motors. Government commitment to electric motor energy efficiency was specifically reaffirmed in 1998, and government and industry have been working toward agreeing to minimum energy performance levels (MEPS) for three-phase induction motors with a rating of 0.75 to 150 kW.

In Australia, the MEPS levels for the various categories of motors will be published in a joint Australian/New Zealand Standard. Performance curves for high efficiency motors (HEMs) will also be defined in the Standard, to ensure future advertising claims meet a specified performance standard.

Regulatory authorities will give mandatory effect to the minimum and high efficiency curves in this Standard. Industry has proposed an implementation date from 1 July 2001 though governments have still to formalise this commencement date.

To complement the proposed regulatory action, the Australian Motor System Challenge is being developed as a voluntary program with the cooperation of Government and industry stakeholders. This program will focus on the benefits of a total system approach to electric motor energy efficiency.

Keywords: MEPS, motors, efficiency, greenhouse

1 Policy and Economic Issues

1.1 Policy Background

Australia's endorsement of the UN Framework Convention on Climate Change provides the basis for action to reduce greenhouse gas emissions. The Prime Ministerial Statement, *Safeguarding the Future: Australia's Response to Climate Change (1997)* stated that "...Australian governments would work to develop minimum energy performance standards ... and, where appropriate, labelling or rating equipment to help consumers with their selection." Australia has signed the Kyoto Protocol to the Convention and if ratified the Protocol will commit Australia to a legally binding limit on its greenhouse gas emissions.

Measure 4.10 in *The National Greenhouse Strategy (1998),* which has been developed by the Commonwealth and State and Territory Governments, formalises Australia's commitment to energy performance codes and standards for commercial and industrial equipment. The combined approach of MEPS and labelling for high efficiency motors will assist in overcoming a number of market barriers that traditionally have impeded emission reductions relative to a business–as–usual (BAU) scenario.

The National Appliance and Equipment Energy Efficiency Committee (NAEEEC), which consists of representatives from the Commonwealth, State and Territory regulators and the New Zealand Government Energy Efficiency and Conservation Authority, coordinates energy efficiency initiatives on behalf of the Australian and New Zealand Mineral and Energy Council (ANZMEC). ANZMEC meets at the Ministerial level on an annual basis.

The Australian Greenhouse Office (AGO) is the lead Commonwealth agency on greenhouse matters, responsible for both the coordination of domestic climate change policy and managing the delivery of new and existing Commonwealth initiatives and programs. The AGO chairs NAEEEC and coordinates activities necessary for a national regulatory approach. The separation of powers between the Commonwealth and States/Territories under the Australian constitution means that the Commonwealth has no regulatory power in relation to energy efficiency.

1.2 Economic Considerations

On behalf of NAEEEC, Energetics Pty. Ltd. conducted a detailed study in 1996 of electric motor use in Australia that provided data for sales, capacity, and electricity consumption in 1995. Electricity use in the Australian industrial/commercial sector was 95.3 TWh, or 68% of total electricity consumption. Consumption by three phase electric motors (0.75 to 150 kW) was

39.6 TWh, or 28% of total electricity consumption. Energy consumption in electric motors was responsible for about 37 Mt of CO_2 emission per annum, or about 9% of Australia's total greenhouse emissions in 1995.

Following further consultations with industry representatives and other stakeholders on the key findings of the Energetics study, a NAEEEC/industry working group recommended a survey to seek feedback from a wider group of industry stakeholders. This survey was undertaken by SRC International in 1998, and confirmed that there was strong support from the stakeholders for a MEPS level that would remove the least efficient 40 percent of motors from the market.

Under National Competition Policy, governments are required to ensure that regulation does not act as a barrier to competition, and a regulatory impact statement (RIS) must be produced to demonstrate overall economic benefit to the economy.

2 The Australian Motor Market

2.1 Stand-Alone Motor Market

The "new" stand–alone motor market (1995) was characterised as follows:

- 130,000 new three phase motors with capacities between 0.75 kW and 150kW are sold annually (excluding motors directly imported by Original Equipment Manufacturers – OEMS);
- 88% of new motors (115,000 units) were imported and 12% (15,000) were made locally;
- twelve major suppliers were responsible for about 92.5% of new sales;
- five of the twelve market a distinct "high efficiency" range, although this does not necessarily imply that the efficiency of these motors is superior to the standard range of other motor suppliers;
- the new motor market was worth $80 million per annum; and
- there was no strong correlation between motor efficiency and price across the various motor suppliers.

2.2 Original Equipment Manufacturers

- OEMs are a key market player – Energetics estimated that 26,000 motors of the range covered by MEPS are directly imported by OEMs each year.
- OEMs are also responsible for a significant proportion of all motor sales in Australia – estimated at 40% of the market.
- OEMS purchase motors from a range of suppliers, and the market is very competitive with price is a key factor in motor selection criteria.

- Exports account for 5% to 60% of OEMs sales – with 15% being a typical figure.

2.3 Rewinders

- Rewinding is very common due to the economics in terms of initial cost, and in some circumstances gives a shorter turnaround time than the purchase of a new motor. In addition, the rewind industry has a niche market supplying "new" motors for end–users with unusual applications. Rewinders are a major part of the motor industry – with a turnover of some $30 million per annum.
- The structure of the rewind industry is virtually the opposite of the new motor market. It is characterised by a large number of players – for example, the Electrical Apparatus Service Association (EASA) represents approximately 800 motor rewinders across Australia. In addition to those rewinders represented by the EASA there are a large number of "small" facilities not members of the Association.
- In theory, rewinding can produce a motor with the same efficiency rating it had when it was new. In reality, the quality of a motor rewind, and hence the impact on efficiency, is highly variable; and depends on the materials and techniques used in reassembling the motor.
- Rewinding practices for motors of 180 W to 400 kW are specified in AS 4307.1. This Standard states that where motor efficiency is stipulated on a motor nameplate the efficiency of the rewound motor tested to AS 1359 will comply with the nameplate efficiency. However, as many rewind facilities do not have the required test gear the Standard is not generally enforced.

2.4 Summary

Additional details of the stand–alone motor market in Australia, derived from the Energetics study, are provided in the table below. There are no comparable estimates available for the OEM market.

Comparison of the average motor efficiency in the market relative to the maximum attainable clearly indicates that there is considerable scope for efficiency gains. The projected reduction in greenhouse gas emissions resulting from increased efficiency in the market via the introduction of motor MEPS and a high efficiency motor endorsement scheme is presented in the next section.

Table 1: The 1995 Electric Motor Market in Australia

Motor Range (kW)	*Annual Sales*	*% of Sales*	*$M Value*	*Ave % Efficiency*	*Max % Efficiency*	*% Capacity Share of Sales*
0.75 to 2.2	61,500	47	13	76.8	86.4	7
3.0 to 7.5	44,000	34	18	85.4	89.0	18
11.0 to 37.0	17,000	13	19	90.7	93.0	27
45.0 to 90.0	6,000	5	20	93.2	95.5	30
110.0 to 150.0	1,700	1	11	94.4	96.2	18
Total	130,200	100	81	n/a	n/a	100

3 Regulatory Objectives and Market Effects

3.1 Market Barriers

Numerous studies have shown that, in theory, end–users stand to gain significant energy and monetary savings from the installation of more efficient motors. In reality, there are a number of barriers which impede the realisation of this potential. Some of the most important are:

- *Lack of Information* – end–users do not have ready access to accurate information on the financial and operating benefits of energy efficient motors.
- *Limited Access to Capital* – even with informed end–users wishing to purchase energy efficient motors, these projects must compete for capital with other potential investments.
- *Lack of Institutional Incentives* – in many organisations the person in the best position to specify efficient motors does not get to keep the monetary benefit of the investment in their budget and hence has little motivation to make the effort.
- *Market Structure* – in many instances end–users are several steps removed from the actual selection of the motors they purchase; notably, in equipment purchased from OEMs and from consultants designing specifying equipment for greenfield sites.
- *Aversion to Downtime and Innovation* – lack of familiarity with efficient motors and misconceptions regarding performance and reliability may cause end–users to default to the "tried and true".
- *Competition on Speed* – end–users desire to have failed motors repaired as quickly as possible can cause rewinders to compete not only on price but also on turnaround time – a factor that can induce practices that damage motor efficiency.

- *Standardised Inventories* – suppliers and distributors may not keep a full range of efficient motors in stock, so in order to save time end–users will default to purchasing standard equipment.

MEPS and the HEM endorsement scheme are ways to assist in overcoming these barriers and transforming the motor market.

Energetic's study provides estimates of the potential of the proposed MEPS regime to reduce greenhouse gas emissions via increased energy efficiency. This study utilised a market assessment to determine key parameters for a BAU scenario, notably:

- sales weighted average efficiencies for discrete motor classes, and
- existing trends regarding improvements in energy efficiency in the absence of intervention.

This information formed the benchmark against which the potential net benefits (and costs) of the proposed motor MEPS was compared. Key findings from this analysis follow.

- The motor MEPS and HEM endorsement scheme targets equipment that accounted for approximately 39.6 TWh, or 28% of Australia's electricity use in 1995.
- This energy consumption translates to some 37 Mt of $CO_{2\ eq}$ emissions annually, or about 9% of Australia's total greenhouse gas emissions.
- For each percentage point improvement in the average energy efficiency of the motor market as a result of MEPS, there is projected to be an annual reduction in greenhouse gas emissions of about 0.370 Mt of $CO_{2\ equivalent}$ (an energy saving of about 0.4 TWh per year) compared with a BAU scenario.
- Cost/benefit calculations have shown that over the longer term, the proposed motor MEPS if implemented in 1999, would have the potential to reduce greenhouse gas emissions by about 7.0 Mt of $CO_{2\ equivalent}$ (cumulative) relative to BAU levels by the year 2030. The corresponding impact on electricity use is a reduction in consumption of 7.6 TWh.

3.2 Market Effects

The Australian Market

Overall, the motor supply industry and other stakeholders appear to view the proposed motor MEPS in a favourable light when it is applied to motors and OEM products within Australia. The majority of suppliers have models that will comply with the standard. End–users will continue to have an adequate range of motors to choose from and the improved performance characteristics of more efficient motors will bring financial benefits.

The Export Market
Motor and OEM products destined for export were viewed somewhat differently. Although many stakeholders philosophically supported OEM export products being included in MEPS, there was also a general concern that this could impinge on the competitiveness of Australian products in overseas markets.

The Rewind Market
It appears that the use of a standard may not be the most effective approach to improving the energy efficiency of the rewind market at this time. An accreditation scheme incorporating training and other assistance to help the industry alter its practices may be more appropriate. This reflects the dispersed nature of the industry and associated difficulties in monitoring compliance, combined with a lack of infrastructure (practices and testing procedures) to support MEPS without affecting competition in the industry.

Appropriateness of the Proposed Efficiency Levels
On balance, establishing the MEPS efficiency ratings at levels that, on average, remove the 40% of the lowest efficiency motors from the market is a reasonable starting point. In general, this approach is supported by the majority of key industry players as a practicable, achievable goal – neither so low as to have no impact, nor so high as to severely limit competition and choice in the marketplace.

3.3 Regulatory Process and Standards Development

Currently, the new motors standard is in draft form. This joint Australian/New Zealand standard consists of two parts: the first part describes the test methods to be used and the second part deals with the MEPS and HEM levels. The manufacturer has the option of two test methods, based on the proposed IEC 61972 and IEC 34-2. Each test method has a corresponding MEPS and HEM value outlined in Part 2 of the standard taking into account the different values obtained by each method. The test methods are currently under debate by the IEC. Australia and New Zealand have decided to finalise their standard and will consider revisions to align with the IEC standard once it has been finalised. The draft standard is open to public consultation at this time and the final version should be published in late 1999.

Once the standard is finalised the next step is to develop regulations to call up the standard. In Australia, each State and Territory is responsible for introducing energy efficiency regulations. The Commonwealth coordinates this process through NAEEEC and ANZMEC. In order to ensure consistency of law, each state will call up the standard which contains all the technical detail necessary for the supplier.

It is expected that market consistency will be assisted by having the requirements for (2 and 4 pole) MEPS and HEMs aligned with the European Eff 2 and Eff 1 curves. Australia and New Zealand will draw the new standard to the attention of energy efficiency regulators in other Asia Pacific economies via APEC energy efficiency fora.

3.4 Australian Motor Systems Challenge (AMSC)

The AGO conducted a series of workshops in May 1999 seeking stakeholder views on the structure of a new government initiative, to be known as the "Australian Motor Systems Challenge" (AMSC). The AMSC will be a voluntary program aimed at encouraging industry to use energy efficient motors and drive components.

The main themes arising from the workshops were the need to:

- Educate and inform motor users by delivering information necessary to understand and capture the benefits of the program;
- Overcome motivational and institutional barriers within organisations by developing tools to allow engineering staff to convince finance officers of the benefits of energy efficiency;
- Address the payback gap with information on financial benefits that energy efficiency can deliver, and consider financial incentives to promote the program.

Following analysis of the feedback of participants at the workshops and of similar overseas programs, the development of an AMSC program has been agreed with industry. The AMSC program will be launched in March/April 2000.

4 Cost/Benefit Analysis

A summary of cost and benefits and the cost–effectiveness of the proposed motor MEPS relative to a BAU scenario is summarised below. It is based on energy and cost saving data derived from the Energetics 1997 report, updated with the administrative costs. The Net present value (NPV) analysis reflects the installation of compliant motors over a 15 year period with benefits continuing for another 15 years for the service life of the motors.

Table 2: Estimated NPV Costs & Benefits of Proposed Motor MEPS.

Discount Rate	New Purchase + Administration Costs $Million			Energy Costs $Million			BCR
	BAU	MEPS	Difference	BAU	MEPS	Difference	
0%	2057	2210	153	91476	90870	606	3.9
5%	1346	1440	96	36992	36742	250	2.6
8%	941	1003	63	17488	17368	120	1.9

5 References

[1] Commonwealth of Australia, "The National Greenhouse Strategy", 1998

[2] Energetics Pty Ltd, "Energy Labelling and Energy Performance Standards: Motors – Stage 1," May 1996

[3] Energetics Pty Ltd, "Energy Labelling and Energy Performance Standards: Motors – Stage 3," September 1997

[4] SRC International Pty Ltd, "Stakeholder Feedback on Electric Motors MEPS", April 1998

[5] SRC International Pty Ltd, "Consultation Paper: Draft Standard for MEPS for Electric Motors", April 1999

[6] Prime Ministerial Statement, "Safeguarding the Future: Australia's Response to Climate Change", November 1997

Results of The Danish "High Efficiency Motors Campaign"

Klaus Malmose

1 Introduction

In Denmark energy consultancy (Demand Side Management activities (DSM)) took its first steps in the middle of the 80'ies.

The energy consultancy was built up around the Danish electric utilities on which the Danish government imposed the implementation of energy savings for funds collected through the electricity bill (approx. 0.0006 DKK/kWh corresponding to approx. 0.001 ECU).

The electric utilities trained energy consultants who specialised in visiting industrial enterprises and counselling on energy consumption, primarily focusing on auxiliary systems like lighting, compressed air, ventilation and heating.

Within these areas large savings could be obtained because each company was focusing on the actual production but also because the cost of energy was a fraction of the total costs.

As energy consultancy gradually became well-known and the industry learned that money could easily be earned just by making a few adjustments or changes, it was more the rule than the exception to focus on energy in connection with the abovementioned auxiliary systems.

It became obvious that energy efficiency is not only about repairing existing plants but also about being energy aware when equipment is purchased.

It is a fact that the largest financial gain is present when new equipment is purchased if focus, as it is very often seen, is not only on the price but rather on the energy consumption of the product in its entire lifetime.

A case in point is High Effiency Motors. Approx. 70 per cent of the electricity consumption of the industry is directly attributable to electric motors. Most of the motors have a high amount of annual operation hours. The pay-back time for a motor with a high conversion efficiency is therefore often only a few months.

Therefore it is astonishing that the sales of High Effiency Motors was close to nil in the first half of the 90'ies.

In the light of this it was decided to initiate a national campaign aimed at the industry.

2 Campaign Strategy

When a campaign is initiated it is very important to have a general view of the players who are going to participate in the campaign. Furthermore, it is important to analyse the market and also to find out which factors come into play when motors are purchased for the industry.

The implicated parties were identified:

- The electric utilities and their consultants
- Suppliers and producers of motors
- End-users - the ones that purchase and are in charge of operation
- Consulting engineers
- OEM suppliers
- The Danish Energy Agency

2.1 The Electric Utilities and their Consultants

The Electric Utilities and their Consultants should to a great extent be the ambassadors of the campaign, especially in the local areas and should influence the end-users of the motors directly.

2.2 Suppliers and Producers of Motors

It would be difficult to complete a campaign without the participation of the suppliers of the motors. The supplier influences the end-users and the OEM-group very much.

2.3 End-user

The campaign was aimed at the end-user - the industry. The end-user was influenced with simple messages.

2.4 Consulting Engineers (Consultant for the End-User)

The Consulting Engineers are often involved when designing industrial plants and are therefore an important group.

2.5 OEM Suppliers

This group supplies products (ventilation, compressed air, pumps etc.) containing motors. A major part of the sale of motors takes place via this group as part of a larger delivery. This is therefore a very important group.

2.6 The Danish Energy Agency

From the beginning, The Danish Energy Agency was a very active participant. The Danish Energy Agency was for instance responsible for the introduction of a subsidy that was organised parallel to the campaign.

After discussions with the above-mentioned groups - interested parties - the strategy of the campaign was determined:

- End-user - the industry - should be motivated to demand High Effiency Motors, primarily by means of obvious financial arguments.
- The motor producers should be "forced" to produce and promote High Effiency Motors primarily on the basis of customer demand.
- The OEM's should be able to see the advantage in advertising with the text "High Effiency Motor inside".
- A public support scheme should make it even more popular to purchase High Effiency Motors.

This strategy is what the advertising business calls a push-pull strategy.

3 Operational Tools

Before initiating the campaign it was important to find out what operational tools should be used in the campaign.

Experiences from other campaigns for instance in connection with the campaign for compact luminous tubes, had shown that a very strong tool could be to promote selected energy efficient equipment and components directly to make a positive list.

Therefore a list of energy efficient motors was made, which turned out to be a very strong tool.

In addition to this list the project group co-operated with the Danish Energy Agency on supporting the campaign by a direct subsidy of motors on the positive list. The subsidy should be simple to calculate and simple to apply for.

4 The Course of the Campaign

In co-operation with the Danish Energy Agency and the motor suppliers criteria for admittance to the positive list was set up.

Together with the Danish Energy Agency a very simple subsidy model was designed: Subsidy = DKK 100 per kW motor power (shaft power) if the motor was on the list.

In co-operation with an advertising agency the creative elements of the campaign was determined. A recurrent theme was chosen and built up around humoristic drawings by a well-known Danish cartoonist and author.

The campaign was officially initiated on 3 September 1996 in connection with an industrial exhibition.

The same day, the first edition of the positive list was sent out to the 4000 largest Danish industrial enterprises together with direct mail. Several full-page advertisements were inserted in national newspapers. A press conference was held and the homepage of the campaign was introduced.

In the months following the initiation of the campaign several meetings and seminars were held by the electric utilities, among others. Articles were printed in the daily press and the trade magazines.

Already in March 1997 a new edition of the positive list was sent out to the 4000 largest industrial enterprises together with new direct mail.

During the almost 3 years in which the campaign ran a total of 6 positive lists were sent out together with direct mail.

Articles with cases from the industrial enterprises that had chosen High Effiency Motors were regularly printed, primarily in the trade magazines.

In periods, a few of the largest motor suppliers took active part in the campaign using it (as was also intended) as a stepping stone to obtaining an even larger market share.

In March this year the positive list was sent out for the last time.
Together with this a special campaign related initiative was taken towards producers and suppliers of compressed air, ventilation and pump systems. The last mail urged the end-users to pay attention to equipment that contains electric motor and to demand High Effiency Motors.
In addition to this a large seminar was held to which the above-mentioned group of suppliers was invited. The purpose was to influence them to use High Effiency Motors in their products and use this in their advertising/marketing strategy.

Consequently, the so-called push-pull was also used in the final phase of the campaign.

The campaign is now over but the positive list still exists and is active on the Internet.

New suppliers regularly apply for admittance on the list.

5 The Result of the Campaign

The objective of the campaign was that during the three-year campaign period at least 25.000 motors out of approx.150.000 motors sold annually should be High Effiency Motors.

This objective should be seen in the light of the sale of High Effiency Motors nearly being nil at the beginning of the campaign, and only very few people were aware of High Effiency Motors.

Besides the concrete and quantifiable objective a large number of objectives dealing with the degree of knowledge about the campaign with the different professional groups and target groups were set up.

In the three-year campaign period approx. 75.000 High Effiency Motors were sold. This is 3 times more than the objective.

An independent analysing agency made analyses of how each major mail and advertising round was received. If the analyses did not show any result the campaign was supposed to change course.

A final analysis of the measures taken in spring this year aimed at the compressed air, ventilation and pump suppliers remains to be carried out.

The result of the analyses of the target group made to date can be summarised as follows:

- Prior to the campaign only 28 per cent had heard about High Effiency Motors. Today 77 per cent can mention a high efficient motor without hesitation.
- Prior to the campaign 18 per cent believed that High Effiency Motors was too expensive. Today only 10 per cent share this opinion.
- Prior to the campaign 18 per cent believed that conversion efficiency is important. Today 26 per cent shares this opinion.
- Prior to the campaign 17 per cent found it of crucial importance to purchase High Effiency Motors. Today 59 per cent do.

Moreover, we know from interviews with several large Danish industrial enterprises that some of them have chosen High Effiency Motors as part of their procurement policy.

Several motor producers have aimed at producing High Effiency Motors. A very large Danish motor producer chose early on to aim at High Effiency Motors exclusively and therefore changed the entire production.

The positive list has gradually become more extensive. Today, it contains both 2, 4 and 6 pole motors. 12 suppliers are represented and approx. 500 motors are on the list.

The list is now on the Internet and is continusly updated.

6 Campaign Barriers

Especially in the beginning of the campaign many of the suppliers were sceptical. This scepticism was primarily aimed at the suppliers themselves. There were accusations about exaggerations of the conversion efficiency in order to ensure a place on the positive list. This self-regulation has naturally also had a positive effect.

From the beginning of the campaign the stage was not set for a control of the conversion efficiency. When early in the campaign a type test certificate was demanded from the motor factory, a few suppliers chose to withdraw their motors from the list.

The Danish Energy Agency has just granted money for making random tests on the motors on the list i.e. an actual measuring of the conversion efficiency carried out in a test bench. It will be interesting to see how the suppliers react to this.

Producers not accepted on the list spent a lot of energy trying to find faults and irregularities in the campaign. However, concurrently with the increasing demand for High Efficiency Motors the criticism almost died away and instead the producers tried to improve their own motors in order to be accepted on the list.
Part of the general criticism (always expressed by the technicians) was that the messages used in the campaign were too ordinary. Saving energy takes a lot more than just using High Efficiency Motors. This is of course true and the campaign management was aware of the fact that all other aspects concerning operation of motors were and behaviour not being the least important of these.

7 Finance, Energy and Environment

Totally, the campaign has cost approx. DKK 7 million (approx. EURO 1 million). This means that the campaign is among the by far most cost efficient measures when comparing the energy savings to the investment.

As earlier mentioned, 75,000 High Efficiency Motors had been sold to the industry by the end of the campaign.

These motors give annual electricity savings of minimum 20,000,000 kWh or 20GWh. The total electricity consumption of motors in the Danish industry is approx. 6,500 GWh.

Based on the present annual electricity production in Denmark the electricity savings achieved are equal to a cut in the Co2 emissions of around 15,000 tons per year.

8 Closing Remarks

The campaign has been a total and unconditional success.

The tools used; the positive list and the subsidy have shown their effectiveness. Moreover, the market forces have also had a major influence on the success.

The recipe for success is:

- Base your campaign on a very simple message
- even if this means sacrificing some technical accuracy

Energy savings in the industry are complex and call for qualified counselling. The first measures should often be taken by the top management or the people financially responsible. Especially with a campaign like this, this group becomes aware that it is worth focusing on the energy consumption. In other words: you might as well pick up the gold that lies right under your nose.

China – Motors: Opportunities and Activities

Denise E. Knight

International Institute for Energy Conservation

1 Introduction

Annual electricity consumption in China in 1997 was approximately 1,100 billion kWh. Motors use over 600 billion kWh per year, representing the single largest electricity-consuming segment in China. Small motors, ranging from .55 kW to 200 kW, account for about 90% of motor energy losses and 70% of annual demand. With an average efficiency of only 89.3%, these motors represent an enormous opportunity for energy savings. It is estimated that, within 10 years, over 34 billion kWh per year could be saved if a minimal efficiency standard was set equivalent to the US EPACT efficiency standards.
Despite wide support for improved motor efficiency from end-users and motor manufacturers, barriers exist which limit the penetration of energy efficient motors. The China government recognizes the energy and environmental benefits of increased motor and motor system efficiency and have identified these areas as energy efficient priorities. Several initiatives are underway to determine the most appropriate course of action to reduce motor electricity losses and to promote the use of energy efficient motors.

2 Market Overview

Motors account for over 70% of industrial electricity consumption and 50% of total electricity consumption in China. The two primary motor categories in China are large/medium motors (> 220 kW) and small motors (.55 kW – 200 kW). Fractional horsepower motors ($\leq$ 1.1 kW) and mini-motors are also substantial in number, but offer less energy and economic benefits due to the light and/or intermittent loading of these smaller motors. The large/medium motor and small motor segments can be further classified as AC (alternating current) and DC (direct current) motors. AC motors account for over 70% of annual production and about 90% of energy losses from new installations. Small motors offer the greatest opportunity for increased efficiency and are the focus of initiatives to capture these savings.

Analyzing the small motor production by size, the weighted energy efficiency of small motors is 89.3%, much lower than the efficiency of large motors (~96%). The greatest opportunity for energy efficiency improvement is the small AC motor segment between 3 kW and 90 kW. In 1997, 28.6 thousand MW of small AC motors were produced and the average annual growth rate has been 3.2% over the past 8 years.

3 Technology Status

There are six types of motors installed or produced in China. The oldest is the J-series motor. In 1984, government policies forbid the use and purchase of these motors. However, it is estimated that about 120,000 MW remain in service, accounting for about 30% of the installed base.

The Y-series motor is the dominant motor with about 98% share of the domestic market, and is considered the "standard". It was locally developed in the early 80's, and accounts for about 70% of total installment. The YX is an energy efficient model of the Y motor. To date, this motor has been unsuccessful in penetrating the market and is not readily available for purchase.

In the early 90's, the Y2 model was developed as an upgrade of the Y-series. It was also locally developed and designed to improve noise and vibration. The Y2 series has a much lower efficiency, lower than the Y-series motor. The Y2E motor is an energy efficient model of the Y2 motor, but efficiency just meets the Y model.

In addition to the motors used in the domestic market, China produces motors that meet NEMA standards, tested under the IEEE methodology. These motors are for export only and are not available in the domestic market.

The YX motor has the highest efficiency, 92% on average, of the domestic motors. When comparing the efficiency of three representative motor sizes: 2.2kW, 18.5 kW, and 37.5 kW, the YX motor is very close to the US EPACT standard motor for efficiency. China has the technology and capability to produce energy efficient motors, but these motors are not being used for domestic applications.

There are two motor standards in China. GB755 specifies motor performance, efficiency, tolerances, and technical requirements. GB 1032-85 defines the appropriate test methodologies. Two test methodologies have been specified for domestic motors. The Y and Yx series motors are tested similar to the IEEE 112E test methodology, and require measurement of miscellaneous or stray losses. The Y2 and Y2e series are generally tested to the IEC methodology and assume the same percentages for stray losses.

The Shanghai Electrical Appliance Research Institute is the government certified testing laboratory and is the only laboratory in China with equipment and certification to test to the IEEE methodology. SEARI has indicated that motors produced by large manufacturers and tested for stray losses have an actual efficiency of 1-2% lower than the estimated efficiency. Motors produced by smaller companies have actual efficiencies of 3-4% lower than estimated. In general, however, there is very limited on-going motor testing and verification.

4 Opportunity

Market transformation of all new motors produced from the current standard, Y-series, to the energy efficient, YX-series, would result in about 34B kWh of savings per year, within 10 years. Although market transformation will not occur immediately, the energy saving potential is enormous and compounds each year as more motors are installed.

At current annual production rates and existing Y series motor efficiency levels, within 10 years the losses from motors will be about 77 million kW. However, 10 million kW could be saved if, instead, energy efficient Yx motors were installed. Further, an additional 5 million kW could be saved, if Yx motors replaced existing J-series motors. At typical motor loading of 3000 hours at 75% load, 34 billion kWh could be saved. This is equivalent to about 17B RMB or US $2.0B per year and approximately 35 million metric tons reduction of CO_2 emissions. The financial and environmental benefits are substantial.

5 Economics

Energy efficient motors are cost effective on the basis of total owning cost. Total ownership costs (or lifecycle costs) included the motor price and the annual energy operating costs. The total ownership costs of three China motor series Y, Y2, and Yx motors were calculated based on typical industry operational parameters and energy cost factors, as follows:

- annual cost of capacity of $26/kW;
- average industrial energy cost of $0.06/kWh;
- typical loading rate of 75%; and,
- 3000 hours of annual hours of operation.

Energy costs represent more than 80% of the motor's total costs. While the first-cost or price of the Yx motor is between 15% and 25% higher than the standard Y motor, the total operating costs are substantially lower.

However, most industries evaluate various options based on the time required to payback the investment. An energy efficient 55 kW motor would payback a 25% price premium investment in less than 13 months, using the same parameters. Payback times vary with hours of operation and loading, but in most cases the energy efficient motor pays back the added investment in less than 2 years, and

then continues to provide energy and economic benefits for the remaining years of operation.
A similar analysis, evaluating the cost effectiveness of replacing a failed motor with a new energy efficient motor, indicates that the investment would payback in about 1.5 years.

6 Market Dynamics

The overall economic growth is slowing and GDP has declined from a peak growth rate of 14% in 1992, to 7.8% in 1998. This has had serious repercussions for industry and motor manufacturers. The motor industry is experiencing slow market growth and over 40% of all manufacturers lost money in 1998. Industrial customers, in an effort to improve their poor financial performance and lacking the resources to pay the price premiums, are purchasing the cheapest equipment that performs the specified function. Therefore, original equipment manufacturers, responding to the needs of their industrial clients, are reducing equipment costs and energy efficiency is not a consideration.
Compounding the industrial financial problems, which are not favorable for energy efficient motors, is a general lack of awareness of energy efficient motors and their benefits. The top three purchase criteria for motors are reliability, price and quality. The potential for motor efficiency and its economic justification, are rarely considered. Industrial leaders, those that recognize the value of efficiency, find it difficult to find energy efficient motors that meet their requirements. Demand is not sufficient to justify manufacturer investment in energy efficient motor production. Those energy efficient motors that are produced for the domestic market, are custom designed and manufactured and have very high prices. The industrial motor *purchaser* is generally not the industrial motor *user*, especially in the large state-owned enterprises, and their objectives and interests differ significantly. Thus, even if the *user* requests an energy efficient motor, the *purchaser* will make the purchase decision based primarily on price.
Most motors for pumps, blowers, and compressors are purchased by original equipment manufacturers (OEMs). Replacement motors and experienced users, will purchase motors directly from the manufacturer. About 50% of the sales are direct end-user purchases and 50% are through OEMs.

7 Competition

Motor manufacturers can be classified into two groups, small enterprises and large state-owned or joint venture companies. Small enterprises are generally, family or county-owned enterprises. They are characterized by inexpensive labor, low land costs, low social security costs, and tax “exemption”. The motors produced by

these manufacturers are low priced and often poor quality. Their competitive advantage is price.
The large state-owned manufacturers have higher costs, but better quality. Many state-owned manufacturers are losing money due to the economic downturn, continued price pressure from the small manufacturers, and mismanagement. The joint venture companies also have higher costs, but substantially better quality products. Joint venture companies and the state-owned enterprises target exports, which are viewed as a more lucrative market. Internationally recognized manufacturers brand Chinese manufactured motors, and the Chinese manufacturer is unknown in the international markets.

8 Relevant In-Country Initiatives

The recognized potential for motor efficiency has resulted in numerous activities to capture the energy, environmental, and economic benefits. In country, both the State Economic and Trade Commission and the State Development and Planning Commission have announced support for energy efficient motors and motor systems. The State Bureau of Quality and Technology Standards (SBQTS) has initiated activities to establish energy efficient motor standards. The Standards and Codes Institute is collecting market information to review the potential for energy efficient motor standards and product certification. And financial institutions, such as the China Energy Conservation and Investment Corporation (CECIC) are positioned to support energy efficient motor leasing and investment opportunities. In addition to these government directed initiatives, there are a number of initiatives supported by the international community that are helping to advance the energy efficient motors programs. The China Electric Power Research Institute is currently funded by GTZ, the German aid organization, to establish motor efficiency standards and a motor testing laboratory. The US Department of Energy, in collaboration with the China government, is implementing the Motor Challenge program, which is focused on motor systems efficiency. Energy Management Centers (EMCs or ESCOs), funded by the World Bank, are still in the early development stage, but are well positioned to implement motor efficiency programs.

9 Strategic Options

The China Energy Efficiency Program (CEEP), funded by the International Copper Association and the Common Fund for Commodities, and implemented by IIEC, is focused on harnessing the in-country efforts and is acting as a catalyst for action. A "Working Team" is being formed of end-users, motor manufacturers,

and government representatives to define energy efficient motors for China, establish marketing and promotional programs and to develop the market.
CEEP is helping to drive an energy efficient motor transformation strategy by initiating actions to address the market barriers. The CEEP strategic goals are to:

- Develop a premium quality motor brand, incorporating efficiency, performance standards, customer service, and product certification and verification. The brand will link quality, efficiency, and environmental benefits;
- Establish a motor component brand to support OEMs with energy-efficient motors;
- Identify financing support for the manufacturing and purchase of energy efficient motors;
- Initiate government policy support and standards; and
- Educate customers on energy efficiency and its benefits to their businesses.

10 Summary

In summary, an enormous opportunity exists for energy efficient motors in China. The capability to produce quality, reliable, efficient motors exists in country and there is high level support and interest for efficient motors and motor systems. A number of in-country initiatives are targeting motor efficiency. CEEP is pulling together these initiatives and helping to drive market transformation.

Proposed Principles of the Future Standard for High Efficiency Induction Motors

Tadeusz Sliwinski

Instytut Elektrotechniki, Warszawa, Poland

Abstract. In the first part of the paper the dependence of the electromagnetic losses of induction motors on the iron-core volume (mass) is presented. Also the influence of the starting parameters on losses is analyzed. In the second part proposals are put forward to the principles of rating high efficiency motors, to the allocation of outputs to frame numbers and to the minimum rated efficiency in the future international standard.

Keywords. Induction motors; energy losses; optimization: standardization of electrical machines.

1 Introduction

Since the invention of induction motors the progress in their production was directed towards minimizing their dimensions, mass and price. This trend was limited by requirements of international standards respective the temperature rise of winding and the starting parameters [7]. Succesive introducing of new classes of insulation (F and H) with higher limits of temperature rise favoured this course of progress. Efficiency was regarded as being of minor importance and until now no restrictions regarding the minimum efficiency values have been fixed in international standards. Also the endeavours made in IEC (SC2B) in the years 1970 - 1980 to establish a world standard on allocation of outputs to motor frames failed. The reason was that different allocations existed before in different countries or group of countries for many years and these traditions could not be overcome.

About 25 years ago from well known reasons a new trend emerged: to bring down the energy losses in the motor. Many manufacturers in different countries started with production of induction motors with efficiency higher than in the case of standard motors, but the efficiency level of these motors is now very different what causes misunderstandings on the market and confusion of the motor users. Having in mind the bad experience with creating an international standard on the allocation of outputs to motor frames which was started to late I am of the opinion that now is the last moment to establish (preferably by IEC) a world-wide standard for high efficiency induction motors. This standard shall contain principles of rating, methods of measuring efficiency, the minimum values of efficiency and if possible the allocation of rated outputs to motor frames.

Generally the motor efficiency may be rised by following means:

1. Applying better materials.
2. Refining the manufacturing methods.
3. Enlarging the iron-core volume in order to reduce the density of energy converted in the motor.
4. Optimizing the shape of the iron-core, all the sheet dimensions and the winding data.
5. Moderating the standardized limits of starting parameters (i.e. reducing the minimum locked rotor torque and rising the locked-rotor apparent power) in relation to required in [7].
6. Reducing the mechanical losses.

The paper deals mainly with problems named in positions 3,4 and 5 as the remaining problems were already treated in many other papers [3, 4, 5, 6]. Here the motors of different shape and dimensions but made from the same materials and using the same manufacturing technology are analyzed. Also the mechanical losses are regarded as being constant at the same output. On the basis of performed analysis proposals concerning the future international standard for high efficiency motors (h.e.m.) are presented.

2 Method of Analysis and Economic Conditions

As the basis for evaluating and comparing different motor solutions from the users point of view the total cost K will be used being the sum of the motor price K_m and the cost K_e of energy losses in the motor during the amortization period of n years. The cost K_e of energy losses in one year is calculated assuming that the motor operates h hours per year with load factor p and with efficiency η_p at this load.

$$K_a = \frac{ph(1-\eta_p)}{\eta_p} P_N c_e$$

P_N is the rated output in kW and c_e the mean price of 1 kWh of energy.
In the case of h.e.m. which may be overloaded without any harm it is reasonable to assume that the mean load factor is p=1, i.e. that the motor operates at rated output. The cost of energy losses K_k in a succesive k year of operation is reduced to the motor installation year taking into account the discount rate s [%]

$$K_k = \frac{K_a}{q^k}$$

with

$$q = 1 + \frac{s}{100}$$

The sum of costs of energy losses in the whole amortization period of n years is

$$K_e = \sum_{k=1}^{k=n} K_k = \frac{K_a}{d}$$

where the amortization rate d is

$$d = \frac{q(q-1)}{q^n - 1}$$

With the assumed values $s = 8\%$ and $n = 5$ years is $d = 0{,}24$.

The cost of reactive energy may be neglected as the power factor of h.e.m. is high. On the stage of designing the motor price is replaced by the cost of main materials: copper wire for stator winding, iron sheets, aluminium for cage and cast iron or aluminium alloy for the frame.

The study was performed using the authors program OSIN-3 for optimal synthesis of induction motors [9]. The objective function in this program is the total cost K which is minimized. In order to receive results independent on local currencies relations of the prices of 1 kg of separate materials to the price of 1 kWh are used. For this study following price relations have been assumed: copper wire - 40, iron sheets - 8,2, aluminium for cage 25,2 and cast iron frame - 12. Constraints in the optimization procedure are:

- maximum temperature rise of the stator winding at rated load
- minimum locked-rotor torque
- maximum locked-rotor apparent power
- maximum jump of temperature rise of stator winding at one start from hot conditions as required by IEC 34-12 [7].

The limiting values of starting parameters have been assumed with a 10% safety margin: locked-rotor torque by 10% greater and locked-rotor apparent power by 10% less than required in 34-12 [7]. As maximum allowed jump of temperature rise during starting 20 K was assumed.

At the given output, the assumed constraints, price relations and operating conditions the program OSIN-3 evaluates all important dimensions and winding data at which the objective function reaches its minimum. The whole calculation takes about 2 minutes.

3 Dependence of Losses on Iron-Core Volume

When designing motors with provided use of definite kinds of materials and manufacturing methods it is very useful to know the dependence of loss on iron-core volume. To analyse this dependence calculations were performed at different numbers *h* of operation hours per year. All motors dimensions were free, no limitations e.g. to the iron-core diameters have been assumed. At each *h* value an optimal solution has been received with a definite iron-core volume and efficiency. In the calculations the same materials and the same manufacturing methods as for the actually produced standard motors have been assumed. The winding losses in stator and rotor have been evaluated at the calculated temperature rise and the additional load losses assumed to be 0,5% of the power input. The results of the calculations for 3 four-pole totally enclosed motors of 0,75 kW, 7,5 kW and 75 kW are presented on graphs (Figure 1, 2, 3). The losses P_{em} are so called „electromagnetic losses" i.e. the sum of all losses but the mechanical losses which do not depend directly on the iron-core volume and may be treated separately. As iron-core volume or more convenient iron-core mass is understood here the volume (or mass) of a cylinder with the outer diameter of the stack and with its length. On the curves the points are marked with the respective numbers of operation hours per year.

The actually manufactured standard motors are designed availing the full admitted temperature rise and with their rather small iron-volume and high losses correspond with the points on the graph in the interval of $h = 100 \div 200$ in the case of small motors and in the interval of $h = 200 \div 500$ in the case of greater motors. It means that these are the optimal operation conditions for the considered motors. When enlarging the iron-core mass above the mass of standard motors a rather great reduction of electromagnetic losses is received. Anyhow further increase of iron-core volume brings small gain on losses and it seems to be unjustified to go beyond of 2000÷4000 hours.

The curves on Figure 1, 2 and 3 were received taking into account the cost of main materials only. Additionally on Figure 4 the results of designing the same 7,5 kW motor as on Figure 2 are presented but taking into account the total manufacturing costs. The curves on Figure 2 and 4 are similar, only the points showing the respective numbers of hours are shifted in the second case towards smaller mass.

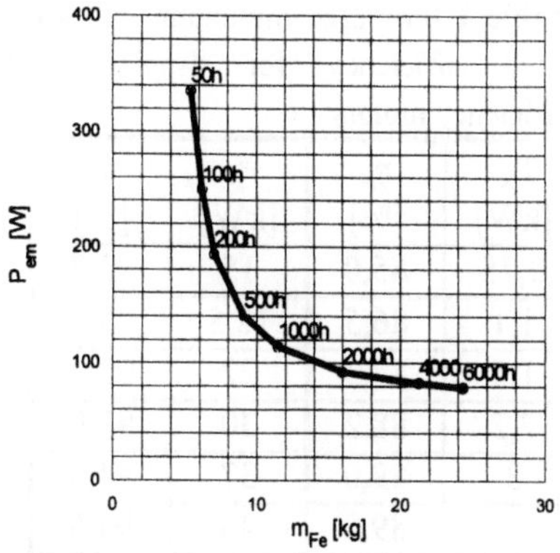

Figure 1: Dependence of electromagnetic losses in the optimal designed 4-pole 0,75 kW induction motors on the mass of iron-core (main material costs taken into account).

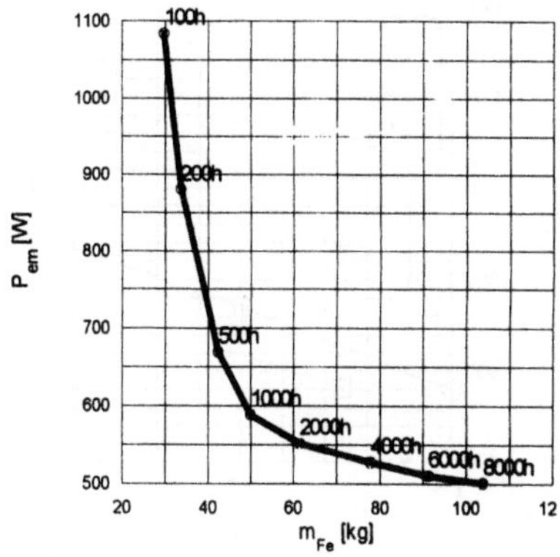

Figure 2: Dependence of electromagnetic losses in the optimal designed 4-pole 7,5 kW induction motors on the mass of iron-core (main material costs taken into account).

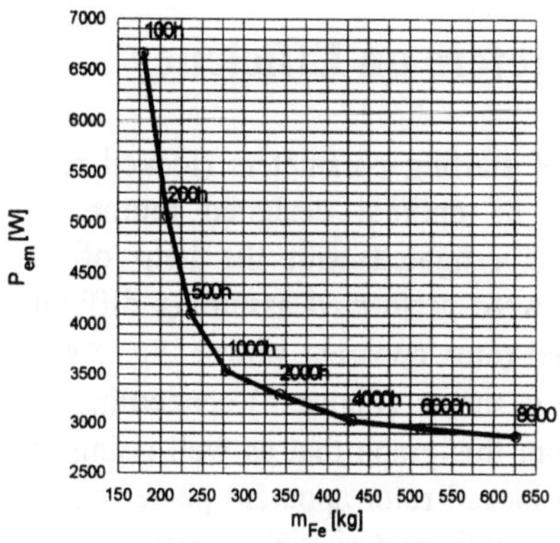

Figure 3: Dependence of electromagnetic losses in the optimal designed 4-pole 75 kW induction motors on the mass of iron-core (main material costs taken into account).

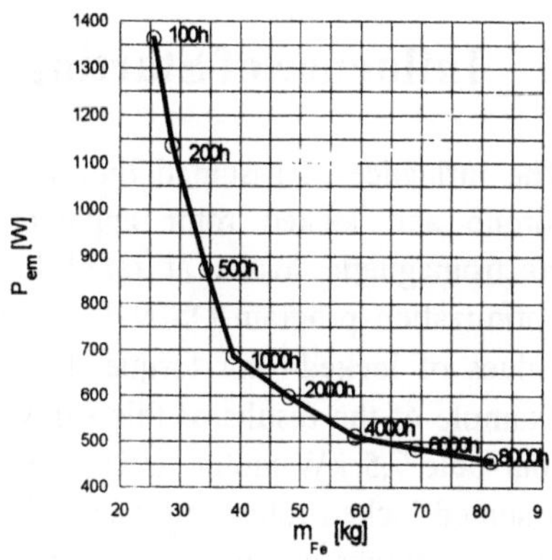

Figure 4: Dependence of electromagnetic losses in the optimal designed 4-pole 7,5 kW induction motors on the mass of iron-core (totall manufacturing costs taken into account).

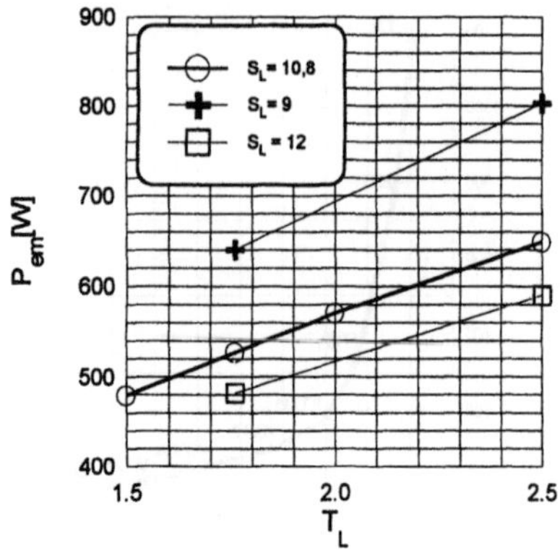

Figure 5: Influence of starting parameters on electromagnetic losses in 4-pole 7,5 kW high efficiency motor.

Table 1: Proposed minimum values of rated efficiency for 4-pole high efficiency motors

P_N [kW]	η_N [%]	P_N [kW]	η_N [%]
0,75	85,0	15	92,7
1,1	86,5	18,5	93,0
1,5	87,3	22	93,4
2,2	88,2	30	93,9
3	89,0	37	94,3
4	89,8	45	94,5
5,5	90,4	55	94,7
7,5	91,3	75	95,0
11	92,2	90	95.1

4 Influence of Starting Parameters on Efficiency

The influence of assumed limiting values of starting parameters (locked rotor torque and locked rotor apparent power) of optimal designed motors on electromagnetic losses at rated load was investigated with the help of the optimization program OSIN-3. Calculations were performed assuming different values of locked-rotor torque T_L and locked-rotor apparent power S_L. As an example of the results of this study a graph is shown (Figure 5) concerning the four-pole 7,5 kW motor. Is is to be seen from this graph that as well rising the assumed locked rotor torque as reducing the locked rotor apparent power causes significant increase of electromagnetic losses in optimal designed motors.
This conclusion is valid also for other cases whenever the limit of one of the starting parameters appears to be active in course of the optimization procedure.

5 Proposals for Future Standard

Until now the increase of efficiency was attained mainly by applying better materials, refining manufacturing methods, correcting the inner dimensions of standard motors and eventually by raising a few the stack length. All this changes were introduced preserving the existing allocation of outputs to motor frames. Such a way [3, 4, 5, 6] is progressive and shall be appreciated very high from technological point of view, but it leads to significant rise of cost and possibilities to continue this course are limited. Another way of rising efficiency is enlarging of the iron-core volume without significant change in used materials and manufacturing processses. This way is not so attractive as the first one but offers also great possibilities to reduce losses what was demonstrated before. It causes also a rise of cost but in certain cases this way may be less expensive as the first one.

In both cases the user must be prepared to pay much more for a h.e.m. than for a standard motor of the same output and it is the main psychological barrier in propagation of h.e.m. It is proposed to overpass this barrier by other rating these motors. On account of considerable losses reduction the temperature rise of h.e.m. is much lower than of standard motors. So it is possible to overload them without any harm. Of course the motor will operate at overload with a little lower efficiency than at rated output but anyhow this efficiency may be not worse than of a standard motor. The overloading capacity of h.e.m.'s is a great advantage when comparing them with standard motors. In the case of standard motors continuous overload is unallowed as these motors are designed normally with the temperature rise on the edge of the limit. The purchaser comparing the price of a h.e.m. with the price of a standard motor is unaware of this advantage.

Calculations and tests show that it is possible to overload the h.e.m. rather high even to the next output in the output range without overpassing the limit of the temperature rise. Also the breakdown torque in relation to the torque at overload remains over the limiting value given in IEC 34-12.

It is proposed to put on the rating plate of h.e.m. two values of output: the rated output at which the efficiency shall be greater than the minimum value fixed in the standard and the value of allowed overload, preferably the next value in the output range. Such a method of rating will suggest the purchaser that a new h.e.m. may replace his old motor with a greater rated output and he will compare the prices of both. Perhaps such a rating system will brake the psychological barrier of the users, promote the h.e.m. and bring greater energy savings.

Additionally to make easier the replacement of standard motors by h.e.m.'s of lower rated output I suggest to apply for them identical motor frames. This will give the possibility to have a larger iron-core and as it was demonstrated in the first part of the paper it will enable to gain additional energy savings.

In such a version the allocation of outputs to motor frames for h.e.m. would be by one step lower than in the actual CENELEC standard [8]. Minimum values of rated efficiency which may be attained in the proposed formulation are presented in table 1. These values are profitable at the defined economic conditions if the h.e.m. operates with rated output 2000 hours or more per year. According to [1] it is often the case.

6 Conclusions

1. The dependence of electromagnetic losses on iron-core volume is helpful in designing h.e.m. and shows how far it is possible and reasonable to reduce the losses by enlarging the iron-core volume depending on the motor operating time.
2. The assumed limit values of locked-rotor torque and locked rotor apparent power influence the losses at rated load. It is suggested to put in the standard for h.e.m. the limit values of locked-rotor torque smaller by 10% and the locked rotor apparent power greater by 10% than the values fixed for standard motors [7].
3. It is an urgent need to establish an international standard (by IEC/TC2 or CENELEC/TC2) for high efficiency motors. This standard shall contain the method for measuring efficiency, the principles of rating, minimum values of efficiency and allocation of outputs to motor frames.
4. It is proposed to put on the rating plate the allowed overload output beside the rated value.
5. Other proposed principles of the future international standard for high efficiency motors are following:
   - the allowed overload output of h.e.m. shall be by one step of the output range [8] higher than the rated output
   - the allocations of outputs to motor frames for h.e.m. shall be by one step lower than for standard motors [8] i.e. the overload output of a h.e.m. and the output of a standard motor shall be identical in the same frame.
   - the minimum efficiency for h.e.m. shall be on a high level (see table 1) availing both better materials and manufacturing methods as larger iron-core volume
6. The proposed solutions in the future standard for h.e.m. may suggest the users to replace his standard motor by a h.e.m. with the same frame having lower rated output but the same maximum output. The price difference between both motors may appear to him more favourable than now between the motors of the same rated output. Such a solution may promote the h.e.m. and favour greater energy savings.

Acknowledgement

The author is grateful to the Polish Committee of Scientific Research (KBN) for granting the project Nr 8T 10A 031 13.

7 References

[1] de Almeida A.T., Fonseca P.(1997): Characterisation of the Electricity Use in European Union and the Savings Potential in 2010. In: de Almeida, Bertoldi P. , Leonhard W. (Ed.): Energy Efficiency Improvements in Electric Motors and Drives. Springer, Berlin, Heidelberg, New York, 19-36

[2] Rosenberg M. (1997): The United States Motor Systems Baseline: Inventory and Trends. ibid., 37-60

[3] Walters D.G. (1997): The Whole Life Efficiency of Electric Motors -UK Developments. ibid., 81-94

[4] Kline D.G. (1997): Energy Efficient Motors. ibid., 95-101

[5] Sjöberg S. (1997): Motor Developments and Energy Efficiency. ibid.,102-112

[6] Watanabe T. (1997): Efficiency Improvements in Induction Motors. ibid.,113-115

[7] IEC 34-12 (1980): Rotating Electrical Machines. Part 12: Starting Performance of Single-speed Three-phase Cage Induction Motors for Voltages up to and Including 690 V.

[8] CENELEC HD 231 S1:1974: Dimensions of Three-phase Induction Motors.

[9] Sliwinski T. (1992): Optimal Synthesis of Induction Motors (in Polish). Zeszyty Naukowe Politechniki Pozna skiej, Nr.40, 31-43.

Brazilian Minimum Efficiency Level for Industrial Motors: A Successful Low Cost Program

George Alves Soares[1], Ronaldo de Paula Tabosa[2], Alexandre Novgorodcev[3]

[1] Centro de Pesquisas de Energia Elétrica - CEPEL, Rio de Janeiro, Brazil
[2] Centrais Elétricas Brasileiras – ELETROBRAS, Rio de Janeiro, Brazil
[3] Intituto Brasileiro de Metrologia – INMETRO, Brasilia, Brazil

Abstract. This paper presents the process and the developed actions that made the atmosphere fertile to the Protocol Signature for establishing the minimum efficiency level for three-phase induction motor. The assumptions on which our strategy was based are showed. The set-up of all aspects led to establishment of two minimum efficiency levels: one for standard motors and other for high efficiency motors. The tables are compared with EPACT 92 and CEMEP tables. The present motor efficiencies of the main Brazilian motor manufacturers are compared with the CEMEP tables. The final comments are presented showing our concerns about CEMEP tables. This strategy has been proved to be successful and low cost. It would be well adequate to the developing countries or part of it could be adjusted for developed countries.

Keywords: Induction Motor efficiency, motor standards, energy conservation

1 Introduction

The most important electrical energy consumer equipment is the electrical motor, indeed electrical energy transducer would be more appropriated. It is responsible for about 50% of the industrial consumption. This great importance of motors in the energetic matrix is a common characteristic of developing and developed countries. Every energy efficiency program must find its way to increase the penetration of high efficiency motors in its market or to increase the efficiency of standard motors. Each country chooses different ways to implement a program to reach these targets.

Brazil as a developing country and owner of a big motor market would require large financial investments to implement rebate strategies to make our market transformation. This paper shows our strategy started on technical efforts addressed to solve or minimize the technical problems related to motor efficiency improvement and it has been transformed in effective market action. The comparison between CEMEP proposal and Brazilian Agreement (BA) for minimum efficiency levels is carried out and the present efficiencies of Brazilian induction motor are discussed.

2 Brazilian Motor Market

The tables 1 and 2 compare the Brazilian market with the European Union and North America ones. They lead us to conclude that Brazil is among the nine biggest of the world, probably the sixth. There are some big original equipment manufacturers (OEMs) in Brazil that are responsible for about 60% of the motor market. This can justify why we are in better position in the motor market rank than we are in the gross national product.

Table 1: Brazil and European Union AC motor market (monetary value) [1].

Country	Market (MECUs)
Germany	677
Brazil	*370*
France	263
United Kingdom	244
Italy	244

Table 2: Brazil and North America average sale of three-phase induction motor [2]

Country	Quantity per year
USA	1,500,000
Brazil	*800,000*
Canada	145,000

3 Brazilian Program on Electric Motors and Drives

3.1 Historical Aspects

In the middle of 80's, two events were significant to change the Brazilian reality: the creation of the Brazilian National Energy Conservation Program (PROCEL) and the creation of the Brazilian Labeling Program (PBE).

PBE was a launched on October 1984 by a protocol signed by the Mines and Energy Ministry, Industry and Trade Ministry and Electric and Electronic Manufacturers Association. National Metrology Institute (INMETRO) is responsible for this program. Its main objective is to inform certified values of the energetic efficiency and energy consumption of the most important equipment to the users. It is a complex program which involves many agents like manufacturers, research institutes, independent laboratories among others. The final results are improvements in the equipment performance and safety. The methodology developed was based on the success of the National Program of the Ethanol as an alternative fuel for vehicles (1975) which induced the mass production of this type of vehicles in four years.

PROCEL was launched on December 1984 by a presidential decree as a joint initiative of the Mines and Energy Ministry and Industry and Trade Ministry. Brazilian Power Sector Holding Company (ELETROBRAS) coordinates it. Its main mission is to decrease electrical energy wastes inside and outside Brazilian power system. Several actions are promoted in all energy conservation related areas. Its present annual budget is about US$ 100 millions. The technological arm of ELETROBRAS is the Electric Power Research Center – CEPEL created in 1974, whose activities are addressed to the continuous improvement of energy services by means of research, energy services and technology transfer. CEPEL is in charge to develop the major technical actions.

The PROCEL electric motor program started in 1987. The first set of projects was done to identify the market and its problems. This stage lasts until 1992. These projects evaluated several aspects as market, motor utilization, standards, technology, motor drives and other related ones. The figure 1 summarizes the evaluated areas.

The results of this set of projects led to the technological planning on December 1991 aiming to minimize or to solve the found problems. In this planning, a project package was elaborated taking account also the experiences of several specialists. Figure 2 shows this package organized in four main areas; motor and ASD performance, educational products development, case study demonstrations and repair analysis.

Other relevant aspect to point out in this section is Brazilian motor standards evolution along the time.

Before 1980, our motors standards were based on the NEMA ones strongly influenced by General Electric activities in Brazil. At the end of 70's started a big movement towards IEC standards, the Brazilian manufacturers wanted to change because IEC standards allowed higher starting currents which can lead to cheaper motors. As Brazilian electric power system is hydraulically based, no one was concerned about energy conservation at that time. In 1981, induction motor specification standard adopted N and H categories, in 1982, IEC test procedures for efficiency were introduced in our test standards. The adoption of IEC frame occurred in 1983 and the Brazilian manufacturers developed a unique standard establishing the correspondence between rated power and frame in 1984. This strategy for producing cheaper motors was successful and the Brazilian manufacturers became domestic market dominants for industrial motor until 200 Hp.

Figure 1: Areas covered for the Evaluation and Analysis Phase.

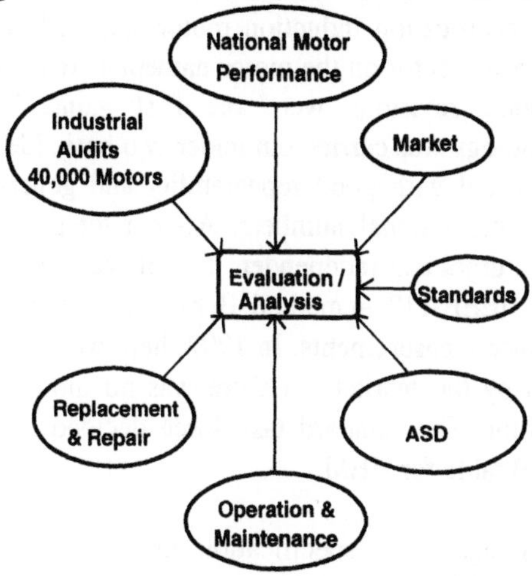

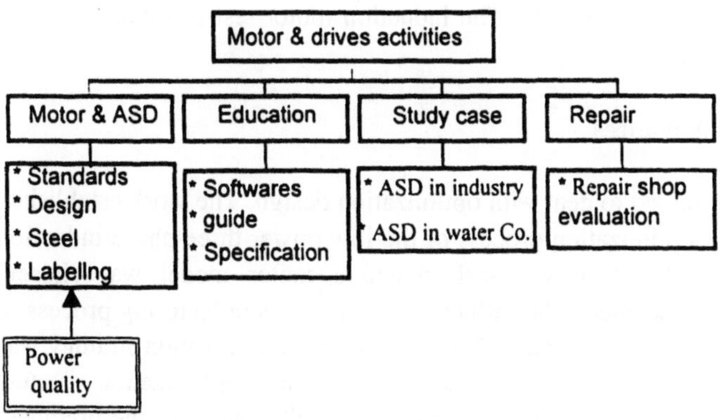

Figure 2: Motor and Drive project package (1992-1998)

3.2 Main Advances Coming from the Projects

The first important task was the creation of a working group composed by manufacturers, INMETRO, PROCEL and CEPEL. It is part of PBE methodology and its actions influence greatly PROCEL projects. This group made all other activities more effective and quicker. Some costs were shared, manufacturers produced a lot of prototypes to the technological projects like optimization design and steel evaluation. The total budget of the projects was US$ 3-4 millions until

1998. The following sections summarize the main advances occurred in the last five years.

3.2.1 Standardization Advances

The first activity was to revise the specification induction motor standard. The inscrtion of the rated efficiency and power factor on the motor nameplate data and efficiency and power factor tolerance revisions were the first gains. The development of efficiency test methodology was carried out under two basic ideas, the methodology should be confident and with good repeatability and generate results that could be compared with international numbers. After a lot of tests using different procedures in manufacturers and independent labs it was decided to follow basically CSA M 390 and IEEE-112 – method B but improving the temperature determinations by resistance measurements. In 1992 there was a high efficiency motor (HEM) line offered by the market but there was no minimum efficiency values for this type of motor. The standard task force decided to be suitable to adopt the 1987 NEMA 12 B table for HEM.

Now the working group is revising again the specification induction motor standard in order to update with IEC 34-1 10th edition and to go further by inserting minimum efficiency table for standard motors and making an upgrade in the HEM minimum efficiency table. The induction motor test procedures is also under revision.

3.2.2 Technical Advances

One project was addressed to deal with optimization design. The work established and implemented an optimization process of the low power three-phase induction motor efficiency in Brazil. A classical induction motor model was chosen, validated by test and discussed. The influences of motor manufacturing process in the motor efficiency were examined. A feasible low cost optimization strategy was developed. Manufacturing motors specified by the optimization results consolidated the whole process. The experimental results validated the predicted ones. Besides this, the optimization opportunities and standards modifications were suggested taking into account the characteristics of the Brazilian motor market.

The first studies identified high no-load losses in the Brazilian motor. Investigations were addressed to evaluate national carbon and silicon steel performances, the steel thermal treatment done by the manufacturers and the standards. The project provided assistance for better thermal treatment, suggested steel standards revisions, for instance, higher permeability level, and concluded that our carbon and silicon steel were similar to the international ones. It also stressed the need for making feasible the mass production of electrical steel that could present better cost-benefit ratio than the silicon steel in some applications.

Large efforts were directed to improve laboratory facilities and instrumentation of the manufacturers and CEPEL, as the independent lab. Periodical verifications among these labs are done to keep the repeatability of the results. Yearly routines for mark verification were required to guarantee the confidence of the declared values. These activities allowed the consolidation of a confident motor efficiency database that can be used for economical evaluations.

The next step was to study the establishment of minimum efficiency level for the three-phase induction motor and some simulations were done based on projects developments.

3.2.3 Educational Aspects

It was identified a lack of information about energy conservation in motor and ASD, the existing documentation was focused in the technology itself not in efficient use. Two books were developed covering all aspects concerning energy conservation in motor and motor drives, booklets were derived from these books. More than 7,000 exemplars were distributed.
Two softwares were developed by CEPEL. The motor software called BDMOTOR has three main modules: the first is a database with more than 500 Brazilian motors, with more than 26 information for each motor. The second is to make economical analysis to help the user to buy a new motor, to replace a motor and to verify it is better to repair or to buy another motor. The other is used to verify if the operating motor is well sized. The ASD software called ECON is designed to be a simple tool for determining the save potential of replacing throttle control to speed control. More than 600 exemplars were distributed. The distribution means are being revised to be more effective.

3.2.4 Market Activities

A motor rebate pilot project is been implemented by one of the biggest utilities in Brazil. It is a too small project, the target is to influence about 2,500 motors and it will be used to analyze how the market and the utility reacts to this kind of activity. Some rebate projects were postponed due to the national power sector restructuring.
The great marketing action has been the concession of annual awards for the most efficient standard and high efficiency motors in the range from 1 to 10 Hp, 2 and 4 poles. This prize has been an effective mechanism to induce technological competition among the manufacturers. The evolution of the efficiency showing its effectiveness is presented in the next section.

3.3 Energy Saving Gains

The energy saving evaluation of any energy conservation action is complex and subject to many variables. PROCEL has done many estimates. This section shows the estimate related to the improvements of the standard motor efficiencies motivated by PROCEL awards.

The table 3 shows the gain of efficiency achieved by the three more relevant Brazilian manufactures of industrial motor for 4 poles units in the range of 1 to 10 HP. These PROCEL Awards gains refer to the period from 1996 to 1999. $\Delta\eta$ means the absolute difference of the efficiency.

Table 3: Gains of efficiencies for 4 poles standard motor from 1996 to 1999 [3].

Power	Manuf. A		Manuf. B		Manuf. C		Average Gain
kW	$\Delta\eta$	%	$\Delta\eta$	%	$\Delta\eta$	%	%
0,75	6,5	9,42	11,2	16,23	7,4	10,48	12,04
1,5	7,5	10,00	8,2	10,66	0,4	0,52	7,06
3,7	3,7	4,46	5,5	6,83	1,1	1,32	4,2
5,5	3,0	3,53	4,4	5,26	2,3	2,67	3,82
7,5	2,2	2,56	3,4	3,99	3,8	4,46	3,67

These awards are addressed to the most significant market segment. The average efficiency gain was about 6%. The effectiveness of this action surprised us and it means that the efficiency starts to be a differential characteristic in the national motor market.

The estimated accumulated energy savings associated to the PROCEL Awards were 354 GWh in 1998 [2].

4 Brazilian Agreement for Minimum Efficiency Level

This section presents BA and makes a comparative analysis of it to the CEMEP proposal and EPACT92. BA covers three-phase squirrel-cage induction motor up to 250 Hp, 2, 4, 6 and 8 poles, horizontal and vertical mountings, face and foot-mounted, duty type S1. The efficiencies are determined using test procedure similar to IEEE-112 – method B. BA establishes one table for standard motor and another for HEM. The table for standard motor has the same values of 1987 NEMA 12-B table and for HEM, the table has the same values of 1993 NEMA 12-10 table.

The agreement is voluntary and the government is commited to make it mandatory in the end of the next 4 years with manufacturers' accordance. The table 4 compares the Brazilian Agreement, CEMEP proposal and EPACT92.

4.1 CEMEP Tables versus Brazilian Motor Efficiency

This section shows the comparison among the present Brazilian motor efficiencies and CEMEP proposal. The three more relevant industrial motor manufacturers are represented in figure 3, 4, 5 e 6. They are responsible for 90% of industrial motor production until 250 Hp. The figure 3 e 4 refer to 2 poles motors. Figure 3 for standard motor and figure 4 for HEM. The figure 5 e 6 refer to 4 poles motors. Figure 5 for standard motor and figure 6 for HEM.

Table 4: Comparison among EPACT92, CEMEP proposal and Brazilian Agreement

	EPACT92	**CEMEP**	**BRAZ. AGR.**
Power Range	1 – 200 Hp	1 - 90 kW	1 – 250 Hp
Voltage	230/460 V	400 V	220 / 380 / 440 V
Coverage	Horizontal and foot-mounted, 2,4 and 6 poles, general purpose, open-closed motor	Horizontal and foot-mounted, 2 and 4 poles, S1	Horizontal and vertical, foot and face mounted, 2, 4, 6 & 8 poles, S1
Test method	IEEE –112 – method B	IEC 34-2	Similar to IEEE-112 – method B
Efficiency region	01	03	02 , for standard motor 1989 NEMA 12-6B table, for HEM 1993 NEMA 12-10 table
Regulation	Law since October 1997	Voluntary, 50% of production will be in eff3 in five years	Voluntary now, it will be mandatory in four years by standard and/or law

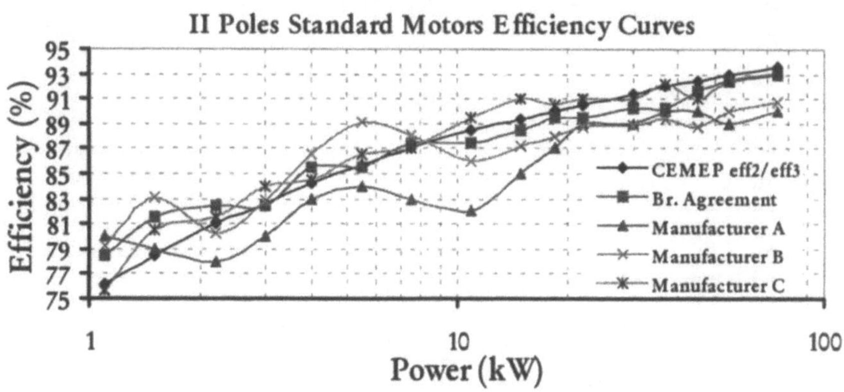

Figure 3: Comparison of Brazilian 2 poles standard motors with CEMEP curves

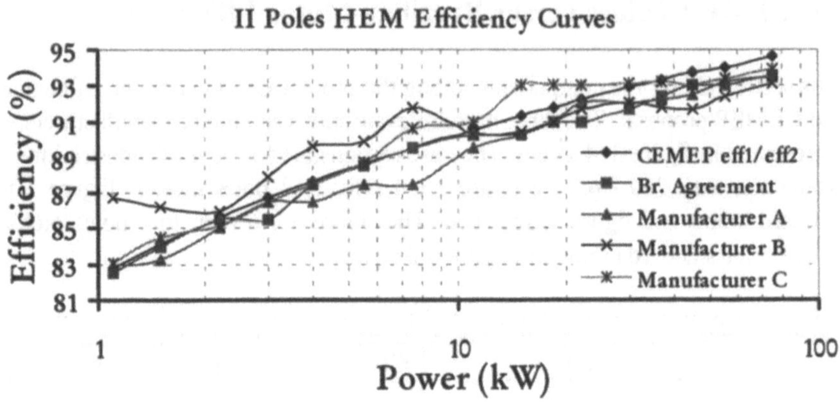

Figure 4: Comparison of Brazilian 2 poles HEM motors with CEMEP curves

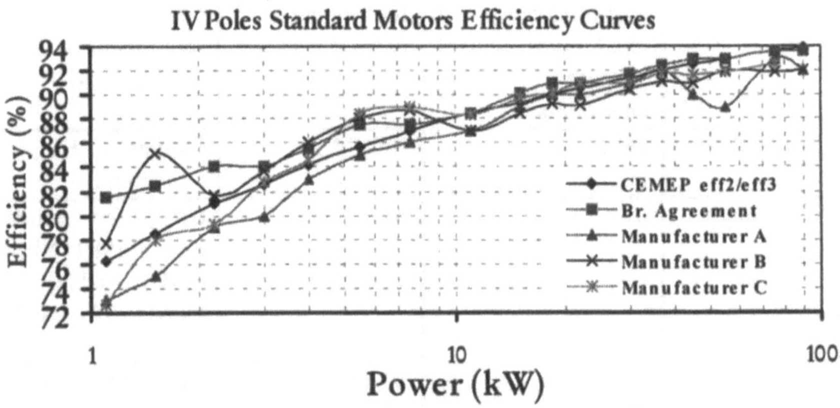

Figure 5: Comparison of Brazilian 4 poles standard motors with CEMEP curves

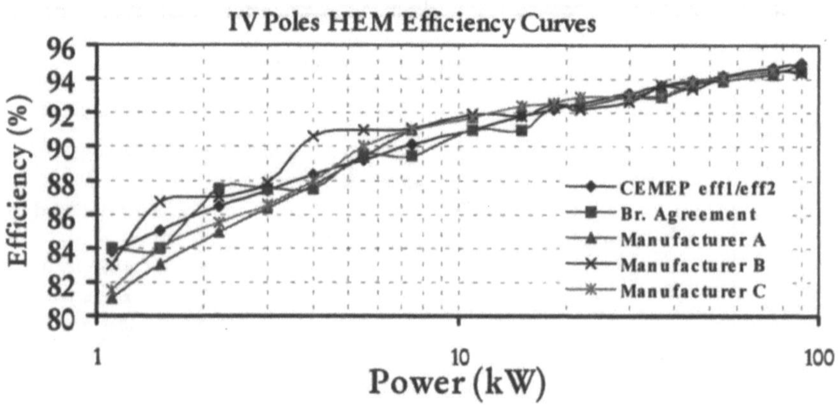

Figure 6: Comparison of Brazilian 4 poles HEM with CEMEP curves

Indeed the CEMEP curves are not directly comparable with Brazilian agreement due to the differences of the test procedures, however it is interesting to establish some comments.

- For standard motors, the CEMEP efficiencies are lower than Brazilian agreement (BA) ones until 10 kW that is most significant market segment and similar or better for greater motors;
- For HEM, the CEMEP and BA efficiencies are similar until 10 kW and for greater power, CEMEP efficiencies are higher for 2 poles and lower for 4 poles;
- Many Brazilian motors already comply the CEMEP curves for 2 poles motors;
- The CEMEP curves also can raise up the standard motors and the HEM until 10 kW for 4 poles motors;
- The BA was developed based on our previous experiences but in accordance with international trends;
- The BA scope is more comprehensive than EPACT92 and CEMEP ones.
- The Brazilian motor standards are under intense revisions and they have been used as an effective tool for energy efficiency;
- Experiences in many countries show that the energy conservation actions in equipments should be followed by some kind of regulation to keep the efficient market. EU is going in another direction.

5 Final Comments

The Brazilian Agreement will be revised yearly, in this way, EU decisions certainly will influence the original proposal. Due to the importance of EU market and of its manufacturers, if EU strategy fails it can cause retrogression in the motor efficiency regulation processes of other countries.

The Brazilian approach is emphatically technical. It has been proved to be successful and low cost. It could be suitable for other developing countries or part of it could be adjusted for developed countries.

6 References

[1] [1] Fonseca, P. and Almeida, A. "Characterisation of the Electricity Use ...in 2010", Energy Efficiency Improvements in Electric Motors and Drives

[2] [2] Information obtained from National Program of Energy Conservation PROCEL.

[3] [3] Information obtained from National Institute of Metrology - INMETRO

The UK Energy Efficiency Best Practice Programme – Lessons Learned

Hugh Falkner

Summary. The UK DETR Energy Efficiency Best Practice Programme (EEBPP) has been actively promoting energy saving in motors and drives for over 10 years. To date, energy savings worth over 45MECU pa in motors and drives alone have been stimulated by the programme – a remarkable achievement considering that it is an education-only initiative, offering no direct financial incentives. Despite the obvious success of the EEBPP, there will always be opportunities to do even better, and so this paper looks to the future by reflecting on some of the lessons that have been learnt.

1 Top 10 Issues

- Energy saving is usually an "also ran" subject, and so should be "piggy backed" on to other higher priority areas where possible.
- Reasons for non-investment in energy saving measures are often non-technical.
- It is vital to keep in close touch with our audience to ensure that we are answering their real needs.
- The temptation to make promotional materials too complicated needs to be avoided.
- Best use should be made of International links to recycle ideas and make the most of existing materials.
- Working with equipment suppliers and trade associations can give excellent gearing of effort, but care is needed.
- Regular structured assessment of the success of any energy saving programme is essential.
- Programmes need to change to reflect changes in the Motors and Drives market
- Careful targeting of different industry sectors
- Care must be taken over all aspects of branding

2 Energy Saving Isn't a Turn on

"Down-sizing" of companies and the need to comply with legislation on Health & Safety, pollution and environmental issues means that energy saving can easily be ignored simply through lack of sufficient time. And these issues are stacked underneath the over-riding drivers of profit, production, maintenance, and market position. It has been tempting to think that just because there is an energy manager we'll have an easy time - but too often he/she has all these other pressures, or may simply not have the "clout" to actually make things happen.

"Energy savings falls below other more important business issues".

Pragmatically, it can often make sense to sell energy saving on the back of factors that do matter, and here promoting VSDs (variable speed drives) and HEMs (Higher Efficiency Motors) on the basis of reduced equipment maintenance and sometimes better productivity appears to be a very promising way to go. Indeed, although our many Case Studies talk about energy savings, in most instances it was actually a maintenance problem or productivity issue that "clinched" the funding for the project!

To illustrate this point, a good practice guide on the repair of motors for minimum efficiency loss has been a particular success. Although the estimated energy saving potential is only about 10% of that from the use of higher efficiency motors, the interest shown is so much greater, that this indirect way of encouraging thinking about motors is having an impact far beyond that anticipated.

3 Making the Case

Our market research shows that the reason most energy saving investments are not made is not for technical reasons, or even lack of funds, but a failure by the project proposer to convince often non-technical management to authorise expenditure. The reasons for this are complex, but essentially projects are not "sold" in a way that is appropriate to those who need convincing. As a start to addressing what we see as a major blockage to our work, we have now introduced light-hearted sessions on just this subject in all of our seminars. This "soft" area marks a key part of our 1999 – 2000 activities, and represents a tacit acknowledgement that we need to spend far more time and effort in these non-technical areas. Sadly, while the overall response to ongoing series of Motor Challenge seminars is excellent, the audience appreciation of this session is the lowest of the day. It seems that engineers still need convincing of the importance of this point.

4 Keeping in Touch

As we all get busier it is too easy to give up actually talking to end users, but this is essential to keep in touch with their real concerns and issues. Talking to colleagues or indeed senior personnel of equipment suppliers or motor-using companies can too easily lead to initiatives which completely ignore the "down to earth" concerns of the engineer or maintenance manager.

A seductively easy "short cut" is to set up advisory committees drawn from industry who represent the people we want to communicate to. However, the drawback with this is that invariably we only attract those with the experience and knowledge to be really "in to" the subject, the less-enlightened who we are really after are very unlikely to have the will or the time to contribute.

The 'information explosion' means it is important to signpost materials which are particularly useful.

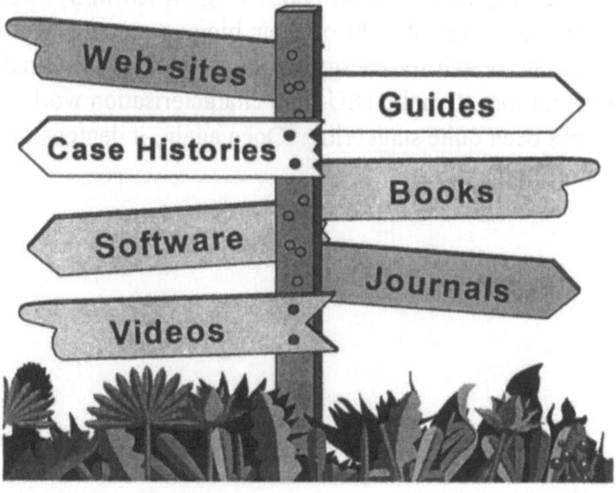

Apart from our own commissioned market research, we get most of our market intelligence from speaking to delegates at our events (but again these are a definite sub-group of those who are pre-disposed to the subject), and in particular at other events where we are in contact with those not necessarily interested primarily in energy saving. Other good sources include “on the ground” salespeople, and keeping an eye on the subjects of articles in the trade press. Equipment suppliers can be an excellent further source, but senior management are sometimes too far removed from day to day concerns to give a good idea of what immediate issues are.

5 Keeping it Simple

For most of our audience, what is required are materials which are authoritative, clear and concise. But to provide such materials is not that simple - it requires a high level of expertise to be able to understand the rules so well in order to condense information in this way. It is a truism that we are all being bombarded with more and more information, and so there is a growing role for the EEBPP to act as “signposts” to other useful documents. “Less is more.”

A very useful test is the 20-second rule – given a maximum initial exposure consideration time of 20 seconds - will a product be filed in the bin, put on the shelf, or in the bag for “evening” reading? This means that for instance we use fewer pages of high quality (and expensive) virgin paper rather than longer documents on recycled paper for all our publications.

The variety of energy saving measures possible, and the 100s or 1,000s of motors on a site, mean that we find too many people are unsure where to start - or start in the wrong place and get fed up with the whole business! In response to this, we developed a simple programme called *Quickstart* ,which in 3-4 hours enables a user to screen and rank typically 90 of their biggest motors in order to decide where to prioritise the search for energy savings. Although originally developed as a very simple tool for internal ETSU field characterisation work, the success of this simple tool has been quite staggering. Once again, it demonstrates clearly the demand for simple products.

Quickstart – a very simple but very well received motor energy saving opportunity screening tool.

- It's simple, crude – but it's quick, and any it's a start
- In just 4 hours without leaving the desk, anybody with a good knowledge of the plant can
  - see which of their top motors use the most energy
  - estimate energy savings from using HEMs, VSDs and switching off!

New technologies, such as CD-ROMs and the Internet, offer new possibilities for the dissemination of large amounts of information in huge quantities at extremely low cost. However, how many plant managers have the time to use the Internet or to look through a CD-ROM? Both fail the 20-second test.

There is sometimes a danger in trying to maintain interest in a subject by exploring issues which are very interesting, (sometimes just to the programme manager!) but ultimately of only real energy small benefit. An example is that of the energy saving options for minimising motors losses:

Action advocated	Usefulness of promoting the action	Desired Response
Specify higher efficiency motors	Clear, simple message, easy to act on.	Implement policy to purchase HEMs
Permanently connect in star very low loaded motors	Clear message, but can be hard to estimate the likely savings, only a few applications likely	Mainly for interest
Downsize over-sized motors	Modern motors maintain good efficiency to 50% load, replacement motor costs money, and time needed to alter mountings etc.	Note for future purchase decisions
Create a repair/replace policy for failed motors based on lifetime costs	Most schemes fail (for administrative reasons), which harms the energy saving drive within the organisation. It is recognised that these decisions are dominated by the need to get plant up and running – energy costs are not the prime concern.	Consider changing your repair/replace procedures in such a way that energy costs are also minimised? But don't attempt anything too ambitious to start with
Implement a site motor inventory on which to base energy saving decisions	Very time consuming, and liable not to be kept up to date	Should appreciate how useful this might be – but it is costly, and should only do as part of a combined maintenance package.
Use energy optimising (voltage reduction) devices	Extensive independent tests have found energy saving claims to be often over-stated, and very sensitive to the load.	If offered "free" in a soft-starter or VSD, might as well try using the feature. Ensure have full understanding of the limitations of the technology before purchasing standalone units.

Being quite realistic about where the achievable energy savings are, and the often very practical issues which could cause difficulty in reaping them, not only focuses effort but gives energy saving programmes a great deal of credibility.

This is even more the case where programmes have been partly critical of products or services offered by sometimes respected commercial organisations.

6 System or the Component?

So where are the remaining savings? Let's have a look at typical system diagram.

Components in a simplified pumping system.

- The first questions are: "Does the job need doing at all?", and "Is this system the right way to do it". The under-used maxim "Efficiency is doing the thing right, effectiveness is doing the right thing" neatly sums this up. These sort of questions are so high level and so application specific that it is often hard to get people to ask them - but the results can be spectacular.

- Second point is to look at the system - i.e. everything on the "real-world" side of the driven machinery - pipework and ducting or insulation issues usually account for a large proportion of the losses. Fixing leaks and mending insulation is always found to be something that people will actually do - it requires no system knowledge, its intuitive, needs little or no cost justification, and does not interfere with the installation- its simple!

- Third point is to consider how well matched the driven machinery is to the system. Here VSDs are very popular in fans and pumps and sequence controllers in compressed air systems. This is because these are effectively "black boxes" which can be retrofited to an existing system without having a detailed knowledge of the system, (although an optimised system is unlikely without detailed knowledge). Some people are also "technophiles" who just like having the latest equipment.

- Last, look at the efficiency of each of the components - the motor, transmission and driven machine. Here the savings may be less than in the system, but where it is a simple purchasing option to buy something more efficient it again requires little technical knowledge. Advice on buying more efficient components is easy to give – although trying to get persuade people to pay more for more efficient components is hard.

The lesson - people will do things that are simple, obvious, or require little thought. Hoping to educate people in proper system design is hard, and anyway it is often not possible to measure the system characteristics sufficiently accurately to properly use this knowledge.

Aim for the stars and people will be put off and won't do anything, but aim for the ceiling and they may not only just rise to the challenge, but then flushed with success go on to reach for something higher – and one day the stars.

7 International

Politically the UK enjoys the position of being both part of Europe, and having a "special relationship" with the US. This is just as true of work in energy savings in motors and drives, where we are always striving to learn from, and share materials with, energy saving programmes in these and other continents. The savings in effort are considerable, although in practice materials do need customising for local audiences, so as to overcome barriers such different currencies, "foreign" case study sites, or simply unfamiliar language. Recent examples include a US Dept of Energy/Hydraulic Institute video now available in Europe through the UK British Pump Manufacturers Associations (sponsored by the EEBPP), and an EEBPP/Association of Electrical and Mechanical Trades Guide on Motor repair now available in the US through EASA.

In terms of policy, the SAVE programme is of paramount importance in leading the way in Europe, and the EEBPP are active and enthusiastic contributors to this programme. This reflects not only the closer political and economic convergence, but also the globalisation of the motors and drives markets. While policies and efficiency standards need clear EU co-ordination, UK programmes still need to be designed for our own markets, as each member state has individual characteristics which makes it hard to come up with a single "best way" of doing things.

Beyond the EU/USA, many countries are seeing the EEBPP approach as "right for them", and we have been very glad to help countries such as Australia, New Zealand, China - to set up similar programmes.

Indeed, at ETSU over half of our work is now for overseas customers, and we find the new perspectives this brings invaluable for our UK work.

While it makes sense to look far afield for new ideas, and to Europe for standards and policy direction, there is still the need for individual programmes to match the markets of each country.

8 Alliances & Partnerships

Perhaps the most important asset that ETSU and the EEBPP has, is its authority and hence credibility brought about by its clear independence from other industry bodies, and in particular equipment and service providers. Against this need to maintain our independence are the opportunities for gearing effort through working more closely with these bodies, and this is an area where we tread very cautiously.

It should always be remembered that Trade Associations of equipment suppliers exist first and foremost to protect the interests of their members as a whole. This can have several pitfalls for the unwary!

- Individuals from companies who are actually supportive of an idea may have to publicly oppose it on behalf of the trade association. This is at best confusing.
- The need to wait for the next meeting of a committee for endorsement of even fairly minor actions can greatly slow down progress.
- Decisions and discussions can be slowed down by large committees being used to debate matters which really only directly concern a few members.
- If a body is discussing ideas then it is essential that either the proposer or somebody who fully understands the proposal and background is present during discussions, otherwise there is no control of the tone or direction of the discussion.
- Trade Associations are always looking for new sources of funding, and will be very happy to produce materials, of a high technical standard, with Government money. However, the need for consensus means that leading edge products or services offered by just one or two companies run the risk of being excluded by the bulk of the membership. Also, issues which suppliers as a whole would rather kept quiet about might simply be left out of materials.
- Large committees with lots of people wanting to make their mark means that documents can easily become far too long.

Having spent many years working quietly with motors and drives suppliers, we have now supplied "like minded suppliers with complete EEBPP tool-kits, which includes bulk copies of literature and a complete set of EEBPP slides on CD-ROM. This has marked an extremely cost-effective way of reaching end users through either third party seminars or direct sales contact - they are increasingly our "foot soldiers". There is now a joint GAMBICA (UK VSD suppliers trade association) / ETSU marketing working group which meet regularly to discuss and progress items of common interest, and where useful undertake joint projects. While this group enjoys considerable success, it is clearly understood that ultimate editorial control of all materials lies with ETSU.

Working with equipment suppliers and Trade Associations can give tremendous gearing for efforts, but care must be taken to be aware of their interests and to always maintain authority and credibility by retaining the final say on all joint activities.

9 Monitoring Progress

The EEBPP is arguably unique in regularly commissioning independent market research to assess the energy savings influenced by its activities. This not only gives the DETR a clear measure of ETSU's performance in running the EEBPP, but in addition gives invaluable information on how successful specific initiatives have been. While not cheap, it gives hard data on which to base future work, which is essential when in so many areas the impact of our work may only take place some time after an activity, and often the actions are not always those which we would have expected. For the record, we have helped to stimulate energy savings of £3M pa in HEMs, and some £30M in VSDs. Of often greater interest than the energy savings which have been made through actions influenced by the EEBPP, is finding out just why other actions have not been taken.

The crude EEBPP spend to carbon savings ratio picture is good, but is actually better than this current figure because even if all our activities were to end immediately, the legacy of our publications, knowledge instilled, and new products developed, will continue to improve the performance without further spend.

As experienced survey clients, we are very aware of the many practical problems, with the issue of "double and multiple counting" being constantly addressed to ensure that energy savings are only counted once, and are attributed to the correct area. For instance, fitting a VSD to a boiler fan could be considered a Motor saving, but it could also be considered a saving to the Chemicals Sector in which the boiler is being used, or to the Boilers and Steam Initiative. There are also colleagues undertaking general management techniques and control work who could also make justifiable claims for the energy savings. These issues can all be resolved, but it requires a lot of patience, willingness to compromise, and a clear view of just what is trying to be achieved. As a measure of how seriously we take this matter, we have an internal "independent " auditor whose sole link to the EEBPP is as energy saving claims manager. Not a job for the faint hearted!

Large EEBPP-funded R&D projects such as the Brook W series motor / Orb low loss steel project allow for simple energy savings accounting by simply receiving annual sales returns against which energy saving claims can be attributed.

In the Motors and Drives portfolio there are now over 26 publications, and clearly the additional impact from further similar materials is becoming progressively more limited. The DETR has responded to this - but it calls for new ways of measuring our success, since it is no longer just a case of counting the numbers of seminars, publications and other forms of deliverables. We are in many ways becoming more entrepreneurial, looking for opportunities that will give us the best £/CO_2 saving ratios - and this in turn calls for more flexibility in our annual plans with the DETR, another area which is being developed.

Such "nuts and bolts" issues regarding the contract between programme funders and managers must be right if changes are going to be made.

10 The Changing Market

The biggest change effecting the EEBPP has been the increasingly competitive nature of both the motors and VSDs markets.

In the UK, buyers are increasingly turning to lower cost, usually lower efficiency, motors from outside Europe. Traditionally these were of lower quality, but the quality gap is closing, and so they now represent a genuinely viable alternative which is attractive to many users. This price erosion is putting further pressure on the Western European producers of higher efficiency motors (Eff class I) who have traditionally dominated the market, and (new EU efficiency classification aside), it seems likely that the average efficiency of motor sold in the UK will decrease. This is a major cause for concern and we, together with sellers of class I motors and other interested stakeholders, have big hopes for the efficiency classification scheme.

The prices of VSDs have fallen in real terms over the last 10 years in the UK, and further price falls are expected. While appearing good news in that it reduces the payback on energy saving investments, it does mean that there is less margin which suppliers can use for helping to assess potential applications. This is made worse by fan/pump drives only needing to be very basic, low cost and hence lower profit, "entry level" models. The challenge is therefore for us to create much simpler tools for assessing applications, and to go "downstream" by giving more of the practical information previously given by suppliers. While classical energy saving texts talk about the need for detailed energy profiling on which to base investment decisions, the cost of such profiling now makes it very much the exception in all but larger drives.

11 Industry Targeting

The EEBPP works on a simple matrix system of technologies and industry sectors, which allows us to target specific industries with motors and drives information tailored to their specific requirements. This approach is at the heart of the EEBPP, and has been a major reason for its success.

Large companies with high energy bills (such as water, steel, paper and pulp and chemicals) are usually pretty "switched on" anyway, and need little technical advice. Such companies will readily take and disseminate very technical information, providing it is of sufficient quality.

Reaching Small to Medium Enterprises (SMEs) is hard to do cost effectively, not least because they are rarely able to give up the time to attend a traditional energy saving seminar. To address this we have a 40-seater mobile seminar unit (MSU) which sets-up on local industrial estates to reach these organisations. We are now reaching further audiences by hiring the MSU out free of charge, complete with speakers, to other organisations who want to give energy saving seminars to local groups. Not only is this cost effective, but the MSU vehicle does a lot for raising the profile of the EEBPP. In response to the particular shortage of time which SMEs have to explore energy saving opportunities, the EEBPP has produced a variety of short summary guides to energy savings, which give both basic information and signpost to other more detailed essential information.

There are also teams of Regional Energy & Environmental Officers based around the country, who are in regular contact with companies on their "patch", giving specific advice tailored to particular needs. Many of these have helped to set up autonomous Regional Energy and Environmental Management Groups, which are simply groups of local organisations who meet on a regular basis to discuss issues of concern, usually with the meeting led by an external speaker. The "buzz" at the meetings of the most successful groups is always uplifting, in large part because the events are arranged by the audience, addressing issues which they are concerned about.

12 Branding and Publicity

One of our biggest regrets within the EEBPP is that generally we have not attracted enough publicity for the EEBPP itself. While the excellent achievements of the EEBPP speak for themselves amongst those in regular contact with the EEBPP, a higher public profile is necessary to gain wider awareness of our work amongst the many other organisations who can help to influence the future of the EEBPP.

Another "image" problem is that there is a proliferation of brand names and logos all being used in the energy saving arena - often on the same materials. To some extent multiple endorsement of materials is a good thing, but too often it just causes confusion, and actually devalues individual brands. Of course, no organisation will actually vote its own brand out of existence, and so we are now judicious over the selection of branding for products.

13 Legislation

Creating legislation for energy efficiency is a time-consuming process, although as we have seen from the US *EPACT* experience, it can be very effective. However, beyond efficiency standards for drive system components, trying to devise workable legislation to improve systems is particularly hard. The EEBPP is though "piggybacking" on to two other initiatives which will give energy saving a higher profile, and this "opportunistic" approach is seen as a useful way forward, although it is hard to gauge the likely success of these routes.

The European **Integrated Prevention of Pollution (IPPC)** directive, in which specific, major EU industries are bench-marked for best practice from a pollution and emissions point of view is an opportunity got energy efficiency. Once introduced, regulated companies will be regularly inspected to check that they are complying with environmental "best practice". Unfortunately, motors and driven systems are generic, and so are not covered in any of the individual sector checklists, but in the UK the EEBPP has provided the Environment Agency with a short appendix on Best Practice in motors and drives. This will then form the basis of at least a rudimentary assessment.

Energy taxes are planned for introduction in the UK in April 2001, and as part of this process key industry sector bodies (in fact those covered by IPPC) have entered into negotiations with the Government to agree specific energy efficiency targets for their industries. The threat of these taxes has already moved the position of energy costs up the agenda in many companies, which is already making companies much more receptive to energy saving messages.

14 Market Transformation Programme

As the EEBPP has developed, it has become much more open, with more time spent in consultation with key stakeholders to help determine future programme directions. An extreme example of this approach is the new DETR Market Transformation Programme where, working closely with key stakeholders, we are jointly developing sector review papers to give a consensus view on the current state of the market in key energy saving products, including motors and VSDs. Future projections on the likely impact of a range of different policy options are then debated and added to these sector review papers. Representing a consensus view, these key documents are becoming the corner-stone against which all future work in these areas will be formulated and assessed, and as such will form the focus for ongoing dialogue with all stakeholders and policy influencers. These Sector Review Papers are regularly updated, and are freely available on the DETR Market transformation programme web-site, **http:\\www.mtprog.com**

This open and clear approach to creating these key papers has been very warmly welcomed by participating stakeholders, to date largely equipment suppliers, and is already sparking much interesting discussion on how the market can be best influenced.

15 The Energy Saving Trust

Another force in the UK Motor and Drives field is the Energy Saving Trust, an independent body funded by the Electricity Industry and Government to promote energy saving in the domestic sector. However, the achievable energy savings from the use of VSDs means that EST has now launched a pilot campaign to encourage the uptake of VSDs Schemes for HEMs and motor repair are also being considered. The success of the low cost branding campaign so far means that consideration is being given to future rebate schemes, with the planned UK Energy Tax being a possible source of funding.

Care is taken throughout both EST and EEBPP work to avoid customer confusion by ensuring that the
programmes compliment each other, with minimum overlap. The essential and clearly understood difference is that the EST is very much about endorsing particular products or services, whereas the EEBPP does not reference particular products.

While it is good to have lots of organisations promoting energy efficiency, each must have a quite clear and specific role to play if there is not to be confusion.

The Energy Saving Trust "Energy Efficiency" Brand

"You don't have to be a genius to save money on your energy bill.
Just look for products and services
where you see the Energy Efficiency logo. They
save or use less energy, and save you money. And because they
use the latest technology, there's no compromise on performance.
This prestigious logo
has been introduced by the Energy Saving Trust, an independent,
government-funded body, working in partnership with industry.
Look for it.
Anything else is a waste of money."

16 Conclusions

The EEBPP has made significant progress in facilitating energy savings. Given the issues in the market place it could be argued that without the involvement of the EEBPP these savings would not have been realised.

Energy efficiency advice and support needs to be carefully structured and targeted. This requires an on-going liaison with those that the Programme serves.

Partnerships – trade and international, can "leverage" the effect of the Programme, as long as they carefully considered. A key strength of the Programme is its independence. Accordingly, the use of "trade" interests has to be in such a way as not to jeopardise this status.

The introduction of legislation and financial measures in the UK will cause energy efficiency increase in importance. The EEBPP is well placed to provide assistance as the demand increases.

The promotion of energy efficiency is an evolving practice. It needs to be very market conscious and work to maximise the effect of its funding.

The New European Database on Efficient Electric Motor Systems: EURODEEM-98.

Stefano Cocchi and Flavio Conti

European Commission – DG Joint Research Centre (JRC)
Environment Institute - Renewable Energies Unit
21020 Ispra (Varese) - Italy

Abstract. EuroDEEM is a Database tool promoted by the European Commission to help and guide decision-makers to manage energy efficiency in Electric Motor Systems (EMS), from the grid to the end-use. At the beginning of 1999 the Joint Research Centre of the European Commission released a completely new version (EuroDEEM-98 V.1.1) with many enhancements. The new software was conceived to consider all energy paths and components, as well as the most interesting energy conservation opportunities applicable to EMS. Although the software is being further developed and refined, it is already now possible to perform a simple energy and economic analysis of a complete electric motor system. This report also describes EuroDEEM's conceptual design, its present features and the improvements underway during 1999, and, in particular, how the end-use Pump Module Help and the related selection tool are being developed.

Keywords: Electricity savings, database, system approach, efficient motor system components.

1 Introduction

In 1996 the European Commission' DG XVII, in the framework of the SAVE programme, published the results of the study *Actions To Promote Energy Efficient Electric Motors* (EEMs), which stressed, among other important issues, the need to improve market transparency for efficient motors and support the market transformation toward EEMs. DG XVII, being well aware of the importance and of the large saving potential of the electric motor systems, well before the conclusions of the above-mentioned study, asked the Joint Research Centre (JRC) to create a database and a selection tool on Electric Motor Systems. This work was carried out during 1996 with the production of an initial version of the EURODEEM database, which was presented for the first time at the Lisbon Conference at the end of 1996. (Cocchi et al., 1996).

The first version of the Electric Motors Database (called EuroDEEM) was a database only for motors, unable to treat the whole EM System. Nevertheless, the motor part is the core for any further developments. The motor database was the starting point for the promotion of information, knowledge, and for dissemination of decision-support tools.

The results obtained from the first version of EuroDEEM can be summarised as follows:

- More than 1800 electric motors loaded into the database,
- A Selection Tool for effective motor choice,
- A Tutorial and Help for easy understanding of high efficiency issues in motor management.
- A World Wide Web page for information dissemination.

The EURODEEM 96 version also suffered from low user-friendliness, almost out-dated motors, with only 14 manufacturers included in the database. There was no fast and reliable procedure for updating and upgrading the database and this was a serious problem since it did not allow the inclusion of new technologies and manufacturers. Therefore, a thorough revision of the EURODEEM structure was necessary and further development was decided in 1997 and a new version conceived.

Obviously, the Electric Motor Systems (EMS) (Figure 1) are formed not only by the motor but also by other important components (e.g. regulation and transmission devices) that can contribute substantially to the efficiency improvement of the whole system. The technology improvements in the field of regulation devices (frequency converters, variable speed drives - VSD, etc.) and in transmission components (belts, gears, etc.) need to be included in the system for proper energy efficient counter-measures (e.g. for avoiding over-sizing). A large number of VSD types and new coupling devices now exist on the market and it is therefore important to include them in the motor database in order to improve the service the database provides to users.

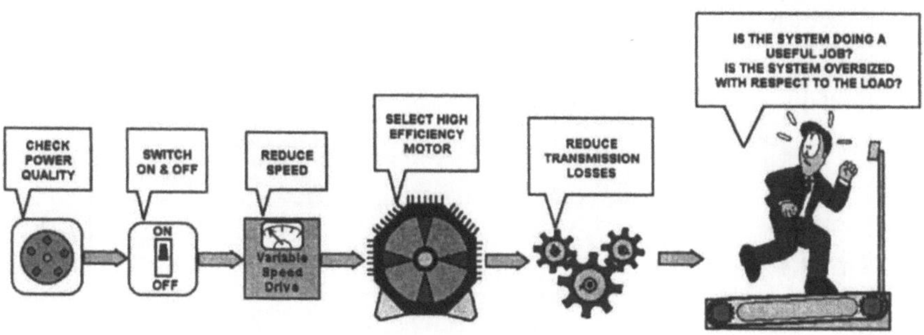

Figure 1: The whole scheme of an electric motor system

For the above reasons, the JRC was committed to continuing the EURODEEM Project.

2 Development Targets of EURODEEM-98

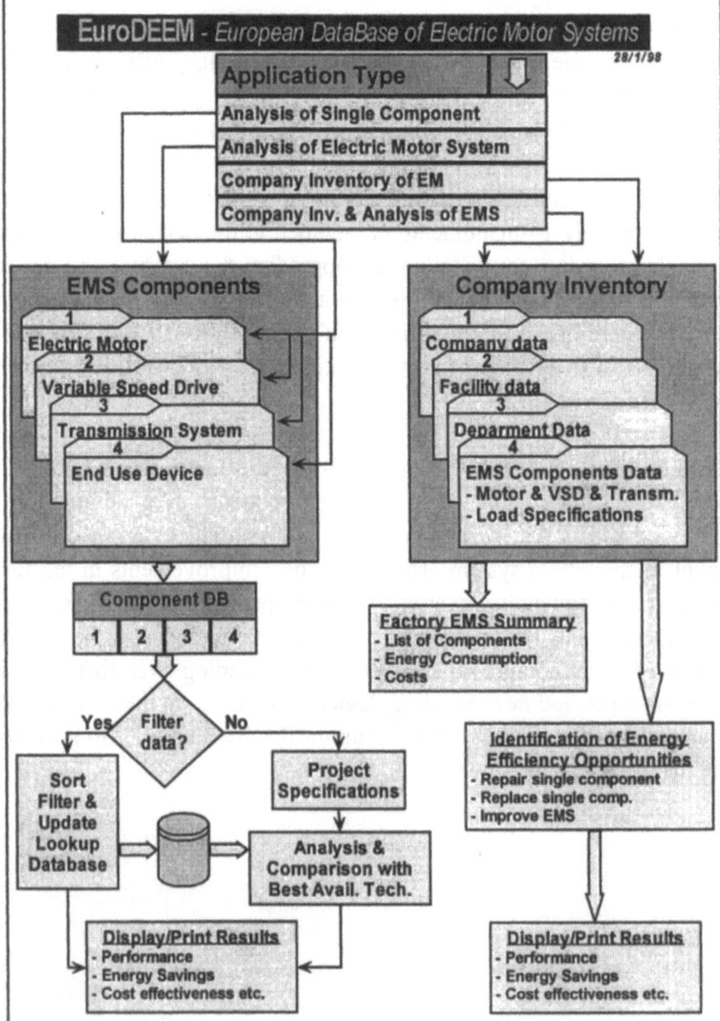

Figure 2: Schematics of the new version of EuroDEEM.

The ultimate goal of the EURODEEM Project is the development of a whole motor system assessment tool, as user-friendly as possible, and able to provide advice and decisional support for electricity efficiency. Therefore, as shown in Figure 1, the main scope of the work is to extend the database of electric motors to all the other motor system components. In this way the tool can reach a full capability of assisting plant energy managers to identify any inefficiencies in the motor system chain and to take optimal decisions.
Last but not least, it was a primary requirement to develop a new interface that could be easily understood by users.

3 Conceptual Structure

To achieve the expected design for the new EuroDEEM software the JRC conceived a conceptual structure shown in Figure 2 and mainly described by the following four types of user applications:

1. Analysis of a single component: The database permits both to query the database for search on each motor system component (e.g. electric motor, VSD, transmission end-use devices). Each component can be searched either directly, by filtering the data in the base, or according to user's load requirement (project specification). In this latter case the user has the possibility to compare various components available on the market and to make a choice based on the cost-effectiveness and/or energy efficiency.
2. Analysis of an Electric Motor System (EMS): This will allow to perform energy and economic analysis of single and combined components of a complete Electric Motor System, (for instance, to assess the overall efficiency of a system chain formed by the motor, the pump and the valve).
3. Company Inventory: The Company Inventory is intended as a tool for archiving all motor system component-related data of a company. The Company Inventory is a database that collects all data regarding Electric Motors, Variable Speed Drives, Transmission devices, Load Description, and Utility rates paid by the company. It permits and helps users to organise, maintain, and manage, at different corporation levels (company, factory, department), inventory, replacements and maintenance of all existing electric motor system components of the user's Company, including the spare stock.
4. Company Inventory Analysis of EMS: It must finally allow the user to perform economic and energy analysis of existing Electric Motor Systems of the Company database inventory, identifying Energy Savings Measures for single and batch components.

Regarding the Electric Motor System Components (left branch in figure 2), four databases are foreseen:

1. The database of Electric Motors, which has been enlarged to take into account some extra data fields, such as Efficiency Class, Certification Type, Warranty, WWW links to manufacturers, etc.
2. The database of Variable Speed Drives (VSDs) (both electronic and mechanical). It must include power ratings, drive efficiency and power factor (only for electronic VSDs) at different speed and load, and frequency ranges.
3. The database of Transmission and coupling devices: Efficiency data is required as a function of load demand.
4. The database of end-use types. This will permit the end-users also to check the sizing of end-use devices with the load requirements and to optimise industrial process.

In addition, two other databases had been created:

5. The Company Inventory database: where company motor data are stored
6. Utilities database: needed for hosting data on tariffs and rates for economic calculation

The relation between the Electric Motor System Database and the Company Inventory must be straightforward. It must be possible, for example, to insert motors (or other components) from the EMS Component database into the Company Inventory.

Finally, there must be an intuitive motor selection tool for choosing best-available motors for a desired application, including those belonging to the Inventory Database. The selection tool should take into account the load profile and type, the utility rates and tariffs, and transmission and control devices, if any. The outcome of the selection should give information on energy consumption and costs, and eventually payback times.

4 Development and Features of EURODEEM-98

4.1 Development Approach

Since the objectives are ambitious, it is very important to rigorously define all possible components and behaviour of an electric motor system. First of all, we have analysed the whole system drive chain, defining the "key nodes" where energy flows and where losses can degrade the system's performance. As shown in Figure 1, almost all EMS components are taken into account:

- Power Quality Issues, namely Reactive Power,
- Control Devices for rational use of electricity
- Variable Speed Drives, both electronical and mechanical,
- Electric motors,
- Transmission and coupling devices,
- End Uses (pumps, fans, compressors, machinery, etc.),

The description of the *hardware* layout of the EMS provides the basis for a reasonable description of the system's performance. Nevertheless, we identified some *software* variables that enable a better analysis of the whole system:

- Load profile, e.g. daily profile of the power requested by the end-use.
- Speed profile, for variable speed end-uses.
- Electricity tariffs, real power and reactive power, day and night values.

Secondly, we made a research of data availability from manufacturers and from catalogues regarding electric motor system components. It would seem easy to retrieve all possible information and data about every system component, in order to have the largest choice. However, some non-technical barriers imposed some serious constraints on the system modelling. Every Electric Motor System component had to be defined according to the availability and the homogeneity of data, but this was not enough. The model must be handled in such a way that all component data match in a harmonised and precise combination. This development is in continuous evolution, since new devices, technologies, data availability, and more detailed models are being introduced into EM systems.

4.2 Technical Features

4.2.1 User Friendliness

Since Version 1 of EuroDEEM suffered from its lack of being user-friendly, a big effort was made in the study and design of the interface itself and its features.
First of all it had to be very intuitive, with only a few commands and text insertion. Secondly, it had to have all the usual Windows 95 features, like "Drag & Drop", "Cut, Copy, and Paste", Help Captions to guide the user, and a familiar layout.
The result of this study was the choice of the usual Windows 95 Explorer as the container of all information, databases, and analysis tools.
The reason for this choice was based on the following ideas:

1. The Explorer has become a very familiar and user-friendly interface on Windows Machines,
2. The Directory - Sub Directory structure of the interface is very useful for containing all database information in different sub-directories.
3. The user can view all information in one single screen shot.
4. All common Windows features (e.g. Drag & Drop, Copy & Paste) can be easily accessed.
5. The user can easily switch between different databases (Electric motors, VSD's, Transmission devices) and can easily browse the content of the Company Inventory (view Electric Motor Systems in each Company Department or Facility).
6. Software upgrades and new databases can be easily added in future.

The screenshot in Figure 3 shows the final layout of EURODEEM-98.

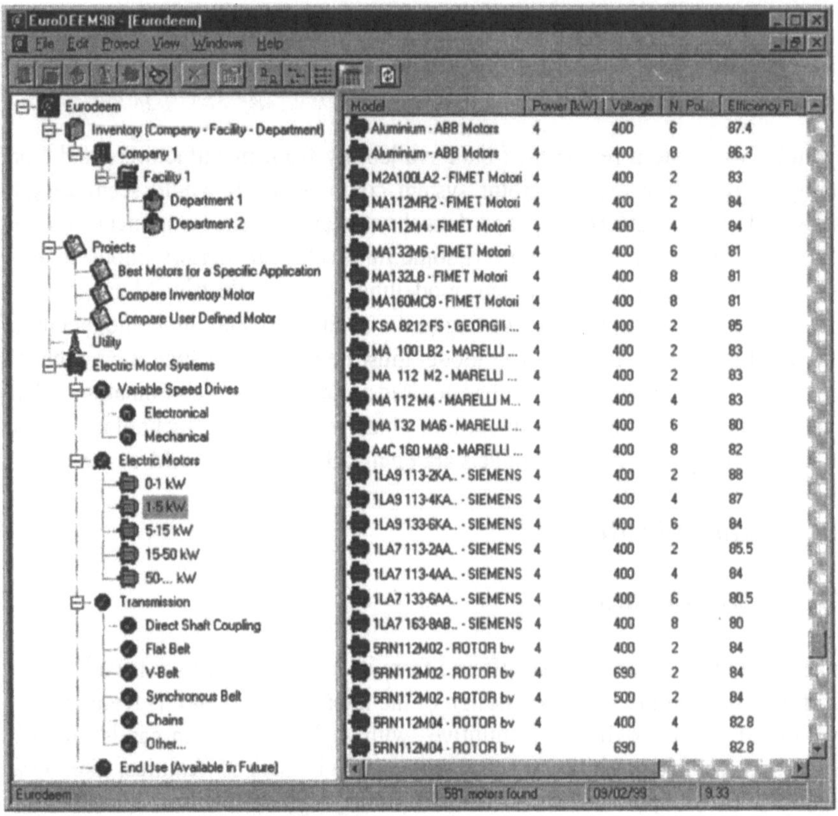

Figure 3 Final layout of EURODEEM-98

4.2.2 Software Development

The upgrade of the first version of EuroDEEM required to design from scratch an interface suitable for the most common operating system, that is Windows 95. The JRC has developed the design of the new interface and its features and has done the software writing with the help of an external software company. It was decided to use Visual Basic 5 as the Programming language in order to be able to apply, if necessary, all minor modifications according to the results of bug-testing, and to implement inside the JRC new features and enhancements.

According to the project design, five new additional databases have been foreseen and developed (Company Inventory, Utility Database, Project Database, Variable Speed Drives, and Transmission Devices), while the existing one for Electric Motors has been modified and improved.

New tools have been added to enhance the usability of the software:

- Catalogue Motor-Find tool, selecting through manufacturers, number of poles, motor power, motor torque, motor frame, etc. (See Figure 4). You can find and store (for vision later) a set of motors, according to general parameters, and/or detailed electric and mechanical data values.

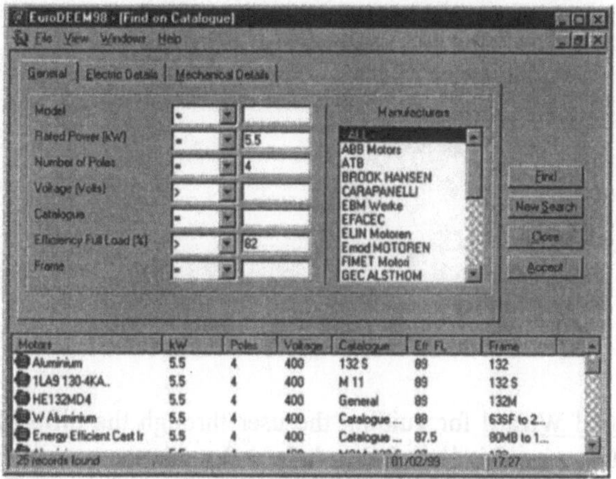

Figure 4: Screenshot of the Catalogue Motor-Find tool

- Inventory Motor-Find tool, selecting through manufacturers, number of poles, motor power, motor torque, motor frame, motor load, energy consumption,

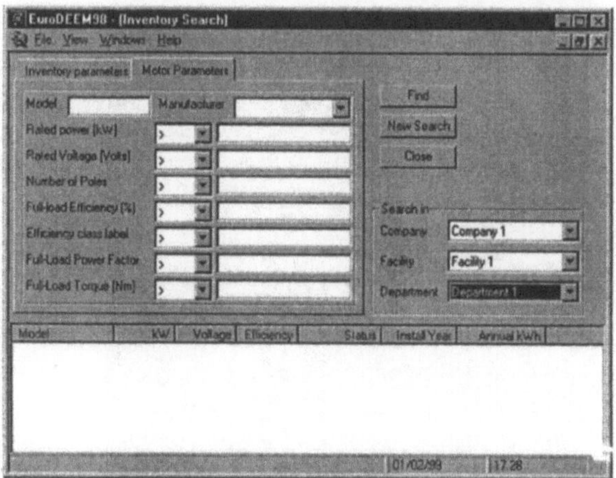

Figure 5: Screenshot of the Inventory-Find Tool.

etc. (see Figure 5). One can search in a Company, Facility, or Department for Motors or Motor Systems that match a specific request (e.g. Overall system efficiency, annual energy consumption, etc.)

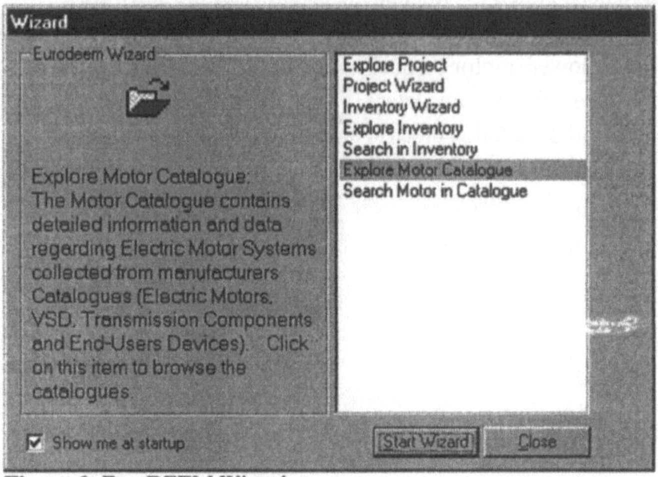

Figure 6: EuroDEEM Wizard.

- A full featured Wizard for guiding the user through the different tools and databases. The wizard is intended to help and guide the user in the selection of the best available motors for specific applications, in replace or rewind analysis for Inventory Motor Systems, in inserting new motor systems into the Inventory Database, and in browsing the Electric Motor System Catalogue and the Company Inventory. See Figure 6.

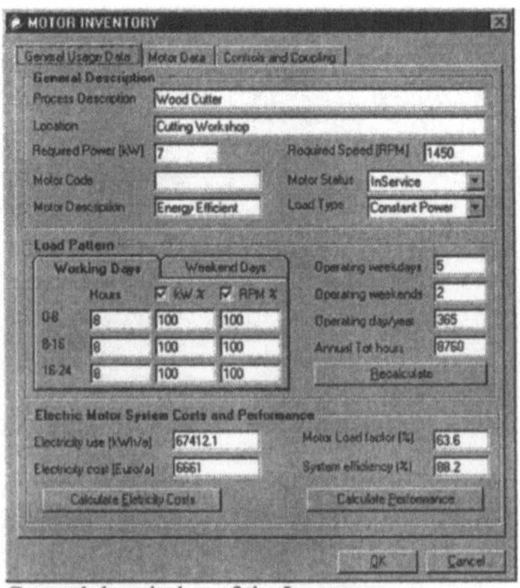

Figure 7: General description of the Inventory.

- Some Analysis Tools are placed in the Company Inventory database for checking and calculating energy losses due to bad control systems, or motor oversizing (system efficiency and motor load factor). See Figure 7.

- Finally, a clear, fast and reliable Procedure For Updating The Data in the base has been set-up in order to keep the Catalogue Database updated. Only JRC staff can maintain, update, and modify the Motor Catalogue Database by the. The JRC staff, in order to avoid any kind of manipulation and responsibility on the data supplied, has developed an automatic loading procedure of the files received from manufacturers. In this way manufacturers are the only persons responsible for the data shown in EURODEEM.
- The user can view the EURODEEM data either by receiving a CD-ROM or downloading EURODEEM from the web site

  http://iamest.jrc.it/projects/eem/eurodeem.htm ,

  but in no way he/she can alter or modify the catalogue data. This is done in order to prevent the circulation of incorrect data. The user can only insert data for the specific use of motor components in the Company Inventory database.

4.3 Non-Technical Issues

4.3.1 Reliability of Energy Analysis

The EURODEEM Software allows ranking the electric motors according to the least energy requirement for supplying a given load. The calculation of the energy consumption requires, besides the load pattern specification, also the availability for each motor assessed, of a given number of motor parameters, such as the Nominal Power at Full Load (Pn), and the efficiency value curve as a function of the Load Factor (LF).
Problems arise when motor parameters are unavailable, partially or completely. This is a frequent case for Inventory old motors, whose nameplate parameters are unavailable. In these cases the program cannot compute the energy consumption and, hence, the energy savings.
But also for new motors, several catalogues do not specify efficiency values below 50% part load. Also in these cases the evaluation of the energy consumption could be very inaccurate.
For these reasons it has been decided to improve the interpolation routine with algorithms that allow obtaining part load efficiency values. The routine calculates estimates of motor losses due to Constant Losses + I^2R Losses + Negative Linear (Core) Losses.
The type of efficiency curve identified for interpolation is the following:

$$\text{Part Load Efficiency } \eta = \mathbf{1/\,(1 + (p_1/LF + p_2{*}\,LF)}$$

where the parameter values p_1 and p_2 actually vary with the motor power level.

4.3.2 Data Collection Issues

The most difficult task of EuroDEEM was to persuade motor manufacturers in supporting the project, by means of making available in a harmonised way all expected data.
Although motor catalogues may contain a huge amount of data, they are insufficient for the purpose of the project. First of all, the catalogue data values were not always in the same units. Secondly, some catalogues did not have efficiency figures (Full Load and Part Load). Finally not all catalogues available were updated.
The JRC and DG XVII have organised several meetings inviting manufacturer representatives to show the advantages of being a EuroDEEM supporter. The efforts in this promotion campaign gave good results and, through a human-error free procedure for collecting and updating data, permitted the database to store more than 3500 motors, from 24 manufacturers.
For what concerns pump manufacturers, they were less confident in providing their data, and basically for a couple of reasons:

1. There is already good software (free and commercial) available in the market for selecting pumps, but generally only a few brands, if not only one, are considered in their databases.
2. Since pump systems are definitely more complicated than just simple motor systems, many pump manufacturers provide extra services such as pipe sizing, pump selection, system optimisation, etc.

Obviously, EuroDEEM does not want to compete with already established software, but it wants to provide a system vision of energy efficiency opportunities. EC-JRC experts had several meetings with pump manufacturers and finally an agreement was reached on the structure and layout of the pump module, and on a draft set of pump data to be included.

4.3.3 Efficiency Classes of EMs

Every action of energy efficiency is basically a replacement of an obsolete or inefficient device or system with a more advanced and efficient one. This implies that efficient devices are identified according to the values of certain parameters and, possibly, labelled in pre-defined classes of efficiency.
An ad hoc study group set up by the DG XVII, where representatives of the European Committee of Manufacturers of Electrical Machines and Power Electronics (CEMEP) were present, has treated the problem of identifying efficiency classes for electric motors as a function of motor nominal power. During 1999 an agreement had been reached and three efficiency classes has been defined for 2 and 4 pole electric motors. EuroDEEM has already included a data

field for the efficiency class and the corresponding class assignment will be made in the next release. The following graph (Figure 8) shows the 4-pole motor efficiencies compared to the curves, which define the 3 efficiency classes.

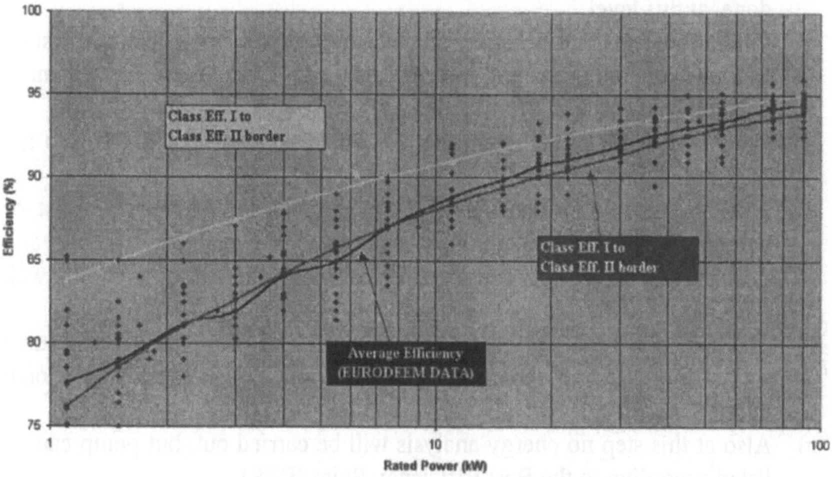

Figure 8: 4-Pole Motor Efficiency Distribution (EuroDEEM data) compared to Final Energy Efficiency Classification Curves for 4 Pole Motors – 50 Hz, according to IEE 34

5 Development of Pump Device Application

Although many software databases and tools are already existing on the market (PumpBase - Tahoe design USA, Flowserve PumpSel - USA, Wilo SELECT, etc.), they are related only to a single manufacturer production line. The existing databases do not include data from many manufacturers.

One of the goals of EuroDEEM is the dissemination of information and management tools for promoting energy efficiency & savings regarding the use of electric motor driven systems in the industrial sector. Thus it is important to rely on the European market pump data for providing to end-users a wide range selection.

The EuroDEEM Project has planned to include a Pump database and Selection Tool as follows (see Figure 9):

1) Development of a <u>*"fast & easy"* type/application selection tool</u>
   a) The tool will permit the user to find a list of manufacturers that produce pumps matching a specific pump type or application.
   b) The output of the query will give a list of manufacturers, each with a list of pumps together with their description (type and application, head and flow ranges, drive type, construction materials, national distributors, web links, email, etc.)

c) The idea behind this option is to have all possible pump manufacturers and their pumps in the database in order to let the end user look-up manufacturers that produce a specific pump type or a specific application within a specific range of head and flow. No energy analysis will be done, at this level.
d) The amount of data required for this easy tool is small, but the system can provide only general information with links where to get more detailed analysis and information..

2) Development of an intermediate tool for selecting a pump according to the user's specifications (Figure 10).
   a) This tool should provide a list of manufacturers, each with a list of pumps together with their description (type and application, head and flow ranges, drive type, construction materials, national distributors, web links, email, etc.).
   b) The selection is based on Total Dynamic Head (TDH) and flow, static suction head, pump speed, pump application, pump type and optional pump features.
   c) Also at this step no energy analysis will be carried out, but pump can be listed according to the Best Efficiency Point (BEP).
3) Development of a tool for selecting a cost-effective pump according to the user's specifications.
   a) This tool is based on the previous one, but is more complex and sophisticated. This selection tool has to provide a list of most suitable and cost-effective pumps for a specific use. It must be capable of performing energy analysis, power requirements, and it must get linked to the motor and VSD database for suggesting possible energy savings.
   b) The end-user must insert the description of the system to be driven by a pump, by means of design TDH and flow (or system curve), NPSHA and static suction head, total static head, liquid type, duty profile, pump speed, orientation, application, type, and optional pump features.
   c) The data that are required for this type of application are more than those required from the previous case. The data required are all contained in pump catalogues: Head at different flow points for various impeller size, Efficiency and power at different points, NPSH at different points, type of pump, applications, orientation, drive type, features, insulation class, materials, shaft seal, motor included, etc.
   d) The goal of this application is to provide to the end-user a tool for Company Inventory management (pump replacement and/or energy saving opportunities through VSD, etc.), as well as a tool for selecting suitable and cost-effective pumps for a specific application.

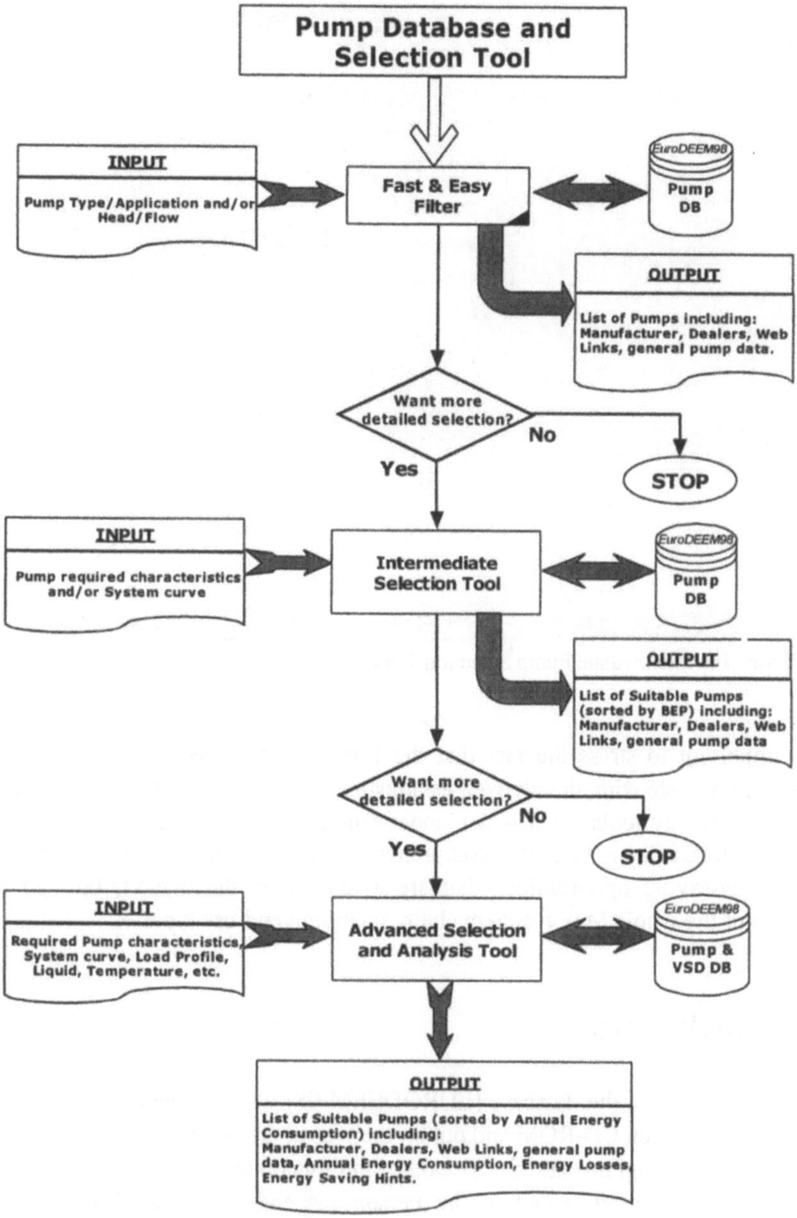

Figure 9: Schematics of the Pump Selection tool and Database.

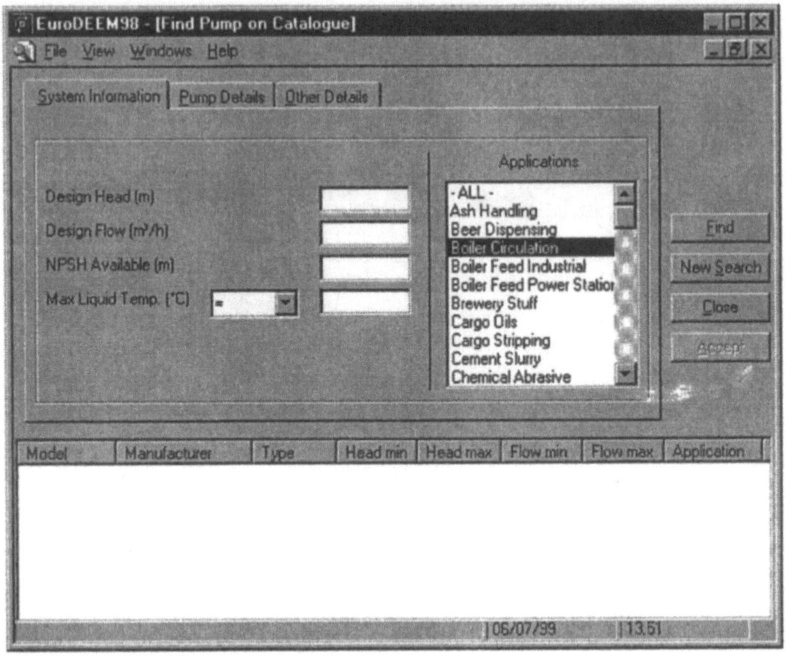

Figure 10: Intermediate Pump Selection Tool.

It is important to stress the fact that the EuroDEEM Project does not want to repeat or compete with already well established tools, but it wants to provide easy and user-friendly tools for a wider understanding of energy related issues. The goal of EuroDEEM is to give a comprehensive and somehow analytical view of the energy-saving opportunities that are available on the market, through the analysis of the whole Motor-system chain, including end-use types.

6 Conclusions

The new version of the database (EURODEEM 98) was completed at the end of 1998, distributed on CD-ROM and put on the web at the beginning of 1999.
Software block diagrams and description of the functional and visual performance were prepared and presented to the person responsible at DG XVII, and to CEMEP experts during various meetings.

Since EURODEEM must be a tool for supporting motor technical experts in decision-making, the collaboration from Electric Motor manufacturers for obtaining technical data, with National Institutions and industrial users for disseminating and using the results of the project had been considered from the beginning of the project of basic importance. Informal advisory group and contacts had been set up for receiving suggestions and remarks aimed at improving the database development. Permanent contacts with CEMEP, Europump, ETSU and various EM experts have been established.

The EuroDEEM product seems to receive a large consensus and this is an encouragement to pursue in its improvement and development. Dissemination actions are also needed in the very near future in order to involve more and more users, whose critical feedback and suggestions are essential for the development of future releases of EuroDEEM.

7 Bibliographic Reference

Motor study group (October 1996): Actions To Promote Energy Efficient Electric Motors, European Commission, DG XVII

Cocchi S., Conti F., Tanner R., (1996): The European database on Efficient Electric Motor Systems, Lisbon Conference 1996.

Evaluation of the Design Options and Cost Impact of Improving Induction Motor Efficiency

F.Parasiliti and M. Villani

Department of Electrical Engineering, University of L'Aquila,
67040 Poggio di Roio, L'Aquila, Italy
Tel. : +39 0862 434433, Fax : +39 0862 434403, e-mail : rock@ing.univaq.it

Abstract. Aim of the paper is the evaluation of the manufacturing cost to move a motor from one efficiency class to the next one, with reference to the European Commission and CEMEP (European Committee of Manufacturers of Electrical Machines and Power Electronics) classification scheme: the analysis deals with low voltage three-phase, 4 pole, induction motors in the 0.76÷90 kW power range. For each movement, the low cost design has been chosen, by considering several design strategies.
The results of the study concern both the increase of Manufacturing Costs and Total Costs of the motor, and the percentage variation of active material costs respect to the "Reference motors" that represent typical commercial motors.
The study was carried out in co-operation with several european Manufacturers that provided technical and economical data of commercial induction motors.

Keywords: Efficiency, induction motors, optimisation, design, cost.

1 Introduction

Developing new ranges of motors represents additional cost for the Manufacturers and should be useful to evaluate the increase of manufacturing cost of new design and the impact on the motor production.
Useful improvements in efficiency can be made modifying existing designs by adding more active material and using lower loss steel, either within existing frame size: additional tooling costs for these options are of less importance.
Further efficiency improvement can be achieved by a new design of stator and rotor shape but this solution requires higher cost of tooling for such modern designs above all if outside stator diameter is modified respect to the conventional design.
Aim of the study has been the evaluation of the manufacturing cost increasing when the existing motor is moved from one efficiency class to the next one with reference to the EC/CEMEP classification: the analysis concerned low voltage

three-phase induction motors, 4 pole, 50 Hz, 400 V, TEFC, in the 0.75÷90 kW power range.
Seven power ratings have been chosen and particularly: 1.5, 3, 7.5, 15, 22, 45 and 75 kW.
The present study has been carried out thanks to the co-operation with several EU motor Manufacturers which were submitted a detailed form to, in order to collect technical and economical data on their commercial motors.
The EC/CEMEP induction motor classification scheme foresees three levels as shown in Table 1 and Figure 1.

Table 1: The EC/CEMEP induction motor classification scheme (4 pole)

kW	Class Eff III	Class Eff II	Class Eff I
1.5	< 78.5	≥ 78.5	≥ 85.0
3	< 82.6	≥ 82.6	≥ 87.4
7.5	< 87.0	≥ 87.0	≥ 90.1
15	< 89.4	≥ 89.4	≥ 91.8
22	< 90.5	≥ 90.5	≥ 92.6
45	< 92.5	≥ 92.5	≥ 93.9
75	< 93.6	≥ 93.6	≥ 94.7

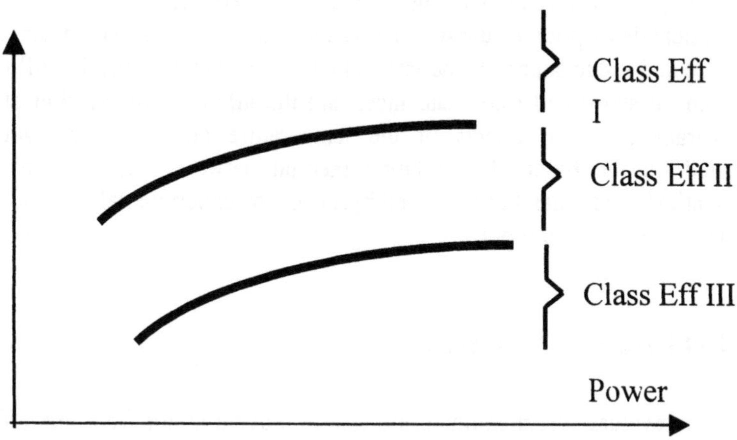

Figure 1: Efficiency levels

For each movement a low cost motor design was involved by means of an optimisation procedure developed by the authors [1,2,3,4,5]; the optimisation was formulated as constrained maximisation of the objective function "rated efficiency" expressed in terms of the motor design variables.
The design optimisation can be dealt with as a mathematical problem as follows:
find $X = (x_1, x_2, \ldots x_n)$ such that

$F(X)$ is a maximum and
$G_j(X) \geq 0$ for $j = 1, 2, \ldots m$,

where:

$F(X)$	is the objective function;
X	is the set of independent variables;
$G_j(X)$	are the constraint functions (inequality constraints);
n	is the number of independent variables;
m	is the number of constraint functions.

The constraints are the physical and performance constraints of the design, that is its specifications. For an induction motor typical constraints are: starting torque, starting current, breakdown torque, rated power factor, stator winding temperature rise, rotor bars temperature rise, rated slip, slot fullness, efficiency at partial load.
The independent variables concern the stator and rotor stampings, the stator windings, the stack length and the air-gap length. Each variable was varied between an upper and a lower limit according to the Manufacturers suggestions, in order to obtain a final optimised design whose dimensions are consistent with the frame of the same power initial motor. Some of the variables like number of slots, coil pitch, stator slot opening etc. were assumed fixed.
The physical description of the motor is reduced to equivalent parameters such as resistance and inductances: the adopted model takes into account the influence of saturation on stator and rotor reactances and the influence of the skin effect on rotor parameters. The effects of the temperature on motor resistances are computed on the basis of a detailed thermal network. The validity of the mathematical model has been verified by means of experimental tests on several three-phase induction motors.

2 Preliminary Analysis

In order to evaluate the manufacturing cost to move a motor from one efficiency class to the next one, two "Reference motors" have been introduced, for each power, that should represent typical commercial motors in the Class Eff III and Class Eff II respectively.
These motors, labelled R0 and RM, present efficiencies equal to average values of the motors in Class Eff III and Eff II available in EU [6, 7]:

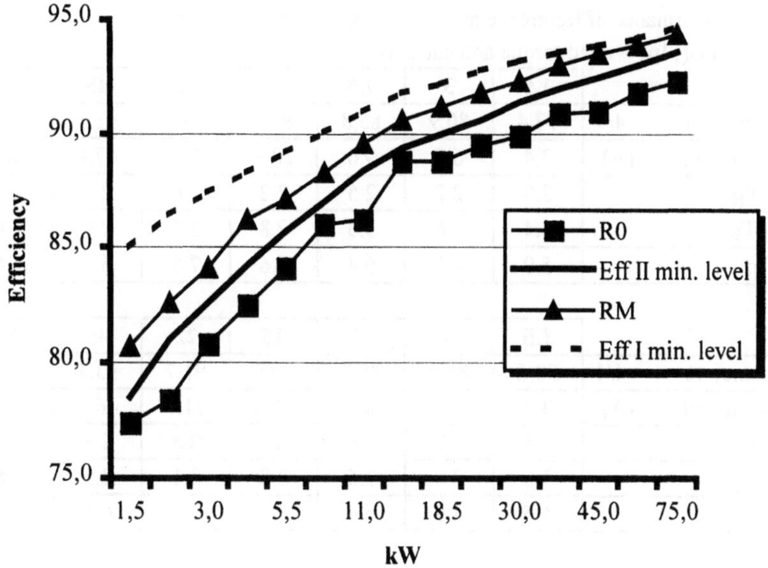

Figure 2: Minimum efficiency values for the Classes Eff II and Eff I, and average efficiencies for R0 and RM motors available in the market, versus power.

R0 → Reference motor in the Class Eff III, whose efficiency is the average of Class Eff III motors available in the EU market (in 1998);
RM → Reference motor in the Class Eff II, whose efficiency is the average of Class Eff II motors available in the EU market (in 1998).

Figure 2 shows the minimum efficiency values for Classes Eff I and II, and the average efficiencies for R0 and RM available in the market, versus power.
In order to compare the costs of these "Reference motors" with the new higher efficiency motors ones, the authors designed 14 motors (one for each considered power) with the same efficiency of R0 and RM: they have been designed by means of the optimisation software, starting from the design data of existing motors provided by European Manufacturers. Their performance are shown in Table 2, with reference to the efficiency, starting current (I_{ST}), starting torque (T_{ST}) and breakdown torque (T_{BR}), and are very close to those of typical designs.

Starting from the Reference motors, two distinct movements have been studied (see Figure 3) and particularly :

Mov1 → From R0 to the minimum efficiency value of Class Eff II;
Mov2 → From RM to the minimum efficiency value of Class Eff I.

Table 2: Performance of Reference motors corresponding to R0 and RM
(T_R, I_R, = rated torque and current)

R0		**1.5**	**3**	**7.5**	**15**	**22**	**45**	**75**
Efficiency %:	4/4	77.4	80.9	85.8	88.6	89.6	91.1	92.3
Phase current	(A)	3.4	6.8	9.0	28.4	22.8	47.6	78.6
T_{ST}/T_R		2.0	2.2	2.5	2.2	2.6	2.5	2.9
T_{BR}/T_R		2.4	2.6	3.5	2.8	3.2	2.6	2.8
I_{ST}/I_R		5.0	4.7	6.4	6.6	7.5	6.2	7.4

RM		**1.5**	**3**	**7.5**	**15**	**22**	**45**	**75**
Efficiency %:	4/4	80.6	84.2	88.2	90.6	91.7	93.4	94.4
Phase current	(A)	3.2	6.7	8.8	27.8	21.7	46.9	77.4
T_{ST}/T_R		2.5	2.2	2.6	2.2	2.5	2.5	2.9
T_{BR}/T_R		3.0	2.8	3.5	2.8	3.1	2.6	2.8
I_{ST}/I_R		5.5	4.9	6.4	6.6	7.3	6.1	7.4

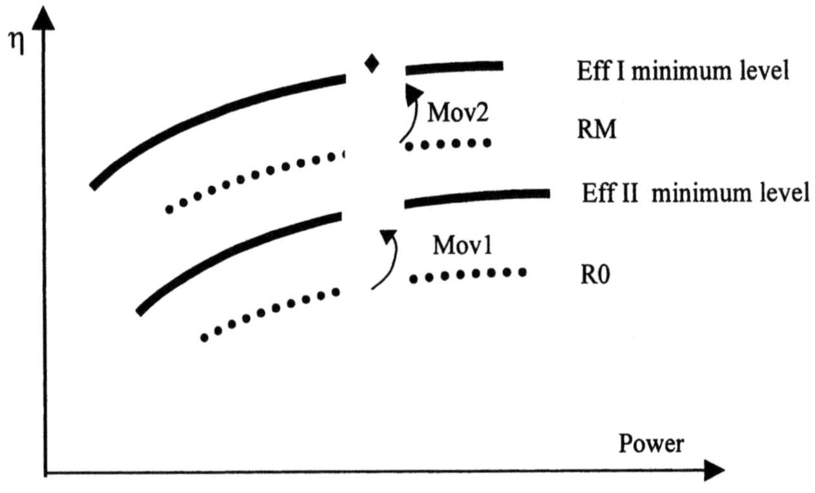

Figure 3: Movements from reference motors

For each "movement" several design strategies have been considered and the lowest cost one has been chosen, with reference to the Manufacturing Cost of motor:

$$\text{Manufacturing Cost} = (\text{Active material Cost}) + (\text{Structure Cost}) + (\text{Direct Labour Cost}) \quad (1)$$

The "Active material Cost" includes copper, aluminium and steel, while the "Structure Cost" includes the other components of the motor (housing, bearings, shaft ...) ; the "Direct Labour Cost" is directly related to motor manufacturing.

In addition to the Manufacturing Cost, the Total Cost (TC) of motor has been introduced that is :

$$\text{Total Cost} = \text{Manufacturing Costs} + \text{Other Costs} \tag{2}$$

where "Other Costs" include services, auxiliaries, financial interests and personnel expenses.

In the present study the following hypothesis have been introduced :

- "Other Costs" are constant during each "movement" ;
- New "Tooling Costs" (i.e. new lamination tooling, new housing die-casting,) have not been taken into account.

The economical data obtained from EU motor Manufacturers have allowed to extrapolate the average costs of active materials and Direct labour, used in the present study for the estimation of the Manufacturing Cost; these costs are shown in Table 3. Moreover, in order to evaluate correctly the increase of Total Cost respect to the Reference motors, it has been necessary to know a priori the percentage influence of each voice in the Total Cost of motor; thanks to the co-operation with Manufacturers, the subdivision shown in Table 4 has been adopted.

Table 3: Average costs of active materials, structure and direct labour

<table>
<tr><th>kW</th><th>1.5</th><th>3.0</th><th>7.5</th><th>15</th><th>22</th><th>45</th><th>75</th></tr>
<tr><td>Materials (Euro/kg)
Cu
Al
Steel</td><td colspan="7">
3.10
1.85
0.65</td></tr>
<tr><td>Structure (Euro)</td><td>17.4</td><td>27</td><td>48</td><td>55.7</td><td>74.8</td><td>304</td><td>500</td></tr>
<tr><td>Direct Labour (Euro/h)</td><td colspan="7">20</td></tr>
<tr><td>hour/kg (related to active materials)</td><td>0.05</td><td>0.045</td><td colspan="3">0.035</td><td colspan="2">0.023</td></tr>
</table>

Table 4: Percentage subdivision of costs (in Total Cost)

	$P \leq 22$ kW	$P > 22$ kW
Material Cost (Active material + Structure)	45 %	45 %
Direct Labour	20 %	10 %
Other Costs	35 %	45 %

3 Design Strategies

The movement from one efficiency class to the next one can be achieved by choosing several design strategies each of one presents different cost for the Manufacturers. A good choice should tend to a lower manufacturing cost that does not affect heavily the tooling costs. For this reason three levels have been introduced each of one has different cost of tooling (Table 5) and foresees several design strategies (labelled 0, 1, ... 9): the chosen design strategy affects the number of independent variables for the optimisation procedure.
For each movement the motor design has been optimised by testing gradually each design strategy.

Table 5: Design strategies

1st level : No tooling cost (NTC)	0	New stator winding
	1	New Steel
	2	(0) + (1)
	3	(2) + New Stack length (*)
2nd level : Low tooling cost (LTC)	4	New Stator lamination and/or Rotor Lamination + New Stator Winding
	5	(4) + New Steel
	6	(5) + New Stack length (*)
3rd level : High tooling cost (HTC)	7	(4) + New Outside Stator Diameter and/or Inner Stator Diameter + New Housing
	8	(7) + New Steel
	9	(8) + New Stack length

(*) alternative existing housing, as required and available.

It is important to underline that the variables “Stack length”, “Outside Stator Diameter” and “Inner Stator Diameter” have been varied with reference to “normalised values” only, according to the Manufacturers suggestions : in this way no additional cost for casting process will occur.
The first level concerns the change of steel (e.g. low loss steel), stack length and stator winding : in this case the stator and rotor slots dimensions and the inner and outside stator diameter are unchanged. The challenge is to physically fit more active material into the motor to reduce iron and copper losses.
The cost of tooling for the new designs is effectively the same of the traditional design since the need for costly new lamination punch tools or stator housing tools are avoided.
The second level foresees also the change of lamination (stator and rotor slots dimensions) and this requires a renewal of the lamination tooling even if the inner and outside stator diameter are unchanged.

For both levels, if the stack length of the new design does not exceed of about 10% the old motor one, an alternative existing housing could be used.

The third and more expansive level requires to change all motor dimensions, inner and outside stator diameter included. In this case a final option is the use of a larger frame size with modification of the existing housing: the shaft height can be maintained by using lower profile feet, alternatively in flange mounting types a standard flange can accommodate a longer motor body.

3.1 Steel Grade

The electrical steel affects heavily the motor performance, and its right choice combined with the design optimisation of the motor allows to achieve better results and higher efficiency [8].
The choice of a "suitable" electrical steel depends on several aspects such as cost, workability, 'business tradition' and storehouse demands. In this study five "fully processed" commercial steels have been considered and tested (labelled with S1…S5) in addition to a typical steel used by several Manufacturers in their designs and labelled S0 : all the alternative electrical steels present lower loss respect to S0 as shown in Figure 4 and two products (S1 and S5) have higher permeability.
The percentage cost increase respect to S0 varies from 15% for S1 to 75% for S4 and S5 (see Figure 5).

3.2 New Stator Winding

The new stator winding presents, respect the reference motor one, a different number of turns per phase and wire size: the winding distribution is unchanged.

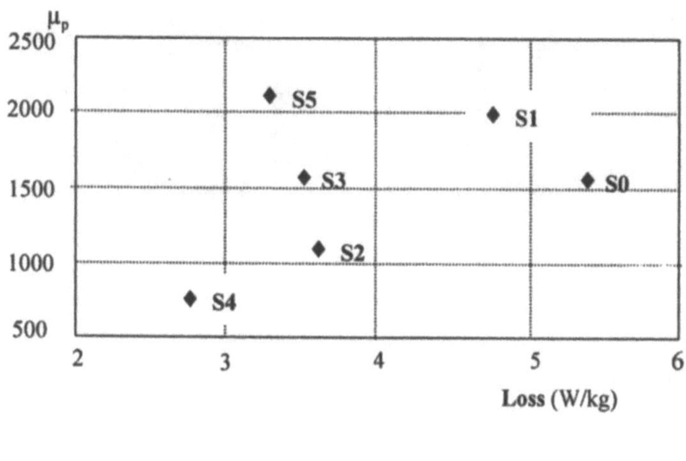

Figure 4: Permeability and loss at 1.5 T of the considered steels

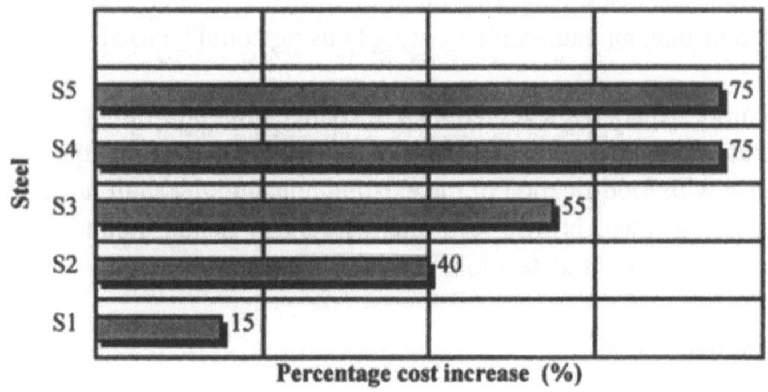

Figure 5: Percentage cost increase of new steels respect to S0

Moreover, the slot fullness has been constrained taking into account the motor sizes and the insulation thickness of stator slots, according to the Manufacturers suggestions.

3.3 Stack Length and Diameters

The Stack length and the Inner and Outside Stator Diameters (Figure 6) affect the manufacture of motor and its cost. Particularly :

Stack length → Housing dimension and Casting process
Outside Stator Diameter → Housing dimension
Inner Stator Diameter → Casting process

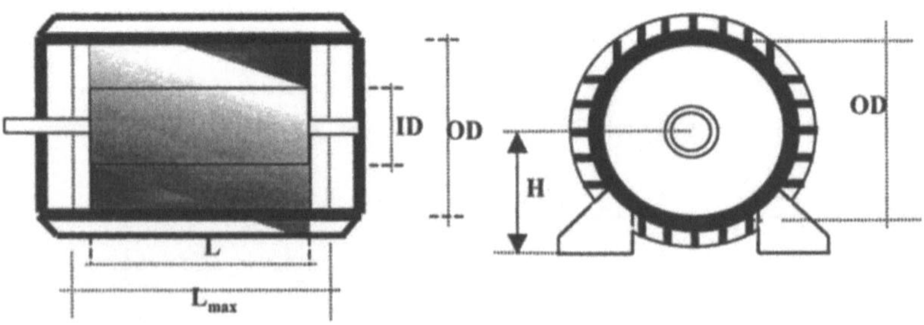

Figure 6: Motor dimensions

Since only "normalised values" have been considered, this has given rise to a limited number of combinations and consequently has affected the final design. The possible combinations depend on each manufacturer. Those considered in the present study correspond to actual situations: they are realistic but not necessarily typical. Table 6 and Table 7 show two figures with a "low number" and an "high number" of combinations. The design optimisations have been carried out taken into account these imposed values, i.e. the variables "Stack length", "Outside Stator Diameter" and "Inner Stator Diameter" have been varied in a "discrete way".

Table 6: Example of low number of combinations

1.5 kW – 4 pole H = 90	**A**	**B**	**C**
Outside St. Diameter		$(OD)_1$	
Stack length	L_1	L_2	L_3
Inner St. Diameter		$(ID)_1$	
Housing	90L	100L	100L

Table 7: Example of high number of combinations

15 kW – 4 pole H = 160	**A**	**B**	**C**	**D**
Outside St. Diameter	$(OD)_1$		$(OD)_2$	
Stack length	L_1	L_2	L_1 L_2 L_3	L_1 L_2 L_3 L_4
Inner St. Diameter	$(ID)_1$		$(ID)_2$	$(ID)_3$
Housing	160L	180M (*)	180L (*)	200L (*)

(*) The use of this housing requires some changes on the "housing feet" and "fins height" in order to satisfy the shaft height.

4 Results and Conclusions

Table 8 presents the results of the study and particularly the percentage increase of Total Cost (ΔTC) respect to the Reference motors. Moreover, for each movement the percentage variation of losses and active material costs have been included. Figure 7 shows a reasonable extrapolation to the full considered power range.

- For all power ratings the first movement has not required any additional tooling cost since the increase of efficiency has be reached combining the new stator winding with a low loss steel (S1) (for the 75 kW better results has be obtained with S2). The increases of total cost vary between 2 and 7%.
- The second movement needs a low tooling cost since the stator and rotor slots dimensions have been modified. For 1.5, 3 and 45 a low loss steel at high permeability has been adopted. The new motors have the same outside

and inner stator diameter of the reference motors and this has not affected the cost of casting process and housing. In case of the 7.5 and 75 kW motors, alternative existing housings have been used, without lamination changes for 75 kW. The total costs vary between 5 and 15%, decreasing for upper powers.

- The cost evaluation is related to a movement towards the lower limit of the upper efficiency class. Better efficiencies could imply lightly higher cost.
- The power ratings considered in the study are in the upper band of the standard shaft height ranges. For the other power ratings the movement towards the upper class could be easier.

The new designs satisfy fully the efficiencies at ¾ that are higher than the full load ones.

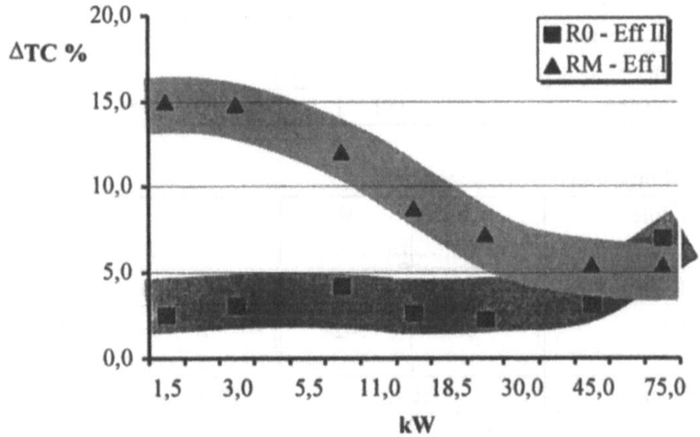

Figure 7: Total cost increase respect to the reference motors

Table 8: Percentage variations of Losses, Active material costs and Total costs respect to the Reference motors (NTC = No tooling cost ; LTC = low tooling cost ; HTC = high tooling cost).

(0),(1),...(9) labels of design strategies (see Table 5). S1,...S5 types of steel (see Figure 4).

1.5 kW	*Efficiency*		*Design strategy*	*ΔLosses %*			*ΔCosts %*			*ΔTC*
	¾	**4/4**	*and Steel*	Cu	Al	Fe	Cu	Al	Fe	%
R0 → Eff II	80.0	**78.4**	NTC → (2) ; S1	-2	-2	-23	3	-	15	2.0
RM→ Eff I	84.9	**84.8**	LTC → (5) ; S5	-28	4	-39	46	5	75	15.0

3 kW	*Efficiency*		*Design strategy*	*ΔLosses %*			*ΔCosts %*			*ΔTC*
	¾	**4/4**	*and Steel*	Cu	Al	Fe	Cu	Al	Fe	%
R0 → Eff II	84.1	**82.7**	NTC → (2) ; S1	-18	-	-17	17	-	15	3.1
RM→ Eff I	87.8	**87.2**	LTC → (6) ; S5	-18	-12	-34	8	5	88	14.8

7.5 kW	*Efficiency*		*Design strategy*	*ΔLosses %*			*ΔCosts %*			*ΔTC*
	¾	**4/4**	*and Steel*	Cu	Al	Fe	Cu	Al	Fe	%
R0 → Eff II	86.8	**87.1**	NTC → (3) ; S1	-18	-9	-11	8	7	22	4.9
RM→ Eff I	89.8	**90.0**	LTC (*) → (6) S3	-26	-30	10	13	26	21	12.6

(*) alternative existing housing required

15 kW	*Efficiency*		*Design strategy*	*ΔLosses %*			*ΔCosts %*			*ΔTC*
	¾	**4/4**	*and Steel*	Cu	Al	Fe	Cu	Al	Fe	%
R0 → Eff II	90.2	**89.4**	NTC → (2) ; S1	-2	-	-26	2	-	15	2.6
RM→ Eff I	92.4	**91.8**	LTC → (5) ; S2	-12	-2	-35	12	-	40	8.7

22 kW	*Efficiency*		*Design strategy*	*ΔLosses %*			*ΔCosts %*			*ΔTC*
	¾	**4/4**	*and Steel*	Cu	Al	Fe	Cu	Al	Fe	%
R0 → Eff II	91.1	**90.5**	NTC → (0) ; S0	-18	-	-2	19	-	-	1.7
RM→ Eff I	93.2	**92.5**	LTC → (5) ; S3	-3	-18	-26	8	14	35	7.7

45 kW	*Efficiency*		*Design strategy*	*ΔLosses %*			*ΔCosts %*			*ΔTC*
	¾	**4/4**	*and Steel*	Cu	Al	Fe	Cu	Al	Fe	%
R0 → Eff II	92.9	**92.5**	NTC → (2) ; S1	-20	-5	-27	22	-	15	3.1
RM→ Eff I	94.6	**93.9**	NTC → (2) ; S5	-21	-2	5	25	-	13	4.5

75 kW	*Efficiency*		*Design strategy*	*ΔLosses %*			*ΔCosts %*			*ΔTC*
	¾	**4/4**	*and Steel*	Cu	Al	Fe	Cu	Al	Fe	%
R0 → Eff II	93.8	**93.6**	NTC → (2) ; S2	-27	3	-28	38	-	40	7.1
RM→ Eff I	94.7	**94.6**	NTC (*) → (3) S4	-13	3	-3	16	4	7	5.4

(*) alternative existing housing required

4.1 Design of Improved Efficiency Motors without Additional Tooling Costs

The additional tooling and other factory costs associated with making higher efficiency motors, compared to standard efficiency motors, will vary between companies.

However, the biggest differences will be between those companies which base high efficiency motors on existing standard designs, and those which use completely new designs.

Within the constraints of an existing design, it should be useful to evaluate the upper limit to how much efficiency can be improved.

For this, starting from the references motors, the stack length, stator winding and steel grade have been modified in order to evaluate the maximum achievable efficiency without affecting the tooling costs. Table 9 and Figure 8 show the obtained results.

- Starting from R0, Class Eff II is always achievable..
- Starting from RM, Class Eff I minimum level has been achieved (but not exceeded) for the higher powers only.
- Starting from R0, the efficiency can be improved by an amount equal approximately to 2/3 of the Eff II band.
- The highest efficiencies have been achieved by choosing the expensive "high permeability" steel S5 and "low loss" steel S4 (for the 75 kW motor). This affects heavily the total cost of the motor, mainly starting from R0.
- Manufacturer having a motor close to the lower limit of Class Eff II will be able to produce a motor in the upper part of the Class Eff II without new tooling costs.
- The choice to base the new higher efficiency motors on existing standard designs precludes future efficiency improvements.

Table 9: Maximum achievable efficiency from R0 and RM without additional tooling costs

kW	from R0		from RM	
	η	ΔTotal Cost %	η	ΔTotal Cost %
1.5	79.1	7.8	82.9	9.0
3	83.4	11.7	86.4	8.0
7.5	88.7	17.0	88.9	7.1
15	90.3	19.7	91.7	17.9
22	92.0	25.7	92.6	14.0
45	92.8	14.3	93.5	4.2
75	94.3	11.7	94.6	≈ 0

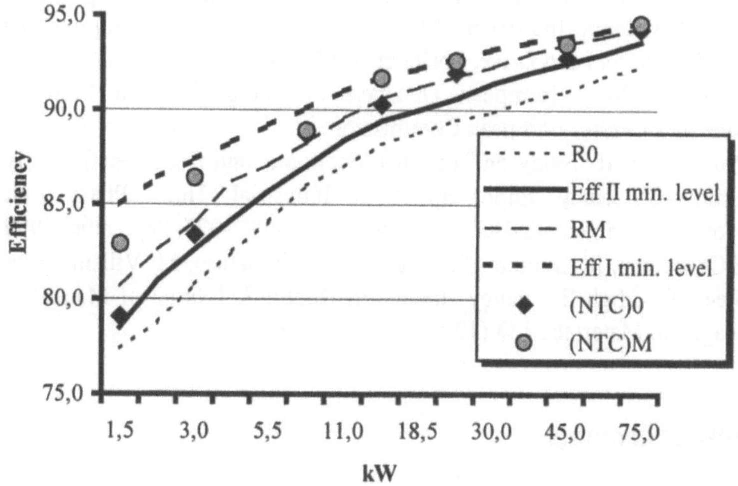

Figure 8: Highest achievable efficiency designs without additional tooling cost respect to efficiency classes:

♦ $(NTC)_0$ = highest efficiency from R0

● $(NTC)_M$ = highest efficiency from RM

5 References

[1] E. Chiricozzi, F. Parasiliti, M. Villani: "On the Use of an Optimization Technique for Small Induction Motor Design", International Conference on "Evolution and Modern Aspects of Induction Machines" - Torino, Italy, July 1986, pp. 295-303.

[2] E. Chiricozzi, F. Parasiliti, M. Villani: "Optimization of the Manufacturing Costs of 3-phase Induction Motor: Economic Analysis" (in Italian) - "L'Energia Elettrica" n. 2, 1991, pp. 67-72.

[3] F. Parasiliti, M. Villani: "Optimized Design of a Double Cage Induction Motor", ICEMA '93, Adelaide, Australia, September 1993, Vol. 1, pp. 81-86.

[4] F. Parasiliti, M. Villani: "Low Cost Optimization of an Induction Motor Line", Proceedings of the Tenth Scientific-Technical Conference on "Alternative Current Electrical Drives" (ACED), Ekaterinburg (Russia), March 1995, pp. 38-41.

[5] E.Chiricozzi, F.Parasiliti, M.Villani, "Experience in Design Optimization of High Efficiency Induction Motors", Energy Efficiency Improvements in Electric Motors and Drive, Springer, 1997, pp.116-137.
[6] EURODEEM – European Database of Energy-Efficient Motors, Joint Research Centre, European Commission.
[7] Project Report "Study on Technical/Economic and Cost/Benefit Analysis of Energy-Efficiency Improvement in Industrial Three Phase Induction Motors", Contract EC-DG XVII No. 4.1031/D/97-027, September 1999.
[8] E.Chiricozzi, R. Gubiotti, F. Parasiliti, G. Vigliaturo, M. Villani, "Electrical Steels for High Efficiency Induction Motors", Journal of Magnetism and Magnetic Materials, 133 (1994), pp. 533-535.

Acknowledgements

The authors would like to thank the following for their contribution:

CEMEP (European Committee of Manufacturers of Electrical Machines and Power Electronics);
Prof. A.De Almeida, University of Coimbra, Portugal;
Prof. D.Walters, EDS Consultant, Doncaster, United Kingdom;
Dr. F.Conti and S.Cocchi, Joint Research Centre, Ispra, Italy;
Dr. H. Falkner, ETSU, Harwell, United Kingdom.
Dr. J. Haataja, University of Lappeenranta, Finland.

This work was developed in the ambit of the SAVE Program and supported by European Commission. Contract EC-DG XVII No. 4.1031/D/97-027.

"Life-Cycle Costs" of Industrial Electric Drives in the Process Industry - Energy Consumption and Economics of Electric Drives

Kurt Bieniek

BASF Aktiengesellschaft, D-67056 Ludwigshafen, Germany

Abstract. For electric motors in the Process Industry, the purchase price is only a small fraction of the overall "life-cycle costs". The most important criteria are reliability, noise and efficiency. The engineering expense for assessing energy consumption must be kept in accordance to the saving potential.
In the German chemical industry, reliability has attained an encouragingly high level thanks to applying the "VIK recommendations".
Standardised sizes ("European Standard Motor") guarantee the interchangeability. That gives also a good basis for introducing improved motors, e.g. with less noise or with better efficiency.
Today's market offers already a high level of efficiencies: Improvements will justify only small price increases.
Despite their large number, small motors of less than 1 kW consume only a small percentage of the total electricity demand in a plant, and for large motors (above 100 kW) energy losses are already evaluated in detail. For the "Standard Motors" approximately between 1 and 100 kW, a standardised classification and labelling of the motor efficiencies will be a useful tool for selecting the optimal motors for their applications. To ensure interchangeability, the motor sizes and installation dimensions must remain unchanged.

Keywords. Electric drive, induction motor, life-cycle costs, efficiency, energy-efficient motor ("EEM")

1 Introduction

For electric drives in the process industry, the purchase price is only a small fraction of the overall "life-cycle costs". The most important criteria for the selection of motors are reliability, noise and energy consumption or efficiency.
This paper deals with the assessment of life-cycle costs of electric drives in the process industry, especially regarding the expense of engineering.

The US and the Canadian governments prescribed minimum efficiencies for EEMs (Energy-Efficient Motors) in the range of 0.7 to 140 kW and banned the sale of motors not conforming with these minimum efficiencies (in the USA laid down in the "Energy Policy Act of 1992" <1> having taken effect on October 24 1997). Stimulated by this, the European interest focussed on EEMs resulting in an agreement between the manufacturer association CEMEP and the European Commission for motors to be grouped into three efficiency bands in 1998 and 1999. Minimum efficiencies are set for the upper two bands ("ClassI" and "ClassII"). But what will be the benefits of such efficiency improvements, and what are the main obstacles to introducing EEMs?

2 Electric Drives in the Process Industry

Its robustness, reliability, ease of installation and good efficiency has made the squirrel-cage induction motor the most common type of drive in the process industry. They can be divided roughly into three power ranges:

- The large number of small motors up to about 1 kW driving auxiliaries, small handling or transport equipment and so on.
- Standard motors mainly between 1 and 100 kW, consuming about 30 to 50% of the industrial electrical energy. They are mostly selected according to the machine to be driven. Due to oversizing and / or varying duty the average load of these motors is about 60 % of the rating.
- Low-voltage "trans-standard" motors above approx. 100 kW and high-voltage motors consuming about 30 to 50 % of the industrial electrical energy (in the example of the Ludwigshafen plant of BASF (Figure 1), this percentage is even higher.). Their average load is closer to the rating, and the operating time is longer in most cases.

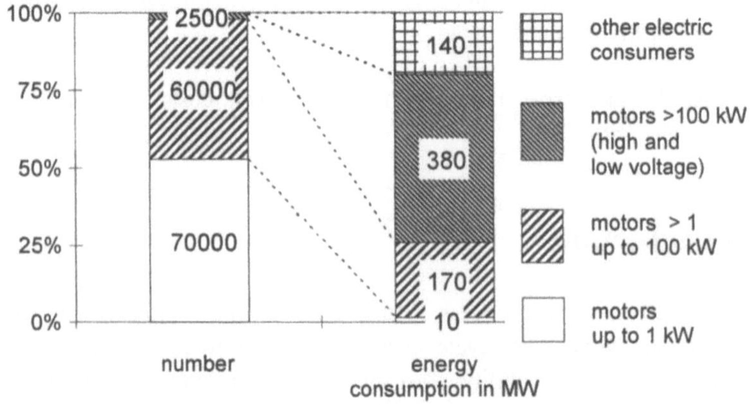

Figure 1: Numbers and energy consumption of electric drives of the BASF plant Ludwigshafen (1993)

Converter-fed variable-speed drives are becoming more and more important. Compared to direct-on-line motors they allow improved process control and adaptation to the actual load and thus a substantial improvement of the efficiency of the machine to be driven in cases of varying loads.

3 Life-Cycle Costs

Life-cycle costs mean the sum of the costs of a drive throughout its entire "life", from engineering / procurement via installation and operation / maintenance up to de-installation and final disposal. Figure 2 shows the typical costs of a line drive, Figure 3 gives some log-log curves from 0.1 to more than 100 kW for the costs incurring over 10 years of operation.

In the range of standard motors, the **costs for engineering and procurement** (called "transaction costs" by economists) depend only marginally on the power and may for small sizes even become higher than the purchasing price, a typical effect with small material. For this reason and in order to reduce the number of interfaces in the engineering process, engineering of these motors is highly standardised and simplified and integrated into process engineering. Introducing new engineering steps or even involving motor experts would significantly increase the expense and - in many cases more important - the time delay for the engineering process. For the "trans-standard" range and the high-voltage motors, higher expense and more time for engineering are accepted and therefore motor experts are involved.

The **purchasing prices** are subject to fierce competition. The prices increase less than proportionally to the power, and the ratio of price to output power approaches a rather low value for high powers. Where the same output power is available, high-voltage motors are distinctly more expensive than low-voltage motors.

The **costs for installation** (switchgear, control, wiring, transport, installation and commissioning) do not depend on the power up to about 5 kW and rise in line with the weight and the current for larger motors, but less than proportionally to the output power. The costs for installation of high-voltage motors are higher, mainly because of the more expensive switchgear.

For the **energy consumption** we have to estimate the averages in the process industry: In the "standard motors" range from 1 to 100 kW, the motors average load percentage is about 60 % of their rating for a period of about 2000 hours per year. The load of 50 to 70% of the rating results from oversizing and / or from varying process loads. The operation time of "only" 2000 hours per year despite the plant operating more than 8000 hours per year ("24 hours a day") results from the terms of operation, mainly from installed or uninstalled spares and from non-continuous operation. Since most of such motors are kept on stock, the load and the operating time of individual motors cannot be predicted. Above 100 kW, the load conditions are mostly known better and can therefore be considered in a way that for a higher loaded and / or longer running motor the energy consumption is evaluated higher.

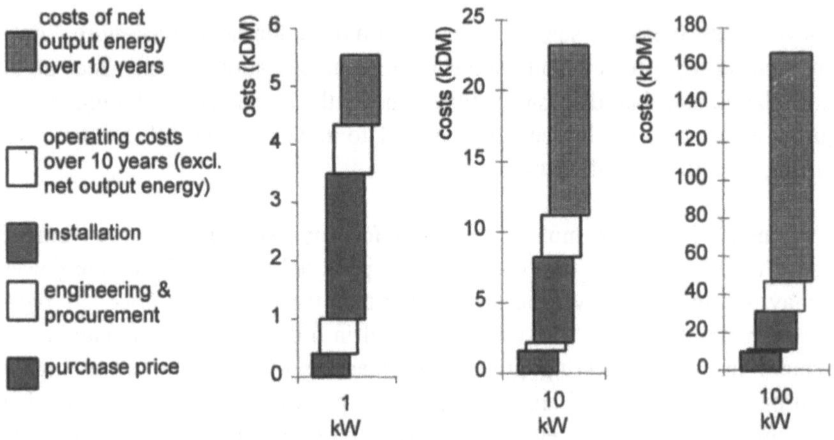

Figure 2: "Life-cycle costs" of typical electric drives of 1, 10 and 100 kW rated output over ten years of operation

As **operating energy costs** we regard only the costs of the **energy losses**. The costs of the output energy the motor delivers to the machine to be driven cannot be influenced by the selection of the motor. Nevertheless, the output energy and / or the energy consumption of the machine to be driven becomes the main field of energy conservation at powers of 10 kW and more. This requires a good knowledge of the machine and of the process. In many cases, variable-speed drives offer a high potential for energy conservation.

The **maintenance costs** can be assessed basing on failures and repair statistics: The reliability of motors of the German chemical industry has attained an encouragingly high level thanks to the application of the "VIK recommendations" <2> (VIK = Association of Industrial Cogenerators and Energy Consumers in Germany). Based on an average failure interval MTBF ("Mean Time Between Failures") of standard motors of VIK quality between 15 and 25 years we can assess an adequate figure for repair.

Besides the cost for repairing the drive (i.e. repairing the faulty motor or replacing it by a new motor) a failure may cause much higher **costs due to the loss of production**. They can be estimated with the formula

CF = PF x (CR + DF x CT)

where the symbols mean:

CF	=	costs of production failure
PF	=	probability of failure
CR	=	cost of restarting
DF	=	duration of failure
CT	=	cost per time of loss of production.

The formula shows the importance of reliability (i. e. the inverse value of the probability of failures). For that reason the German chemical process industry uses almost exclusively motors fulfilling the VIK recommendations <2>. In order to detect defects prior to a failure, important larger machines are subject to preventive maintenance like monitoring bearing vibrations and temperatures. In order to avoid the necessity of stopping and restarting the process, vital machines have stand-by spare machines (mainly as a precaution against machine failures but also effective in cases of motor failures). The costs for stopping and restarting can become extremely high for plants continuously running. But also in "normal" cases, the costs resulting from a failure would become dramatically higher than all the costs of a drive described in this paper. For that reason, reliability is by far the most important quality of a motor in the process industry.

The costs of de-installation and disposal of motors when closing down the plant are of no importance in most cases. Sometimes even some small proceeds may be obtained from selling the scrap.

The costs of variable-speed drives follow - at a higher level - to a comparable course, i.e. an approximately constant "base" cost up to about 1 kW and a less-than-proportional increase of costs vs. the power. The major saving potential lies in the reduction of the power of the machine to be driven, the net output power. Figure 3 shows for line drives that above approx. 5 kW the cost of the net output power exceeds the sum of all operation-independent costs.

Up to about 40 kW, standard solutions for engineering, procurement and installation are applied, for powers up to about 10 kW - the range of the largest numbers of drives - "integral motors" with integrated converter and speed control will offer further relief. After more than 20 years of spreading use of converters and of variable-speed drives their reliability has reached a high level: In the chemical process industry the MTBF has grown up to about 10 years.

4 Selection and Dimensioning of Electric Drives

Figure 3 compares the operation-independent costs with the operating costs (including the costs of the energy losses) and with the costs of the net output energy. The operating costs are based on the average load of 60 % of the rating over a period of 2000 hours per year. The values for 500 kW are fictional: in the process industry drives of that range would not be installed as direct-on-line but either as a high-voltage line motor or as a variable-speed drive. The figure shows that the original purchase price is of minor importance compared with other operation-independent costs; its share is between 10 and 50 %. More important are the costs for installation and switchgear and those for engineering and procurement, the so-called "transaction costs".

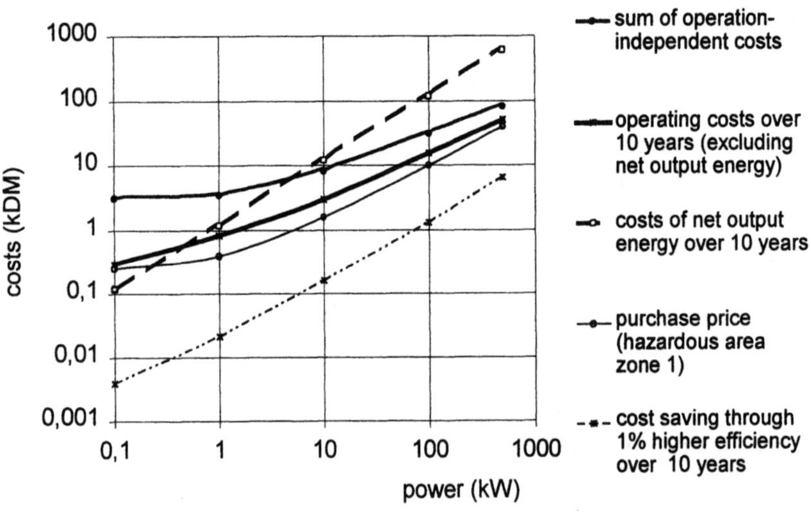

Figure 3: Costs of line drives 0.1 to 500 kW over ten years

The maintenance costs and the failure rate have attained a favourably low level thanks to high-reliability specification and selection of the motors. Lately attention has been focussing on low noise level requirements as laid down in occupational safety and health regulations as well as in environmental protection regulations. The most important criteria for selecting a motor are in order of importance: reliability, noise and efficiency. Easy handling and ease of repair and / or the availability of spare motors are presupposed. The standardised motor sizes with the allocation of power output and installation dimensions ("European Standard Motor") <3> guarantee interchangeability of motors of different series and different manufacturers. This is also a good basis for introducing improvements, such as motors producing less noise or offering better efficiency.

Since above 1 kW the costs of the energy losses over a period of three to five years already exceed the purchase prices, the energy losses or efficiencies are assessed when selecting a motor and comparing prices. The bottom curve reflects the accumulated energy costs savings over 10 years if the efficiency were 1 % higher. This curve lies by a factor of 20 (for 1 kW) to 7.5 (for 100 kW) below the purchase price. That means that based on a return-of-investment time of 4 years an efficiency improvement of 1 % could "justify" a price increase of 2 % (for 1 kW) to 5 % (for 100 kW) of the purchase price for an average motor without presuming increased load or operating time. For higher- or longer-loaded motors even higher prices would be acceptable. Of course, reliability, noise and operability must not deteriorate.

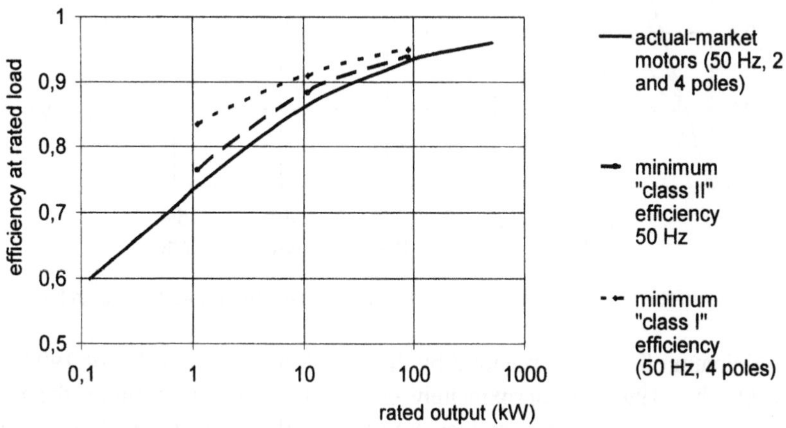

Figure 4: Full-load efficiencies of actual-market motors vs. rated output (totally-enclosed fan-cooled standard induction motors, 50 Hz, 2 and 4 poles) compared to the limits for "Class I" and "Class II"

However, the assessment of efficiency or energy losses must be easy for the non-expert since the selection of the motor should remain part of the general material procurement or of the plant engineering process. Involving extra steps to be performed by experts would inflate the engineering costs and might even cause time delays. For that reason the introduction of standardised efficiency classes will be a very effective tool allowing assessment without the need to interpret catalogue data from the suppliers.

Already standard motors available on today's market offer favourable efficiencies (Figure 4) - especially at partial load shown in Figure 5. The efficiency has its maximum at about 75 % of the rated load. It remains close to rated efficiency from 100 % down to 50 %, for large motors even down to 30 % of the rated load. For comparison, Figure 4 shows the class limits agreed between CEMEP and the EC for "Class I" (highest line) and for "Class II" (medium line). Compared to the actual market values, the motor losses are reduced by about 1/10 for "ClassII" and by about 1/3 for "Class I" raising the efficiency by 1 % and 8 % (for 1 kW) or by 0 % and 1 % (for 90 kW) respectively. Motors meeting the "ClassII" level are already available on the market, "Class I" motors are - at least as standard motors - still rare.

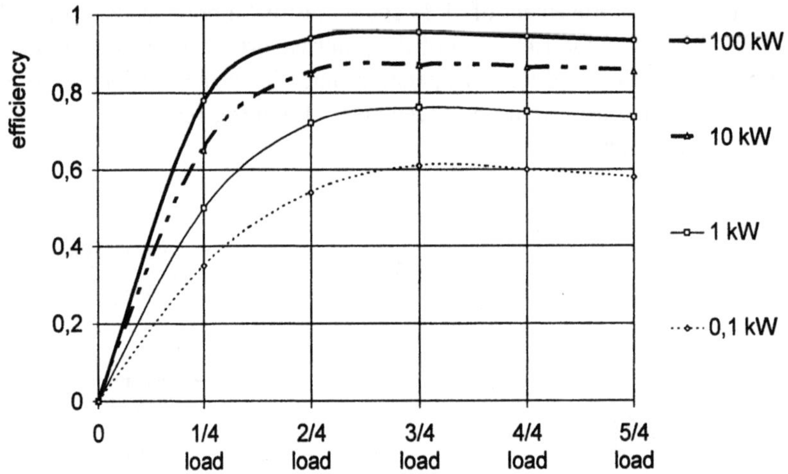

Figure 5: Course of the efficiencies of actual-market standard motors over the relative load

The minimum efficiencies required by the U.S. "Energy Policy Act of 1992" <1> since October 1997 lie approximately on the "Class II" limit curve, the higher frequency (60 Hz) results in an increase and the different method of measurement in a decrease of efficiency. Generally, limited efficiency improvements can be made at a reasonable cost of production <4>.

The current method of determining efficiency according to the summation-of-losses method with assumed 0.5 % stray-load losses distorts the assessment of efficiencies: in comparing efficiencies, the stray-load losses become a very important factor. Since an improved efficiency level demands a highly sophisticated production technique and quality, all losses will have to be verified by type and / or sample tests. Recently IEC / CENELEC have started to develop a new standard for determining the efficiency of "energy-efficient motors" < 5 >.

The rated power of the motor is selected according to the torque required. It does not need to exactly "meet the point": The favourable behaviour of the efficiency at partial load (Figure 5) results in an equivalent efficiency of an "oversized" motor, in some cases the larger motor at partial load may even have a higher efficiency!

5 Energy Conserving Drives

"Energy-Efficient Motors" (EEMs) of different efficiency levels are or will be available. It is absolutely important that they are interchangeable with the commonly used standard motors in terms of outputs and dimensions as well as in terms of operating qualities like reliability, noise etc. . Due to increased material and production costs, EEMs are more expensive than current standard motors. This will result in a selection of the efficiency level according to the operation characteristics: the higher and / or the longer the motor is loaded the higher the price that is acceptable for efficiency improvement. For that reason, larger motors (above 100 kW) have already very high efficiency levels which will not be easy to exceed. Motors in the standard range from 1 to 100 kW are bought in larger numbers and in many cases the operation characteristic cannot be predicted for each individual motor. This is why average loads and operating times are assumed when selecting the motors.

There are four important obstacles to introducing EEMs:

1. The liberalisation of the electric energy market distorts the calculation base: energy costs depend less on the amount of energy consumed than on other technical or economical factors.
2. EEMs are "new products" and the users have to gain confidence in reliability etc.,
3. The operating cost gain must exceed the engineering expense for selecting the motors ("transaction costs") in an appropriate "return of-investment interval".
4. A large number of drives are already installed. Replacing the installed motors by EEMs might reduce future energy consumption but at high expenses. Moreover, production and installation of the new motors requires additional energy. However, it may be very effective to replace a motor by an EEM instead of refurbishing the old motor.

Optimising the machine to be driven offers a large energy conservation potential. For example, pumps or fans show significant efficiency deterioration off the optimal operating range. A proper selection and design of the machine reduces the net output power offering much higher cost savings than reducing the motor losses. Another important means to reduce energy consumption is to avoid no- or nearly-no-load operation of such machines, e. g. by starting and stopping the machines according to the flows required: a machine stopping will always consume less energy than any kind of high-efficiency machine running!

In many cases, in the design stage of a plant the flows and pressures required cannot be determined accurately enough or the flows and pressures in the operating plant differ from the design data resulting in an operation off the optimal points. In such cases correcting the impeller may be useful to reduce the energy consumption. As already explained, the reduced drive power does not require to change the motor size, the motor will keep about the same efficiency level.

Variable-speed converter drives offer unique possibilities of controlling the process by enabling operation at every combination of speed and torque within a wide range, e. g. variable-speed drives of extruders, mixers / agitators, positive-displacement pumps or drive-and-brake control drives of centrifuges. The main advantage is that variable-speed drives can follow variable loads or adapt to an unknown load without operating the machine to be driven too far off its optimal range.

Formerly used devices like mechanical or hydraulic variable-speed transmissions have low efficiencies, mainly at partial load, and high wear resulting in high maintenance costs. Besides, the transmission must be installed directly on the machine to be driven. For variable-speed converter drives the motor - in most cases a standard cage induction motor - can be coupled directly to the machine. The additional energy losses in converter and motor are about 2 % to 5 % of the rated output. The failure rates of variable-speed drives are low.

What is more, for many machines like centrifugal pumps or compressors every deviation from the optimal operating range results in significant efficiency reductions. Furthermore, any output surplus has to be dissipated e. g. in a throttling valve. If the flows and pressures vary or cannot be predicted accurately enough, speed control is the most simple, the easiest and the most effective method of optimising efficiency and energy consumption <6>.

6 Conclusion

For electric motors in the Process Industry, the purchase price is only a small fraction of the overall "life-cycle costs". The most important criteria are reliability, noise and efficiency. The engineering expense for assessing energy consumption must be kept in accordance to the saving potential.

For the "Standard Motors" approximately between 1 and 100 kW, a standardised classification and labelling of the motor efficiencies will be a useful tool for selecting the optimal motors for their applications. To ensure interchangeability, the motor sizes and installation dimensions must remain unchanged.

7 Bibliographic References

[1] Public Law, 102 - 486 - October 24, 1992: "Energy Policy Act (EPAct) of 1992"

[2] "VIK-Empfehlung Drehstrom-Asynchronmotoren; Technische Anforderungen", VIK, Essen (Germany), April 1999
("VIK Recommendation for Three-Phase AC Induction Motors: Technical Requirements", available in German, yet to be translated)

[3] HD 231 S1: "Dimensions of Three-Phase Induction Motors", CENELEC Harmonisation Document, 1978/1980/1981

[4] David Walters: "Energy-Efficient Motors - Saving Money or Costing the Earth?", IEE Power Engineering Journal, Vol. 13, No. 1 (Feb. 1999) pp 25-30 & No. 2 (April 1999) pp 44-48

[5] IEC Draft 2G / 102 / CDV: "Methods for Determining Losses and Efficiency of Three-Phase Cage Induction Motors",
CENELEC / IEC committee draft for vote (CDV) to become IEC 61972

[6] K. Bieniek, N. Gröning: "Controlling the Output of Centrifugal Pumps by Means of Electronic Speed Control",
KSB Technische Berichte vol. 22e (1986, ISSN 0179-8715), Frankenthal (Germany), pp 16-31

IEA Hi-Motors Competition

Haataja J[1]. and Härkönen H[2]

[1] Lappeenranta University of Technology, Finland
[2] MOTIVA Energy Information Center, Finland

Abstract. IEA Hi-Motors Competition was arranged to find electric motors with energy efficiency higher than in any motor before and get them to the marketplace. The competition goals were to establish electric motors with losses about 25-50% lower compared to average motors. The 3-phase, 4-pole, 1500 r/min, most common motor type on the market, was selected to be the competition product. The IEA Hi-Motors Competition challenge was introduced to motor manufacturers in October 1996. The call for tenders - the competition start - happened in June 1997 and it ended by the deadline for entries on 15 August 1998. The competition jury consisting of members from United Kingdom, Denmark, Sweden and Finland and assisted by motor experts, thoroughly evaluated the entries. The entry motors were tested using direct method measurements; by the entry makers as a part of their entry documentation and also by the jury tests. Two entries from ABB Motors met the tough requirements of the competition, and were awarded with "IEA DSM Award of Excellence". The smaller entry motor, 5.5 kW, represents the lower end of the power range. The efficiency level at full load had to be at least 90.4% using IEC 34-2 direct method, and the tested entry motor reached the value 90.5%. The bigger entry motor, 75 kW, met the efficiency value 96.0%, when the mandatory requirement was 95.8%. The project has been carried out as an international pilot project within IEA DSM Agreement Annex III, Co-operative Procurement of Innovative Technologies. The project manager has been MOTIVA, Energy Information Centre, Finland assisted by motor experts from Lappeenranta University of Technology.

Keywords. Energy efficiency, induction motor

1 Background

Electric motors consume a major part of the electric energy in the modern industrial society. Because of its simplicity and robustness the three-phase squirrel-cage induction motor is the prime mover of the modern industry. Most of the research in the field of energy efficient motors (EEMs) has concentrated on motor sizes from 1 kW to about 110 kW. These motors are very important both in industry and commerce operating pumps, compressors, fans, machine tools and air conditioning equipment. Generally motors constitute of about 65 percent of the whole industrial energy consumption. It is estimated that about 6 % of the electric energy consumed in induction motors is directly converted to heat.

For many years the trend of motor design was to minimise the motor cost. This was partly done at the cost of the motor efficiency. With the continuing increase of the cost of electric energy since the energy crisis of 1973 and the pressure caused by the environmental aspects, motor manufactures and consumers have become increasingly concerned with the energy conservation potential of the energy efficiency motors.

Motor efficiency can be improved by increasing the volume of the active materials (electrical steel and current conducting material), by using more expensive technologies (better electrical steel, higher copper slot fill-factor, increased amount of the stator and rotor slots, etc) and by optimising motor design with respect to efficiency. This is, however, not a straightforward process because of several limitations which are set for induction motors by international standards. High efficiency motors have often problems in limiting the starting current to an acceptable value. These problems can be overcome by sophisticated motor design. Also standards have been modified so that a larger inrush current for energy efficient motors is allowed.

The existing energy efficient motors still have not been widely adopted despite of their many benefits. There are plenty of barriers that have prevented the adoption of energy efficient technologies. The main barrier for a motor manufacturer to improve motor efficiency is that there are not enough customers for EEMs. That is mainly caused by the market structure. About 60 % of all motors are purchased directly by OEMs (original equipment manufacturer), and only 40 % by end-users.

In the IEA Hi-Motors Competition two goals are set. Firstly: Motor manufacturers are urged to develop high efficiency induction motors. Secondly: The customers are guided to demand and buy motors better than previously. A major manufacturer meeting took place in October 1996 in P-Lisbon in connection with a conference “Energy Efficient Improvements in Electrical Motors and Drives”.

1.1 Technical Requirements

The goal of the competition was to establish motors with losses of 25-50 per cent lower compared to average motors. The 3-phase, 4-pole, 1500 r/min motor, the most common motor type on the market, was selected as the competition product.

The motor must be three phase squirrel cage induction motor and designed according to relevant IEC standards and specifications for use at 690 / 400 V,

50Hz in the power range from 0.18 to 110 kW. The replacement of a primary motor should be possible without constructional changes in motor bed.

The average motors used in industry have slightly higher efficiency than the average motors. That is why in the Hi-Motors technology procurement project the mandatory efficiency level is based on a 15 percent and the desired efficiency level on a 35 percent decrease in average motor losses compared to the best motors available today. All the values are shown in Figure 1.

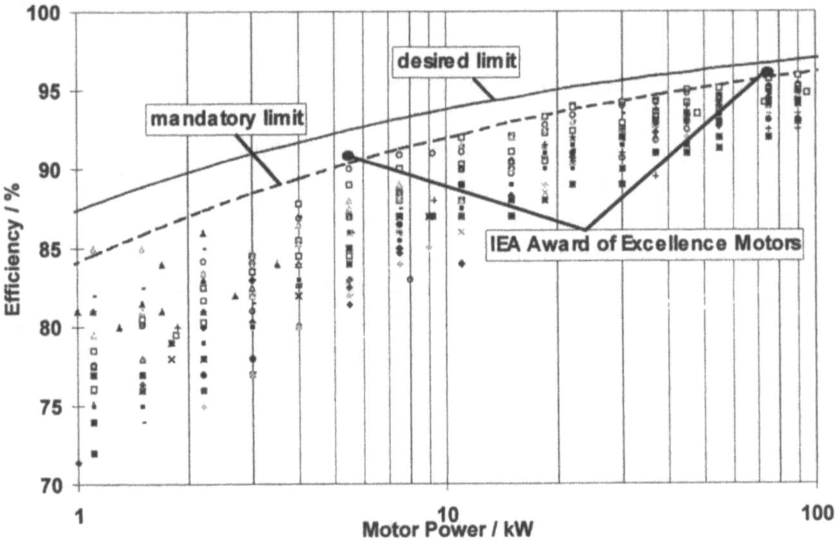

Figure 1: The desired and mandatory limits of IEA Hi-Motors Competition. The dots (• Δ – etc.) in the graph are efficiency values of major motor manufacturers´ motors available on the market. These values are taken from the EURODEEM database.

2 Two Motor Types from ABB Were Tested

Within the deadline of the competition, 15 August 1998, two entries were received, both from ABB Motors. There were no other entries.

In the tests carried out by Lappeenranta University of Technology and followed by international experts and jury members, the ABB 5.5 kW entry motor reached the efficiency value of 90.5 per cent, while the ABB 75 kW entry motor reaching the value of 96.0 per cent. Thus the already very tough mandatory requirements were even exceeded.

The smaller entry motor, 5.5 kW aluminium frame motor represents the lower end of the power range while the bigger entry motor was a 75 kW cast iron frame motor.

The two entries from ABB Motors were awarded the "IEA DSM Award of Excellence" at an international IEA Workshop in London on 24 February 1999.

2.1 Brilliant Motor Efficiency Was Achieved

The ABB motors designed for the competition were the motor M2AA132, 5.5 kW 400 V, 50 Hz, and the motor M2BA280, 75 kW 400 V, 50 Hz. The efficiency requirements of the competition were tough: 90.4 per cent for the smaller motor and 95.8 per cent for the bigger motor, using IEC 34-2 direct method. The efficiencies of the entry motors are shown in Table I.

Table I. Efficiencies of IEA Hi-Motors Competition entry motors

ABB 5,5 kW Motor

Output / pu	Hi-Motors mandatory requirement / % [1]	Direct IEC 34-2 / % [1,4]	IEEE 112 B / % [3,4]	Indirect IEC 34-2 / % [2]
0,5	89.4	89.9	89.7	90.1
0,75	90.4	90.4	89.8	90.4
1,0	90.4	90.5	89.7	90.4
1,25	89.4	90.1	89.1	90.0

ABB 75 kW Motor

Output / pu	Hi-Motors mandatory requirement / % [1]	Direct IEC 34-2 / % [1,4]	IEEE 112 B / % [3,4]	Indirect IEC 34-2 / % [2]
0,5	94.3	95.4	95.3	95.5
0,75	95.8	96.0	96.1	96.2
1,0	95.8	96.0	96.1	96.3
1,25	95.3	95.8	95.9	95.8

[1] Direct Method IEC 34-2

[2] Indirect Method IEC 34-2, summation of losses

[3] IEEE 112 Method B

[4] Measured by Lappeenranta University of Technology

3 Longer Lifetime through Reduced Losses

Efficiency is a measure of how well a motor converts electrical energy to useful work. The motor loses energy as heat. The reduced losses of the motor will result in a lower temperature rise in the motor. This will in turn effect the lifetime of the winding insulation and the bearing and bearing grease lifetime. The lifetime of the winding insulation will be four times longer for the smaller entry motor and eight times longer for the bigger entry motor compared to the standard motors, and the grease intervals are doubled for the entry motors compared to the standard motors.

Improvements in entry motor efficiency were achieved by using different materials in the stator package, by changing the number of slots in stator and rotor and by using special stator winding technique.

3.1 Higher Efficiency Means Lower Energy Consumption

Generally motors constitute of about 65 per cent of the whole industrial energy consumption. With the above mentioned improvement in motor efficiency which ABB motors achieved in the competition, increase of 1.6 per cent in efficiency means that a 75 kW motor uses 1.3 kW less energy, and 4.0 per cent increase in efficiency means that a 5.5 kW motor uses 0.2 kW less energy. On annual basis these savings are of great importance and especially when the motors are in frequent use.

The pay-back time of a high efficient motor is relatively short in regard to the fact that electricity consumed by a motor during its life time costs one hundred times more than the motor itself.

The development of high efficiency motors copes with the Kyoto Climate Agreement on reducing CO_2 emissions. Thereby less energy production will be needed, resulting in reduced harmful environmental impacts. When choosing high efficiency motors we also make a choice for the environment.

To achieve these goals and to help the market introduction of high efficient motors, the most influential purchasers have been gathered into a buyer group containing more than 30 major companies from Finland representing paper, pulp and process industry.

4 Conclusion

Energy efficient motors have a large saving potential. The most remarkable saving potential is found within the power range below 37 kW. 80 % of the total saving potential is in this power range. EE-motors are cost-effective in all power ranges.

Even if there are high efficiency motors available in the market now new advanced motor materials, improved motor manufacturing processes and design tools still allow a challenge to increase efficiency in an economical way.

More information on the IEA Hi-Motors Competition can be found at the web site http://info.lut.fi/ente/sahko/Hi-Motors/Intro.htm.

LCC-based Guidelines on Procurement of Energy Intensive Equipment in Industries

(Save Contract XVII/4.1031/Z/98-037)

Mans Johansson

1 Background

In procurement work, the factor of decisive importance in the selection of equipment and supplier has so far usually been the initial cost. Delivery time, quality, service, terms of payment and maintenance cost have also been included in the evaluation. But the energy efficiency, i.e. how efficiently energy, usually electricity, is put to use, has come into the picture more rarely. However, this is often vitally important, particularly on equipment with long operation times combined with long useful life, such as ventilation and lighting systems, pumps, air compressors and refrigeration machines.

The unit within a company or organisation that are entrusted with procurement and installation of machines and equipment, is sometimes not the same as those subsequently entrusted with their operation and maintenance. The different units generally also have different budgets, which could make it difficult to establish a common view on these matters. In many cases, the building owner/purchaser may not be the same company as the user of the premises and equipment. As a result, the equipment purchased may be cheap to buy but expensive to operate and maintain. This, in turn, leads to higher **Life Cycle Cost (LCC)**, i.e. a higher total cost viewed over the total utilisation time of the equipment.

However, at the procurement stage, the life cycle cost has unfortunately not yet assumed a more prominent role. If a different procurement methodology was adopted, in which concepts such as **overall view, far-sightedness** and **life cycle cost** are of decisive importance, better scope would be available for assigning priorities to profitable long-term efforts in the field of energy, environment and cost.

Against this background the Association of Swedish Engineering Industries (VI) has taken the initiative to develop European guidelines for the procurement of energy-efficient equipment in industry.

2 Project Description

2.1 Objectives

The primary goals of the project are:

- to develop EU-adapted guidelines for procurement of energy-efficient equipment in industry, particularly the engineering industry;
- to promote the application and implementation of the guidelines among enterprises in the EU.

Long-term objectives of the project are:

- to change attitudes and procedures in the procurement of energy-intensive equipment in industry;
- to define a readily usable system of procurement based on LCC calculation;
- to create a set of guidelines that make it possible to check whether an item of equipment installed fulfils agreed performance specifications;
- to encourage manufacturers in member countries of the EU to develop energy-optimised systems and products;
- to reduce the consumption of energy in industry and thus also the environmental impact of energy production.

How should LCC-Guidelines be used?

The LCC-Guidelines should serve as a practical means of assistance in:

- planning, project design and procurement of both construction-related equipment and production equipment;
- comparisons between various investment options;
- measurement and control;
- calculation of any performance bonuses or penalties at final inspection or trial operation.

Scope of products and systems included.

The first edition of the LCC-Guidelines will apply to the following categories of products and systems:

- Air treatment systems, including electric motors;
- Pump systems, including electric motors;
- Refrigeration installations, including electric motors.

2.2 Partners in the Project

Contractor:
The Association of Swedish Engineering Industries (VI)

Partners:
AREA, Air Conditioning & Refrigeration European Association
EUROPUMP, European Association of Pump Manufacturers
EUROVENT/CECOMAF, European Committee of Air Handling and Refrigeration Equipment Industries

2.3 Project Execution

A joint management group (JMG) is established consisting of representatives from the partner organisations.

Stage 1 - Feasibility study

At this stage information is gathered from members of the partner organisations. Are there any experiences of life cycle cost calculation? Are there any national rules that must be taken care of in the project? Is it really possible to bring about guidelines that could be accepted all over EU?

If the result from stage 1 is positive:

Stage 2 - Development and editing of LCC-Guidelines

The work is carried out by working groups and specialists in the partner organisations.

One important task in the project is the adaptation of the guidelines to other regulations in the EU. Once a first version of the guidelines has been drafted, it will be distributed to selected experts in law, economics and technology for examination. After examination and any revisions, an initial edition of the guidelines will be produced which will be used for implementation among the partner organisations.

Stage 3 - Implementation of LCC-Guidelines among member companies of European industry organisations

During this stage, the first edition of LCC-Guidelines will be distributed to companies of European industry organisations. Examples of implementation activities are seminars, articles in professional journals etc. The first edition of LCC-Guidelines will be in English.

2.4 The Structure of LCC-Guidelines

A preliminary structure of the LCC-Guidelines is as follows:

- A main section that describes the objective and methodology.
- Guidelines for energy efficient project design and procurement, with both system requirements and component requirements for energy-consuming equipment.
- A system of forms for calculating and following up the life cycle energy costs of various components and systems. The forms contain all of the information needed by a tenderer to make his calculations.

2.5 Project Status

The project started at 1 January 1999.

The feasibility study is finished (aug-99) and indicates a large interest in the field of procurement utilising LCC.

The project is now in stage 2 Development of the guidelines.

2.6 Project Data

Total project cost	752 k ECU
Support from SAVE	300 k ECU

Project time 27 months.

A New Pay-as-You Save Scheme for the Retrofitting of Variable Speed Drives to Fans and Pumps

Edward Kirk[1], Bob Oglesby[2]

[1] Control Techniques
[2] SWALEC Energy Services

1 British Industry Cannot Afford to Save Money - or Can it?

Almost 2/3 of the UK's industrial electricity consumption is accountable to the running of electric motors and drive systems, at a cost to consumer of £3 billion/per year. Calculations show that as much as 40% of this energy is not used to perform the function required and is therefore wasted due to ineffective control.

A major proportion of industrial electricity consumption (≅ 40%) is utilised as the motive power for pumps and fans. The majority of these motors are diiven at constant speed by squirrel cage machines, and for reasons of low original cost, variation in system output is frequently achieved by throttling or damping in the system an inefficient mechanism at best. In addition to this, a substantial amount of this energy is wasted, because most fan and pump systems are oversized. Consequently, significant amounts of energy are expended unnecessarily, and the operating cost of the system can be as much as 50% more than it should be.

The energy benefits of applying variable speed drives to centrifugal pumps and fans have been understood for over 20 years, but the technique has really only become viable with the decrease in cost of larger power ac drives. The reason for this is that existing fans and pumps are almost invariably driven by ac motors and the retrospective fitting of an inverter to an existing motor is usually quite simple and relatively cheap. Nowadays, many pump/and fan installations over 25kW would pay back the cost of fitting variable speed control in less than two years.

1.1 Oversized and Under Efficient

Frequently, existing pump and fan systems are larger, sometimes far larger, than they need be. This is brought about by the need to err on the positive side at all stages of the design process, as an undersized system cannot easily be made larger, but an over-large system can always be turned down. Building cooling systems, for example, need to be designed to cope with the maximum seasonal and daily conditions and encounter these infrequently. In the past, dampers and throttles were often installed to effect crude control of air and cooling water flow, usually because of the low installed cost. Time moves on and, as have all things electronic, variable speed drives have become very much cheaper per kW and thankfully are being adopted widely in new installations.

1.2 Energy Savings Benefits Unrealised

However, take up for existing systems is scandalously low considering the huge energy benefits to be gained. Even when as often happens, a two year (why only two years, why not three, when the equipment has a 10 year life?) payback or less can be illustrated, the proposal to fit drives often fails to go ahead. There are three reasons for this.

1.2.1 First, lack of resource at engineering level, where the proposal starts and all too often dies. There may be a saving to be had, but the 'it isn't broke, doesn't need fixing' attitude means the project stays in the pending tray rather than moving into the action tray. Engineering's job is to keep the wheels turning, not necessarily to reduce running costs.

1.2.2 Second, the main cash saving incentive resides with the finance department, which often doesn't know about the potential for energy saving, because no application is made for capital funding. They also don't understand the technologies of fans, pumps or drives and frequently find it difficult to believe such substantial savings can be made.

1.2.3 Third, there is the reluctance of British business and commerce to dip into the capital budget to save in the revenue budget. Effectively, they believe they cannot afford to save money! Curiously, the appointment of Energy Managers in many large businesses does not seem to have stimulated the uptake of energy saving drives, despite the best efforts of ETSU at all their seminars.

2 Case Study Highlights Benefits

The short sightedness of British Industry in all the above respects is highlighted by a recent case study which shows conclusively the benefits of installing variable speed drives.

2.1 Energy Saving Case Study

Recently, Control Techniques retrofitted a variable speed drive (V.S.D.) to the AHU[1] supply in Armitage Shanks, a leading UK supplier of ceramic bathroom

---

[1] Air Handling Unit

utilities. The AHU in question was one of the 20 identical units used to supply constant air volume to a network of ducting for the purpose of drying out porous moulds used during the manufacture of toilet bowls.

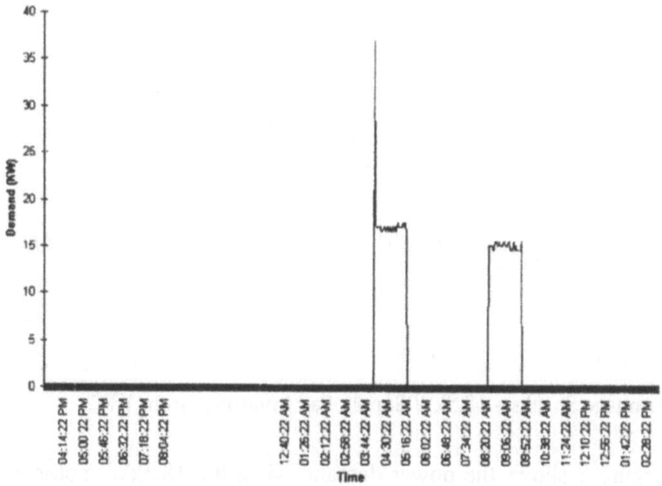

Figure 1: Armitage Shanks – 18.5kW Supply Fan (Star/Delta Starter)

2.2 One of the main issues was to reduce the starting current/torque as the fans are started at 2.00 am in the morning and the resulting "belt screech" was creating significant noise pollution. It was also thought there was sufficient air volume to allow a slight reduction in fan speed, (down to 45 Hz) and thereby save energy. The VSD supplied was a Unidrive 3402VTC (18.5kW) packaged in a standard IP54 enclosure. The unit was fitted to run at a constant 45 Hz and was monitored along side another unit running directly from the mains supply at 50 Hz from a Star/Delta starter.

2.3 Figure 1 shows the peak kW when starting from the Star/delta starter. The peak power demand reached as high as 38kW during starting. During normal running, the power consumed was close to the rated value of the motor.

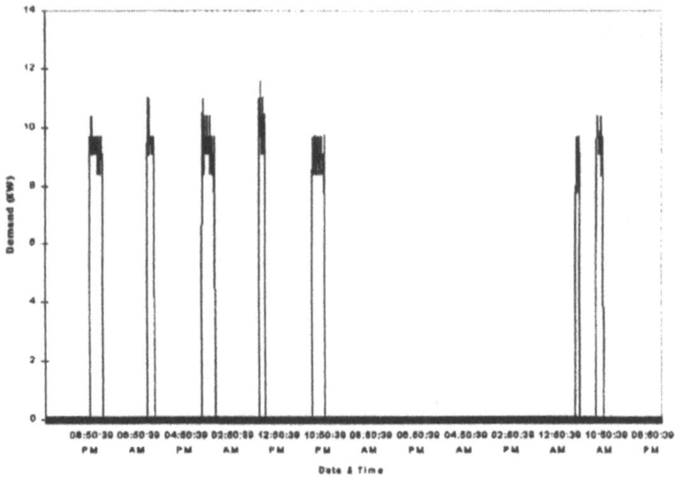

Figure 2: Armitage Shanks – 18.5kW Supply Fan Controlled from VTC

2.4 Figure 2 shows the power demand when the 18.5kW motor was started under inverter control. The power demand was much lower since the acceleration time of the load was controlled. The power measured was lower than the power rating of the motor. This was because the drive was set to run the motor at a lower speed, 45 Hz instead of its rated frequency of 50Hz. The problem of "screeching" drive belts when starting the existing motors has been entirely overcome by controlled starting.

2.5 During the investigation period, the systems ran for an average of six hours per day, this had been used along with the following information in order to calculate the annual running costs and payback period:-

1. Existing fixed speed motor absorbed an average of 16.56kW power
2. Unidrive VTC controlled motor absorbed an average of 9.06kW power
3. Daily run time - 6 hours
4. Weekly run time - 30 hours
5. Total operating period - 46 weeks/annum
6. Day Tariff - GBP0.045/kWh
7. Night Tariff - GBP0.02/kWh
8. Percentage of run time on Day and Night tariff - We had used a figure of 10% for the Night Tariff although the investigation period indicated that only 6% of the total run time occurred during the Night tariff period.

2.6 Figure 3 shows the energy consumption of the existing fixed speed motor and VSD controlled motor. The amount of energy of saving was significant.

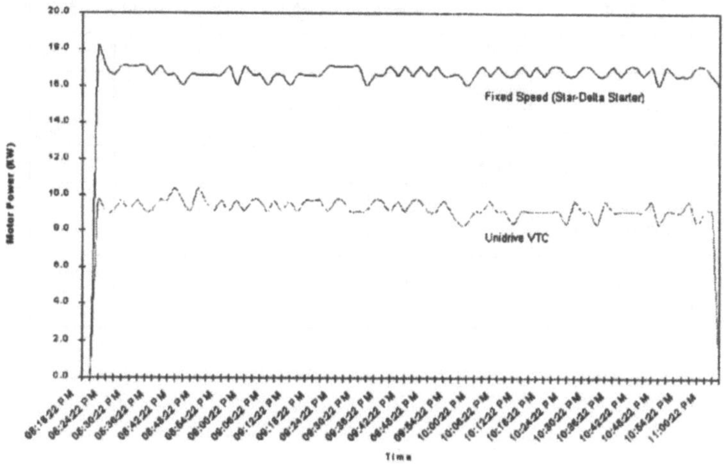

Figure 3: Comparison of Existing Fixed Speed Motor with Unidrive VTC

2.7 The investigation showed that there was significant real-time energy saving obtained from controlling the supply fan with the inverter running at a slightly reduced frequency of 45 Hz; airflow was reduced slightly to approximately 90% with no detrimental effect to the performance. The overall running cost was reduced from GBP0.33/hr to GBP0.18/hr, providing the potential for excellent savings. With the hourly running costs calculated the payback period was dependent upon the amount of time the system would run. Based upon the readings taken and our assumption of six hours per day the payback period was found to be 4.52 years. The calculation is summarized in table 1.

2.8 Figure 4 depicts the running cost differential between fixed speed and the interter control with respect to time in months. Although the star-delta system was cheaper to install, inverter control could provide substantial savings in the long term.

Table 1 Summary of Energy Savings (based on 6 hour/day runtime)

Plant	Motor Rating	Existing kWh	Projected kWh	Existing Cost (GBP)	Projected Cost (GBP)
AHU Supply Fan Day Tariff Night Tariff	 18.5kW 18.5kW	 20,568 2,285	 11,253 1,250	 925.56 61.70	 506.39 33.75
	Totals	22,853	12,503	987	540
	Savings				447
	Capital Costs				2,020
	Payback (Yr.)				4.52
High Tariff- GBP0.045/kWh	1242 hrs				
Low Tariff- GBP0.027/kWh	138 hrs				

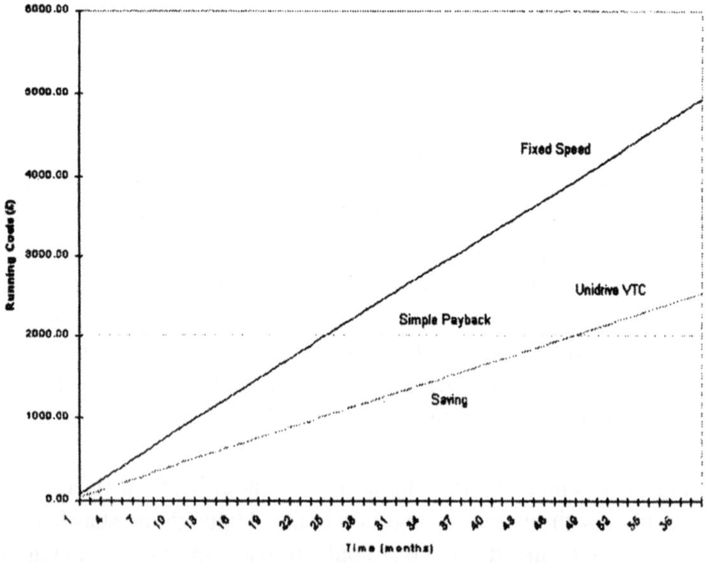

Figure 4: Running Cost Comparison between Fixed Speed and VTC

The projected calculation demonstrates that with an increased average runtime of between 8 hours and 10 hours per day, the payback would reduce to 3.3 years and 2.75 years respectively. This might be further reduced to just under two years by running the fan at 40 Hz with an average daily runtime of 8 hours.

2.9 Conclusion

Even though this application is limited to an 18.5kW motor, it is clearly demonstrated that there is significant scope for energy savings as a consequence of variable speed control of the motor. For larger power applications, where the installed cost per kW of variable frequency inverters reduces dramatically, the cost justification period reduces equally dramatically, and can be halved.

The Driving & Vehicle Licensing Authority in Swansea fitted a 160kW inverter to one of its air conditioning chilled water circulating pump motors, realising savings of over £1,000 per month and a cost justification period of around only18 months.

Oxford University fitted variable speed control to two 25kW air handling units in the Students Union, replacing star / delta starters. The result was an annual saving of over 164,000kWh and a payback period of around 2.5 years. In addition there was a significant improvement in air quality and temperature control in the Union environment.

Two new fan control installations at JRB Recovery and Owens Corning, totalling over 1.5MW, will realise annual energy savings in excess of £180,000.

In addition to the self financing benefits of variable speed control, these applications also demonstrated significant additional benefits and flexibility arising from controlled starting and running - reduced starting torque, reduced stress on the motor and the driven fan or pump; Reduced cavitation in pumps, improved power factor; reduced maximum demand, reduced fan and pump acoustic noise and reduced maintenance. The financial benefits of these improvements is in addition to the energy saving and have not been quantified or allowed for in the justification, but alone could reduce the payback period by a quarter.

3 An Easy Way to Pay

A new initiative means that there is now a hope of overcoming the reluctance to spend on energy saving drives. Control Techniques, working in partnership with South Wales Electricity (SWALEC), have successfully pioneered an innovative mechanism for the supply of energy saving drives to industry and commerce. CT survey the installation, act as project manager and supply the drives. SWALEC fund the total initial project cost, recover the funding in stages over the pay back period, and take a share of the energy savings. This spreads the cost, reducing the capital commitment and links the cost more closely to the savings.

The Impact of Efficiency Legislation on the US Motor Market

R. B. Boteler

Director Marketing U. S. Electrical Motors

Abstract: The purpose of this paper is to summarize the impact of Energy Policy act of 1992 [Epact92] motor efficiency regulations and what the net affect has been on the motor manufacturers and motor users in the United States. This paper presupposes that the reader has been briefed and has a though understanding of Epact92 regulations covering integral horse power motors. This paper will address those issues that have caused both manufacturers and users to make changes to how motors are manufactured and applied to motor driven systems.

1 Efficiency Levels in the Market Today

Motors available in the market today are categorized into one of three different levels of efficiency. Each of these levels can be seen in [Figure 1]. The lowest level is standard efficient. This level is not defined by a NEMA standard, it represents the pre-Epact92 level and it varies by manufacturer. The middle line is the new Epact92 efficiency level known as energy efficient. This level is defined in a NEMA standard [12-10] the actual efficiency levels have been written onto the Epact92 motor regulations. The highest level of efficiency shown on this chart is known as premium efficient. Premium efficient is not presently defined by a NEMA standard, each motor manufacturer has set their own level for this efficiency.

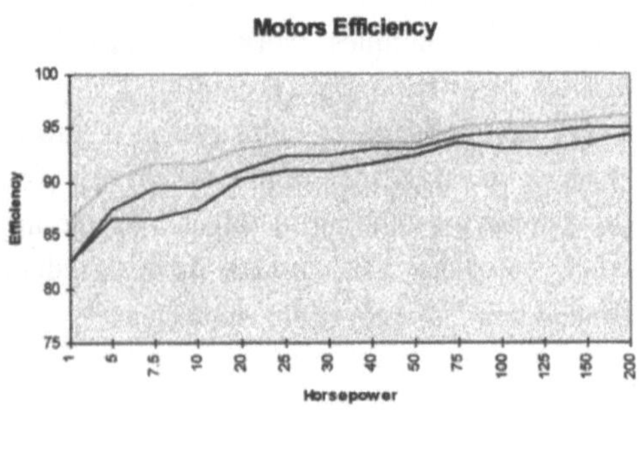

Figure 1

2 Motor Manufacturer

The issues facing motor manufacturers can be put into two distinct areas. The first is the design and manufacture of electrical components that meet the minimum efficiency requirements. Each manufacturer began the redesign of standard efficient products from their own perspective or base design. Some manufacturers already had electrical designs that included many ratings that met or exceeded the regulatory levels required by the federal law. While others produced products that required significant redesign to meet the new levels. Some manufacturers found that they not only needed to add electrical material, but in some cases they were forced to change frame lengths or other mechanical features to accommodate larger rotor and stator stack lengths. [See Figure 2]

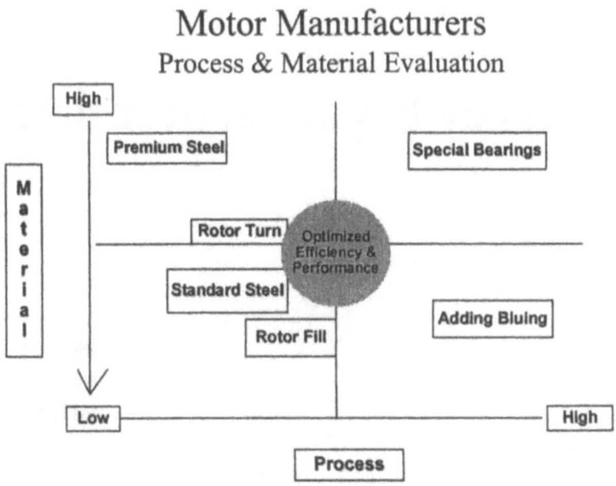

Figure 2

Epact92 implementation was negotiated by the National Electrical Manufacturers Association [NEMA] member manufacturers to allow approximately five years for this redesign to be accomplished. This period of time was adequate for the motor manufacturers to redesign and test their core products to meet the new efficiency standard. The issue that has created confusion and some resourcing problems for the manufacturers is a delay in the Department of Energy's release of proposed rules that will provide clearly defined rules distinguishing all motors. A clarification document was proposed by the NEMA task-force in the form of a product matrix. This matrix was then incorporated into a DOE proposed rule. It is from this document of product definitions that motor companies have been guiding their finished motor designs.

The lesson learned from this experience is to have clear and explicit product definitions completed as early in the process as possible.

3 Market Transition

A second issue that all motor manufacturers faced, was how rapidly would the market make the transition to Epact92 product [Figure 3] and what would the affect be on motor inventories and other products such as premium efficient motors. Manufacturers normally maintain an inventory of finished motors that is equal to a fairly constant percent of their sales. For this discussion we will assume 20% of a 100Million dollar sales company. [Figure 4] shows an example of how one manufacturer managed their inventory through the change to Energy Efficient Epact92 product.

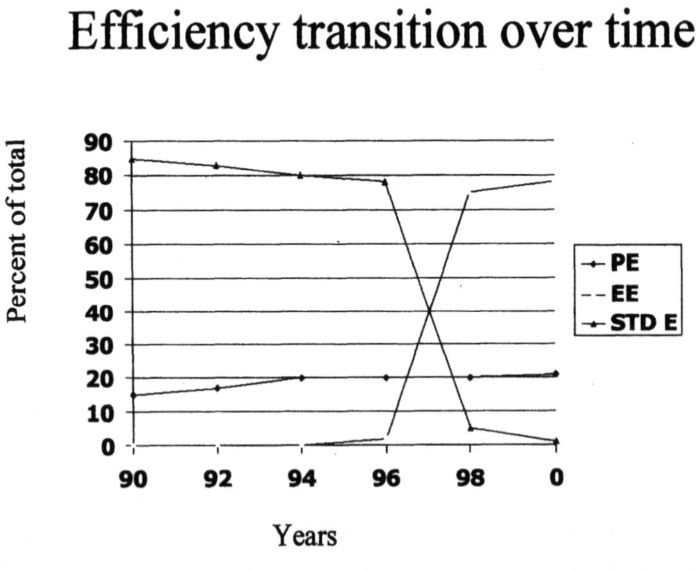

Figure 3

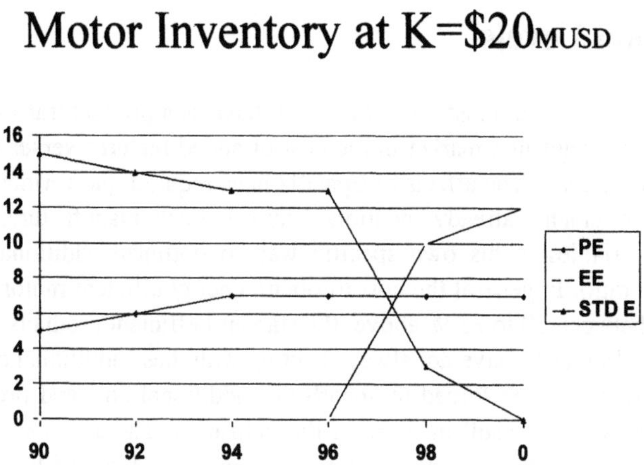

Figure 4

All manufacturers were faced with making assumptions of how customers would react to the new products. Would there be demand for the standard efficient product that remained in inventory? Should a manufacturer build standard efficient product to maintain a reserve of pre-Epact products? Manufacturers were concerned that customers would switch quickly leaving the manufacturer with old standard efficient product that was no longer wanted by motor users? We now know that the motor customer shifted to the new Epact92 product from the standard efficient product very rapidly.

The motor manufacturers were told by the Dept of Energy that the lighting ballast manufacturers had experienced an "over night" transformation of their market, with the acceptance of electronic ballast from manual ballast occurring very quickly. Some motor companies shifted their production early to build inventory of the new product while others built an inventory of standard efficient product. In some cases manufacturers still have a significant amount of this unsold inventory.

Lessons learned do not underestimate the speed with which a regulated market can change and recognize that each additional efficiency levels will increase inventory requirements and potentially decrease motor availability in the market place.

4 Market Levels

One of the concerns that most manufacturers have as a product transitions from a free market to a regulated market in the cost of added features versus their ability to recoup added costs. The affects of Epact92 have required the various design and process improvements already mentioned above. Each manufacturers has dealt with these variables in his own specific way to optimize additional costs and performance ratio. In general the cost to obtain energy efficient motors has meant an increase between 10 to 25 % above the standard efficient products. The selling prices in the US market have not always kept up with these additional costs. Motor manufacturers have been forced to absorb this additional cost and profit margins have come down as a result in some of the industries. The acceptance of energy efficient and premium efficient motors has been very high with end users of motors. The factory operators who must pay utility bills are often persuaded to use more efficient products when they evaluate motor cost as a cost of ownership over the life of the motor. Original equipment manufacturers are most interested in the lowest cost motor solution that will adequately operate their equipment of system. It is for this reason that motor regulation originally was considered appropriate.

5 Labeling of Product

Manufacturers of motors have included the nominal efficiency of a motor on the motor name plate for several years. Epact92 requires this information and will eventually include a manufacturers compliance number and special energy mark for compliance. NEMA motor manufacturers believe that a compliance number issued to each manufacturer, upon verifying the accuracy of the manufacturers computer simulation program, should be included on the name plate of each motor. Because all motors are now regulated, regardless of country of origin, the compliance number concept was proposed by the Energy Management Taskforce as a means of documenting and monitoring the shipments of motors from any where in the world. The EMTF believes the issuance of compliance numbers will make the task of regulating motor efficiency more accurate and reduce the potential for errors in enforcement. The added efficiency mark [ee] will help distinguish motors covered by Epact92 once they enter the market.

6 Catalogs and Performance Information

To communicate the Epact92 motor data to motor customers the manufacturers were instructed by DOE to provide efficiency information in all catalogs printed after October 24th of 1997. None of this information format was dictated by DOE. Manufacturers were only to include the specific nominal efficiency [Figure 5] as set by the law with each motor rating. This change was made and with few exceptions has not been a major problem for the motor companies.

Manufacturers Price Page Example

GENERAL PURPOSE
THREE PHASE, TEFC HOSTILE DUTY
ENERGY EFFICIENT

Typical 184T frame

APPLICATIONS:
Designed for pulp & paper plants, saw mills, mines, foundries, chemical plants, waste management facilities, and other process-related industries requiring protection within harsh operating conditions.

FEATURES:

- Fully comply with EPACT '92 and NRCan efficiency standards
- Suitable for wye-delta start on 250T frame and larger
- Dual voltage ratings suitable for part wind start on lower voltage
- Full 50 and 60 Hz operating data on nameplate
- Double shielded bearings on 140T - 360T frame, open on 400T - 440T frame
- Lubed for life on 140T frame
- Regreasable bearings on 250T frame and larger
- Class F insulation system, class B rise at full load (4)
- 40°C ambient continuous duty
- Corrosion resistant mill and chemical duty paint
- Stainless steel nameplate and zinc plated hardware
- NEMA Design B performance (4)
- Cast iron frame construction 180T frame and larger, steel frame construction 140T frame
- Cast iron endshields
- All cast iron upgrade kit available
- Field convertible to F-2 mounting - 180T frame and larger
- Field convertible C & D flange kits
- Lifting provisions on 180T frame and larger
- Condensation drains with plastic plugs
- Shaft slinger on pulley end for IP 54 protection
- "CE" mark on nameplate
- 320 & 360 frame models shown do not accept canopy caps

HP	RPM	FRAME	VOLTS	CATALOG NUMBER	MODEL	TYPE	LIST $	DS SYM	SF(1)	FULL LOAD EFF	FULL LOAD PF	SHIP WGT	NOTES
1	1800	143T	200	H1E2H	T336	FCT	$260	3FX	1.15	82.5	76.2	65	
	1800	143T	208-230/460	H1E2D	T332	FCT	$260	3FX	1.15	82.5	76.2	65	3
	1800	143T	575	H1E2G	T334	FCT	$260	3FX	1.15	82.5	76.1	65	
	1200	145T	200	H1E3H	T356	FCT	$325	3FX	1.15	80.0	64.1	70	
	1200	145T	208-230/460	H1E3D	T350	FCT	$325	3FX	1.15	80.0	65.1	70	3
	1200	145T	575	H1E3G	T352	FCT	$325	3FX	1.15	80.0	64.8	70	
	900	182T	200	H1E4H	T368	FCT	$489	3FX	1.15	74.0	52.7	100	
	900	182T	208-230/460	H1E4D	T369	FCT	$489	3FX	1.15	74.0	52.5	100	3
	900	182T	575	H1E4G	T371	FCT	$489	3FX	1.15	74.0	53.1	100	

Figure 5

7 Compliance and Enforcement Test Procedures

The need for clear well defined compliance and enforcement test procedures is the next big hurdle for motor efficiency regulation. In North America the IEEE test method 122 method B or CSA 390 has been adopted by the US, Canada, and Mexico in each country's motor regulation. Most of the motor manufacturers will actually use CSA 390 in their labs. The law allows organizations such as NVLAP and CSA to provide lab accreditation services to the manufacturers to assure labs accuracy and procedures are followed. The EMTF has proposed specific procedures to the Department of Energy to use in determining a motor manufacturer's products compliance. Manufacturers have asked that these procedures be setup in such a way that the need for on going motor testing is kept to a minimum number of units. The proposed rules will require that each manufacture test a minimum of [25] motors to verify the accuracy of it's computer simulation models. The accuracy of these tests will require the mean of the manufacturers sample meets the law's nominal efficiency requirement less 5% losses or one half of on NEMA efficiency band. Manufacturers spent a great deal of time and thought to document this procedure and communicate it to the Department of Energy in the hopes that this critical element would provide a clear measure from which all manufacturers can be judged.

The motor manufacturers in the US have found the need to accredit their labs to be a necessary yet costly issue. Most of them have gone through an Epact process determined by NVLAP [National Voluntary Lab Accreditation Program]. This accrediting body has also established " memos of understanding" with several other international organizations that can also provide lab accreditation to foreign manufacturers.

8 Motor User's and Original Equipment Manufacturers [OEM]

The motor market can be divided into two distinct categories. The first group of motor buyers is the original equipment manufacturer. This type of customer is predominantly focused on purchasing the lowest cost motor that will deliver the necessary performance to drive the OEM's load. The OEM was the primary target of Epact92. It was believed that this type of customer would not switch from the lower cost standard efficient motor to a more costly high efficient design unless a regulation or law was enacted to mandate a change. The motor user, that person or company that actually used the motor and paid for power to operate was more likely to switch to a higher efficient design. In the US the majority of these large processing companies have switched new motor purchasing from standard efficient motors passing the Epact92 level product and moving all the way to a "premium" efficient design. Most OEMs have accepted Epact92 standards. In the first year of Epact92 regulation there have been a few cases of motor manufacturers dealing with small to medium size equipment manufacturers using a design "C" motor to "avoid" Epact92. The Department of Energy has been unable or unwilling to make any significant effort to revise their interpretation of Epact92 in such a way that this practice would be stopped.

Lesson learned - the definition of "covered product" must take into consideration the fact that some manufactures and OEMs will use the definitions to avoid compliance if some motor designs are left unregulated.

9 Agency Files

There is an issue with manufacturers of equipment to make certain all of their various product files be up dated to include new model numbers and in some cases different performance. UL has been very accommodating in providing assistance to equipment manufacturers, making sure they are aware of filing procedures and additional testing required to maintain Underwriter Labatory [UL] recognition. Any efficiency program whether voluntary or legislated must take the timing, cost, and testing of agency certification into account. Equipment builders and listing agencies must become active participants in the motor efficiency issue to assure as orderly a transition as possible.

10 Motor Performance Issues

Motors that are designed for a higher level of efficiency may run at slightly higher RPM as a result of their lower losses. In the US, the pump manufacturers working closely with the motor manufacturers recognize this possibility and have adjusted the pump design to correct any variation. Initially several motor manufacturers were very concerned that the performance of energy efficient motors would cause major complications when motors were applied. This has not been the case, while a few isolated applications may have needed adjustments to address motor performance changes the over all technical transition for motor users has been free of performance related problems.

11 Conclusion

Motor legislation in the US has taken nearly eight years to enact. Since 1992 the amount of time and money spent by the motor industry to understand, design and build the necessary products has been significant. The US motor industry and the US Department of Energy have developed a good working relationship, that has allowed both sides of this issue to be heard. Should the European Community decide to proceed with motor regulation or voluntary programs , the lessons of the US manufacturers can serve as a guide to help reduce the costs and time required to achieve a more efficient motor population.

12 Bibliographic References

[1] Galowin, Lawerence (1995): Efficiency of Electric Motors, NIST handbook 150-10

[2] US Department of Energy (1996) : Section 10 CFR Part 431 , Energy Efficient Program for Certain Commercial and Industrial Equipment: Test Procedures, Labeling, and Certification Requirement for Electric Motors; Proposed Rule

Contract Energy Management of Compressed Air

Craig Ibbetson

Head of Marketing, Dalkia Utilities Services plc, Staines, TW18 4BQ. UK

Abstract. Compressed air is a utility that is expected to be constantly available and is free at the point of use. Both of these statements are far from reality; an adequate supply will only be ensured by the correct level of investment in equipment (with on-going maintenance), and compressed air is an expensive form of energy.

It is estimated that the approximately 10% of the UK industrial electrical demand is consumed by compressed air systems, and that up to 30% of this energy is wasted. The source of the wastage is from a number of areas including mis-matching of production capacity versus demand, inadequate maintenance, ageing and inefficient equipment, losses and misuse.
For a large user of compressed air the value of the wasted energy can be significant. However, rectification may be difficult due to the need for skills that are not possessed in-house, the need for investment that is not available, and maintaining optimum system performance requires management attention that could be focussed on the core activity.

For these reasons many companies choose to outsource the management of energy and utility plant to a third party. In the UK most of these suppliers are known as "Contract Energy Management" companies. They will assume responsibility for operation and maintenance, and provide guaranteed performance of the equipment to meet the end users needs. In many cases they will also fund investment of new and more energy efficient plant in exchange for a long-term contract to operate the plant on behalf of the end user. These companies provide all of the skills, resources and expertise to enable the end user to concentrate on their core business under the security of a long term guarantee that ensures the appropriate performance measures are achieved.

1 Introduction

The use of compressed air in industry is widespread. This readily available and (apparently) cheap utility is used for all sorts of applications including: moving materials, moving machines, cleaning, breathing air supply, instrumentation supply, dilution of vapours in pipe work. The wide use of compressed air is promoted by its ease of use and availability. It is very common for compressed air systems to grow substantially with time causing there to be substantial differences between the original design requirements and the current operational performance. These differences can have a significant impact on the cost, energy efficiency and operational effectiveness of a compressed air installation.

2 Simplicity vs Cost

It should be recognised that compressed air is an inefficient way of providing energy. The relative cost can be many times that of electricity or steam

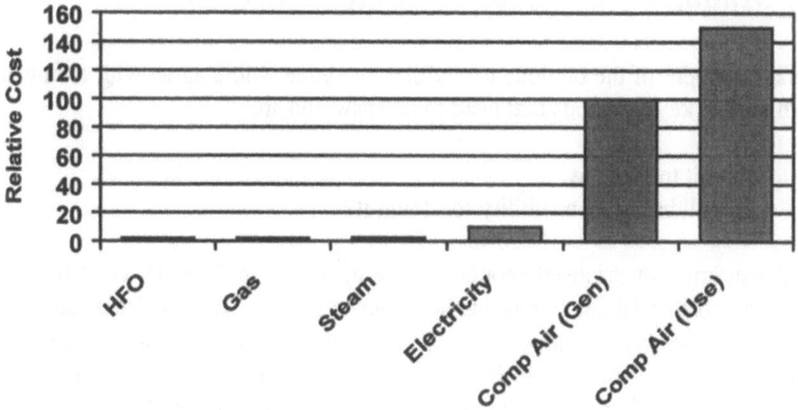

Figure 1: Relative cost of energy

The reason why compressed air is so widely used is because of its ease of installation and use. This has resulted in many installations with energy considerations taking second place.

Research in the UK suggests that the power requirement for compressed air in industrial installations represents 7.5 million MWh, this Figure is equivalent to 10% of the UK industrial electricity demand. It has also been estimated that approximately 30% savings in energy demand are achievable by good design and operational practice.

Most of the cost of compressed air is associated with energy (see Figure 2)

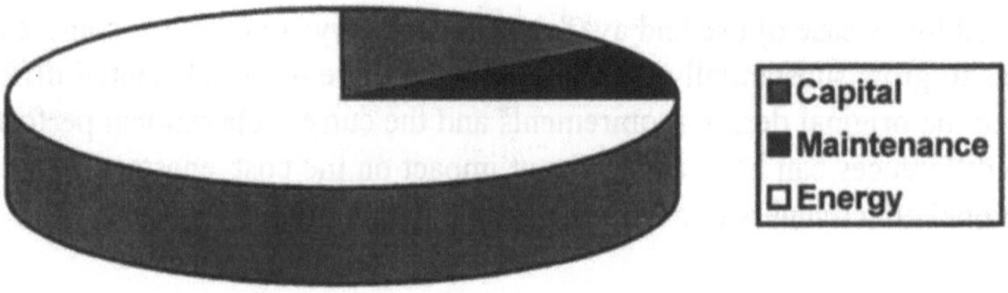

Figure 2: Cost of compressed air

3 Managing the Cost of Compressed Air

It is a statement of the obvious to say that everyone would agree with the need to do this. The key questions that need to be answered are:

- how is it to be done
- who will have responsibility for doing it?

Most western industrialised economies are operating in financial conditions that mean the cost of labour is one of the major expenses. A significant part of the labour force has been removed over the last few years, in some cases this has also removed some technical expertise. In the UK many of the people in industry who have had responsibility for the management of energy have had to manage their activity with less people and resources.

In order to fill this gap different external service suppliers have been able to provide their skills to the host company. Typically these kinds of companies are consultants, equipment manufacturers, energy suppliers (with energy management activity) or specialist energy management companies.

Depending on the need of the company these companies are able to meet some or all of the apparent need (see Table 1)

Table 1: Typical service element/supplier capability matrix

Service element	Consultant	Eqpt Supplier	Energy Management Company
Review of existing installation	✓	✓	✓
Operate and maintain Compressed air generation			✓
Take total system responsibility			✓
Operate and maintain all energy plant and utilities			✓
Invest in new equipment for improved performance			✓
Warranty performance			✓
Deliver continuous improvement of energy and utility costs			✓
Demonstrate track record			✓

In most countries the energy management companies are best placed to provide the package which has the widest scope. As compressed air is a relatively small part of the site energy use it usually makes sense to look at the total energy use in order to maximise the potential savings.

4 Commercial Aspects

An agreement for management of the compressed air system should deliver the following benefits:

- Focus on core activity
- Budgetary control
- Guaranteed levels of performance
- Management information
- Continuous improvement
- Investment for lowest lifetime cost
- Transfer of risk

The contract form should be agreed by the two parties to provide an appropriate structure to meet the end users' objectives. It is particularly important that the objective is clearly understood by both the supplier and user, in order to ensure the end result is achieved. A longer term (typically greater than 5 years) contract will provide an incentive for the supplier to invest in order to maximise savings and deliver lowest lifetime cost supply. If investment by the supplier is needed, a long term contract will allow the recovery of the investment to be made at a rate that the end user will find attractive. In some cases the supplier will offer warranty on the existing plant so that in the event of breakdown the risk is with the supplier.

Contractual terms are arranged so that it meets the end users business needs. Usually this consists of a two-part tariff structure to reflect the fixed and variable costs, however it is possible to construct fully fixed or fully variable tariffs. The supplier is incentivised to deliver good performance by a form of sharing of benefits that flow from performance levels that exceed the agreed benchmarks. It is quite normal that the contract allows the benchmark levels to be reviewed and adjusted to reflect the passage of time and changing business need. As mentioned before it is in both parties interests to make the scope of supply as large as is practical. For the supplier a large scope of supply will enable investment in the necessary people, plant and systems that will deliver the required performance. Put simply if the contract is too small there will be insufficient profit to make it worth investing for the long term. If the end users objective is to reduce cost over a sustained period the interest of supplier and user must be aligned.

5 Summary

- Identify the objective
- Identify the potential suppliers in terms of capability and relevant experience
- Discuss and agree the required approach with preferred supplier(s)
- Agree the risk and reward structure
- Make the scope as large as possible/feasible
- Review performance

The principles described in this paper apply to all energy and utility outsourcing. Dalkia has been applying this approach over a large number of years. Today Dalkia manages installations for 2400 industrial customers, 2300 hospitals, 8000 commercial premises and 12000 public buildings in 26 countries around the world.

Electric Motor Management Scheme

Richard Blandford

Southern & Redfern Ltd, Bradford, England

Abstract. The paper explores traditional repair situations and draws a comparison with a structured scheme, which commences with an audit of all stock and plant motors held by a typical user. A nominated repairer is selected, repair and overhaul specifications defined and a manufacturer of new motors chosen. Storage of motors, tracking of repairs and consideration for replacement of motors with high efficiency types, are covered in a practical yet thought provoking paper.

After the successful implementation of a Motor Management Scheme consideration is given to ongoing Condition Monitoring and methods of saving energy.

1 Introduction

This paper is not intended to be a technical appraisal of how or how not to repair electric motors. Rather, it is a basic "Motor Management Scheme" based upon proven practical and commercial experience, which can be used as a framework to improve the management of electric motor repairs, stock control, and replacement of motors that cannot be repaired. The definition of a "Motor Management Scheme" for the purposes of this paper has been defined as:-
"Co ordination of the Supply of New and Refurbished Motors within a Planned Energy Saving, Maintenance and Repair Scheme."

2 Typical Repair Scenario

Experience tells us that many companies do not have a structured approach to electric motor repairs. Typically we find that records have not been kept of how many times a motor has been repaired, or by whom; a day shift engineer for example will use repairer 'A' whilst a night shift engineer will use repairer 'B'. The motor may be old and inefficient, yet it is still repaired, with no plan to replace it after a given number of repairs. No specification has been given to the repairer's, so each repair is to a different standard / specification. This can cause problems, particularly in the case of a "copy wind", where bad practices can easily be replicated.

If adequate stocks of replacement motors are not available to meet a breakdown situation, then the faulty motor may have to be repaired instantly, and if this is during the night, or at weekends the cost could be high, and out of proportion to the actual value of the motor. However the cost of "plant downtime" cannot always be measured in relation to the cost of a motor repair, therefore this practice is often commonplace. If a replacement motor were available, is it "fit for purpose"? When was the motor last powered-up, is it electrically and mechanically complete and is it the right motor for the job?. Many times the answer is NO! but because of time constraints, a quick fix is necessary.

In addition to the plant engineer's time, the cost of administration should not be ignored. Orders have to be placed with each repairer, spare parts purchased, often in a hurry, again resulting in increased costs.

Having explored a typical repair scenario how can we embark upon a structured "Motor Management Scheme"?

3 Implementation

The implementation of a management scheme can be carried out by the user, or by a nominated repairer. Often best results are obtained by a combination of each since the user is familiar with his plant, and location of motors.

The first task should be to carry out a formal audit of all motors. Motors should be identified as follows:-

- Manufacturer / Serial Number.
- Voltage / Frequency.
- KW / HP.
- Type eg. TEFC / Foot Mounted / Ex.
- Any special ID or unique reference.
- Plant Location.

If a motor is identified as an out and out special type, then a winding specification should be obtained from the manufacturer.

Motors held in stock should be logged in an identical manner, with an assessment made of which type of motor is required for each application or plant location. If the plant has say 50 of motor type 'A', then a sufficient number of type 'A' motors should be held in stock to cover any breakdown situation, without the need for out of hours repairs. If the motors are of a standard type then the dependency on in-house stocks is reduced, however if a motor is a special non-standard type then a suitable replacement should be available at all times.

The relationship between stock and plant motors should be reviewed regularly, since plant may change, new machines or processes maybe installed which can introduce new motor types or sizes. If possible ensure that when new plant is purchased, motors are specified which conform to the users exact requirements. In some instances this could invoke a "premium", however any increase will often be saved in later years, when breakdowns do occur.

Preferably records should be kept on a PC based system, and a number of proprietary software packages are available. Alternatively a schedule can be prepared on a simple spreadsheet which will ease the entry of additions and deletions to the motor inventory, yet provide ongoing stock records.
Having prepared a list of all motors then an assessment can be made with regard to the selection of a suitable repairer, if this has not yet been carried out.

4 Choice of Repairer

The repairer should be of sufficient size and have the experience and ability, to carry out repairs on the full range of motors identified in the audit. The workshop should be fully equipped for repair and testing in order to meet any repair situation. If this cannot be achieved, a suitable subcontractor should be identified who can carry out any "specialist" repairs. If the user operates 24 hours every day, 365 days per year, the repairer should be able to react to this situation. A maximum response time should be agreed, and the repairer should ensure that all his personnel attending to breakdowns are familiar with the exact location of the user's plant. Knowledge of the exact route and travelling times can often save precious time in an emergency situation. Having to explain the route to someone unfamiliar with a user's address can be frustrating at 2.00am on a Sunday morning, and can waste valuable time.

If the repairer has to carry out any site work, ensure he is familiar with both plant layout and any Health & Safety requirements. If Health & Safety induction is needed make sure that all repairer's engineers, who could be involved, have been through the induction process, prior to the commencement of any repair contract. Identify delivery or collection points, publish lists of all key personnel with emergency telephone numbers, etc.

If the user insists on the issue of an order number prior to the commencement of any repair, ensure that the repairer understands this situation. If communications break down, the repair may not be carried out, leading to possible expensive downtime.

The repairer should have transport able to carry the users range of motors; ensure that the weights and dimensions of the full range of motors are checked against the type of vehicles operated by the repairer.

The use of mobile telephones by all personnel involved in the repair scenario can also save time in breakdown situations.
If the user operates, for instance, a chemical plant, where motors may be contaminated, the motors sent for repair should be de-contaminated by the user and a suitable "de-contamination certificate" issued. The repairer should not handle motors, which have not been de-contaminated.

The repairer should be ideally involved in ongoing training, which could be allied to the types of motors and control systems the user proposes to install in the future. Membership of organisations such as AEMT, Association of Electrical and Mechanical Trades can be beneficial. In order to gain membership, companies are required to meet agreed standards, and undertake formal training and certification for such things as Ex repairs.

In addition, ISO9000 accreditation will provide the basis for formal procedures and dedicated work instructions, some of which can be developed to meet the exact needs of a particular type of repair. ISO9000 accreditation can be combined with known standards such as BEMA / AEMT Code of Practice. Many repairers are also nominated Authorised Repair Agents (ARA's) for individual motor manufacturers, which can help familiarity with the users nominated supplier of motors.

Looking to the future, the repairer should be working towards meeting the requirements of the AEMT Repair Shop Guide for Rewinding Induction Motors with Minimum Efficiency.

Finally the user should insist on the repairer working to a minimum "specification" for both rewinding and overhauling motors. A typical specification is shown below. Insist on good quality spare parts eg. bearings, seals.

- BEMA / AEMT Code of Practice.
- AEMT Repair Shop Guide for rewinding induction motors with minimum loss of efficiency.
- Clean all parts.
- Strip and record winding details. (Rewind only)
- Rewind to manufacturers specification or "Copy Wind".
- Clean windings and revarnish (Overhaul only).
- Clean rotor and check for defects.
- Fit new bearings and seals of proprietary brand (Drive / Non Drive End).

- Reassemble.
- Paint to clients agreed colour.
- Record No load test.
- Vibration test analysis.
- Ensure all practices are carried out by experienced personnel.
- AEMT Ex Code of Practice (if flameproof)

A "Test Certificate" should be issued for the repair of each motor.

The repairer must of course have the knowledge to repair each and every motor, and also a system to track every motor involved in the repair and replacement chain.

5 Storage / Stock Rotation

Stocked motors can be located at either the user's or the repairer's premises and should be stored in a clean and dry area. If the motors are held at the repairer's premises then the user should ensure that the repairer can provide a suitable service to ensure that the motors are accessible at all times. Agreement should be reached to ensure that the motors are fully insured and if the motors are held at the user's premises then the motors should be accessible to the repairer. In most situations the user will remove motors requiring repair from the plant, and install it's replacement.

Each motor should be suitably identified with details as shown in Para.3, sometimes a unique reference can also be used to tag each motor. As a motor is taken from stock, its replacement should be put into storage in strict rotation e.g. last in - last out. If a faulty motor is scrapped then a suitable replacement should be purchased, added to the stock and recorded in the motor inventory. If the scrapped motor has a high usage, then consideration should be given to replacement with a high efficiency motor.

The following procedure should be adopted for maintenance, storage and stock rotation.

- Megger test
- Clean
- Spin shaft
- Check general appearance
- Grease polished surfaces
- Record condition and usage
- Power up annually (If not used)
- Check tag and identification reference

- Ensure that stock motors are rotated e.g. last in - last out.
- Update records

6 Commercial Considerations

We now have to consider how the Motor Management Scheme can be of commercial benefit to both repairer and user.
Fixed prices for both Rewind and Overhaul should be established for the full range of motors, and reviewed on an annual basis. The prices should be based upon the specification detailed in Para.4. An agreed price structure for new motors and spare components should be determined. In some instances the repairer or the user will keep in stock a range of bearings, fans and terminal boxes etc.

A range of optional services could also be identified and rates agreed for such things as Emergency Call-Out, Normal and Out of Hours Working, Collection and Delivery.

The question of Replace or Repair is often contentious; however the following guidelines can be used:-

Depending on the price of a new motor, consider scrapping motors up to 5.5kW which require repair.

If a motor has been repaired 3 times or its repair costs 66% of new price, consider a new motor rather than repair.

If replacing a motor having a high usage, consider a high efficiency model.

Any motor being repaired should meet the requirements of the agreed specification, and be repaired to standards such as the AEMT guidelines mentioned earlier.

Agreement should be reached between repairer and user, as to the costs associated with preliminary work carried out on a faulty motor, which consequently cannot be repaired.

The benefits to the repairer in a Motor Management situation, is that any repair work is guaranteed to be carried out by the nominated repairer, overhead costs are reduced since the repairer is not in an ongoing bidding situation, each time a repair or plant shutdown occurs.

Year-on-Year improvements to the condition of motors involved in a management scheme could mean fewer repairs for the repairer, but in the long term the reduced administration costs encountered by the repairer should enable him to put additional effort into winning more Management Contracts.

Since much technical and commercial data has been agreed within the framework of a Management Scheme, the opportunity exists to standardise such things as test certificates, delivery notes, and in particular invoices. Electronic Data Interchange can also reduce administration costs and timescales, leading to greater efficiency in the overhead costs of both supplier and user.

7 Condition Monitoring

We now have a successful Motor Management Scheme operating - good house keeping is now an important ingredient in maintaining the condition and efficiency of our motors. Keeping the plant and associated drives and equipment clean, and in prime condition will help to optimise production and avoid failures. Many plants have essential drives, and consideration should be given to the provision of Condition Monitoring to these key drives. A new generation of intelligent motor starters can be used in a dual role: motor control and condition monitoring.

Generally a motor will fail because of:-

- Heat.
- Ageing.
- Humidity and Dirt.
- Mechanical Breakdown / Vibration.

Protection methods include motor current and winding temperature and these parameters can be enhanced by the display of statistical information:-

- Phase Current in Amperes and % FLC.
- Phase Imbalance.
- % Thermal Capacity Utilised.
- Time - to - Trip.
- Time - to - Reset.
- Cause of last Five Trips.

This type of condition monitoring can be carried out locally or remotely at the repairer's premises via modem link.

By use of condition monitoring, trends in motor wear can be estimated and essential motors can be maintained at regular intervals, or during planned plant shutdown periods.

If breakdowns can be avoided, then repair costs will be reduced and unscheduled plant shutdowns avoided.

8 Energy Savings

In the long term a well maintained and repaired motor will help reduce energy costs, but other methods of energy savings can also be introduced. The use of Adjustable Frequency Inverter Drives on pumps and fans in particular, can result in a reduction of power consumption. An Inverter can ensure that an induction motor and it's driven equipment, will operate along its optimum operative curve and maintain efficiency. The benefits thus gained far outweigh the small additional loss caused by the inverter. Conversion from fixed to variable speed can also reduce energy in areas other than electrical energy, giving additional savings from the extended life of bearings and seals.

9 Summary

It is hoped that the paper will evoke some thought as to how an existing or proposed Motor Management Scheme can be improved or implemented. The overriding success of such a scheme can only be achieved by co-operation between user, repairer and motor manufacturer. Each has a key, but differing role to play, if any Motor Management Scheme is to succeed. It may be that some reorganisation of existing working practices needs to take place; however the benefits will be enjoyed by all participants.

- Consider a Motor Management Scheme.
- Apply Condition Monitoring to Essential Motors.
- Fit VSD's to Fans and Pumps.
- Insist on Rewinding Induction Motors with a Minimum Loss of Efficiency - AEMT Workshop Guide.

MIX
Papier aus verantwortungsvollen Quellen
Paper from responsible sources
FSC® C105338

If you have any concerns about our products,
you can contact us on
ProductSafety@springernature.com

In case Publisher is established outside the EU,
the EU authorized representative is:
Springer Nature Customer Service Center GmbH
Europaplatz 3, 69115 Heidelberg, Germany

Printed by Libri Plureos GmbH
in Hamburg, Germany